NON-CIRCULATING MATERIAL

S0-AND-659

Organization of the sections in the chapters on "Physical data of semiconductors"

The data on the physical properties of semiconductors are generally arranged in subsections of the order given below. Since there is some arbitrariness in the assignment of a property to the seven subsections, physically related properties may appear in different subsections in some cases.

(The detailed subclassification could frequently not be retained because of the small amount of data for some compounds dealt with in this volume. The data have been then merely grouped under two headings "Structure" and "Physical properties", in some cases they have not been classified at all.)

Structure

Static properties of the lattice as structure, space group, lattice parameters, phase transitions, chemical bond. Density, melting point, thermal expansion are sometimes listed under "Further properties" or "Lattice properties". Further static properties are given in "Lattice properties".

Electronic properties

Information and data about electronic and excitonic energy states as well as electron and hole parameters (band structure, density of states, energy gaps, transition energies, core levels, effective masses (sometimes also presented in "Transport properties"), g-factors, other band parameters).

Impurity and defect states

Basic data on shallow and deep states (trap levels, nature of defects, g-factors of defects; for the influence of impurities on other properties, see also the respective subsections; data on diffusion are also listed under "Transport properties" or "Further properties").

Lattice properties

Static and dynamic properties of the lattice (phonon dispersion relations, phonon frequencies, sound velocities, elastic and other moduli, Grüneisen parameter, dielectric constants (sometimes also listed under "Optical properties"); for Debye temperature, heat capacity, see also "Further properties"; for structure, density, lattice parameters, chemical bond etc., see subsection "Structure").

Transport properties

Electronic transport parameters (conductivities, carrier concentrations, mobilities, Seebeck coefficient, Nernst coefficient etc.; for thermal conductivity, see also "Further properties").

Optical properties

Optical spectra, optical constants, parameters obtained from optical measurements if not already listed in the subsections "Electronic properties" and "Impurity and defect states" (absorption, reflection, (photo-) luminescence and photoconductivity, decay times, refractive index, dielectric constants, Raman scattering etc.).

Further properties

Thermal, magnetic properties, data more completely presented in other Landolt-Börnstein volumes (thermal conductivity, thermal expansion, magnetic susceptibility, magnetic transition temperatures, magnetic moments, melting point, density, heat capacity, Debye temperature etc.).

LANDOLT-BÖRNSTEIN

Numerical Data and Functional Relationships
in Science and Technology

New Series

Editors in Chief: K.-H. Hellwege · O. Madelung

Group III: Crystal and Solid State Physics

Volume 17

Semiconductors

Editors: O. Madelung · M. Schulz · H. Weiss†

Subvolume h

Physics of Ternary Compounds

M. Böhm · G. Huber · A. MacKinnon
O. Madelung · A. Scharmann · E.-G. Scharmer

Edited by O. Madelung

Springer-Verlag Berlin · Heidelberg · New York · Tokyo

Landolt, Hans Heinrich

LANDOLT-BÖRNSTEIN

Zahlenwerte und Funktionen
aus Naturwissenschaften und Technik

Neue Serie

Gesamtherausgabe: K.-H. Hellwege · O. Madelung

QC
61
L332
G.3
...
17.h

Gruppe III: Kristall- und Festkörperphysik

Band 17
Halbleiter

Herausgeber: O. Madelung · M. Schulz · H. Weiss †

Teilband h

Physik der ternären Verbindungen

M. Böhm · G. Huber · A. MacKinnon
O. Madelung · A. Scharmann · E.-G. Scharmer

Herausgegeben von O. Madelung

NON-CIRCULATING MATERIAL

ALLE ZEIT WACHT · 1842

Springer-Verlag Berlin · Heidelberg · New York · Tokyo

542549

LAMAR UNIVERSITY LIBRARY

CIP-Kurztitelaufnahme der Deutschen Bibliothek

Zahlenwerte und Funktionen aus Naturwissenschaften und Technik / Landolt-Börnstein. – Berlin; Heidelberg; New York; Tokyo:
Springer. Teilw. mit d. Erscheinungsorten Berlin, Heidelberg, New York. – Parallelt.: Numerical data and functional relationships
in science and technology.

NE: Landolt, Hans [Begr.]; PT. Landolt-Börnstein, N.S./Gesamthrsg.: K.-H. Hellwege; O. Madelung. Gruppe 3, Kristall- und
Festkörperphysik. Bd. 17. Halbleiter/Hrsg.: O. Madelung ... Teilbd. h. Physik der ternären Verbindungen/M. Böhm ... Hrsg.
von O. Madelung. – 1985.

ISBN 3-540-13507-3 Berlin, Heidelberg, New York, Tokyo
ISBN 0-387-13507-3 New York, Heidelberg, Berlin, Tokyo

NE: Hellwege, Karl-Heinz [Hrsg.]; Madelung, Otfried [Hrsg.]; Böhm, Manfred [Mitverf.]

This work is subject to copyright. All rights are reserved, whether the whole or part of
the material is concerned specifically those of translation, reprinting, reuse of illustrations,
broadcasting, reproduction by photocopying machine or similar means, and storage in data
banks.

Under § 54 of the German Copyright Law where copies are made for other than private
use a fee is payable to 'Verwertungsgesellschaft Wort' Munich.

© by Springer-Verlag Berlin-Heidelberg 1985

Printed in Germany

The use of registered names, trademarks, etc. in this publication does not imply, even in
the absence of a specific statement, that such names are exempt from the relevant protective
laws and regulations and therefore free for general use.

Typesetting: Universitätsdruckerei H. Stürtz AG, Würzburg; printing: Druckhaus Langen-
scheidt KG, Berlin; bookbinding: Lüderitz & Bauer-GmbH, Berlin

2163/3020-543210

Contributors

M. Böhm, I. Physikalisches Institut der Universität Gießen, 6300 Gießen, FRG

G. Huber, Institut für Angewandte Physik der Universität Hamburg, 2000 Hamburg, FRG

A. MacKinnon, Blackett Laboratory, Imperial College, London SW7 2BZ, UK

O. Madelung, Fachbereich Physik der Universität Marburg, 3550 Marburg, FRG

A. Scharmann, I. Physikalisches Institut der Universität Gießen, 6300 Gießen, FRG

E.-G. Scharmer, Institut für Angewandte Physik der Universität Hamburg, 2000 Hamburg, FRG

Vorwort

Der vorliegende Band befaßt sich mit den physikalischen Eigenschaften ternärer Halbleiter.

Im ersten Kapitel werden die tetraedrisch gebundenen ternären (und quasibinären) Verbindungen behandelt, die sich durch Substitution aus den II–VI- und III–V-Verbindungen in gleicher Weise ableiten lassen, wie diese aus den elementaren Halbleitern der IV. Gruppe. Dieses Kapitel schließt also eng an die Bände 17a und 17b an.

Die beiden folgenden Kapitel enthalten Daten über halbleitende Verbindungen der Übergangs-Elemente und Seltenen Erden. Hier zeigt sich deutlich die Schwierigkeit der Klassifikation einer Gruppe von Festkörpern und ihrer Einordnung in einen Band dieser Serie. Denn eine große Anzahl dieser Verbindungen sind auf Grund ihrer magnetischen oder ferroelektrischen Eigenschaften bereits in den Bänden 4, 12 und 16 ausführlich dargestellt worden. Ein Weglassen dieser Halbleiter-Familien unter Verweis auf die genannten Bände hätte jedoch die mit den Bänden 17a bis 17i beabsichtigte Gesamtübersicht über das Halbleitergebiet empfindlich beeinträchtigt. Die für den Halbleiterphysiker wichtigen Eigenschaften dieser Substanzgruppen wurden hier also nochmals zusammengestellt und durch neueres Material ergänzt.

Das letzte Kapitel schließlich gibt einen Überblick über weitere ternäre Halbleiter. Eine vollständige Aufzählung aller in der Literatur als „Halbleiter" bezeichneter Substanzen war nicht möglich und erschien auch nicht sinnvoll. So gibt dieses Kapitel nur einen Eindruck über die Vielfalt weiterer Substanzgruppen, an denen Halbleitereigenschaften gefunden wurden.

Mit diesem Band wird die systematische Behandlung der kristallinen Halbleiterfamilien beendet. Im abschließenden Band 17i werden Eigenschaften der amorphen und der organischen Halbleiter dargestellt sowie einige spezielle Halbleiterphänomene behandelt, die sich in den systematischen Aufbau der Bände 17a bis 17h nicht einfügen ließen.

Mein Dank gilt den Autoren für die kritische Aufarbeitung der Daten dieses Teilbandes. Es war nicht immer einfach, für die behandelten Substanzen eine geschlossene Darstellung der Halbleitereigenschaften und damit verbundener Phänomene zu geben, ohne allzugroße Überschneidungen mit den bereits in anderen Landolt-Börnstein-Bänden präsentierten Daten zuzulassen.

Mein Dank gilt auch der Landolt-Börnstein-Redaktion, insbesondere Herrn Dr. W. Polzin, der diesen Teilband redaktionell betreut hat sowie Frau I. Lenhart und Frau R. Lettmann für die sorgfältige Überprüfung der Manuskripte und Druckfahnen. Dem Springer-Verlag danke ich für die verständnisvolle Zusammenarbeit bei der Fertigstellung.

Marburg, Oktober 1984   **Der Herausgeber**

Preface

This volume presents the physical properties of ternary semiconductors.

In the first chapter the tetrahedrally bonded ternary (and quasi-binary) compounds have been treated. These compounds are related to the II–VI and III–V compounds in the same way as those compounds to the semiconducting elements of the fourth group of the periodic table. Thus this chapter is closely connected with volumes 17a and 17b.

The two following chapters contain data of semiconducting compounds of transition elements and rare earths. Here the difficulty is clearly visible to classify groups of solids according to their properties and to assign these groups to the different volumes of this series. Many of the transition element and rare earth compounds have magnetic or ferroelectric properties and have accordingly been treated in the respective volumes 4, 12 and 16. To leave out these groups of semiconductors and to refer solely to the volumes cited would have impaired considerably the program to give a complete review of the field of semiconductors within the volumes 17a to 17i. Thus the properties of these groups have been treated here to such an extent as a semiconductor physicist would be interested in.

The last chapter finally gives a review about further semiconducting ternary compounds. An exhaustive presentation of all substances called "semiconductors" anywhere in the literature was not possible. So this chapter solely gives an impression about the variety of semiconducting substances.

With this volume the systematic treatment of the families of crystalline semiconductors is closed. The final volume 17i will contain the presentation of the properties of amorphous and organic semiconductors as well as of several semiconductor phenomena which did not fit into the systematic arrangement of volumes 17a to 17h.

I wish to thank the authors for their critical preparing of the data in this subvolume. It often was a hard work for them to give a comprehensive presentation of the semiconductor and related properties of the ternary compounds without repeating too much of the material presented in other Landolt-Börnstein volumes.

Thanks are also due to the editorial staff of Landolt-Börnstein, especially to Dr. W. Polzin who was in charge of the editorial preparation of this subvolume and to Mrs. I. Lenhart and Mrs. R. Lettmann for their careful checking of the manuscripts and galleys. I am also grateful to the Springer-Verlag for their patient care and experienced help in the final preparation.

Marburg, October 1984   **The Editor**

Table of contents

Semiconductors

Subvolume h: Physics of ternary compounds
(edited by O. MADELUNG)

A. Introduction

1. List of symbols

In the following list frequently used symbols are specified. The references in the last column refer to the introductory part A in subvolume 17a (cited as 17a/followed by the number for the respective section) or to that section of part B of this subvolume where the quantity is defined or introduced the first time (cited as 17h/followed by the number of the respective section). The units listed in the second-last column are the most frequently used units. In the tables of part B data are generally given in the units of the original paper. To facilitate a conversion from CGS units to SI units or vice versa conversion tables are presented in section 3 below.

Symbol	Property	Unit	Introduced in section
a, b, c	lattice parameters	Å, pm	
b	electron-hole mobility ratio (μ_n/μ_p)	–	
$B(B_S, B_T)$	bulk modulus (adiabatic, isothermal)	$N\,m^{-2}$	17a/2.1.2, eq. (A.17)
B	Nernst coefficient	$cm^2\,K^{-1}\,s^{-1}$	17a/2.2.1
$\boldsymbol{B}$	magnetic induction	T, G	
$c_{(Fe_{2+})}$	concentration (of Fe^{2+} ions)	cm^{-3}	
c_{lm}	elastic moduli (stiffnesses)	$N\,m^{-2}$	17a/2.1.2, eq. (A.19)
c_p	specific heat capacity	$J\,K^{-1}\,g^{-1}$	
$C_p(C_v)$	heat capacity at constant pressure (volume)	$J\,mol^{-1}\,K^{-1}$	
d	distance	cm, Å	
d	thickness of a sample	cm	
d	density	$g\,cm^{-3}$	
d_{exp}	experimental density	$g\,cm^{-3}$	
d_X, d_{th}	X-ray density	$g\,cm^{-3}$	
d_{opt}	optical density (log I_0/I)	–	
d_{ij}	nonlinear dielectric susceptibility	$cm\,V^{-1}$	17a/2.2.2
$D_{n(p)}$	diffusion coefficient for electrons (holes)	$cm^2\,s^{-1}$	
e	elementary charge ($|e| = 1.6021892(46) \cdot 10^{-19}\,C$)	C	
e	polarization vector	–	
$\boldsymbol{E}$	electric field strength	$V\,cm^{-1}$	
E	energy	eV	
$E_{0,1,2\ldots}$	energies of critical points in optical spectra	eV	17a/2.1.1
$E(\Gamma_6)\ldots$	energy of band edge of type $\Gamma_6\ldots$	eV	17a/2.4.1
$E_{[hkl]}$	Young's modulus (measured in [hkl] direction)	$dyn\,cm^{-2}$	17a/2.4.2, eq. (A.100)
$E_{a(d)}$	energy of acceptor (donor) state measured from the respective band edge	eV	17a/2.3.1
E_A	activation energy (of conductivity or other temperature or pressure dependent properties)	eV	
E_b	binding energy	eV	
E_c	coercive field	$V\,cm^{-1}$	
$E_{c(v)}$	band edge of conduction (valence) band	eV	17a/2.1.1, eq. (A.1)
E_F	Fermi energy	eV	
E_g	energy gap	eV	17a/2.1.1
E_{gx}	excitonic energy gap	eV	17a/2.2.2
$E_{g,th}$	energy gap extrapolated to 0 K (thermal energy gap)	eV	17a/2.1.1
$E_{g,dir}$	direct energy gap	eV	17a/2.2.1
$E_{g,ind}$	indirect energy gap	eV	17a/2.2.1
$E_{g,pseu}$	pseudodirect energy gap	eV	17h/10.1
E_r	recombination center energy level	eV	17h/10.1
E_t	energy of trap level	eV	
ΔE	width of valence band	eV	17h/10.1

Symbol	Property	Unit	Introduced in section
f	frequency	Hz	
$g(E)$	density of states	$cm^{-3}\,eV^{-1}$	17a/2.1.1, eq. (A.8)
g^*	effective g-factor	–	
g_c	g-factor of conduction electrons	–	
g_v	g-factor of holes (valence band)	–	
h	Planck constant ($h = 6.626176 \cdot 10^{-34}$ Js)	Js	
$\boldsymbol{H}$	magnetic field strength	$A\,cm^{-1}$, Oe	
$\boldsymbol{i}$	current density	$A\,cm^{-2}$	
$I_{(lum,\,R)}$	intensity (of luminescence, Raman intensity)	$cm^{-2}\,s^{-1}$	
I_{rel}	relative intensity	–	
I_{ph}	photocurrent	A	
k	extinction coefficient (absorption index)	–	17a/2.2.2, eq. (A.41)
k	Boltzmann constant ($k = 1.380662\,(44) \cdot 10^{-23}\,J\,K^{-1}$)	$J\,K^{-1}$	
$\boldsymbol{k}$	wave vector of electrons	cm^{-1}	17a/2.1.1
K	absorption coefficient	cm^{-1}	17a/2.2.2, eq. (A.38)
$L,\,l$	length, carrier diffusion length	cm	
$\Delta l/l$	linear thermal elongation		
m_0	electron mass ($m_0 = 9.109534\,(47) \cdot 10^{-28}$ g)	g	
m^*	effective mass	m_0	17a/2.1.1
m^{**}	polaronic mass	m_0	17a/2.1.1, eq. (A.10)
m_c	conductivity effective mass	m_0	17h/10.1
$m_{n\,(p)}$	effective mass of electrons (holes)	m_0	17a/2.1.1, eq. (A.1)
m_{ds}	density of states mass	m_0	17a/2.1.1, eq. (A.9)
$m_\parallel$	longitudinal effective mass	m_0	17a/2.4.1, eq. (A.76)
$m_\perp$	transverse effective mass	m_0	17a/2.4.1, eq. (A.76)
n	(real) refractive index	–	17a/2.2.2, eq. (A.37)
$n_{a,b,c}$	refractive index in a, b, c direction	–	
n_o	refractive index for ordinary ray	–	
n_e	refractive index for extraordinary ray	–	
Δn	birefringence	–	
n	electron concentration (also carrier concentration in general)	cm^{-3}	
n_i	intrinsic carrier concentration	cm^{-3}	17a/2.1.3, eq. (A.20)
$n_{a\,(d)}$	acceptor (donor) concentration	cm^{-3}	
n_t	trap concentration	cm^{-3}	
N	count rate	–	
P	Peltier coefficient	V	17h/10.2
p	pressure	bar	
p_S	sulfur pressure	bar	
p_{tr}	(phase) transition pressure	bar	
p	hole concentration	cm^{-3}	
p_{ij}	elastooptic constants	–	17a/2.4.2
p_{eff}	effective (paramagnetic) magnetic moment	μ_B	
p_m	magnetic moment per formula unit	μ_B	
$P_{(s)}$	(spontaneous) polarization	$C\,m^{-2}$	17h/10.2
$\boldsymbol{q}$	wave vector of phonons	cm^{-1}	
r	radius, distance	Å	
r_{ij}^{X}	linear electrooptic constant at constant stress	$cm\,V^{-1}$	17h/10.1
R	resistance	Ω	
R	reflectance, reflectivity	–	17a/2.2.2, eq. (A.42)
R_H	Hall coefficient	$cm^3\,C^{-1}$	17a/2.2.1
S	spin quantum number	–	
S	Seebeck coefficient, thermoelectric power	$V\,K^{-1}$	17a/2.2.1, eq. (A.42)
s_{ml}	elastic compliances	$m^2\,N^{-1}$	17a/2.1.2, eq. (A.19)
t	time	s	

Symbol	Property	Unit	Introduced in section
T	transmission	–	
T	temperature	K, °C	
T_C	Curie temperature	K, °C	
T_f	firing temperature	K, °C	
T_m	melting temperature	K, °C	
T_N	Néel temperature	K, °C	
T_p	hot pressing temperature	K, °C	
T_{perit}	peritectic (decomposition) temperature	K, °C	
T_{sub}	substrate temperature	K, °C	
T_{tr}	transition temperature	K, °C	
u	position parameter	–	
U	voltage	V	
U_H	Hall voltage	V	
U_{ph}	photovoltage	V	
v_{dr}	drift velocity	cm s^{-1}	
$v_{l(t)}, v_{L(T)}$	velocity of longitudinal (transverse) waves	cm s^{-1}	17a/2.1.2
$V_{(m)}$	(molar) volume	cm^3 (mol^{-1})	
x, y, z	fractional coordinates of atoms in unit cell	–	
X_{ik}	stress tensor (6×6) (in the literature often labeled T_{ij})	bar	17a/2.1.2
X_k	stress vector (6-component)	bar	17a/2.1.2
Z	figure of merit	deg dB^{-1}	17h/10.2.2, Fig. 25
Z	number of formula units in unit cell	–	
α	linear thermal expansion coefficient	K^{-1}	17a/2.1.2, eq. (A.17)
$\alpha_{a,b,c}$	linear thermal expansion coefficient in a, b, c direction	K^{-1}	
β	volume thermal expansion coefficient (3α)	K^{-1}	17a/2.1.2, eq. (A.17)
γ	Grüneisen constant	–	17a/2.1.2, eq. (A.18)
Γ	center of Brillouin zone	–	17a/2.4.1, Fig. A.4
δ	Mössbauer isomer shift	m s^{-1}	17h/10.1
Δ	quadrupole splitting	m s^{-1}	17h/10.1
Δ_{cf}	crystal field splitting energy	eV	17h/10.1
Δ_{so}	spin-orbit splitting energy	eV	17h/10.1
ε_0	permittivity of free space ($\varepsilon_0 = 8.85418782\,(5) \cdot 10^{-14}$ F cm^{-1})	F cm^{-1}	
ε	dielectric constant	–	
ε_{ij}	component of dielectric constant tensor	–	
$\varepsilon_{1(2)}$	real (imaginary) part of dielectric constant	–	17a/2.2.2, eq. (A.39)
$\varepsilon(0)$	low frequency dielectric constant	–	17a/2.2.2, eq. (A.53)
$\varepsilon(\infty)$	high frequency dielectric constant	–	17a/2.2.2, eq. (A.53)
ζ	reduced wave vector coordinate	–	
θ_F	Faraday rotation	deg cm^{-1}	17a/2.2.2
θ_K	Kerr ellipticity	deg	17h/10.3
Θ_a	asymptotic Curie temperature	K	
Θ_C	ferroelectric transition temperature	K, °C	17h/10.2
Θ_p	paramagnetic Curie temperature	K	
Θ_{pe}	paraelectric Curie temperature	K	17h/10.4
Θ_D	Debye temperature	K	17a/2.1.2
$\kappa(\kappa_{L,el})$	thermal conductivity (lattice, electronic contribution)	W cm^{-1} K^{-1}	17a/2.2.1, eq. (A.36)
κ	compressibility ($= 1$/bulk modulus)	Pa^{-1}	
$\kappa_{v(l)}$	volume (linear) compressibility	Pa^{-1}	
λ	wavelength	cm	
μ_B	Bohr magneton ($\mu_B = 9.274078(36) \cdot 10^{-24}$ JT^{-1})	J T^{-1}	
$\mu_{n(p)}$	electron (hole) mobility	cm^2 V^{-1} s^{-1}	17a/2.2.1, eq. (A.25)
μ_{dr}	drift mobility	cm^2 V^{-1} s^{-1}	17a/2.2.1

Symbol	Property	Unit	Introduced in section
μ_{H}	Hall mobility	$\mathrm{cm^2\,V^{-1}\,s^{-1}}$	17a/2.2.1, eq. (A.27)
$\mu_{a,b,c}$	mobility in a, b, c direction	$\mathrm{cm^2\,V^{-1}\,s^{-1}}$	
μ_{imp}	mobility in impurity band	$\mathrm{cm^2\,V^{-1}\,s^{-1}}$	17h/10.1
ν	frequency	$\mathrm{s^{-1}}$	
$\nu_{\mathrm{TO(LO)}}$	frequency of transverse (longitudinal) optical phonon	$\mathrm{s^{-1}}$	
$\bar{\nu}$	wavenumber	$\mathrm{cm^{-1}}$	
$\bar{\nu}_{\mathrm{R}}$	Raman wavenumber	$\mathrm{cm^{-1}}$	
π_{ik}	piezoresistance coefficients	$\mathrm{cm^2\,dyn^{-1}}$	17a/2.1.1, eq. (A.32)
ϱ	resistivity	$\Omega\,\mathrm{cm}$	
$\varrho_{a,b,c}$	resistivity in a, b, c direction	$\Omega\,\mathrm{cm}$	
$\varrho_{\mathrm{d,(i)}}$	dark resistivity (ϱ under illumination)	$\Omega\,\mathrm{cm}$	17h/10.1
$\varrho_{\mathrm{H}}^{\mathrm{a}}$	anomalous Hall resistivity	$\Omega\,\mathrm{cm}$	17h/10.2
$\Delta\varrho/\varrho_0$	magnetoresistance	–	17a/2.2.1, eq. (A.30)
$\sigma_{(\mathrm{i})}$	(intrinsic) conductivity	$\Omega^{-1}\,\mathrm{cm^{-1}}$	17a/2.2.1
σ_{ij}	conductivity tensor components	$\Omega^{-1}\,\mathrm{cm^{-1}}$	17a/2.2.1, eq. (A.24)
$\sigma_{a,b,c}$	conductivity in a, b, c direction	$\Omega^{-1}\,\mathrm{cm^{-1}}$	
σ_{d}	dark conductivity	$\Omega^{-1}\,\mathrm{cm^{-1}}$	
σ_{ion}	ionic conductivity	$\Omega^{-1}\,\mathrm{cm^{-1}}$	
σ_{ph}	photoconductivity	$\Omega^{-1}\,\mathrm{cm^{-1}}$	
σ_{m}	molar magnetization (magnetic moment per mol)	$\mathrm{G\,cm^3\,mol^{-1}}$	17h/10.2
$\tau_{(n,p)}$	relaxation time, decay time, rise time, lifetime of carriers	s	
φ_{K}	Kerr rotation angle	deg	17h/10.3
Φ	(electron) irradiation dose	$\mathrm{electrons/cm^2}$	17h/10.1
χ_v	magnetic volume susceptibility	–	
χ_{g},χ	magnetic mass susceptibility	$\mathrm{cm^3\,g^{-1}}$	
χ_{m}	magnetic molar susceptibility	$\mathrm{cm^3\,mol^{-1}}$	
ω	circular frequency	$\mathrm{rad\,s^{-1}}$	
$\hbar\omega$	photon energy	eV	
$\hbar\omega_{\mathrm{ph}},\hbar\omega_q$	phonon energy	eV	

2. List of abbreviations

a	acceptor
a	amorphous
ac	alternating current
arb.	arbitrary
av	average
AF	antiferromagnetically ordered spin system
bcc	body centered cubic
bct	base centered tetragonal
BZ	Brillouin zone
calc	calculated
cub	cubic
crit	critical
CB	conduction band
CDW	charge density wave
d	donor
dc	direct current
dir	direct
DTA	differential thermal analysis
e	electron
el	electronic
eff	effective
epr, EPR	electron paramagnetic resonance
esr, ESR	electron spin resonance
exp	experimental
EMF, emf	electromotive force
EXAFS	extended X-ray absorption fine structure
fcc	face centered cubic
F	ferromagnetic ordered spin system
FE	ferroelectric
h	hole
hcp	hexagonal close packed
hex	hexagonal
HT	high temperature
i	intrinsic; sometimes used for illuminated or incident
ind	indirect
ion	ionic, ionization
ir, IR	infrared
In_{Cd}^{3+}	In ion on Cd site
l, L	longitudinal; sometimes abbreviation for "laser"
l, L, liq	liquid
latt	lattice
LA	longitudinal acoustic
LO	longitudinal optical
LCAO	linear combination of atomic orbitals
Ln	lanthanides
LT	low temperature
m	as subscript: per mol, per molecule
mon	monoclinic (mostly subscript)
magn	magnetic
max	maximum
min	minimum
M, Me	metal
$Mn_{Mn}^{\cdot}$	Mn ion on Mn site, positively charged
oct	octahedral (sometimes o is used as subscript)
orth	orthorhombic

List of abbreviations (continued)

p	pseudocubic (subscript of lattice parameters)
ph	photo-, phonon- (subscript)
pol	polaron
P	paramagnetic
rel	relative
rh	rhombohedral
RDF	radial distribution function
RE, R	rare earth
REEL	reflection electron energy loss spectroscopy
RT	room temperature
s	surface, sometimes used for scattered
s.s.	solid solution
t	tetrahedral
tetr	tetragonal
th	thermal; sometimes for theoretical
tot	total
tr	transition (subscript for phase transition parameters)
t, T	transverse
TA	transverse acoustic
TO	transverse optical
uv, UV	ultraviolet
vac	vacuum, sometimes abbreviation for vacancy
V_{Ga}	vacancy on Ga site
VB	valence band
w	wurtzite
WDS	wavelength derivative spectroscopy
WMR	wavelength modulated reflectance
X	anion (e. g. S, Se, Te)
XPS, XPE	X-ray photoelectron spectroscopy
$\perp$, $\parallel$	perpendicular, parallel to a crystallographic axis

3. Conversion tables

A. Conversion factors from the SIU system to the CGS-esu and the CGS-emu systems.

Quantities	Symbols	SIU	CGS-esu (non-rationalized)	CGS-emu
bulk modulus	B	$Pa\,(=N\,m^{-2})$	$10\;dyn\,cm^{-2}$	$10\;dyn\,cm^{-2}$
magnetic induction	B	$T\,(=Wb\,m^{-2})$ $=Vs\,m^{-2}$	$10^{-6}/3$ esu	10^{4} G
molar heat capacity at const. pressure	C_p	$J\,K^{-1}\,mol^{-1}$	$10^{7}\;erg\,K^{-1}\,mol^{-1}$ $(=0.239\;cal\,K^{-1}\,mol^{-1})$	$10^{7}\;erg\,K^{-1}\,mol^{-1}$ $(=0.239\;cal\,K^{-1}\,mol^{-1})$
elastic moduli (stiffnesses)	c_{lm}	$N\,m^{-2}\,(=Pa)$	$10\;dyn\,cm^{-2}$	$10\;dyn\,cm^{-2}$
density	d	$kg\,m^{-3}$	$10^{-3}\;g\,cm^{-3}$	$10^{-3}\;g\,cm^{-3}$
piezoelectric strain coefficient	d_{ik}	$C\,N^{-1}$ $(=m\,V^{-1})$	$3\cdot10^{4}$ esu	10^{-6} emu
strain tensor	e_{ik}	dimensionless	1 (dimensionless)	1 (dimensionless)
piezoelectric stress coefficients	e_{ik}	$C\,m^{-2}$	$3\cdot10^{5}$ esu	10^{-5} emu
Young's modulus	E	$N\,m^{-2}\,(=Pa)$	$10\;dyn\,cm^{-2}$	$10\;dyn\,cm^{-2}$
electric field strength	E	$V\,m^{-1}$	$10^{-4}/3$ esu	10^{6} emu
piezoelectric strain coefficients	g_{ik}	$m^{2}\,C^{-1}$	$10^{-5}/3$ esu	10^{5} emu
molar free energy change	ΔG	$J\,mol^{-1}$	$10^{7}\;erg\,mol^{-1}$ $(=0.239\;cal\,mol^{-1})$	$10^{7}\;erg\,mol^{-1}$ $(=0.239\;cal\,mol^{-1})$
piezoelectric stress coefficient	h_{ik}	$V\,m^{-1}\,(=N\,C^{-1})$	$10^{-4}/3$ esu	10^{6} emu
hardness	H	$N\,m^{-2}\,(=Pa)$	$10\;dyn\,cm^{-2}$	$10\;dyn\,cm^{-2}$
magnetic field strength	H	$A\,m^{-1}$	$12\pi\cdot10^{7}$ esu	$4\pi\cdot10^{-3}$ Oe
molar enthalpy change	ΔH	$J\,mol^{-1}$	$10^{7}\;erg\,mol^{-1}$ $(=0.239\;cal\,mol^{-1})$	$10^{7}\;erg\,mol^{-1}$ $(=0.239\;cal\,mol^{-1})$
current density	i	$A\,m^{-2}$	$3\cdot10^{5}$ esu	10^{-5} emu
elastoresistance coefficients	m_{ik}	dimensionless	1 (dimensionless)	1 (dimensionless)
pressure	p	$Pa\,(=10^{-5}\,bar)$	$10\;dyn\,cm^{-2}$ $(=1.019\cdot10^{-5}\,kg\,cm^{-2}$ $=7.5\cdot10^{-3}\,Torr)$	$10\;dyn\,cm^{-2}$ $(=1.019\cdot10^{-5}\,kg\,cm^{-2}$ $=7.5\cdot10^{-3}\,Torr)$
dielectric polarization	P	$C\,m^{-2}$	$3\cdot10^{5}$ esu	10^{-5} emu
pyroelectric coefficient	p_i	$C\,m^{-2}\,K^{-1}$	$3\cdot10^{5}$ esu K^{-1}	10^{-5} emu K^{-1}
elastooptic constant (in cubic crystals)	p_{ij}	dimensionless	1 (dimensionless)	1 (dimensionless)
piezooptic constant (in cubic crystals)	q_{ij}	dimensionless	1 (dimensionless)	1 (dimensionless)
Hall coefficient	R_H	$m^{3}\,C^{-1}$	$10^{-3}/3$ esu	10^{7} emu
linear electrooptic constant	r_{ij}	$m\,V^{-1}$	$3\cdot10^{4}$ esu	10^{-6} emu
elastic compliances	s_{ml}	$m^{2}\,N^{-1}$	$10^{-1}\;cm^{2}\,dyn^{-1}$	$10^{-1}\;cm^{2}\,dyn^{-1}$
molar entropy change	ΔS	$J\,K^{-1}\,mol^{-1}$	$10^{7}\;erg\,K^{-1}\,mol^{-1}$ $(=0.239\;cal\,mol^{-1})$	$10^{7}\;erg\,K^{-1}\,mol^{-1}$ $(=0.239\;cal\,mol^{-1})$
stress tensor	X_{ij}	$N\,m^{-2}\,(=Pa)$	$10\;dyn\,cm^{-2}$	$10\;dyn\,cm^{-2}$
thermal conductivity	κ	$W\,m^{-1}\,K^{-1}$ $(=J\,m^{-1}\,s^{-1}\,K^{-1})$	$10^{5}\;erg\,cm^{-1}\,s^{-1}\,K^{-1}$	$10^{5}\;erg\,cm^{-1}\,s^{-1}\,K^{-1}$

(continued)

Quantities	Symbols	SIU	CGS-esu	CGS-emu
			(non-rationalized)	
Conversion factors (continued)				
dielectric constant	ε	dimensionless	1 (dimensionless)	1 (dimensionless)
piezoresistance tensor coefficients	π_{ik}	$m^2\,N^{-1}$	$10^{-1}\,cm^2\,dyn^{-1}$	$10^{-1}\,cm^2\,dyn^{-1}$
piezooptic constant (in cubic crystals)	π_{ik}	$m^2\,N^{-1}$	$10^{-1}\,cm^2\,dyn^{-1}$	$10^{-1}\,cm^2\,dyn^{-1}$
resistivity	ϱ	$\Omega\,m$	$10^{-9}/9$ esu	10^{11} emu
conductivity	σ	$\Omega^{-1}\,m^{-1}$	$9\cdot 10^9$ esu	10^{-11} emu
magnetic volume susceptibility	χ_v	dimensionless	$\dfrac{1}{4\pi}$ (dimensionless)	$\dfrac{1}{4\pi}$ (dimensionless)
magnetic mass susceptiblity	χ_g	$m^3\,kg^{-1}$	$10^3/4\pi$ esu $cm^3\,g^{-1}$	$10^3/4\pi$ emu g^{-1} $(=10^3/4\pi\,cm^3\,g^{-1})$
magnetic molar susceptibility	χ_m	$m^3\,mol^{-1}$	$10^6/4\pi$ esu $cm^3\,mol^{-1}$	$10^6/4\pi$ emu mol^{-1} $(=10^6/4\pi\,cm^3\,mol^{-1})$

Lattice parameters a, b, c are mostly given in Å.

B. Energy conversion

Energy: $E = eV = h\nu = hc\,\bar{\nu}$ $1\,V\,As = 1\,J = 10^7\,erg = 2.38845\cdot 10^{-4}\,kcal$

Energy and equivalent quantities	E	V	ν	$\bar{\nu}$
	J	V	Hz, s^{-1}	cm^{-1}
$1\,J \triangleq$	1	$6.2415\cdot 10^{18}$	$1.50916\cdot 10^{33}$	$5.03403\cdot 10^{22}$
$1\,V \triangleq$	$1.60219\cdot 10^{-19}$	1	$2.41797\cdot 10^{14}$	$8.06547\cdot 10^{3}$
$1\,s^{-1} \triangleq$ $(=1\,Hz)$	$6.62619\cdot 10^{-34}$	$4.13550\cdot 10^{-15}$	1	$3.33564\cdot 10^{-11}$
$1\,cm^{-1} \triangleq$	$1.98648\cdot 10^{-23}$	$1.23979\cdot 10^{-4}$	$2.99792\cdot 10^{10}$	1

Error: In this volume, experimental errors are given in parentheses referring to the last decimal places. For example, 1.352(12) stands for 1.352 ± 0.012 and 342.5(21) stands for 342.5 ± 2.1.

B. Physical data of semiconductors VI

10 Ternary compounds

10.1 Tetrahedrally bonded ternary and quasi-binary compounds

10.1.0 Introduction, general remarks on structure and properties

The compounds presented in this chapter all obey the so-called Grimm-Sommerfeld rule. That is the average number of valence electrons per atom is four in a tetrahedral structure [26G]. For example: C: $(1 \cdot 4)/1 = 4$, GaAs: $(1 \cdot 3 + 1 \cdot 5)/2 = 4$, ZnSe: $(1 \cdot 2 + 1 \cdot 6)/2 = 4$, ZnGeAs$_2$: $(1 \cdot 2 + 1 \cdot 4 + 2 \cdot 5)/4 = 4$, AgGaSe$_2$: $(1 \cdot 1 + 1 \cdot 3 + 2 \cdot 6)/4 = 4$, $\square$-CdGa$_2$Se$_4$: $(1 \cdot 0 + 1 \cdot 2 + 2 \cdot 3 + 4 \cdot 6)/8 = 4$, $\square$-In$_2$Te$_3$: $(1 \cdot 0 + 2 \cdot 3 + 3 \cdot 6)/6 = 4$, where $\square$ signifies a vacancy. Such a vacancy is counted as an atom of zero valency. The above procedure can be generalized to give a large number of possible ternary, quaternary, quintinary and even more complicated multinary compounds [64P].

Structure: Compounds of this type generally adopt structures related to zincblende or wurtzite. The few exceptions and some unusual variations will be discussed under the relevant compounds. The simplest case is that where the several cations, including vacancies, are randomly distributed on the zincblende or wurtzite sublattices. In this case X-ray pictures look like somewhat diffuse examples of zincblende or wurtzite. Most ternary compounds adopt such a structure at higher temperatures and many, especially the more complicated ones, have only been prepared in this phase. It is, however, impossible in many cases to distinguish between such a disordered zincblende or wurtzite structure and simple mixed crystals such as CuZnAlSe$_4$ or solid solutions of CuAlSe$_2$ and ZnSe. Disordered phases which can be decomposed in this way will not be listed separately.

Most I–III–VI$_2$ and II–IV–V$_2$ compounds (sections 10.1.2 and 10.1.3) take the *chalcopyrite* structure (Fig. 1), named after the mineral CuFeS$_2$. This structure (space group D_{2d}^{12}–$\bar{I}42d$) contains two formula units in the primitive cell (four formula units in the conventional trigonal cell). The atoms are distributed as follows [64P, 75S, 65W]:

compound ABC$_2$: A on $(0\,0\,0)$, $(0\frac{1}{2}\frac{1}{4})$, B on $(0\frac{1}{2}0)$, $(0\frac{1}{2}\frac{3}{4})$, C on $(x\frac{1}{4}\frac{1}{8})$, $(\bar{x}\frac{3}{4}\frac{1}{8})$, $(\frac{3}{4}x\frac{1}{8})$, $(\frac{1}{4}\bar{x}\frac{7}{8})$.

Thus the structure is completely defined by the lattice parameters a and c and the parameter x. Most of the $\square$–II–III$_2$–VI$_4$ compounds (section 10.1.6) take one of the two structures of Fig. 2: *defect stannite* (space group D_{2d}^{11}–$\bar{I}42m$) and *defect chalcopyrite* (space group S_4^2–$\bar{I}4$) [64P, 65W]. The $\bar{I}4$ structure is sometimes referred to as *thiogallate* since it is adopted by the A-Ga$_2$S$_4$ compounds. In defect stannite (D_{2d}^{11}–$\bar{I}42m$) the atoms are distributed as follows:

compound $\square$-AB$_2$C$_4$: $\square$ on $(0\,0\,0)$, A on $(0\,0\frac{1}{2})$, B on $(0\frac{1}{2}\frac{1}{4})$, $(\frac{1}{2}0\frac{1}{4})$, C on $(x\,x\,z)$, $(\bar{x}\,\bar{x}\,z)$, $(\bar{x}\,x\,\bar{z})$, $(x\,\bar{x}\,\bar{z})$;

whereas in defect chalcopyrite (S_4^2–$\bar{I}4$) the distribution is as follows: $\square$ on $(0\,0\,0)$, A on $(0\frac{1}{2}\frac{1}{4})$, B on $(0\,0\frac{1}{2})$, $(\frac{1}{2}0\frac{1}{4})$, C on $(x\,y\,z)$, $(\bar{x}\,\bar{y}\,z)$, $(\bar{y}\,x\,\bar{z})$, $(y\,\bar{x}\,\bar{z})$.

Thus the $\bar{I}42m$ structure has two free parameters, "x" and "z", and $\bar{I}4$ has three, "x", "y" and "z".

In the ideal zincblende-like structure the lattice constants are related by $c = 2a$ and the free parameters have the values $x = y = \frac{1}{4}$; $z = \frac{1}{8}$. The values of x, y and z are difficult to measure accurately. Only in the chalcopyrites are the published values generally accepted so that they will be given only for these systems.

In the case of the II–IV–V$_2$ chalcopyrites (section 10.1.3) there is a well established relationship between "x" and the tetragonal compression $(2 - c/a)$. The tetravalent element (Si, Ge, Sn) preserves its ideal tetrahedral coordination [75S]. Thus:

$$x = \tfrac{1}{2} - (c^2/32\,a^2 - 1/16)^{1/2}.$$

This tendency exists for the corresponding trivalent element in I–III–VI$_2$ compounds (section 10.1.2) but is not nearly so strong.

A third structure commonly adopted by II–III$_2$–VI$_4$ compounds is *spinel* (Fig. 3) [65W].

This is not so closely related to zincblende but is a structure midway between zincblende and rock-salt. The atoms are distributed as follows:

compound AB$_2$C$_4$: A on (000), $(\frac{1}{4}\frac{1}{4}\frac{1}{4})$, B on $(\frac{1}{8}\frac{7}{8}\frac{3}{8})$, $(\frac{1}{8}\frac{5}{8}\frac{1}{8})$, $(\frac{3}{8}\frac{7}{8}\frac{1}{8})$, $(\frac{3}{8}\frac{5}{8}\frac{3}{8})$, C on $(\frac{1}{8}\frac{1}{8}\frac{3}{8})$, $(\frac{1}{8}\frac{3}{8}\frac{1}{8})$, $(\frac{3}{8}\frac{1}{8}\frac{1}{8})$, $(\frac{3}{8}\frac{3}{8}\frac{3}{8})$, $(\frac{1}{8}\frac{5}{8}\frac{3}{8})$, $(\frac{1}{8}\frac{7}{8}\frac{1}{8})$, $(\frac{3}{8}\frac{5}{8}\frac{1}{8})$, $(\frac{3}{8}\frac{7}{8}\frac{3}{8})$.

This is the so called *normal spinel* structure. The "A" atoms are on tetrahedral sites, as in zincblende, whereas the "B" atoms are on octahedral sites, as on rocksalt.

An alternative is for one half of the "B" atoms to go on to the tetrahedral sites and half on the octahedral sites. The "A" atoms are then all on the octahedral sites. We then have an *inverse spinel*.

It is also possible to find variations between these two extremes, so called *partially inverted spinels*.

The atoms on tetrahedral sites are more covalently bonded whereas those on octahedral sites are more ionic. Thus in II–III$_2$–VI$_4$ compounds, like CdInS$_4$, the more ionic II–VI bond causes the type II atom (Cd) to go on to the octahedral site, giving inverse spinel. On the other hand the somewhat larger In atoms prefer the larger octahedral sites. Result: a compromise, partial inversion.

General remarks on band structure and dispersion relations: Since chalcopyrite, defect stannite and defect chalcopyrite have the same Bravais lattice and similar space groups, they have many properties in common. Fig. 4 shows the Brillouin zone of fcc (zincblende) and two Brillouin zones of chalcopyrite. Each point in the chalcopyrite Brillouin zone maps on to 4 different points of the zincblende zone. Thus the Γ point corresponds to Γ, X and two W points.

A useful starting point for a description of the electronic and vibrational properties of these compounds is the band structure or dispersion curves of the zincblende parent compound represented in the smaller Brillouin zone of the ternary [81M] (see Fig. 4). These "folded back" curves can be labelled with the irreducible representations to which they correspond in both the chalcopyrite etc. or zincblende space groups (I$\bar{4}$2d and F$\bar{4}$3m, respectively). A further analysis treats the differences between the binary and ternary compounds as perturbations. These are usually of three types:

a) Cationic asymmetry: the simple fact that there is more than one cation.

b) Anionic displacement: the anions are no longer on sites with tetrahedral symmetry. They may move slightly from the ideal places (the parameters x, y, z).

c) Tetragonal compression: the lattice is no longer cubic. $c \neq 2a$.

Consideration of these perturbations in several orders of perturbation theory leads to a division of possible physical processes into four classes:

allowed strong: allowed in zincblende and ternary (compound);

allowed weak: allowed in first order perturbation theory;

allowed very weak: allowed in higher order;

forbidden: forbidden to all orders.

Very weak processes are difficult to observe in practice.

In the chalcopyrite system, but not in defect stannite or defect chalcopyrite, matrix elements of the type $\langle\Gamma|\mathscr{H}|X\rangle$ are zero by symmetry. Thus all physical processes involving transitions $\Gamma - X$ or mixing of Γ and X like states are very weak [75S, 81M]. Thus in chalcopyrite like ZnGeP$_2$ where the corresponding zincblende compound GaP has an indirect band gap between Γ and X, the chalcopyrite retains the character of an indirect band gap compound, although the gap is formally direct. Such cases are referred to as *pseudodirect*. (In fact the minimum of the conduction band of GaP is along the Δ direction close to the X-point, but this does not affect the general principle.)

Tetragonal crystals are usually optically anisotropic. That is absorption etc. experiments carried out with $E \| c$ give different results from those with $E \perp c$.

The Γ_{15} state at the top of the valence band of zincblende splits into Γ_4 and Γ_5 in the ternary; the p states split into p$_z$ and p$_x$, p$_y$ [81M]. When the conduction band is Γ_1, s-like states, only the $\Gamma_{4v} \rightarrow \Gamma_{1c}$ transition is seen for $E \| c$ and $\Gamma_{5v} \rightarrow \Gamma_{1c}$ for $E \perp c$. Unfortunately this simple picture is somewhat complicated by the effects of spin-orbit coupling which splits the Γ_{5v} state. The possible transitions (A, B, C) are as in Fig. 5.

Hopfield's quasi cubic model for the splitting of p-like states in the presence of a noncubic crystal field and spin-orbit splitting gives the formula:

$$E_{1,2} = -\tfrac{1}{2}(\Delta_{so} + \Delta_{cf}) \pm \tfrac{1}{2}[(\Delta_{so} + \Delta_{cf})^2 - \tfrac{8}{3}\Delta_{so}\Delta_{cf}]^{1/2} \qquad [60H]$$

where Δ_{so} is the spin-orbit splitting in a cubic crystal field and Δ_{cf} is the crystal field splitting of the valence bands in the absence of spin-orbit coupling. The two solutions E_1 and E_2 give the separation of the two Γ_7 states (Fig. 5) from the Γ_6.

In chalcopyrites it is found that the crystal field splitting is negative, that is Γ_4 lies above Γ_5 and that this splitting is due almost entirely to the tetragonal compression. It compares very well with that expected from the corresponding zincblende compound under uniaxial pressure [75S] (See also Fig. 10α of section 10.1.2 for general features of the band structure in I–III–VI$_2$ compounds.)

On the other hand in the vacancy compounds the crystal field splitting is positive, Γ_4 lies below Γ_5 and is not correlated with the tetragonal compression. In this case the $\langle \Gamma | \mathscr{H} | X \rangle$ matrix elements are non-zero and the Δ_{cf} is due to the interaction of the Γ_{15} states of zincblende with the X_5 states which are quite close in energy [81M].

Effective masses: In the tetragonal system the energy bands near $k = 0$ have the form:

$$E = E_0 \pm \frac{\hbar^2}{2}\left(\frac{k_x^2 + k_y^2}{m_\perp} + \frac{k_z^2}{m_\parallel}\right)$$

where $m_\perp$ and $m_\parallel$ are the effective masses for $k \perp c$ and $k \parallel c$, respectively. From these two other effective masses are often defined: a density of states effective mass $m_{ds}^3 = m_\perp^2 m_\parallel$ and a conductivity effective mass

$$\frac{1}{m_c} = \frac{1}{3}\left(\frac{1}{m_\parallel} + \frac{2}{m_\perp}\right)$$

Lattice properties: Phonons are also treated by the "folding back" method. In some chalcopyrites, however, the simple perturbation theory is rendered invalid by accidental degeneracies. In the vacancy structure it is necessary, to take account of the loss of three degrees of freedom in the creation of vacancies [81M].

References for 10.1.0

26G   Grimm, H.G., Sommerfeld, A.: Z. Phys. **36** (1926) 36.
60H   Hopfield, J.J.: J. Phys. Chem. Solids **15** (1960) 97.
64P   Parthé, E.: Crystal Chemistry of tetrahedral structures. New York: Gordon and Breach, **1964**.
65W   Wyckoff, R.W.G.: Crystal Structures Vol. 2. New York: Interscience, **1965**.
69B   Berger, L.T., Prochukhan, V.: Ternary diamond-like Semiconductors. New York: Consultants Bureau, **1969**.
75S   Shay, J.L., Wernick, J.H.: Ternary Chalcopyrite Semiconductors: growth, electronic, properties and applications. Pergamon, **1975**.
81M   Miller, A., MacKinnon, A., Weaire, D.: Beyond the Binaries – the chalcopyrite and related semiconductor compounds, in Solid State Physics. Academic Press, **1981**.

10.1.1 III$_2$–VI$_3$ compounds

Ga$_2$S$_3$, Ga$_2$Se$_3$, Ga$_2$Te$_3$, In$_2$S$_3$, In$_2$Se$_3$, In$_2$Te$_3$.

No semiconducting properties have been reported for the compounds Al$_2$S$_3$, Al$_2$Se$_3$, Al$_2$Te$_3$, Tl$_2$S$_3$, Tl$_2$Se$_3$, Tl$_2$Te$_3$; Al$_2$Se$_3$ however has a rich chemical literature [1827W].

10.1.1.0 Structure

III$_2$–VI$_3$ compounds generally form tetrahedral structures since when vacancies are counted as zero valent atoms, they obey the Grimm-Sommerfeld rule:

$$\square\text{-III}_2\text{-VI}_3 = (1 \cdot 0 + 2 \cdot 3 + 3 \cdot 6)/6 = 4$$

(see section 10.1.0 for a general discussion).

It should however be noted that the trivalent element (Al, Ga, In) is generally overstoichiometric. The most important structural properties are summarized in the following table (there is some confusion about the terminology of the different phases in the literature.)

Substance	Structure, space group	Lattice parameters	Remarks	Ref.
α-Al$_2$S$_3$	ordered vacancies, wurtzite type: D$_{6h}$ (?), hexagonal 6 molecules/unit cell	$a = 6.423\,(3)$ Å $c = 17.83\,(2)$ Å	stable at $T = 300$ K; $d = 2.32$ g cm^{-3}	51F1
β-Al$_2$S$_3$	disordered vacancies, wurtzite type: D$^4_{6h}$—P6$_3$/mmc or C$^4_{6v}$—P6$_3$mc	$a = 3.579\,(4)$ Å $c = 5.829\,(7)$ Å	stable at $T = 1270 \cdots 1370$ K in the presence of Al$_4$C$_3$; $d = 2.495(15)$ g cm^{-3}	51F2
γ-Al$_2$S$_3$	ordered vacancies, wurtzite type: trigonal 6 molecules/unit cell	$a = 6.47$ Å $c = 17.26$ Å	stable at $T = 1270 \cdots 1370$ K; $d = 2.36$ g cm^{-3}	52F
δ-Al$_2$S$_3$	defect spinel type: D$^{19}_{4h}$—I4$_1$/amd	$a = 7.026$ Å $c = 29.819$ Å	stable at $T = 1270 \cdots 1370$ K; $p = 2 \cdots 65$ kbar	70D2
ε-Al$_2$S$_3$	defect spinel type: O^7_h—Fd3m	$a = 9.94$ Å	stable at $T = 673$ K; $p = 40$ kbar	73R1
ζ-Al$_2$S$_3$	defect spinel type: O^7_h—Fd3m vacancies disordered on tetrahedral sites	$a = 9.93$ Å	stable at $T = 300$ K with 2at% As; $d = 2.80$ g cm^{-3}	63S1
Al$_2$Se$_3$	disordered vacancies, wurtzite type: C$^4_{6v}$—P6$_3$mc	$a = 3.890\,(4)$ Å $c = 6.30\,(1)$ Å	stable at $T = 300$ K; $d = 3.91$ g cm^{-3}; possible ordering of vacancies [52G]; possible high temperature phase [53G]	54S
Al$_2$Te$_3$	defect-spinel type: O^7_h—Fd3m	$a = 10.45$ Å	high pressure phase ($p = 60$ kbar, $T = 870$ K); $d = 4.51$ g cm^{-3}	73R1
	disordered vacancies, wurtzite type: C$^4_{6v}$—P6$_3$mc	$a = 4.08$ Å $c = 6.94$ Å	structure identified by comparison with Al$_2$Se$_3$; $d = 4.54$ g cm^{-3}	59M
α-Ga$_2$S$_3$	ordered vacancies, wurtzite type: C^4_s—Bb	$a = 11.094$ Å $b = 9.578$ Å $c = 6.395$ Å $\gamma = 141°15'$	prepared by the reaction 3H$_2$S + 2Ga(OH)$_3$ = Ga$_2$S$_3$ + 6H$_2$O at $T = 1020$ K	76C
	disordered C$^2_6$—P6$_1$ or D$^3_6$—P6$_5$			55H

(continued)

table (continued)

Substance	Structure, space group	Lattice parameters	Remarks	Ref.
β-Ga$_2$S$_3$	disordered vacancies, wurtzite type: C_{6v}^4—P6$_3$mc	$a = 3.678\,(5)$ Å $c = 6.016\,(6)$ Å	prepared as above at $T = 820$ K; $d = 3.65$ g cm^{-3}	49H1
γ-Ga$_2$S$_3$	disordered vacancies, zincblende type: T_d^2—F$\bar{4}$3m	$a = 5.17$ Å	prepared as above at $T = 873$ K; $d = 3.63$ g cm^{-3} (N.B. [49H1] calls this α-Ga$_2$S$_3$)	49H1
α-Ga$_2$Se$_3$	disordered vacancies, zincblende type: T_d^2—F$\bar{4}$3m	$a = 5.418\,(1)$ Å	$d = 4.92$ g cm^{-3}	49H1
β-Ga$_2$Se$_3$.	ordered vacancies, zincblende type: D_{4h}^{20}—I4$_1$/acd	$a = 23.235$ Å $c = 10.828$ Å	stable with excess Se	77K1
			$d = 4.91$ g cm^{-3}	80T
Ga$_2$Te$_3$	disordered vacancies, zincblende type: T_d^2—F$\bar{4}$3m	$a = 5.874\,(5)$ Å $\quad\;\; 5.899$ Å	$d = 5.57$ g cm^{-3}	49H1 77B
	ordered vacancies, zincblende type: orthorhombic $8 \times$ Ga$_2$Te$_3$/unit cell	$a = \;\; 4.17$ Å $\simeq a_0/\sqrt{2}$ $b = 23.60$ Å $\simeq 4a_0$ $c = 12.52$ Å $\simeq 3a_0/\sqrt{2}$	doubtful (a_0: lattice constant of ZnS subcell)	63N
	wurtzite like structure		see transport data	56H
α-In$_2$S$_3$	disordered defect spinel type: O_h^7—Fd3m tetrahedral sites full, octahedral sites: In and vacancies	$a = 5.36$ Å	stable at $T = 693\cdots1023$ K (see also Fig. 9) stable at $T = 300$ K with excess In [70D1]; may be badly crystallized β-In$_2$S$_3$ [73S, 74A1, 76K, 75E] $d = 4.63$ g cm^{-3}; probably contains voids	49H2 49H2
β-In$_2$S$_3$	ordered defect spinel type: D_{4h}^{19}—I4$_1$/amd octahedral sites full, tetrahedral 2/3; see Fig. 2	$a = \;\; 7.618\,(1)$ Å $c = 32.33\,(1)$ Å	stable up to $T = 693$ K, but see above; $d = 4.613$ g cm^{-3} (N.B. α–β-transition temperature therefore uncertain)	70L, 76K
γ-In$_2$S$_3$	layered structure (—S—In—S—In—S—$\cdots$) S close packed, In octahedral, D_{3d}^3—P$\bar{3}$m1	$a = 3.8$ Å $c = 9.04$ Å	stable above $T = 1023$ K annealing produces α or β phase [78B1]; stable at $T = 300$ K with As or Sb; As, Sb in tetrahedral positions between layers [73D1, 76D] $d = 4.75\,(8)$ g cm^{-3}, with As, $d = 4.80\,(8)$ g cm^{-3}, with Sb	70D1 76D

(continued)

table (continued)

Substance	Structure, space group	Lattice parameters	Remarks	Ref.
In_2Se_3	see below			
α-In_2Te_3	ordered vacancies, zincblende type: T_d^2—$F\bar{4}3m$ $9 \times In_2Te_3$/unit cell	$a = 18.486\,(20)$ Å	contains a Te atom with no first neighbors; probably mixture of 2 phases (see below) $d = 5.79\,\mathrm{g\,cm^{-3}}$	54I, 60Z1, 71Z 60Z1
α-In_2Te_3–I	ordered vacancies, zincblende type, defect famatinite: D_{2d}^{11}—$I\bar{4}2m$	$a = 6.173\,(10)$ Å $c = 12.438\,(73)$ Å	really In_3Te_4 with less In; has been prepared as single crystal (N.B. disobeys the Grimm-Sommerfeld rule.)	75K, 57G
α-In_2Te_3–II	ordered vacancies, zincblende type: C_{2v}^{20}—Imm2, see Fig. 3	$(a_0 \simeq 6.163$ Å$)$ $a = a_0/\sqrt{2}$ $b = a_0 3/\sqrt{2}$ $c = a_0$	not yet prepared as single crystal	59W, 75K, 78B2
β-In_2Te_3	disordered vacancies, zincblende type: T_d^2—$F\bar{4}3m$	$a = 6.163$ Å	phase change: $T = 790 \cdots 890$ K, stable at lower temperature with excess In (almost always the case) Phase change slow, goes through partially ordered transition state [76B] $d = 5.73\,\mathrm{g\,cm^{-3}}$	49H2, 60Z1 60Z1

In_2Se_3: The literature on the structure of In_2Se_3 is confused and in some respects contradictory. Therefore only those features will be reproduced here on which there is reasonable agreement.

Below 923 K the structure is composed of layers, which are only weakly bound to each other. Each layer contains planes of Se and In in the sequence —Se—In—Se—In—Se— where the Se atoms in each plane build a triangular lattice with lattice constant: $a = 4.025\,(25)$ Å at RT.

The three Se planes are stacked in ABA (hcp) or ABC (fcc) sequence, each layer having a thickness of 9.6 (1) Å. The In atoms occupy either the tetrahedral or octahedral sites or, more probably, some intermediate, non-ideal positions.

The layers are stacked in AAA [78L1], ABAB [61S1, 61S2, 71P, 76P1, 78L2], or ABC [66O] sequence, giving a lattice vector perpendicular to the layers of $c = 9.6$ (1) Å, 19.2 (2) Å or 28.5 (3) Å (at RT), respectively. (Numbers estimated from above references.)

The space group generally has a 3-fold or 6-fold axis in the c direction.

In addition the layers may be distorted, especially at low temperatures giving rise to a superlattice structure. (N.B. There is some confusion about the nomenclature of the phases.)

Phases (see also Figs. 7, 8): $a' = 7.1$ Å $\approx a\sqrt{3}$; $a' = 16$ Å $\approx 4a$

Substance	T [K]	Remarks	Ref.
α'-In_2Se_3:	< 148	large superstructure; possibly charge density wave	74L, 75L
α-In_2Se_3:	148 $\cdots$ 473	see "Transport properties" below	
	472 $\cdots$ 475	1st order phase transition	74C
β-In_2Se_3:	473 $\cdots$ 923	can be supercooled, possibly to room temperature; see "Transport properties"	78L2
γ-In_2Se_3:	923 $\cdots$ 1023	cubic (non layered) structure; probably decomposition phase In_5Se_3	61S1, 61S2, 79P
δ-In_2Se_3:	1023 $\cdots$ 1163	monoclinic (non-layered) structure; probably decomposition phase In_6Se_7	79P

For a recent review, see [79P].

Solid solutions:

Composition	Structure, space group	Lattice parameters	Remarks	Ref.
$(Ga_2Se_3)_x(In_2Se_3)_{1-x}$:				
$1 \geq x > 0.88$	defect zincblende type	$a = 5.426\,(3)$ Å $a = 5.453\,(3)$ Å	extra diffraction line	80T
$0.88 > x > 0.75$	2 phase region			
$0.75 > x \geq 0.60$	γ_2-region: see below			
$0.60 > x > 0.55$	2 phase region			
$0.55 > x > 0.02$	γ_1-region: like γ-In_2Se_3			
$0.02 > x > 0.005$	$\gamma_1 + \alpha$-In_2Se_3			
$x = 0.6$	γ_2-phase hexagonal structure	$a = 6.864\,(3)$ Å $c = 19.324\,(5)$ Å	stable at $T = 300$ K	80T
	zincblende type	$a = 5.5478\,(10)$ Å	after heating above $T = 1100$ K and slow cooling	80T
$x = 0.4$	γ_1-phase D_6^2—$P6_122$ or D_6^3—$P6_522$	$a = 6.98\,(5)$ Å $c = 18.96\,(5)$ Å	stable down to $T = 0$ K	80T
$x = 0.25$	structure as at $x = 0.4$	$a = 7.043\,(2)$ Å $c = 19.130\,(10)$ Å	stable down to $T = 0$ K	80T
$(AgInTe_2)_{3x}(In_2Te_3)_{2(1-x)}$:				
$x = 0.0$	zincblende like	$a = 6.16$ Å	β-form	75M
0.1	zincblende like			
0.2	thiogallate (defect chalcopyrite)			
0.25	thiogallate			
0.3	thiogallate			
0.4	thiogallate			
0.5	thiogallate			
0.7	chalcopyrite			
0.8	chalcopyrite			
0.9	chalcopyrite			
1.0	chalcopyrite			
$(In_2Te_3)_x(Bi_{0.5}Sb_{1.5}Te_3)_{(1-x)}$:				
	space group: D_{3d}^5—$R\bar{3}m$ phase diagram: Fig. 5	see Fig. 4 for compositional dependence of lattice parameter	up to 33 mol% In_2Te_3	73A
$(In_2Te_3)_x(BiTe_{3-y}Se_y)_{(1-x)}$:				
	phase diagram: Fig. 6			
$(In_2Se_3)_x(Sb_2Se_3)_{(1-x)}$:				
	phase diagram: Fig. 7			
α-$InSbSe_3$	orthorhombic	$a = 9.45$ Å $b = 13.9$ Å $c = 3.92$ Å		77G
β-$InSbSe_3$	monoclinic	$a = 6.86$ Å $b = 3.93$ Å $c = 9.38$ Å	monoclinic angle not given	77G

(continued)

Solid solutions (continued)

Composition	Structure, space group	Lattice parameters	Remarks	Ref.
$(In_2Se_3)_x(Sb_2Te_3)_{(1-x)}$:				
	space group: D_{3d}^5—$R\bar{3}m$	see Fig. 30a		72B1
	phase diagram: Fig. 8			
$3\,In_2Se_3$		$a=$ 4.04 Å		72B1
$\cdot\,2Sb_2Te_3$		$c=29.35$ Å		
$(In_2S_3)_xSe_{3(1-x)}$:				
	phase diagram: Fig. 9			
$(In_2Se_3)_xS_{3(1-x)}$:				
	phase diagram: Fig. 10			

Physical property	Numerical value	Experimental conditions	Experimental method remarks	Ref.

10.1.1.1   Gallium sulfide ($\square$ -Ga_2S_3)

Electronic properties

There are no measurements reported for γ-Ga_2S_3.

α-Ga_2S_3:
energy gap:

		T [K]		
E_g	3.438 eV	1.6	transmission and reflectivity,	77M2
	3.437 eV	4.2	see Fig. 11	
	3.424 eV	77	(no polarization dependence given)	
dE_g/dT	$-3.8\cdot10^{-4}$ eV K^{-1}	1.6$\cdots$4.2		77M2
	$1.8\cdot10^{-4}$ eV K^{-1}	4.2$\cdots$77		

other transition energies (in eV):

		T [K]		
E	3.388	1.6	transmission and reflectivity	77M2
	3.424	1.6	probably excitonic	
	3.387	4.2	(no polarization dependence	
	3.423	4.2	given)	
	3.375	77		
	3.410	77		

β-Ga_2S_3:
energy gap (in eV):

		T [K]		
E_g	2.5$\cdots$2.7	290	optical absorption	59G
	2.6$\cdots$2.9	77.3		
	2.48	290	photoelectric effect	59G

Lattice and further properties

For infrared and Raman spectra in α- and β-Ga_2S_3, see Fig. 12; for dielectric constant, see [68K].

β-Ga_2S_3:
phonon frequencies:

ν_{TO}	$9.72\cdot10^{12}$ s^{-1}	RT	infrared reflectivity	77M1
	$10.44\cdot10^{12}$ s^{-1}		(no polarization dependence given)	
			(see also section 10.1.1.7 (Ga_2S_3—Ga_2Se_3))	

Physical property	Numerical value	Experimental conditions	Experimental method remarks	Ref.
γ-Ga$_2$S$_3$:				
dielectric constants:				
$\varepsilon(0)$	7.5	RT	infrared reflectivity	73M
$\varepsilon(\infty)$	5.8			
phonon frequencies:				
ν_{TO}	$9.45 \cdot 10^{12}$ s^{-1}	RT	infrared reflectivity, see Fig. 13	73M
ν_{LO}	$10.8 \cdot 10^{12}$ s^{-1}			
melting point:				
T_m	1360 K	$p = 1.7 \cdot 10^{-2}$ bar		76C

10.1.1.2 Gallium selenide (☐-Ga$_2$Se$_3$)

Electronic properties

β-Ga$_2$Se$_3$:

energy gaps (in eV):

		T [K]		
$E_{g, dir}$	2.1···2.2	77.3	optical absorption	59G
	1.9···2.1	290		
	1.75···1.9	290	photoelectric effect	59G
$E_{g, ind}$	2.04	77	electroabsorption	72A
	1.96	300		
	1.98	300	optical absorption	76P1
$E_{g, th}$	2.01	360···430	conductivity vs. temperature	76P1

optical transition energies from reflection spectra (in eV):

E_g	2.00	$T = 300$ K	see also Fig. 31	72B3
E_1	3.08			
E_2	3.88			
E_3	4.69			
E	2.28	77 K	electroabsorption, indirect transition	72A

Lattice and further properties

For IR reflectivity and Raman spectrum of α-Ga$_2$Se$_3$, see Fig. 14.

linear expansion coefficient:

α	$10.2(64) \cdot 10^{-6}$ K^{-1}		β-Ga$_2$Se$_3$	80T

melting point:

T_m	1005(3) K			74S

10.1.1.3 Gallium telluride (☐-Ga$_2$Te$_3$)

Electronic properties

energy gaps:

$E_{g, th}$	1.55 (3) eV	$T = 293$ K ··· 1063 K	conductivity vs. temperature	56H
E_g	1.22 eV	273 K	optical reflection	56H
	1.23 (3) eV	273 K	photoemission	55G
			reinterpreted	56H

Transport properties

The transport is intrinsic even with up to 9at% Cu or other elemental impurities including excess Ga [77S]. Even after irradiation the conductivity is unchanged [73K]. No magnetoresistance was measureable [56H].

Physical property	Numerical value	Experimental conditions	Experimental method remarks	Ref.
conductivities (in $\Omega^{-1}\,cm^{-1}$, $T=273$ K):				
σ	$1.1 \cdot 10^{-7}$			73K
	$2.1 \cdot 10^{-4}$		with 20 at% Cu becomes a two phase system	77S
	$9.9 \cdot 10^{-2}$		with 1 at% Cu_2Te	
	$1.6 \cdot 10^{-3}$		with 1 at% Cu Te	

For ac conductivity, see Fig. 15. A switching effect with memory has been reported [79A2].

Further properties
linear expansion coefficient:

α	$8.3\,(3) \cdot 10^{-6}\,K^{-1}$	$T=300$ K		61W

melting point:

T_m	1063 K			80T

10.1.1.4 Indium sulfide ($\square$-In_2S_3)

If not noted otherwise the data refer to n-type β-In_2S_3.

Electronic properties
energy gaps (in eV):

		T [K]		
$E_{g,\,ind}$	1.1	300	Hall effect	65R
$E_{g,\,dir}$	2.03	300	absorption (probably forbidden at extreme point) (no polarization dependence given)	
$dE_{g,\,dir}/dT$	$-7 \cdot 10^{-4}\,eV\,K^{-1}$	$77\cdots360$		
$E_{g,\,th}$	2.30		field emission	70L
γ-In_2S_3:As:				
E_g	1.88	300	absorption	73D2
$E_{g,\,th}$	1.38 (15)		conductivity vs. T	
γ-In_2S_3:Sb:				
E_g	1.44	300	absorption	73D2
$E_{g,\,th}$	1.42 (15)		conductivity vs. T	
further optical transition:				
E	2.45 eV	$T=300$ K	absorption	65R
structure in valence band (energy below E_F, in eV):				
E	1.5		X-ray photoemission	78I
	3.9			
	5.9			

Transport properties
resistivity: Fig. 18; Hall effect: Fig. 19; thermoelectric power: Fig. 20

resistivity:
γ-In_2S_3:As:

ϱ	$9\,(5) \cdot 10^7\,\Omega\,cm$	$T=300$ K	(no orientation dependence given)	73D2

(continued)

Physical property	Numerical value	Experimental conditions	Experimental method remarks	Ref.
resistivity (continued)				
γ-In$_2$S$_3$:Sb:				
ϱ	3 (2) $\cdot$ 10^4 Ω cm	$T = 300$ K		73D2
Optical properties				
dielectric constant:				
$\varepsilon(0)$	13.5	$T = 300$ K		63G2
$\varepsilon(\infty)$	6.5			
refractive indices:				
γ-In$_2$S$_3$:As:				
n	3.06 (15)	$T = 300$ K		73D2
γ-In$_2$S$_3$:Sb:				
n	2.84 (15)	$T = 300$ K		73D2

trap distribution from luminescence and field emission: Fig. 16; IR absorption: Fig. 17

Further properties
thermal expansion: Fig. 21

melting point:

T_m	1363 K			75D1

10.1.1.5 Indium selenide ($\square$-In$_2$Se$_3$)

Electronic properties:
energy gaps (in eV):

		T [K]		
E_g	1.2$\cdots$1.5	RT, $E \perp c$	α- and β-phases, absorption and photoconductivity	71B
	1.39	RT	thin film	72P
$E_{g, ind}$	1.10 (5)	100, $E \perp c$	α-phase, absorption	71M
	1.0	300		
	0.94 (5)	480		
	1.01	300, $E \perp c$	β-phase, absorption	
$E_{g, dir}$	1.16	300, $E \perp c$	β-phase, absorption	71M
$E_{g, th}$	1.41	$E \parallel c$	β-phase, conductivity	71B
	0.62		film on pyrex, deposition $T < 470$ K	72P
	0.88		film on mica, deposition $T < 470$ K	
	1.34		film on pyrex, deposition $T > 493$ K	
	0.96		film on mica, deposition $T > 493$ K	

Lattice properties
phonon frequencies (in 10^{12} s^{-1}):

ν_{TO}	2.723	RT, $E \perp c$	α-phase, infrared reflectivity	78K
	4.82			
	5.624			
ν_{LO}	2.835	RT, $E \perp c$	α-phase, infrared reflectivity, Kramers-Kronig analysis	78K
	5.286			
	6.564			

There is a large background signal, in quenched samples, possibly due to the mobility of the In ions [78K].

Physical property	Numerical value	Experimental conditions	Experimental method remarks	Ref.
dielectric constants:				
$\varepsilon(0)$	16.68	$E \perp c$	infrared reflectivity	78K
$\varepsilon(\infty)$	9.53	$E \perp c$		

Transport properties

In the α-phase the conductivity is metallic, although the optical gap is greater than 1 eV. This is interpreted as being due to shallow donors. The β-phase shows intrinsic semiconducting behavior. The σ vs. T curves (Fig. 22) show a large hysteresis due to the supercooling of the β-phase. See also 10.1.1.7 (Solid solutions).

conductivity (α-phase):				
σ	$2.7 \, \Omega^{-1} \, \text{cm}^{-1}$	RT, $E \perp c$	undoped	71B
	$0.8 \, \Omega^{-1} \, \text{cm}^{-1}$	RT, $E \parallel c$	undoped	
Hall coefficients (in cm^3 C^{-1}) **(α-phase):**				
R_H	-4.5	RT	undoped (no orientation given)	71B
	-2.7		In doped	
	-12.0		Se doped	

Pressed pellets have activated conductivity for all temperatures [71C].

mobility of charge carriers (α-phase):				
μ_n	$10 \, \text{cm}^2 \, \text{V}^{-1} \, \text{s}^{-1}$	$T = 300$ K	Hall effect	74R

The magnetoresistance (Fig. 23) is negative [74R].

Further properties

linear expansion coefficient:

α	$12.4 \cdot 10^{-6} \, \text{K}^{-1}$	$T = 290 \cdots 470$ K	parallel to layer	75G
	$11.5 \cdot 10^{-6} \, \text{K}^{-1}$	(α-phase)	perpendicular to layer	

lattice heat capacity: Fig. 24

melting point:				
T_m	1163 K			76C

10.1.1.6 Indium telluride ($\square$-In$_2$Te$_3$)

All data for α-In$_2$Te$_3$ are for the cubic form (see section 10.1.1.0 "Structure").

Electronic properties

energy gaps (in eV):

		T [K]		
E_g	1.121	82	β-In$_2$Te$_3$, absorption	72V
	1.0	295		
	1.026	290	α-In$_2$Te$_3$, absorption	60P2
	1.02	354	β-In$_2$Te$_3$, absorption	
	1.15 (5)	≈ 0	absorption (phase not stated)	69K
			The optical gap is probably direct [75H], see Fig. 25; see also 10.1.1.7 (Solid solutions).	
	0.94		α-In$_2$Te$_3$, photoemission	60P2
	0.92		β-In$_2$Te$_3$, photoemission	
			E_g reported as indirect in [60P2]	
$E_\text{g, th}$	1.02		U–I characteristic	77M3
	1.12 (5)		α-In$_2$Te$_3$, Hall conductivity	60Z2
	1.55 (10)	> 743	conductivity	61W
$\text{d}E_\text{g}/\text{d}T$	$-5.6 \cdot 10^{-4} \, \text{eV K}^{-1}$	$82 \cdots 295$	β-In$_2$Te$_3$, absorption	72V
	$-3.4 \cdot 10^{-4} \, \text{eV K}^{-1}$	$\approx 0 \cdots 300$	α-In$_2$Te$_3$, Hall effect	60Z2

Physical property	Numerical value	Experimental conditions	Experimental method remarks	Ref.
effective masses (in m_0):				
		T [K]		
m_n	0.70	334	α-In$_2$Te$_3$, Hall effect	60Z2
	0.71	417		
	0.67	556		
m_p	1.23	334	α-In$_2$Te$_3$, Hall effect	60Z2
	1.27	417		
	1.20	556		
Lattice and optical properties				
sound velocity:				
v_l	$2.56 \cdot 10^5$ cm s^{-1}		longitudinal	60Z3
refractive index, dielectric constant:				
n	3.4 (3)	$\lambda = 2.2$ μm	α-In$_2$Te$_3$	60Z2
$\varepsilon(0)$	16		α-In$_2$Te$_3$	60Z2
imaginary part of the dielectric constant: Fig. 26				
Transport properties				
There is little difference between the properties of the two phases α and β.				
intrinsic electron concentration (in cm^{-3}):				
n_i	$7.68 \cdot 10^{12}$	$T = 334$ K	α-In$_2$Te$_3$,	60Z2
	$4.1 \cdot 10^{13}$	417 K	T dependence of Hall constant	
	$1.6 \cdot 10^{15}$	556 K		
resistivity:				
ϱ	$1.8 \cdot 10^6$ Ω cm		before and after quenching; conductivity vs. T: Fig. 27	75D3
mobility (in cm^2 V^{-1} s^{-1}):				
μ_n	5$\cdots$70	$T = 0 \cdots 300$ K	Hall effect	60Z2
	32	RT	before quenching from 823 K (i.e. α-In$_2$Te$_3$)	75D3
	28	RT	after quenching from 823 K (i.e. β-In$_2$Te$_3$)	
Hall coefficient (in cm^3 C^{-1}):				
		T [K]		
R_H	$-1.6 \cdot 10^7$	334	α-In$_2$Te$_3$, see also Fig. 27	60Z2
	$-2.8 \cdot 10^5$	417		
	$-7.9 \cdot 10^3$	556		
thermoelectric power (Seebeck coefficient):				
S	-400 μV K^{-1}	$T = 714$ K	α-In$_2$Te$_3$	60Z2
	-960 μV K^{-1}	333 K		

doping: Mg, Cd, Cu, Hg, Sb, Sn, Zn, Si, Ge show no doping effect for concentrations up to 1 at% [79D, 60Z2]. Bi gives n-type, I gives p-type behavior [60Z2]. The conductivity and photoconductivity are insensitive to irradiation by γ-rays or neutrons [76A].

trap levels:				
E_t	0.3 eV		photoconductivity	77M3
	0.34 eV			79A1

Physical property	Numerical value	Experimental conditions	Experimental method remarks	Ref.
donor levels:				
E_d	0.04 eV		σ vs. T	77M3
	0.03 eV			
	0.24 eV			
Further properties				
linear thermal expansion coefficient:				
α	$1.15\,(10)\cdot10^{-4}\,\text{K}^{-1}$	$T=290\cdots470$ K		61W
thermal conductivity:				
κ	$11.22\cdot10^{-3}\,\text{W cm}^{-1}\,\text{K}^{-1}$	$T=300$ K	α-In$_2$Te$_3$	60P1,
	$6.95\cdot10^{-3}\,\text{W cm}^{-1}\,\text{K}^{-1}$	300 K	β-In$_2$Te$_3$	60Z3
	$7.87\cdot10^{-3}\,\text{W cm}^{-1}\,\text{K}^{-1}$	100 K	β-In$_2$Te$_3$	
melting point:				
T_m	940 K			60Z3

10.1.1.7 Solid solutions

(Ga$_2$Se$_3$)$_x$(In$_2$Se$_3$)$_{1-x}$
linear expansion coefficients (in 10^{-6} K^{-1}):

α_a	13.8 (8)	$x=0.6$;	hexagonal	80T
α_c	7.8 (4)	$T=300\cdots1100$ K		
α	11.10 (8)	$x=0.6$;	cubic	80T
		$T=300\cdots1100$ K		
		(after		
		quenching)		
α_a	4.70 (7)	$x=0.4$;		80T
α_c	38 (4)	$T=300\cdots1100$ K		
α_a	5.3 (5)	$x=0.25$;		80T
α_c	32 (1)	$T=793\cdots1003$ K		

melting points:

T_m	1143 K	$x=0.6$	cubic	80T
	1103 K	$x=0.4$	hexagonal	80T

(In$_2$Se$_3$)$_x$(Bi$_2$Se$_3$)$_{1-x}$
Transport properties: see Fig. 28

(AgInTe$_2$)$_x$(In$_2$Te$_3$)$_{1-x}$
Physical properties (at RT) for some compositions [75M]: (Polarization or orientation dependence not given.)

x	$E_{g,\,dir}$ eV	$E_{g,\,ind}$ eV	ϱ Ω cm	$\varepsilon(0)$	$\varepsilon(\infty)$
0.0	1.14	1.04	10^6	16.0	11.8
0.1	1.06	0.98	10^6	17.0	12.2
0.2	1.04	0.95	10^5	16.7	12.5
0.25	1.18	1.00	10^5	15.0	11.0
0.3	1.12	0.95	10^5	16.1	11.9
0.4	1.06	0.93	10^5	16.5	12.3
0.5	0.97	0.90	10^5	17.0	12.8
0.7	0.95	0.88	10^4	17.2	13.0
0.8	0.98	0.90	10^4	16.9	12.7
0.9	1.05	0.94	10^4	16.5	12.3
1.0	1.12	0.99	10^5	16.1	11.9

Physical property	Numerical value	Experimental conditions	Experimental method remarks	Ref.

For energy gap vs. composition: see Fig. 29.

$(In_2Se_3)_x(Sb_2Te_3)_{1-x}$: conductivity: Fig. 30b; thermoelectric power: Fig. 30c

$(Ga_2Se_3)_x(Cu_2Te_3)_{1-x}$:
conductivities (in $10^{-2} \, \Omega^{-1} \, cm^{-1}$):

			fraction Cu_2Te_3 (in mol%)	
σ	9.9	$T = 300$ K	99	77S
	0.1		98	
	4.2		95	
	1.0		90	
	1.26		85	

$(Ga_2Te_3)_x(Cu_2Te_3)_{1-x}$:
conductivities (in $\Omega^{-1} \, cm^{-1}$):

			fraction Cu_2Te_3 (in mol%)	
σ	$1.6 \cdot 10^{-3}$	$T = 300$ K	99	77S
	$7 \cdot 10^{-6}$		98	

$(Ga_2S_3)_x(Ga_2Se_3)_{1-x}$: optical reflection: Fig. 31

$(In_2Te_3)_x(Tl_2Te_3)_{1-x}$: electrical conductivity: Fig. 32

10.1.1.8 References for 10.1.1

1827W	Wöhler, F.: Poggendorfs Ann. Phys. Chem. **11** (1827) 160.
49H1	Hahn, H., Klinger, W.: Z. Anorg. Chem. **259** (1949) 135.
49H2	Hahn, H., Klinger, W.: Z. Anorg. Chem. **260** (1949) 97.
51F1	Flahaut, J.: Compt. Rend. **232** (1951) 334.
51F2	Flahaut, J.: Compt. Rend. **232** (1951) 2100.
52F	Flahaut, J.: Ann. Chim. (Paris) **7** (1952) 632.
52G	Geiersberger, K., Galster, H.: Angew. Chem. **64** (1952) 81.
53G	Giersberger, K., Galster, H.: Z. Anorg. Allg. Chem. **274** (1953) 289.
54I	Inuzaka, H., Sugaike, S.: Proc. Jpn. Acad. **30** (1954) 383.
54S	Schneider, A., Gattow, G.: Z. Anorg. Allg. Chem. **277** (1954) 49.
55G	Gorgonova, N.A., Grigoreva, V.S., Konovalenko, B.M., Ryvkin, S.M.: Zh. Tekh. Fiz. **25** (1955) 1675.
55H	Hahn, H., Frank, G.: Z. Anorg. Allg. Chem. **278** (1955) 333.
56H	Harbeke, G., Lautz, G.: Z. Naturforsch. **11a** (1956) 1015.
57G	Gaines, R.V.: Am. Mineral. **42** (1957) 766.
59G	Gross, E.F., Navikov, B.V., Razbirin, B.S., Suslina, L.G.: Opt. i. Spektrosk. **6** (1959) 569.
59M	Mirgalovskaya, M., Skudanova, E.: Izv. Akad. Nauk SSSR, Otd. Tekh'-Nauk Metall. Toplivo **4** (1959) 148.
59R	Rooymans, C.J.M.: J. Inorg. Nucl. Chem. **11** (1959) 78.
59W	Wooley, J.C., Pamplin, B.R., Holmes, P.T.: J. Less-Common Met. **1** (1959) 362.
60P1	Petrusevich, V.A., Sergeeva, V.M., Smirnov, I.A.: Fiz. Tverd. Tela **2** (1960) 2894; Sov. Phys.-Solid State (English Transl.) **2** (1961) 2573.
60P2	Petrusevich, V.A., Sergeeva, V.M.: Fiz. Tverd. Tela **2** (1960) 2881; Sov. Phys.-Solid State (English Transl.) **2** (1961) 2562.
60Z1	Zaslavskii, A.I., Sergeeva, V.M.: Fiz. Tverd. Tela **2** (1960) 2872; Sov. Phys.-Solid State (English Transl.) **2** (1961) 2556.
60Z2	Zhuse, V.P., Sergeeva, V.M., Shelykh, A.I.: Fiz. Tverd. Tela **2** (1960) 2858; Sov. Phys.-Solid State (English Transl.) **2** (1961) 2545.
60Z3	Zaslavskii, A.I., Sergeeva, V.M., Smirnov, I.A.: Fiz. Tverd. Tela **2** (1960) 2885; Sov. Phys.-Solid State (English Transl.) **2** (1961) 2565.

61S1	Semelitov, S.A.: Kristallografiya **6** (1961) 200; Sov. Phys.-Crystallogr. (English Transl.) **6** (1961) 158.
61S2	Semelitov, S.A.: Doklady Akad. Nauk SSSR **137** (1961) 584; Sov. Phys. Doklady (English Transl.) **6** (1961) 189.
61W	Wooley, J.C., Pamplin, B.R.: J. Electrochem. Soc. **108** (1961) 874.
62K	King, G.S.D.: Acta Crystallogr. **15** (1962) 512.
63G1	Goodyear, J., Steigman, G.A.: Acta Crystallogr. **16** (1963) 946.
63G2	Garlick, G.F., Springford, M., Checinska, H.: Proc. Phys. Soc. **82** (1963) 16.
63N	Newman, P.C., Cundall, J.A.: Nature (London) **200** (1963) 876.
63S1	Schäfer, H., Schäfer, G., Weiss, A.: Z. Anorg. Allg. Chem. **325** (1963) 77.
63S2	Slavona, G.K.: Zh. Neorg. Khim. **8** (1963) 1161.
65R	Rewald, W., Harbeke, G.: J. Phys. Chem. Solids **26** (1965) 1309.
66O	Osamura, K., Murakami, Y., Tomiie, Y.: J. Phys. Soc. Jpn. **21** (1966) 1848.
67N	Neuman, H.B.: Z. Naturforsch. **22a** (1967) 1012.
68K	Kurtz, S.K.: IEEE J. Quantum Electron. **4** (1968) 578.
69K	Koshkin, V.M., Karas, V.R., Gal'chinetskii, L.P.: Fiz. Tekh. Poluprov. **3** (1969) 1417; Sov. Phys.-Semicond. (English Transl.) **3** (1970) 1186.
70D1	Diehl, R., Nitsche, R., Ottermann, J.: Naturwissenschaften **51** (1970) 670.
70D2	Donohue, P.C.: J. Solid State Chem. **2** (1970) 6.
70G	Gukhman, G., Kostova, M.G., Nikolava, C.M.: C. R. Acad. Bulg. Sci. **23** (1970) 1067.
70L	Lutz, H.D., Häuseler, H.: Z. Naturforsch. **26a** (1970) 323.
71B	Bidgin, D., Popovic, S., Celustka, B.: Phys. Status Solidi (a) **6** (1971) 295.
71C	Celustka, B., Bidgin, D., Popovic, S.: Phys. Status Solidi (a) **6** (1971) 699.
71K	Koshkin, V.M., Gal'chinetskii, L.P., Korin, A.I.: Fiz. Tekh. Poluprovodn. **5** (1971) 1983; Sov. Phys.-Semicond. (English Transl.) **5** (1972) 1718.
71M	Mushinskii, V.P., Kobolev, V.I., Andronck, I.Ya.: Fiz. Tekh. Poluprovodn. **5** (1971) 1251; Sov. Phys.-Semicond. (English Transl.) **5** (1971) 1104.
71N	Nieke, H., Sandt, P.R.: Ann. Physik **26** (1971) 235.
71P	Popovic, S., Celustka, B., Bidgin, D.: Phys. Status Solidi (a) **6** (1971) 301.
71Z	Zaslavskii, A.I., Kartenko, N.F., Karachentseva, Z.A.: Fiz. Tverd. Tela **13** (1971) 1562; Sov. Phys.-Solid State (English Transl.) **13** (1971) 2252.
72A	Askerov, I.M., Gadzhiev, V.A., Guseinova, E.S., Tagiev, B.G.: Phys. Status Solidi (b) **50** (1972) K113.
72B1	Belotskii, D.P., Legeta, L.V.: Izv. Akad. Nauk SSSR, Neorg. Mater. **8** (1972) 1160; Inorg. Mater. (USSR) (English Transl.) **8** (1972) 1017.
72B2	Belotskii, D.P., Legeta, L.V.: Izv. Akad. Nauk SSSR, Neorg. Mater. **8** (1972) 1908; Inorg. Mater. (USSR) (English Transl.) **8** (1972) 1677.
72B3	Bakhyshev, A.E., Musaeva, L.G., Akhundov, G.A.: Phys. Status Solidi (b) **54** (1972) K77.
72P	Persin, M., Persin, A., Celustka, B., Ettinger, B.: Thin Solid Films **11** (1972) 153.
72V	Verkelis, I.Yu.: Fiz. Tverd. Tela **14** (1972) 1676; Sov. Phys. Solid State (English Transl.) **14** (1972) 1445.
73A	Abrikosov, N.K., Makareeva, E.G.: Izv. Akad. Nauk SSSR, Neorg. Mater. **9** (1973) 2103; Inorg. Mater. (USSR) (English Transl.) **9** (1974) 1859.
73B	Buck, P.: J. Appl. Crystallogr. **6** (1973) 1.
73D1	Diehl, R., Nitsche, R., Carpentier, C.-D.: J. Appl. Crystallogr. **6** (1973) 497.
73D2	Diehl, R., Nitsche, R.: J. Cryst. Growth **20** (1973) 38.
73F1	Finkman, E., Tauc, J.: Phys. Rev. Lett. **31** (1973) 890.
73F2	Finkman, E., Tauc, J., Kershaw, R., Wold, A.: Phys. Rev. **B11** (1973) 3785.
73K	Koshkin, V.M., Gal'chinetskii, L.P., Kulik, V.N., Minkiev, B.I., Ulmanis, U.A.: Solid State Commun. **13** (1973) 1.
73M	Musaeva, L.G., Akhundov, G.A., Bakhyshev, A.E., Gasanli, N.M.: Phys. Status Solidi (b) **60** (1973) K1.
73R1	Range, K.J., Hübner, H.-J.: Z. Naturforsch. **28b** (1973) 353.
73R2	Rustamov, P.G., Melikova, Z.D.: Izv. Akad. Nauk SSSR, Neorg. Mater. **9** (1973) 1673; Inorg. Mater. (USSR) (English Transl.) **9** (1974) 1492.
73S	Shafizade, R.B., Efendiev, E.G., Aliev, F.I.: Kristallografiya **18** (1973) 660; Sov. Phys.-Crystallogr. (English Transl.) **18** (1973) 417.
74A1	Aliev, F.I., Efendiev, E.G., Shafizade, R.B.: Izv. Akad. Nauk Az. SSR, Ser. Fiz. Tekh. i. Mat. Nauk **4** (1974) 23.

74A2	Abrikosov, N.K., Makareeva, E.G.: Izv. Akad. Nauk SSSR, Neorg. Mater. **10** (1974) 1204; Inorg. Mater. (USSR) (English Transl.) **10** (1974) 1032.
74C	Chistov, S.F., Boryakova, V.A., Grinberg, Ya.Kh.: Izv. Akad. Nauk SSSR, Neorg. Mater. **10** (1974) 1531; Inorg. Mater. (USSR) (English Transl.) **10** (1974) 1319.
74L	van Landuyt, J., van Tendeloo, G., Amelinckx, S.: Phys. Status Solidi (a) **26** (1974) K103.
74R	Romeo, N.: Phys. Status Solidi (a) **26** (1974) K187.
74S	Suzuki, H., Mori, R.: Jpn. J. Appl. Phys. **13** (1974) 417.
75D1	Diehl, R., Nitsche, R.: J. Cryst. Growth **28** (1975) 306.
75D2	Demidenko, A.F., Koshchenko, V.I., Grinberg, Ya.Kh., Borgakova, V.A., Gastev, S.V.: Izv. Akad. Nauk SSSR, Neorg. Mater. **11** (1975) 2141; Inorg. Mater. (USSR) (English Transl.) **11** (1975) 1839.
75D3	Dmitriev, Y.N., Kulik, V.N., Gal'chinetskii, L.P., Koshkin, V.M.: Fiz. Tverd. Tela **17** (1975) 3685; Sov. Phys. Solid State (English Transl.) **17** (1975) 2396.
75E	Efendiev, E.G., Aliev, F.I., Shafizade, R.B.: Izv. Akad. Nauk Az. SSR, Ser. Fiz. Tekh. i Mat. Nauk **1** (1975) 23.
75G	Gasanov, G.Sh., Mamedov, K.P., Suleimanov, Z.Y., Bagirov, S.B.: Izv. Akad. Nauk Az. SSR **4** (1975) 65.
75H	Hughes, O.H., Nikolic, P.M., Doran, C.J., Vujatovic, S.S.: Phys. Status Solidi (b) **71** (1975) 105.
75K	Karakostas, T., Economou, N.A.: Phys. Status Solidi (a) **31** (1975) 89.
75L	van Landuyt, J., van Tendelov, G., Amelinckx, S.: Phys. Status Solidi (a) **30** (1975) 299.
75M	Mirgorodskii, V.M., Radautsan, S.I., Tsurkan, A.E.: Opt. Spektrosk. **39** (1975) 519; Opt. Spectrosc. (USSR) (English Transl.) **39** (1975) 291.
76A	Anan'ina, D.B., Bakumenko, V.L., Grushka, G.G., Zaitov, F.A., Kurbatov, L.N., Shalyapina, L.M.: Izv. Akad. Nauk SSSR, Neorg. Mater. **12** (1976) 2074; Inorg. Mater. (USSR) (English Transl.) **12** (1976) 1699.
76B	Bleris, G.L., Karakostas, T., Stoemenos, J., Economou, N.A.: Phys. Status Solidi (a) **34** (1976) 243.
76C	Collin, G., Flahaut, J., Guittard, M., Loireau-Lozach, A.M.: Mater. Res. Bull. **11** (1976) 285.
76D	Diehl, R., Carpentier, C.-D., Nitsche, R.: Acta Crystallogr. **B32** (1976) 1257.
76K	Kundra, K.D., Ali, S.Z.: Phys. Status Solidi (a) **36** (1976) 517.
76M	Mills, K.C.: High Temp.-High Pressure **8** (1976) 225.
76P1	Persin, M., Celustka, B., Popovic, S., Persin, A.: Thin Solid Films **37** (1976) L61.
76P2	Popovic, S., Celustka, B., Bidgin, D.: Phys. Status Solidi (a) **33** (1976) K23.
77B	Burlaku, G.G., Markus, M.M., Tyrzin, V.G.: Izv. Akad. Nauk SSSR, Neorg. Mater. **13** (1977) 820.
77G	Guliev, T.N., Magerramov, E.V., Rustamov, P.G.: Izv. Ak. Nauk SSSR, Neorg. Mater. **13** (1977) 627; Inorg. Mater. (USSR) (English Transl.) **13** (1977) 514.
77K1	Khan, M.Y.: J. Appl. Crystallogr. **10** (1977) 70.
77K2	Karakostas, T., Flevaris, N.F., Vlachavas, N., Bleris, G.L., Economou, N.A.: Acta Crystallogr. **A34** (1978) 123.
77M1	Musaeva, L.G., Khomutova, M.D., Gasanly, N.M.: Fiz. Tverd. Tela **19** (1977) 1766; Sov. Phys. Solid State (English Transl.) **19** (1977) 1030.
77M2	Mushinskii, V.P., Palaki, L.I., Chebotaru, V.V.: Phys. Status Solidi (b) **83** (1977) K149.
77M3	Mamedov, A.S., Mamedov, K.P., Gasanov, G.Sh., Bagirov, S.B., Nifkiev, G.M.: Izv. Akad. Nauk SSSR, Neorg. Mater. **13** (1977) 1987; Inorg. Mater. (USSR) (English Transl.) **13** (1978) 1592.
77S	Sheikh-Zananova, R.N., Ponomarev, V.F.: Izv. Akad. Nauk SSSR, Neorg. Mater. **13** (1977) 1308; Inorg. Mater. (USSR) (English Transl.) **13** (1977) 1056.
78B1	Bartzokas, D., Manolikas, C., Spyridelis, J.: Phys. Status Solidi (a) **47** (1978) 459.
78B2	Bleris, G.L., Karakostas, T., Economou, N.A., de Ridder, R.: Phys. Status Solidi (a) **50** (1978) 579.
78G	Gadzhiev, A.R., Niftiev, G.M., Tagiev, B.G.: Fiz. Tekh. Poluprovodn. **12** (1978) 174; Sov. Phys. Semicond. (English Transl.) **12** (1978) 100.
78I	Ihara, H., Abe, H., Endo, S., Irie, T.: Solid State Commun. **28** (1978) 563.
78K	Kambas, K., Spyridelis, J.: Mater. Res. Bull. **13** (1978) 653.
78L1	Lucazeau, G., Leroy, J.: Spectrosc. Acta **34A** (1978) 29.
78L2	Likforman, A., Carré, D., Hillel, R.: Acta Crystallogr. **B34** (1978) 1.
79A1	Anan'ian, D.B., Bakumenko, V.L., Bonakov, A.K., Grushka, G.G., Kurbatov, L.N.: Fiz. Tekh. Poluprovodn. **13** (1979) 961; Sov. Phys. Semicond. (English Transl.) **13** (1979) 561.

79A2   Aliev, F.I., Niftiev, G.M., Pliev, F.I., Tagiev, B.G.: Fiz. Tekh. Poluprovodn. **13** (1979) 579; Sov. Phys. Semicond. (English Transl.) **13** (1979) 340.

79D   Drabkin, I.A., Moizes, B.Y., Sanfirov, Y.B.: Fiz. Tekh. Poluprovodn. **13** (1979) 134; Sov. Phys. Semicond. (English Transl.) **13** (1979) 77.

79P   Popovic, S., Tonejc, A., Grzeta-Plenkovic, B., Celustka, B., Trojko, R.: J. Appl. Crystallogr. **12** (1979) 416.

80T   Tonejc, A., Popovic, S., Grzeta-Plenkovic, B.: J. Appl. Crystallogr. **13** (1980) 24.

10.1.2 I–III–VI₂ compounds

(included are I–Fe–VI$_2$ compounds)

10.1.2.0 Structure and energy gaps

Nearly all of the I–III–VI$_2$ compounds adopt the chalcopyrite structure (space group D_{2d}^{12}—I$\bar{4}$2d). At high temperatures there is a phase transition to a disordered zincblende-like structure, whereas at high pressures a transition to a NaCl like structure is common, such as also occurs in zincblende compounds.

The most important structural properties are summarized in the following tables.

chalcopyrite structures (RT values):

Substance	a [Å]	c [Å]	c/a	x	d [g cm^{-3}]	Ref.
CuAlS$_2$	5.32 (1)	10.430 (15)	1.960 (1)			75S2
					3.45	53H
CuAlSe$_2$	5.61 (1)	10.92 (6)	1.95 (1)			75S2
					4.69	53H
CuAlTe$_2$	5.96 (1)	11.77 (3)	1.97			75S2
					5.47	53H
CuGaS$_2$	5.35 (2)	10.46 (2)	1.960 (2)	0.2539		75S2
					4.38	77B4
CuGaSe$_2$	5.61 (1)	11.00 (2)	1.960 (4)			75S2
					5.57	77B4
CuGaTe$_2$	6.00 (1)	11.93 (2)	1.985 (5)			75S2
					5.95	77B4
CuInS$_2$	5.52 (1)	11.08 (6)	2.00 (1)	0.2296		75S2
					4.74	77B4
CuInSe$_2$	5.78 (1)	11.55 (2)	2.00	0.224		77H1
					5.77	77B4
CuInTe$_2$	6.17 (1)	12.34 (2)	2.00			75S2
					6.10	77B4
CuTlS$_2$	5.58	11.16	2.00			75S2
					6.13	79G2
CuTlSe$_2$	5.83	11.60	1.99			75S2
					7.08	53H
AgAlS$_2$	5.70 (1)	10.26 (2)	1.80			75S2
					3.93	53H
AgAlSe$_2$	5.96 (1)	10.76 (3)	1.85 (4)			75S2
					4.99	53H
AgAlTe$_2$	6.30 (1)	11.84 (1)	1.88			75S2
					6.15	53H
AgGaS$_2$	5.75 (1)	10.29 (2)	1.790 (1)	0.2908		75S2
					4.70	72A2
AgGaSe$_2$	5.98 (1)	10.88 (1)	1.820 (3)			75S2
					5.70	72A2
AgGaTe$_2$	6.29 (1)	11.95 (1)	1.90			75S2
					6.08	72A2

(continued)

chalcopyrite structures (continued)

Substance	a [Å]	c [Å]	c/a	x	d [g cm^{-3}]	Ref.
AgInS$_2$	5.82 (1)	11.17 (2)	1.92			75S2
					4.97	53H
AgInSe$_2$	6.095 (15)	11.69 (3)	1.92 (1)			75S2
					5.82	72A2
AgInTe$_2$	6.43 (3)	12.59 (4)	1.96			75S2
					6.05	77B4
(AgTlSe$_2$	9.70	8.25	(space group D_{3d}^1—$P\bar{3}1m$)			65I)
(AgTlTe$_2$	3.92	15.22	(space group D_{2d}^9—$I\bar{4}m2$)			64I)
(CuFeS$_2$	5.29 (1)	10.41 (1)	1.97 (1)		4.18	72A2)
(AgFeSe$_2$	6.58	8.96	(tetragonal)			58Z)

For further values of lattice parameters, see x = 0 and x = 1 values of the solid solutions in Figs. 3···6, 9 and 10.

high-temperature phases (at $p = 1$ bar):

Substance	T [K]	Structure	Remarks	Ref.
CuInSe$_2$	1083	disordered zincblende	bulk material	77H1
	870	disordered zincblende	thin films	78S3
CuInTe$_2$	950	disordered zincblende		80T
CuTlSe$_2$	<680			75S1
CuTlTe$_2$	<650			75S1
AgInSe$_2$	1047.5	see Fig. 1		76K5
		disordered zincblende		77J
		$a = 6.00···6.06$ Å		
AgInTe$_2$	910 (10)	see Fig. 2		76K5
CuFeS$_2$	830	disordered zincblende		75S1
	930	?	2nd order transition	

high-pressure phases (at $T = 300$ K):

Substance	Transition pressure [bar]	Structure	Remarks	Ref.
CuAlS$_2$	$1.5 \cdot 10^5$	disordered NaCl		76J
CuGaS$_2$	$1.5 \cdot 10^5$	disordered NaCl		76J
CuInS$_2$			no transition up to $5 \cdot 10^4$ bar	77J
CuInSe$_2$	$8 \cdot 10^4$	disordered NaCl		77S1
	$6 \cdot 10^4$	$a = 5.47$ Å		77J
CuInTe$_2$	?	$a = 6.12$ Å		77J
AgGaS$_2$	$4.2 \cdot 10^4$	?		80C
	$1.16 \cdot 10^5$	disordered NaCl or α-NaFeO$_2$ structure		
AgGaSe$_2$			no transition up to 10^5 bar	77J
AgInSe$_2$	see Fig. 1			
	$2.4 \cdot 10^4$	disordered NaCl	diamond anvil	77J
	(see also Fig. 32, conflicting data)	$a = 5.69$ Å	$p = 1$ bar (metastable)	
		5.60 Å	$2.4 \cdot 10^4$ bar	
		5.55 Å	$5 \cdot 10^4$ bar	
		5.51 Å	$8 \cdot 10^4$ bar	
			$d = 6.42$ g cm^{-3}	

(continued)

high-pressure phases (continued)

Substance	Transition pressure [bar]	Structure	Remarks	Ref.
$AgInSe_2$		α-$NaFeO_2$ type $a=$ 3.9467 Å $c=19.884$ Å	$p=1$ bar (metastable) (see Fig. 1 for T-p diagram)	77J
$AgInTe_2$	see Fig. 2 $3.0\cdot 10^4$	disordered NaCl $a=6.02$ Å 5.96 Å	diamond anvil $p=1$ bar (metastable) $3.8\cdot 10^4$ bar	77J

other phases:

Substance	Temperature	Structure	Remarks	Ref.
$AgInS_2$	RT	wurtzite like, orthorhombic $a=6.954$ Å $b=8.264$ Å $c=6.683$ Å	Li doped (6%), stable at all $T<T_m$	74S
	RT	$a=7.00$ Å $b=8.28$ Å $c=6.70$ Å	quenched from temperatures slightly below T_m (pseudowurtzite platelets)	

Solid solutions:

Substance	Lattice parameters	Remarks	Ref.
$Cu_{1-x}Ag_xAlX_2$ (X=S, Se, Te)	Fig. 3		
$Cu_{1-x}Ag_xGaX_2$ (X=S, Se, Te)	Fig. 4		
$Cu_{1-x}Ag_xInX_2$ (X=S, Se, Te)	Fig. 5		
$MAlS_{2x}Se_{2(1-x)}$ (M=Cu, Ag)	Fig. 6a		
$CuGaS_{2x}Se_{2(1-x)}$	Fig. 6b		
x=0.75	$a=5.412$ Å, $c=10.599$ Å		73R
x=0.5	5.478 Å, 10.718 Å		
x=0.25	5.511 Å, 10.882 Å		
$AgGaS_{2x}Se_{2(1-x)}$	Fig. 6b	phase diagram: Fig. 7	
$CuInS_{2x}Se_{2(1-x)}$	Fig. 6c	phase diagram: Fig. 8	
x=0.5	$a=5.653$ Å, $c=11.264$ Å		73R
$AgInS_{2x}Se_{2(1-x)}$	Fig. 6c		
$CuGa_{1-x}In_xS_2$	Fig. 9		
$CuGa_{1-x}In_xSe_2$	Fig. 10		
x=0.4	$a=5.680$ Å, $c=11.264$ Å		80G1
x=0.7	5.736 Å, 11.448 Å		
$AgGa_{1-x}In_xSe_2$			
x=0.2	$a=6.012$ Å, $c=11.028$ Å		80G1
x=0.3	6.023 Å, 11.133 Å		
x=0.6	6.053 Å, 11.359 Å		
x=0.8	6.076 Å, 11.533 Å		

energy gaps:

The electronic energy gaps of the I–III–VI$_2$ compounds at room temperature are summarized in the following table. Typical features of the band structure are shown in Fig. 10α. For further values and details of electronic structure, see under the appropriate compound:

Substance	Energy gap eV	Type		Further data in section	Ref.
CuAlS$_2$	3 ?	(pseudodirect or indirect)		10.1.2.1	77R
	3.49	direct			77R, 75S1
CuAlSe$_2$	2.65	direct	$(T=110 \text{ K})$	10.1.2.2	77Y
CuAlTe$_2$	2.06	direct		–	75S1
CuGaS$_2$	2.43	direct		10.1.2.3	75S1
CuGaSe$_2$	1.68	direct		10.1.2.4	75S1
CuGaTe$_2$	1.0···1.24	direct		10.1.2.5	79N2
CuInS$_2$	1.53	direct		10.1.2.6	75S1
CuInSe$_2$	1.04	direct	$(T=77 \text{ K})$	10.1.2.7	75S1
CuInTe$_2$	1.06	direct		10.1.2.8	77T3, 77T4
AgAlTe$_2$	2.27	direct		–	75S1
AgGaS$_2$	2.64	direct		10.1.2.9	76T
AgGaSe$_2$	1.80	direct		10.1.2.10	76T, 77S3
AgGaTe$_2$	1.32	direct		10.1.2.11	75S1
AgInS$_2$	1.87	direct	(chalcopyrite)	10.1.2.12	74S
	2.081	direct	(orthorhombic)		
AgInSe$_2$	1.24	direct		10.1.2.13	75S1
AgInTe$_2$	0.95	thermal	$(T=0 \text{ K})$	10.1.2.14	75S1

Physical property	Numerical value	Experimental conditions	Experimental method remarks	Ref.

10.1.2.1 Copper aluminum sulfide (CuAlS$_2$)

Electronic properties

band structure: Fig. 10β

energy gap:

$E_{g, \text{dir}}$	(A)	3.49 eV	$T=300 \text{ K}$	electroreflectance; for transitions A, B, C, see Fig. 5 of 10.1.0; see also Fig. 11	75S1
	(B, C)	3.62 eV			

excitonic energy gap (in eV):

$E_{g x}$	(A)	3.54	$T=78 \text{ K}$	absorption	73B
	(B, C)	3.71			
	(A)	3.51	20 K	phase-shift-difference reflection (see also Fig. 34)	77Y
	(B, C)	3.57			

splitting energies at Γ:

Δ_{cf}	−0.13 eV	$T=300 \text{ K}$	electroreflectance	75S1
Δ_{so}	0 eV	300 K		

For higher structures in reflectivity, see Fig. 11.
Possible indirect or pseudodirect gap: 3 eV. The valence band has 35% d-like character [75S1].

Impurities and defects

unidentified donor:

E_{d}	0.4 eV	$T=2 \text{ K}$	photoluminescence	75B1
	1.31 eV	110 K	photoluminescence	76Y1

Physical property	Numerical value	Experimental conditions	Experimental method remarks	Ref.
Fe^{3+} on Al^{3+} site (usually present as impurity):				
g^*	2.024	$T = 77$ K	EPR	75S1
$\bar{\nu}$	$11 \cdot 10^3$ cm^{-1}	300 K	absorption peaks	74D
	$16 \cdot 10^3$ cm^{-1}	300 K		

Fe^{2+}:

Created by charge transfer from valence band to Fe^{3+} under illumination at $\lambda = 0.5 \cdots 1.1$ μm [76K1]. Also created by heating in the atmosphere of Al, Cu, Zn, Cd, Ga, Si, Ge, Sn. Fluorescence at $0.5 \cdots 0.7$ μm due to the transition Fe$^{3+} \leftrightarrow$ Fe^{2+} [74D].

g^*	2.187 (5)	$T = 20$ K; $B \parallel c$	EPR (weak)	76K1
	2.037	20 K; $B \perp c$	EPR (weak)	

Ti^{3+} on Al^{3+} site:

g^*	1.9018	$T = 21$ K; $B \parallel c$	EPR	75K2
	1.9784	21 K; $B \perp c$	EPR	

Ni$^+$ on Cu$^+$ site:

g^*	2.051	$T = 20$ K; $B \parallel c$	EPR	75K3
	2.330	20 K; $B \perp c$	EPR	

Al, Si, Cd + Al: no absorption peak at $16 \cdot 10^3$ cm^{-1} (although Fe present) [74D].

Al, Si: absorption peak at $25.64 \cdot 10^3$ cm^{-1} (acceptor) [74D].

Mn: fluorescence at $\lambda = 630$ (30) nm ($T = 300$ K) [74D].

Lattice properties

wavenumbers of infrared and Raman active phonons (in cm^{-1}, at RT):

		Symmetry I$\bar{4}$2d (F$\bar{4}$3m)			
$\bar{\nu}_{LO}/\bar{\nu}_{TO}$	418/446	Γ_4	(Γ_{15})	all are Raman active; Γ_4, Γ_5	75K1
	497/444	Γ_5	(Γ_{15})	are IR active	
	443	Γ_3	(W$_2$)		
	−/432	Γ_5	(W$_4$)		
	315	Γ_1	(W$_1$)		
	284/271	Γ_4	(W$_2$)		
	266/263	Γ_5	(X$_5$)		
	268	Γ_3	(X$_3$)		
	217/216	Γ_5	(W$_3$)		
	137/137	Γ_5	(W$_4$)		
	112/112	Γ_4	(W$_2$)		
	98	Γ_3	(W$_2$)		
	76/76	Γ_5	(X$_5$)		

Transport properties

resistivities and Seebeck coefficients (values at $T = 300$ K; all isotropic):

p-type samples:

ϱ	$10^2 \cdots 10^4$ Ω cm		colorless crystals	74D
S	$20 \cdots 50$ μV K^{-1}			
ϱ	$10^4 \cdots 10^6$ Ω cm		green crystals	74D
S	$20 \cdots 200$ μV K^{-1}			
ϱ	$10^6 \cdots 10^8$ Ω cm		black crystals	74D
S	200 μV K^{-1}			

(continued)

Physical property	Numerical value	Experimental conditions	Experimental method remarks	Ref.
resistivities and Seebeck coefficients (continued)				
Cu, Al, Si doped samples:				
ϱ	$10\cdots10^3\,\Omega\,cm$			74D
S	$5\cdots50\,\mu V\,K^{-1}$			
n-type samples (Cd, Cd + Al doped):				
ϱ	$10\cdots10^4\,\Omega\,cm$			74D
S	$-100\cdots200\,\mu V\,K^{-1}$			
Further properties				
melting point:				
T_m	1570 K			75S1

10.1.2.2 Copper aluminum selenide (CuAlSe$_2$)

Electronic properties

    band structure: Fig. 11 α

energy gap (in eV):				
$E_{g,dir}$ (A)	2.72	$T = 78$ K	absorption	73B
(B)	2.86			
(C)	3.01			
(A)	2.65	110 K	phase-shift-difference spectroscopy	77Y
(B)	2.88			
(C)	3.02			
excitonic energy gap (in eV):				
E_{gx} (A)	2.71	$T = 78$ K	absorption	73B
(B)	2.85			
(C)	3.0			
(A)	2.77	20 K	phase-shift-difference spectroscopy	77Y
(B)	2.85			
(C)	2.96			
splitting energies at Γ:				
Δ_{cf}	$-0.17\,eV$	$T = 78$ K	calculated	73B
Δ_{so}	$0.18\,eV$	78 K		

Impurities and defects

luminescence peaks (in eV):				
emission:				
E	1.80		donor with $E_d = 0.81$ eV	76Y1
	1.97		unexplained	
excitation:				
E	2.61		probably energy gap;	76Y1
	2.99		no polarization dependence	

Optical and further properties

refractive indices:

		λ [μm]		
n_o	2.7797	0.50		75S1
	2.5293	1.00		
	2.4969	1.50		
	2.4851	2.00		
	2.4795	2.50		(continued)

Physical property	Numerical value	Experimental conditions	Experimental method remarks	Ref.
refractive indices (continued)				
		λ [μm]		
n_o	2.4759	3.00		75S1
	2.4733	3.50		
	2.4712	4.00		
	2.4685	4.50		
	2.4659	5.00		
n_e	2.7886	0.50		75S1
	2.5179	1.00		
	2.4852	1.50		
	2.4734	2.00		
	2.4676	2.50		
	2.4638	3.00		
	2.4609	3.50		
	2.4586	4.00		
	2.4559	4.50		
	2.4533	5.00		
melting point:				
T_m	1470 K			75S1

10.1.2.3 Copper gallium sulfide (CuGaS$_2$)

Electronic properties

band structure: Fig. 11 β

energy gap (in eV):

			T [K]			
$E_{g,\,dir}$	(A)	2.53	2	$E \parallel c$	absorption	75S1
	(B, C)	2.65		$E \perp c$		
	(A)	2.502	77	$E \parallel c$	wavelength derivative reflectivity	75T
	(B)	2.627		$E \perp c$		
	(C)	2.638		$E \perp c$		
	(A)	2.46	110	$E \parallel c$	reflection;	75Y
	(B, C)	2.58		$E \perp c$	yellow crystal: stoichiometric	
	(A)	2.49	110	$E \parallel c$	reflection;	75Y
	(B, C)	2.62		$E \perp c$	black crystal: Cu$_2$S rich	
	(A)	2.43	300	$E \parallel c$	electroreflectance	75S1
	(B, C)	2.55		$E \perp c$		

See also under "higher transition energies".

dE_g/dT	(A)	$-2.2\,(2) \cdot 10^{-4}$ eV K^{-1}	80···300	absorption	79H1
	(A)	$-1.5 \cdot 10^{-4}$ eV K^{-1}	<110	modulated phase-shift-difference	78H1
	(A)	$-1.0 \cdot 10^{-4}$ eV K^{-1}	>110		
dE_g/dp	(A)	$3.4 \cdot 10^{-6}$ eV bar^{-1}	300		76J

excitonic energy gap (in eV):

			T [K]			
E_{gx}	(A)	2.501	2	($n=1$)	absorption	75S1,
	(A)	2.523		($n=2$)	phase-shift-difference spectra	77Y
	(B, C)	2.627				75S2
	(A)	2.480	20	($n=1$)	phase-shift-difference spectra	77Y
	(A)	2.495		($n=2$)		
	(B)	2.59				
	(C)	2.60				
	(A)	2.4995	77	($n=1$)	cathodoluminescence and	77S1
	(A)	2.524		($n=2$)	absorption	
	(A)	2.467 (2)	300		cathodoluminescence	78S1
	(B, C)	2.596 (1)			modulated phase difference	78H1

Physical property	Numerical value	Experimental conditions	Experimental method remarks	Ref.

splitting energies at Γ (in eV):

		T [K]		
Δ_{cf}	-0.126	2	absorption	75S1
	-0.13	77	wavelength derivative reflectance	75T
	-0.12	300	electroreflectance	75S1
Δ_{so}	-0.017	77	wavelength derivative reflectance ($\Delta_{so} \leq 0!$) (unresolved in other experiments [75S1, 75Y])	75T

higher transition energies (in eV) (reflectivity peaks at $T = 80$ K):

	$E \parallel c$	$E \perp c$		
E	2.48	–	gap	77R,
	–	2.60	gap	75H
	2.62	–	gap	
	3.84	3.82	E_1-like	
	–	4.20	Cu 3d	
	4.4	–	Cu 3d	
	4.7	4.68	Cu 3d	
	5.14	5.12	Cu 3d	
	5.58	5.56	Cu 3d	
	6.7	6.70	E_1-like	
	7.7	7.7	E_2-like	
	8.9	8.9	E_2-like	
	–	9.7	Ga, Cu 4s	
	11.0	12.0	Ga, Cu 4s	
	12.5	12.5	Ga, Cu 4s	
	14.1	14.2	Ga, Cu 4s	
	15.7	15.7	S 3s	
	18.0	18.0	S 3s	
	21.025	21.025	Ga 3d	
	21.475	21.475	Ga 3d	
	22.05	22.05	Ga 3d	
	22.50	22.50	Ga 3d	
	23.75	23.75	Ga 3d	
	25.50	25.50	Ga 3d	

N.B. E_1, E_2-like, denote that these probably correspond to the transitions denoted E_1, E_2 in zincblende compounds.

conduction band, valence band and core state energies (in eV):

relative to valence band maximum (from photoemission spectra at 300 K)

$E(C_4)$	6.8	conduction band states	77R
$E(C_3)$	5.1	transition from Ga 3d states	
$E(C_2)$	3.5		
$E(C_1)$	2.5		
$E(C_0)$	2.5	conduction band minimum (energy gap)	
$E(V_0)$	0.0	valence band maximum	
$E(V_1)$	-1.6	Cu 3d (Γ_{12})	
$E(V_2)$	-2.1	Cu 3d + S 3p (Γ_{15})	
$E(V_3)$	-3.4		
$E(V_4)$	-7.1	Ga, Cu 4s	
$E(V_5)$	-13.0	S 3s	
$E(d_{5/2})$	-18.8	Ga 3d (core states)	
$E(d_{3/2})$	-19.3	Ga 3d	

Physical property	Numerical value	Experimental conditions	Experimental method remarks	Ref.
effective masses:				
m_n	0.13 m_0		phase-shift-difference	77Y
m_p	0.69 m_0			

density of states of amorphous CuGaS$_2$: Fig. 12

Impurities and defects

energy levels of intrinsic defects (in eV):

		T [K]		
E	1.68 (40)	110	photoluminescence: emission	76Y1
	2.48	110 $E \parallel c$	photoluminescence: excitation	
	1.62	110 $E \perp c$		
	0.55		activation energy of conductivity	75S1
E_d	0.80		unidentified	77T1
	0.07	0	Cu vacancy; ϱ vs. T (Fig. 14b)	75S1
	0.18 (50)	77	luminescence, unidentified	77T1
E	0.021	4.2	IR absorption peak, S on Ga site	79J

luminescence: Fig. 13

color of samples:

yellow: stoichiometric [74Y, 75Y]
black: Cu$_2$S rich [75Y]
        sulfur annealed [76K2]
red: Ga$_2$S$_3$ rich [75Y]
        Fe doped [76K2]
green: Cu$_2$S precipitate [76K2]

doped samples:

Fe doped (Fe^{3+} on Ga^{3+} site):

E_a	0.2 eV	$T = 77$ K	photoconduction	77Z
E_d	0.6 eV	77 K		

Ni doped (Ni$^+$ on Cu$^+$ site):

g^*	1.915	$T = 20$ K, $B \parallel c$	ESR	75K3
	1.324	20 K, $B \perp c$		

Mn doped (Mn^{2+}):

g^*	2.0020 (15)	$T = 77$ K	Mn^{2+} on Ga site	76T
	2.0107 (40)	77 K	Mn^{2+} on Cu site	
	2.029 (5)	77 K	Mn^{2+} on Cu site with vacant neighbor	

Co doped (Co^{2+}):

g^*	2.2147 (10)	$T = 20$ K, $B \parallel c$	ESR	75S2
	2.2326 (20)	$B \perp c$		
$\bar{\nu}$ [cm^{-1}]	6600	$T = 2$ K	absorption bands	76S
	6900			
	7600			
	13300			
	13600			
	13900			
	14300			

Physical property	Numerical value	Experimental conditions	Experimental method remarks	Ref.

Lattice properties

warenumbers $\bar{v}$ (in cm^{-1}) and Grüneisen parameters γ of infrared and Raman active phonons (for $T = 300$ K):

	[75K1]	[80C]	Symmetry I$\bar{4}$2d (F$\bar{4}$3m)		[80C]
$\bar{v}_{LO}/\bar{v}_{TO}$	401/368	393/367	Γ_4	(Γ_{15})	$\gamma_{i_{LO}}/\gamma_{i_{TO}}$	1.4/1.5 (3)
	384/363	385/367	Γ_5	(Γ_{15})		1.3/1.5 (3)
	358	401	Γ_3	(W$_2$)		
	352/332	347/332	Γ_5	(W$_4$)		1.5/1.2 (3)
	312	312	Γ_1	(W$_1$)		1.5 (3)
	281/267	288/286	Γ_4	(W$_2$)		1.7/1.1 (3)
	276/262	283/273	Γ_5	(X$_5$)		1.1/1.4 (3)
	203	238	Γ_3	(X$_3$)		2.6 (6)
	160/156	167/167	Γ_5	(W$_3$)		2.0 (5)
	–	147/147	Γ_5	(W$_4$)		0.8 (2)
	95/95	95/95	Γ_4	(W$_2$)		2.0 (4)
	97	116	Γ_3	(W$_2$)		1.3 (3)
	74/–	75/75	Γ_5	(X$_5$)		−0.8 (2)

Γ_4, Γ_5 are IR active, all are Raman active; see also Fig. 38.

local mode:

$\bar{v}$	175 cm^{-1}	$T = 4.2$ K	infrared: S on Ga site	79J

resonant Raman effect:

excitons observed at resonant enhancement [77S2]:

Phonon symmetry		Exciton energy [eV]		Remarks
I$\bar{4}$2d (F$\bar{4}$3m)		$\parallel, \parallel$	$\perp, \perp$	incident, scattered light polarizations;
				$\parallel = E \parallel \langle 11\bar{1} \rangle$, $\perp = E \parallel \langle 1\bar{1}0 \rangle$
Γ_4	(Γ_{15})	2.50	2.62	resonance
Γ_1	(W$_1$)	2.45		antiresonance
		>2.6		antiresonance
		2.47···2.6		resonance
Γ_4	(W$_2$)	2.50	2.62	resonance
Γ_5	(W$_3$)	2.50	2.62	resonance
Γ_5	(X$_5$)	2.50	2.62	resonance

Transport properties

resistivity, carrier concentration and mobility:

ϱ	1 Ω cm	$T = 300$ K	p-type sample, annealed under	75S1
p	$4 \cdot 10^{17}$ cm^{-3}	300 K	maximum S-pressure;	
μ_p	15 cm^2 V^{-1} s^{-1}	300 K	no anisotropy reported;	
			see also Fig. 14	

Evidence of ionic conduction has been reported in [77T2].

Physical property	Numerical value	Experimental conditions	Experimental method remarks	Ref.

Optical properties

For luminescence spectrum, see Fig. 13.

refractive indices:

coefficients in the formula: $n^2 = A + \dfrac{B}{1 - C/\lambda^2} + \dfrac{D}{1 - E/\lambda^2}$ 79B1

		T [K]		
A	3.9064	290	ordinary index (O)	
	4.3165		extraordinary index (E)	
	4.0984	390	O	
	4.4834		E	
B	2.3065	290	O	
	1.8692		E	
	2.1419	390	O	
	1.7316		E	
C	0.1149 μm^2	290	O	
	0.1364 μm^2		E	
	0.1225 μm^2	390	O	
	0.1453 μm^2		E	
D	1.5479	290	O	
	1.7575		E	
	1.5955	390	O	
	1.7785		E	
E	738.43 μm^2	290	O+E	
	738.43 μm^2	390	O+E	
		λ [μm]		
n_o	2.7630	0.55		75S1
	2.5517	1.00		
	2.5051	2.00		
	2.4945	3.00		
	2.4884	4.00		
	2.4843	5.00		
	2.4774	6.00		
	2.4714	7.00		
	2.4639	8.00		
	2.4539	9.00		
	2.4429	10.00		
	2.4311	11.00		
	2.4171	12.00		
	2.3999	13.00		
n_e	2.7813	0.55		75S1
	2.5464	1.00		
	2.4991	2.00		
	2.4880	3.00		
	2.4816	4.00		
	2.4772	5.00		
	2.4694	6.00		
	2.4621	7.00		
	2.4539	8.00		
	2.4435	9.00		
	2.4311	10.00		
	2.4179	11.00		

Physical property	Numerical value	Experimental conditions	Experimental method remarks	Ref.
linear electrooptic coefficients (at RT) (in 10^{-12} m/V):				
		λ [nm]		
r_{41}	1.76	633		74T
	1.9	1150		
	1.1	3390		
r_{63}	1.35	633		
	1.66	1150		
	1.05	3390		
For nonlinear dielectric susceptibility, see [71B].				
dielectric constants:				
$\varepsilon(0)$	7.6	$T = 300$ K, $E \parallel c$		75S1
	8.9	$E \perp c$		
$\varepsilon(\infty)$	6.1	300 K, $E \parallel c$		77S1
	6.2	$E \perp c$		
Further properties				
diffusion coefficient of S^{2-} ions:				
D	$2 \cdot 10^{-12}$ cm^2 s^{-1}	$T = 1000$ K		77B3
linear thermal expansion coefficient:				
α	$7.7 \cdot 10^{-6}$ K^{-1}		no temperature or orientation given	80B1
Debye temperature:				
Θ_D	356 K	$T \to 0$ K		75A
melting point:				
T_m	1550 K			76K2

10.1.2.4 Copper gallium selenide (CuGaSe$_2$)

Electronic properties

band structure: Fig. 14α

energy gap (in eV):					
$E_{g,\,dir}$	(A)	1.729	$T = 77$ K	wavelength derivative spectroscopy (WDS)	75T
	(B)	1.813			
	(C)	2.016			
	(A)	1.68	300 K	electroreflectance	75S1
	(B)	1.75			
	(C)	1.96			
dE_g/dT	(A)	$-2.1\,(1) \cdot 10^{-4}$ eV K^{-1}	$120 \cdots 300$ K	transmission	78N1
excitonic energy gap (in eV):					
			T [K]		
E_{gx}	(A)	1.695	20　$(n=1)$	phase-shift-difference spectroscopy	77T
	(A)	1.707	20　$(n=2)$		
	(A)	1.695	70　$(n=1)$	luminescence	
	(B)	1.74 (1)	20	phase-shift-difference	
	(C)	1.925 (1)			
exciton binding energy:					
E_b	(A)	$16 \cdot 10^{-3}$ eV			77Y

Physical property	Numerical value	Experimental conditions	Experimental method remarks	Ref.
splitting energies at Γ:				
Δ_{cf}	-0.112 eV	$T = 77$ K	(WDS)	75T
	-0.09 eV	300 K	electroreflectance	75S1
Δ_{so}	0.231 eV	77···300 K	electroreflectance	75T,
				75S1
higher transition energy (see also Fig. 15):				
E	2.8 eV	$T = 300$ K	Cu d-bands to conduction band transition; absorption, unpolarized	78H2

d-like character of valence bands 36% [75S1].

Impurities and defects

transition energies (in eV) involving impurities and defects:

		T [K]		
E	1.59	4	luminescence, donor-acceptor transition	80P1, 81P
	1.53	70	electroluminescence, recombination in acceptors	77P
	1.59			
	1.77	76	photoluminescence, conduction band-acceptor transition	79S1, 78V
	1.68	76	photoluminescence, recombination of free electrons with bound holes	
	1.65	76	photoluminescence, transitions in deeper localized state (impurities or defects)	
	1.52			
	1.30			

donor-acceptor pair:

E_b	0.111 eV	$T = 9···40$ K	binding energy of deeper level	80P1,
	0.0685 eV		binding energy of shallower level	81P
r	73 Å		separation of pair	

acceptor binding energies (in eV):

E_a	0.54	$T = 150···400$ K		79M1
	0.15			
	0.03···0.05		see also [78V]	

activation energy from impurity level:

E_A	0.27 eV			75S1

Lattice properties

wavenumbers of IR active phonons (in cm^{-1}, at RT):

		Symmetry I$\bar{4}$2d	(F$\bar{4}$3m)		
$\bar{\nu}_{LO}/\bar{\nu}_{TO}$	278/254	Γ_4	(Γ_{15})	infrared absorption	77B2
	276/250	Γ_5	(Γ_{15})		
	196/178	Γ_4	(W_2)		
	190/170	Γ_5	(W_4)		

Physical property	Numerical value	Experimental conditions	Experimental method remarks	Ref.
Transport properties (p-type samples)				
conductivities, Seebeck coefficients, carrier concentrations and mobilities:				
		T [K]		
σ	$2.6\ \Omega^{-1}\,\text{cm}^{-1}$	1040	(just below T_m)	70M
S	$0.38\ \mu\text{V K}^{-1}$			
ϱ	$0.1\ \Omega\,\text{cm}$	300		77P
μ_p	$15\ \text{cm}^2\,\text{V}^{-1}\,\text{s}^{-1}$			
ϱ	$0.4\ \Omega\,\text{cm}$	300	thermal treatment changes	78V
p	$3 \cdot 10^{17}\,\text{cm}^{-3}$		electrical properties, see [78V]	
μ_p	$60\ \text{cm}^2\,\text{V}^{-1}\,\text{s}^{-1}$			
ϱ	$4\ \Omega\,\text{cm}$	76		78V
p	$10^{16}\,\text{cm}^{-3}$			
μ_p	$140\ \text{cm}^2\,\text{V}^{-1}\,\text{s}^{-1}$			
p	$8 \cdot 10^{18}\,\text{cm}^{-3}$	76$\cdots$300	Se annealed	79S1
	$1.5 \cdot 10^{18}\,\text{cm}^{-3}$	300	as grown	
	$6 \cdot 10^{17}\,\text{cm}^{-3}$	76	as grown	
	$4 \cdot 10^{17}\,\text{cm}^{-3}$	300	vacuum annealed	
	$4 \cdot 10^{16}\,\text{cm}^{-3}$	76		
μ_p	$20\cdots30\ \text{cm}^2\,\text{V}^{-1}\,\text{s}^{-1}$	76$\cdots$300	Se annealed	79S1
	$0.2\ \text{cm}^2\,\text{V}^{-1}\,\text{s}^{-1}$	76	vacuum annealed	
	$60\ \text{cm}^2\,\text{V}^{-1}\,\text{s}^{-1}$	300	vacuum annealed	
	$60\cdots100\ \text{cm}^2\,\text{V}^{-1}\,\text{s}^{-1}$	76$\cdots$300	as grown	
S	$7.5 \cdot 10^{-2}\ \text{V K}^{-1}$	300		75S1

N.B. No anisotropy reported.

Optical properties
refractive indices:

		λ [μm]		
n_o	2.9580	0.78		75S1
	2.8358	1.00		
	2.7430	2.00		
	2.7273	3.00		
	2.7211	4.00		
	2.7170	5.00		
	2.7133	6.00		
	2.7101	7.00		
	2.7060	8.00		
	2.7021	9.00		
	2.6974	10.00		
	2.6926	11.00		
	2.6872	12.00		
n_e	3.0093	0.78		75S1
	2.8513	1.00		
	2.7510	2.00		
	2.7344	3.00		
	2.7276	4.00		
	2.7232	5.00		
	2.7192	6.00		
	2.7158	7.00		
	2.7111	8.00		
	2.7065	9.00		
	2.7014	10.00		
	2.6981	11.00		
	2.6898	12.00		

For refractive index and nonlinear dielectric susceptibility, see [72B].

Physical property	Numerical value	Experimental conditions	Experimental method remarks	Ref.
Further properties				
linear thermal expansion coefficient:				
α	$13.1\,(14) \cdot 10^{-6}\,\mathrm{K}^{-1}$	$T = 300 \cdots 670$	a axis	80B3
	$5.2\,(7) \cdot 10^{-6}\,\mathrm{K}^{-1}$		c axis	
Debye temperature:				
Θ_D	$262.0\,(7)\,\mathrm{K}$	$T \to 0\,\mathrm{K}$		82B
melting point:				
T_m	$1310 \cdots 1340\,\mathrm{K}$			75S1

10.1.2.5 Copper gallium telluride (CuGaTe$_2$)

Electronic properties

energy gap (in eV, RT values):

$E_{\mathrm{g,dir}}$	1.0		photocond. ⎫ transitions	79N2
	1.1$\cdots$1.17		absorption ⎬ probably due	
	1.24		electrorefl. ⎭ to impurities	
	(A) 1.227		transmission in polycrystalline,	
	(B) 1.280		thin films	
	(C) 1.97			

splitting energies at Γ:

Δ_cf	$-0.08\,(4)\,\mathrm{eV}$	$T = 300\,\mathrm{K}$	transmission in thin polycrystalline	79N2
Δ_so	$0.71\,(4)\,\mathrm{eV}$	$300\,\mathrm{K}$	films	

higher transition energies (in eV, RT values):

E	2.67		transmission in polycrystalline	79N2
			thin films, d-band	
	2.50		reflectivity, polycrystalline samples	80H1
	3.25			
	3.65			
	4.1			
	4.6			
	5.6			
	6.1			
	6.3			
	6.65			
	7.0			
	7.6			
	7.8			
	8.1			
	8.3			
	8.8			

Impurities and defects

acceptor binding energies:

E_a	0.14 eV	$T = 80 \cdots 300\,\mathrm{K}$	Cu-vacancy ?	79N1
	0.45 eV		Cu-vacancy ?	
			E_a from transport measurements	

Lattice properties

phonon wavenumbers (in cm^{-1}):

$\bar{\nu}$	$209.2\,(1)\,\mathrm{cm}^{-1}$	$T = 300\,\mathrm{K}$	thin films	79N2
	$201.4\,(1)\,\mathrm{cm}^{-1}$			
	$166.4\,(5)\,\mathrm{cm}^{-1}$			

Physical property	Numerical value	Experimental conditions	Experimental method remarks	Ref.
Transport properties (p-type samples)				
carrier concentrations, mobilities, conductivities and Seebeck coefficients (at RT):				
p	$6 \cdot 10^{17} \cdots 2 \cdot 10^{18}$ cm^{-3}		epitaxial layers	79S2
μ_p	$4 \cdots 12$ cm^2 V^{-1} s^{-1}		epitaxial layers	
p	$3 \cdot 10^{16} \cdots 10^{18}$ cm^{-3}			79N1
μ_p	$5 \cdots 39$ cm^2 V^{-1} s^{-1}			
p	10^{18} cm^{-3}			75S1
μ_p	50 cm^2 V^{-1} s^{-1}			
σ	11 Ω^{-1} cm^{-1}			
S	$2.7 \cdot 10^{-1}$ V K^{-1}			
N.B. No anisotropy reported.				
Optical and further properties:				
refractive index:				
n	2.83 (9)	$\hbar\omega = 0.5$ eV		80H2
Debye temperature:				
Θ_D	226.2 (8) K	$T \rightarrow 0$ K		82B
melting point:				
T_m	1140 K			75S1

10.1.2.6 Roquesite ($CuInS_2$)

Most published work is on $CuInS_2$ thin films, because of potential photovoltaic applications [76K3, 79G1].

Electronic properties
band structure: Fig. 15α

energy gaps and splitting energies:				
$E_{g, dir}$　(A, B, C)	1.53 eV	$T = 300$ K	electroreflectance, bulk crystal	75S1
$\Delta_{cf}(\Gamma)$	> -0.005 eV			
$\Delta_{so}(\Gamma)$	-0.02 eV			
d-like character: 45% [75S1].				
thin film data:				
$E_{g, dir}$	1.54 (2) eV	$T = 300$ K	absorption, unpolarized	80H3, 78S2
$E_{g, th}$	1.62 eV		$E_g(T) = E_{g, th} - \beta T^2/(T + \alpha)$	77K1
β	$4.3 \cdot 10^{-4}$ eV K^{-1}			
α	231.54 K			
ΔE	2.15 (30) eV	300 K	shift of Cu-K absorption	78B1
excitonic energy gap:				
E_{gx}　(A)	1.536 eV	$T = 2$ K　($n = 1$)	absorption	75S1
(B)	1.554 eV	($n = 1$)		
effective mass:				
m_p	1.3 m_0		Hall effect	80H3

Physical property	Numerical value	Experimental conditions	Experimental method remarks	Ref.
Impurities and defects				
acceptor, donor binding energies (in eV):				
E_a	0.15		σ vs. T	76L
E_d	0.35		σ vs. T	
	0.017		magnetic susceptibility vs. T with antiferromagnetic exchange coupling between donors	80J
	0.004		σ vs. T; Zn, Cd doped	77M2
	0.02		σ vs. T; Cd doped	
	0.025		σ vs. T; Zn doped	
transition energies involving impurities:				
E	1.41$\cdots$1.42 eV		Cu-vacancy to conduction band; p-type only	78S2
g-factor:				
$g^*(Fe^{3+})$	2.022		ESR	75S1
Lattice properties				
wavenumbers of infrared and Raman active phonons (in cm^{-1}, $T = 4.5$ K):				

		Symmetry $I\bar{4}2d$ ($F\bar{4}3m$)		
$\bar{\nu}_{LO}/\bar{\nu}_{TO}$	352/323	Γ_4 (Γ_{15})	Γ_4, Γ_5 are IR active, all are	75K1
	339/321	Γ_5 (Γ_{15})	Raman active	
	314/295	Γ_5 (W$_4$)		
	294	Γ_1 (W$_1$)		
	266/234	Γ_4 (W$_2$)		
	260/244	Γ_5 (X$_5$)		
	$-/140$	Γ_5 (W$_3$)		
	$-/88$	Γ_4 (W$_2$)		
	$-/79$	Γ_3 (W$_2$)		
	$-/67$	Γ_5 (X$_5$)		

Physical property	Numerical value	Experimental conditions	Experimental method remarks	Ref.
Transport properties				
resistivities, carrier concentrations, mobilities:				
n-type (bulk material):				
ϱ	1 Ω cm	$T = 300$ K	annealed under minimum S pressure	80H3
n	$3 \cdot 10^{16}$ cm^{-3}			
μ_n	15 cm^2 V^{-1} s^{-1}			
p-type (bulk material):				
ϱ	5 Ω cm	$T = 300$ K	annealed under maximum S pressure	80H3
p	10^{17} cm^{-3}			
μ_p	15 cm^2 V^{-1} s^{-1}			
n-type (thin films):				
ϱ	$0.1 \cdots 4.9 \cdot 10^4$ Ω cm	$T = 300$ K		75K4,
n	$4.10^{12} \cdots 10^{19}$ cm^{-3}			80H3,
μ_n	$1 \cdots 240$ cm^2 V^{-1} s^{-1}			80H4, 79N3, 76L
p-type (thin films):				
ϱ	$0.8 \cdots 400$ Ω cm	$T = 300$ K		75K4,
p	$10^{13} \cdots 10^{17}$ cm^{-3}			80H3,
μ_p	$0.2 \cdots 95$ cm^2 V^{-1} s^{-1}			80H4, 79N3, 76L

Physical property	Numerical value	Experimental conditions	Experimental method remarks	Ref.

For temperature dependence of resistivity and mobilities, see Figs. 17···19. N.B. No anisotropy reported.

As grown samples or samples annealed in S are p-type; samples annealed in Cd, Zn, In or in minimum S pressure are n-type [80H4, 77M2].

decay times and diffusion lengths from photoconductivity data (at RT):

τ	$0.1···2.4 \cdot 10^{-3}$ s		photoelectromagnetic effect	77K1
τ_n	$1.2 \cdot 10^{-10}$ s		and photoconductivity	79M2
τ_p	$6.5 \cdot 10^{-7}$ s			
L_n	$2.5 \cdot 10^{-5}$ cm			
L_p	$2.5 \cdot 10^{-4}$ cm			

Optical properties

refractive indices:

		λ [μm]		
n_o	2.7907	0.90		75S1
	2.7225	1.00		
	2.6020	2.00		
	2.5838	3.00		
	2.5760	4.00		
	2.5699	5.00		
	2.5645	6.00		
	2.5587	7.00		
	2.5522	8.00		
	2.5448	9.00		
	2.5366	10.00		
	2.5274	11.00		
	2.5166	12.00		
	2.5108	12.50		
n_e	2.7713	0.90		75S1
	2.7067	1.00		
	2.5918	2.00		
	2.5741	3.00		
	2.5663	4.00		
	2.5598	5.00		
	2.5539	6.00		
	2.5474	7.00		
	2.5401	8.00		
	2.5311	9.00		
	2.5225	10.00		
	2.5112	11.00		
	2.4987	12.00		

For refractive index and nonlinear dielectric susceptibility, see [71B]. Photoluminescence spectrum: see Fig. 16.

Further properties

magnetic susceptibility:

$\chi - 83.7 + 185 [1\text{-}\exp(-100/T)]/(T+3) \cdot 10^{-6}$ cm^3 mol^{-1} (T in K, χ in CGS-emu) [80J].

Debye temperature:

Θ_D	273 K	$T \to 0$ K		77B4

melting point:

T_m		$T = 1270···1320$ K		75S1

Physical property	Numerical value	Experimental conditions	Experimental method remarks	Ref.

10.1.2.7 Copper indium selenide (CuInSe$_2$)

Most published data on CuInSe$_2$ concerns thin films, because of possible photovoltaic applications [76K3].

Electronic properties:

band structure: Fig. 19α

energy gap (in eV):

		T [K]		
$E_{g, dir}$	(A) 1.04	77 K	electroreflectance, single crystals	75S1
	(B) 1.04			
	(C) 1.27			
$E_g(T) = E_g(0) - aT^2/(T+b)$			with fitted parameters below;	77H2
$E_g(0)$	0.965 eV		from absorption	
a	$1.086 \cdot 10^{-4}$ eV K^{-1}			
b	97 K			
dE_g/dT	$-1.4\,(2) \cdot 10^{-4}$ eV K^{-1}	150$\cdots$300	absorption	77H2
		200$\cdots$300	photoconductivity	80S2
	$-3 \cdot 10^{-4}$ eV K^{-1}	77$\cdots$300		80A2
	$-1.2\,(1) \cdot 10^{-4}$ eV K^{-1}	100$\cdots$300	absorption	81A1

thin film data:

		T [K]		
$E_{g, dir}$	1.00 (3)	300	absorption, depends on grain sizes	80G2, 80P2, 78H4, 80P2
	0.989 (3)	100	absorption	81A1
	0.967 (3)	300		
	0.982 (3)	100	wavelength modulated absorption	
	0.964 (3)	300		
	2.6	300	absorption	78H4

excitonic energy gap:

E_{gx}	1.037 eV	$T = 4.2$ K	unpolarized	76P3
	1.042 eV	77 K	($E_b = 0.018$ eV)	

splitting energies at Γ:

Δ_{cf}	$+0.006$ eV	$T = 77$ K	positive!	75S1
Δ_{so}	0.23 eV	77 K		

higher transition energies (from reflectance spectra at $T = 300$ K, in eV):

E	2.5		unpolarized	77R,
	2.92		(disagrees with [80H1])	80A1
	3.24			
	3.29			
	3.72			
	4.85			
	5.43$\cdots$5.38			
	6.2			
	6.7			

Physical property	Numerical value	Experimental conditions	Experimental method remarks	Ref.
conduction band, valence band and core states (in eV, RT values):				
relative to upper valence band edge (from photoemission data):				
$E(C_3)$	3.4		conduction band states	77R
$E(C_2)$	2.9			
$E(C_1)$	1.4			
$E(C_0)$	1.1		conduction band minimum	
$E(V_0)$	0.0		valence band maximum	
$E(V_1)$	-0.5		Cu 3d	
$E(V_2)$	-2.1		Cu 3d	
$E(V_3)$	-3.3			
$E(V_4)$	-6.3		light holes or cation s-states	
$E(V_5)$	-13.0		Se 4s	
$E(d_{5/2})$	-17.2		In 4d, core states	
$E(d_{3/2})$	-18.2		In 4d	
effective masses:				
m_n	0.09 (1) m_0	$T = 300$ K	Faraday effect	77W1
m_p	0.09 m_0	300 K	Seebeck effect	80S1
d-like character of the valence bands: 34% [75S1].				
Impurities and defects				
intrinsic defects:				
E_d (probably Se vacancies)	0.012 eV		various experiments at different temperatures (E_d: donor binding energy)	79I, 79N4, 78N2, 80B3
	0.065 (2) eV	$T = 4.2$ K	luminescence	76P3
	0.22 (3) eV		various experiments at different temperatures	79I, 78N2, 80B3, 80S2
E_a (probably Cu vacancies)	0.012 eV		various experiments at different temperatures (E_a: acceptor binding energy)	79N4, 80B3
	0.020$\cdots$0.028 eV		transport	79I
	0.090 (5) eV		photoconductivity vs. T	78S3, 80S2
	0.40 (3) eV		various experiments at different temperatures	79N4, 76P3, 80B3, 78S3
E	0.94 (1) eV	4.2 K	luminescence; E: donor-acceptor transition energy; shifts to higher energy with increased excitation intensity [76P3]	76P2
extrinsic defects:				
Cd, Zn, Br, Cl are donors [76Y2]. For electroluminescence spectrum, see Fig. 20.				
Fe^{3+}:				
g^*	2.040 (3)	$T = 4.2$ K, $B \parallel c$	EPR ($n_{Fe} = 10^{17}\cdots 10^{18}$ cm^{-3})	80B3

Lattice properties

Infrared reflectivity and Raman scattering have been measured by [78R, 76G1, 78N3], two photon absorption by [81S]. These results are however contradictory.

Physical property	Numerical value	Experimental conditions	Experimental method remarks	Ref.
Grüneisen parameter:				
γ	1.61	$T = 80 \cdots 300$ K	thermal conductivity	80S1

Transport properties

carrier concentrations, resistivities, mobilities, Seebeck coefficients:

p-type samples (usual due to Cu vacancies or overstoichiometric Se. [78K1]):

		T [K]		
p	$10^{16} \cdots 10^{18}$ cm^{-3}	300	see Fig. 21	79N3, 76P3, 78S3, 79I
ϱ	$10^{-2} \cdots 3 \cdot 10^3$ Ω cm	300		80G3, 77H1, 75D
μ_p	$< 1 \cdots 200$ cm^2 V^{-1} s^{-1}	300		77H1, 76P3, 78S3, 79I, 80G2
S	0.45 mV K^{-1}	90		75D

n-type samples (due to Se vacancies created by annealing [78K1]):

		T [K]		
n	$2.6 \cdot 10^{15}$ $\cdots 3 \cdot 10^{17}$ cm^{-3}	300	see Figs. 22 $\cdots$ 24	78N2, 79I
ϱ	$0.05 \cdots 1.6$ Ω cm	90		75S1, 75D
μ_n	$1 \cdots 1150$ cm^2 V^{-1} s^{-1}	300	($= 600$ cm^2 V^{-1} s^{-1} at 100 K [81A1])	76P3, 70M, 78N2, 76K4, 79I
S	-0.17 mV K^{-1}	90		75D

N.B. No ansisotropy reported for transport properties of n- and p-type samples.

magnetoresistance: Fig. 26

There is evidence of impurity conduction in a band 0.006 eV below the conduction band [79I].

Optical properties

refractive index:

n	2.79 (9)	$T = 300$ K	$\hbar\omega = 0.5$ eV $\}$ polycrystalline	80H2
	2.8 (1)	300 K	$\lambda = 10$ µm $\int$ films	80S2
dn/dT	$1.2 (7) \cdot 10^{-4}$ K^{-1}		$\lambda = 3 \cdots 10$ µm	80S2

dielectric constants (at 300 K):

$\varepsilon(0)$	13.6 (8)		capacitance	79L
	15.2	$E \parallel c$	infrared	78R
	16.0	$E \perp c$		
$\varepsilon(\infty)$	8.5	$E \parallel c$	infrared	78R
	9.5	$E \perp c$		

Physical property	Numerical value	Experimental conditions	Experimental method remarks	Ref.
decay times (in s): (from photoconductivity data, see Fig. 25 and Fig. 41)				
τ	$5 \cdot 10^{-8}$	$T = 77 \cdots 260$ K		80A2
	10^{-3}	$77 \cdots 300$ K		76K1
	$4(3) \cdot 10^{-3}$	$77 \cdots 300$ K	as deposited (p-type) thin films	77K1
	$12(3) \cdot 10^{-3}$		n-type	
	$0.4(3) \cdot 10^{-3}$		annealed in Se (p-type)	
carrier lifetimes:				
τ_n	$0.6 \cdot 10^{-9}$ s	$T = 300$ K	photoelectromagnetic effect	77M3
τ_p	10^{-6} s			
Further properties				
linear thermal expansion coefficients:				
α	$11.4 \cdot 10^{-6}$ K^{-1}	no temperature given	a axis	77H1
	$8.6 \cdot 10^{-6}$ K^{-1}		c axis	
Debye temperature:				
Θ_D	222 K	$T \to 0$ K		77B4
	207 K	$T \to 0$ K		80S1
melting point:				
T_m	1260 K			75S1

10.1.2.8 Copper indium telluride (CuInTe$_2$)

Electronic properties:

energy gap (in eV):

$E_{g,dir}$	(A, B)	1.064 (14)	$T = 290$ K	electroreflectance	77T4, 77T3
	(C)	1.67			
		1.12 (3)	77 K		
dE_g/dT		$-2.4(10) \cdot 10^{-4}$ eV K^{-1}	$77 \cdots 290$ K		77T4
$E_g(T) = E_g(0) - \beta T^2/(T+\alpha)$				with fitted parameters below;	77K1
$E_g(0)$		1.05 eV		from photoconductivity	
β		$4.6 \cdot 10^{-4}$ eV K^{-1}		thin film	
		$3.2 \cdot 10^{-4}$ eV K^{-1}		bulk	
α		185.78 K			

d-band hybridisation: 30% [77T3, 77T4].

splitting energies at Γ:

Δ_{cf}	0	$T = 290$ K	electroreflectance	77T4
Δ_{so}	0.61 (1) eV	290 K		

higher transition energies (in eV, RT values):

E	2.5	$\Gamma_7(\Gamma_{15})$	electroreflectance; from Cu d levels	79T2
	2.7	$\Gamma_6(\Gamma_{15})$		
	2.9	$\Gamma_7(\Gamma_{15})$		
	3.15	$\Gamma_6(\Gamma_{12})$		
	4.0	$\Gamma_7(\Gamma_{12})$		
	2.40		reflectivity, unpolarized	80H1
	3.20			
	3.85			
	4.7			
	4.9			
	5.1			

(continued)

Physical property	Numerical value	Experimental conditions	Experimental method remarks	Ref.
higher transition energies (continued)				
E	5.7			80H1
	6.0			
	6.3			
	6.55			
	7.0			
	7.45			
	7.8			
	8.1			
	8.7			
effective masses:				
$m_{p\parallel}$	$0.66\,(10)\,m_0$		free carrier effects in	80R
$m_{p\perp}$	$0.85\,(10)\,m_0$		infrared reflectivity	

Lattice properties

wavenumbers of infrared active phonons (in cm^{-1}, RT values):

		Symmetry		
$\bar{\nu}_{LO}/\bar{\nu}_{TO}$	180/167	Γ_5 (E modes)	infrared reflectivity on single crystals	80R
	179/168	Γ_4 (B$_2$ modes)		
	162/158	Γ_5 (E modes)		
	157/154	Γ_4 (B$_2$ modes)		
	60.5/59		infrared reflectivity on polycrystals	79R
	58.5/58			
$\bar{\nu}$	172, 157, 126, 50	E modes ($E\perp c$)	infrared reflectivity	81H2
	170, 101, 46	B$_2$ modes ($E\parallel c$)		

Transport properties

p-type samples due to overstoichiometric Te (Auger electron spectroscopy) [78K1, 80T].
n-type samples due to understoichiometric Te (Auger electron spectroscopy) [78K1, 80T].

carrier concentrations, mobilities, resistivities (at $T = 300$ K):

p	$2\cdot10^{18}\cdots5\cdot10^{18}$ cm^{-3}			80R
μ	$20\cdots35$ cm^2 V^{-1} s^{-1}			
	10 cm^2 V^{-1} s^{-1}			81H2
ϱ	$3.5\cdot10^{-3}\,\Omega$ cm		stoichiometric, as grown	78D
μ	$120\cdots150$ cm^2 V^{-1} s^{-1}			
ϱ	$700\cdots6300\,\Omega$ cm		Zn annealed (Zn-donor)	78D
μ	$6.3\cdots20$ cm^2 V^{-1} s^{-1}			
ϱ	$0.044\cdots0.1\,\Omega$ cm		Te deficient	78D
μ	$26\cdots120$ cm^2 V^{-1} s^{-1}			

N.B. No anisotropy reported.

thin film data:

p-type samples for $T_{sub} > 450$ K (T_{sub}: substrate temperature) [77K2].
n-type samples for $T_{sub} < 450$ K [77K2].

μ_n	$20\cdots30$ cm^2 V^{-1} s^{-1}	$T = 300$ K	polycrystalline samples	77K2
μ_p	$5\cdots12$ cm^2 V^{-1} s^{-1}	300 K		

Optical properties
refractive index:

n	3.05	$T = 300$ K	$\lambda \to \infty$	78D
	2.71 (9)	300 K	$\hbar\omega = 0.5$ eV, polycrystalline film	80H2

Physical property	Numerical value	Experimental conditions	Experimental method remarks	Ref.
dielectric constants:				
$\varepsilon(0)$	10.5 (8)	$T = 300$ K, $E \perp c$	infrared reflectivity	80R
	12.9 (8)	$E \parallel c$		
$\varepsilon(\infty)$	8.7 (5)	300 K, $E \perp c$		80R
	11.0 (5)	$E \parallel c$		
	9.6	300 K, $E \perp c$		81H2
	9.0	$E \parallel c$		
decay time (from photoconductivity):				
τ	$0.4\,(3) \cdot 10^{-3}$ s	$T = 77 \cdots 300$ K	as deposited film (n-type)	77K1
	$3.2 \cdot 10^{-6}$ s		annealed in Ar	
Further properties				
Debye temperature:				
Θ_D	191.4 K	$T \rightarrow 0$ K		77B4
	195.1 (7) K	$T \rightarrow 0$ K		82B
melting point:				
T_m	1050 K			75S1

10.1.2.9 Silver gallium sulfide (AgGaS$_2$)

Electronic properties

energy gap (in eV):

			T [K]		
$E_{g,\,dir}$	(A)	2.73	77	electroreflectance	75S1
	(B, C)	3.01			
		2.746	77	absorption	76G2
		2.638	300		
		2.70	300	absorption	76J
dE_g/dT		$2 \cdot 10^{-4}$ eV K^{-1}		cathodoluminescence	75S1
		$-4.85 \cdot 10^{-4}$ eV K^{-1}		absorption	76G2
dE_g/dp		0.22 eV/Pa	300	absorption	76J

excitonic energy gap (in eV):

			T [K]			
E_{gx}		2.605	300		absorption	76G2
		2.713	77			
	(A)	2.698	2	$(n=1)$	$E_g = 2.726$ eV	76S
	(A)	2.716	2	$(n=2)$		
	(B)	2.98	2	$(n=1)$	$E_g = 3.02$ eV	

splitting energies at Γ:

Δ_{cf}	-0.28 eV	$T = 77$ K			75S1
Δ_{so}	0	77 K			

d-like character of valence bands: 20% [76S].

higher transition energies (in eV, RT values):

E	3.5 (3)		UV reflectivity and REEL (REEL = Reflection Electron Energy Loss Spectroscopy) heavy hole band (S3p)		75H
	4.5		UV		
	4.8		UV		
	5.7		UV		(continued)

Physical property	Numerical value	Experimental conditions	Experimental method remarks	Ref.
higher transition energies (continued)				
E	6.5 (2)		light hole band (S3p) UV and REEL	75H
	7.9		UV	
	9.8		UV	
	14.0		surface plasmon, REEL	
	18.5		bulk plasmon, REEL	
	24.4		Ga 3d, REEL	

Disagrees with [80A1] at $T = 80$ K.

Impurities and defects
unidentified states:

		T [K]		
E_d	0.11 eV	4.2	ESR	78B3
g^*	2.0321 (20)	$B \parallel c$		78B3
	1.973 (2)	$B \perp c$		
			Only seen under continuous excitation.	
E_d	0.1100 (5) eV	20···300	thermally stimulated current;	80B4
	0.20 eV		optical ionization energy, associated with Ni^{3+} impurities	
	0.368 eV		electroluminescence	77G
E_a	0.071 eV	300	absorption	76G2
	0.050 eV	77		
	0.058 eV	77		
	0.061 eV	77		
	0.071 eV	77		

Ni:
Ni$^+$:

g^*	2.645	$B \parallel c$	ESR	75K3
	2.220	$B \perp c$	ESR	
Ni^{2+}:			donor state	78B3
Ni^{3+}:				
g^*	2.14 (1)	$B \parallel c$	ESR	75K3
	4.52 (2)	$B \perp c$	ESR	
E_a	1.3 (1) eV	20···300	optical ionization energy; see above for associated donor	80B4

Fe:
Fe^{3+}:

g^*	2.00	$B \parallel c$	ESR	78B3
	1.42	$B \perp c$		
	5.98	$B \perp c$	only seen after annealing in S, possibly Fe^{3+}—X, where X is probably interstitial Ag$^+$ [78B3]	
E_a	0.5 (1) eV	20···300	optical ionization energy	80B4
	0.46 (5) eV	20···300	thermally stimulated current	

photoluminescence spectrum: Fig. 27, cathodoluminescence spectrum: Fig. 28

Physical property	Numerical value	Experimental conditions	Experimental method remarks	Ref.

Lattice properties

wavenumbers v of infrared and Raman active phonons (in cm^{-1}) and Grüneisen parameters γ_i (at $T = 300$ K):

		Symmetry I$\bar{4}$2d (F$\bar{4}$3m)	γ_i		
$\bar{v}_{LO}/\bar{v}_{TO}$	392/368	Γ_5 (Γ_{15})	0.81	Γ_4, Γ_5 are IR	77L, 74H, 75L, 80C
	393/367	Γ_4 (Γ_{15})	0.85	active, all	74H, 75L, 80C
	340/321	Γ_5 (W$_4$)	–	are Raman	74H, 75L, 77L
	334	Γ_3 (W$_2$)	–	active	75L, 75K1
	224				74H
	295	Γ_1 (W$_1$)	1.02		74H, 75L, 80C
	240/213	Γ_4 (W$_2$)	1.79/1.49		74H, 75L
	215/213				80C
	230/223	Γ_5 (W$_4$)	2.49/1.49		74H, 75L
	190.5	Γ_3 (X$_3$)	–		75L
	160				74H
	161/157	Γ_5 (W$_3$)	0.56/0.41		77L, 74H, 75L, 80C
	95/95	Γ_5 (W$_4$)	0.52		
	65/65	Γ_4 (W$_2$)	0.0		74H, 75L, 80C
	54	Γ_3 (W$_2$)	–		
	36/36	Γ_5 (X$_5$	−4.4 (soft mode)		77L, 75L, 80C

For a discussion of the influence of ordered vacancies on the lattice vibration spectrum, see [82T1].

wavenumbers of 2nd order modes (in cm^{-1}):

$\bar{v}$	517	$\gamma_i = 1.25$	$T = 300$ K	80C
	85	–		77L

wavenumbers of phonons in high-pressure phases (in cm^{-1}):

phase transition at $p = 4.2 \cdot 10^4$ bar:

$\bar{v}$	17	$\gamma_i = 35$	$T = 300$ K (near transition)	80C
	335	1.6		

phase transition at $p = 1.16 \cdot 10^5$ bar:

$\bar{v}$	373	$\gamma_i = 0.43$	$T = 300$ K	80C
	363	0.66		
	358	0.66		
	350	0.24		
	106	−1.6		
	211	−1.4		
	($= 2 \times 106$ cm^{-1})			

elastic moduli (in 10^{11} dyn cm^{-2}):

c_{11}	8.79 (5)	75G,
c_{33}	7.58 (5)	75K1
c_{44}	2.41 (5)	
c_{66}	3.08 (5)	
c_{12}	5.84 (5)	
c_{13}	5.92 (5)	

See also [81V].

Transport properties

resistivity:

ϱ	$\approx 10^8 \, \Omega$ cm	$T = 300$ K	(orientation not specified)	80B4

There have been reports of ionic conduction with Ag$^+$ as the active ion [77T2, 77W2].

Physical property	Numerical value	Experimental conditions	Experimental method remarks	Ref.

Optical properties
refractive indices:
coefficients in the formula $n^2 = A + B/(1 - C/\lambda^2) + D/(1 - E/\lambda^2)$.

	n_o	n_e		
A	3.6280	4.0172		76B3
B	2.1686	1.5274		
C	$0.1003 \cdot 10^{-12}$ m^2	$0.1310 \cdot 10^{-12}$ m^2		
D	2.1753	2.1699		
E	$950 \cdot 10^{-12}$ m^2	$950 \cdot 10^{-12}$ m^2		
	valid for $\lambda = (0.49 \cdots 12.0)$ μm			
		λ [μm]		
n_o	2.7148	0.49		75S1
	2.5137	0.633		75B2
	2.4582	1.00		75S1
	2.3902	1.15		75B2
	2.4164	2.00		75S1
	2.4080	3.00		
	2.3506	3.39		75B2
	2.4024	4.00		75S1
	2.3953	5.00		
	2.3908	6.00		
	2.3827	7.00		
	2.3757	8.00		
	2.3663	9.00		
	2.3548	10.00		
	2.3417	11.00		
	2.3266	12.00		
	2.3076	13.00		
n_e	2.7287	0.49		75S1
	2.5574	0.633		75B2
	2.4053	1.00		75S1
	2.4495	1.15		75B2
	2.3637	2.00		75S1
	2.3545	3.00		
	2.4057	3.39		75B2
	2.3488	4.00		75S1
	2.3419	5.00		
	2.3369	6.00		
	2.3291	7.00		
	2.3219	8.00		
	2.3121	9.00		
	2.3012	10.00		
	2.2880	11.00		
	2.2716	12.00		

dielectric constants (at $T = 300$ K):

$\varepsilon(0)$	8.21	$E \parallel c$		75S1
	8.5	$E \parallel c$		74H
	8.51	$E \perp c$		75S1
	10.0	$E \perp c$		74H
$\varepsilon(\infty)$	5.50	$E \parallel c$		75S1
	5.7	$E \parallel c$		74H
	5.90	$E \perp c$		75S1, 74H

Physical property	Numerical value	Experimental conditions	Experimental method remarks	Ref.
electrooptic coefficients:				
$r_{41}^{\mathbf{X}}$	$4.0\,(2) \cdot 20^{-12}$ m V^{-1}	$\lambda = 633$ nm,	(superscript **X**: constant stress)	75B2
$r_{63}^{\mathbf{X}}$	$3.0\,(1) \cdot 13$ m V^{-1}	$T = 300$ K		
nonlinear dielectric susceptibility:				
d_{36}	$21\,(10) \cdot 10^{-12}$ m V^{-1}	$\lambda = 10.6\,\mu$m and 3.39 μm		75B2
Further properties				
Debye temperature:				
Θ_{D}	255 K	$T \rightarrow 0$ K		75A
melting point:				
T_{m}	$1220 \cdots 1320$ K			75S1

10.1.2.10 Silver gallium selenide (AgGaSe$_2$)

Electronic properties

energy gap (in eV):

			T [K]		
$E_{\mathrm{g,\,dir}}$	(A)	1.81	10	wavelength modulated reflectance (WMR)	77S3
	(B)	2.02			
	(C)	2.28			
	(A)	1.83	77	WMR and electroreflectance	75S1,
	(B)	2.03			77S3
	(C)	2.29			
	(A)	1.80	300	WMR and absorption	76J, 77S3
	(B)	1.98		WMR	77S3
	(C)	2.28		WMR	77S3
$\mathrm{d}E_{\mathrm{g}}/\mathrm{d}T$	(A)	$0.8 \cdot 10^{-4}$ eV K^{-1}	$10 \cdots 300$	WMR	77S3
	(B)	$1.9 \cdot 10^{-4}$ eV K^{-1}			
	(C)	0.0 eV K^{-1}			
$\mathrm{d}E_{\mathrm{g}}/\mathrm{d}p$		$5.3 \cdot 10^{-6}$ eV bar^{-1}	300		76J

d-like character of valence bands.

excitonic energy gap (in eV):

			T [K]			
E_{gx}	(A)	1.815	2	($n=1$)	absorption	75S1
	(A)	1.818	77	($n=1$)	cathodoluminescence	77S1
	(A)	1.801	300	($n=1$)		77S2
	(A)	1.826	2	($n=2$)	absorption	75S1
	(B)	2.02	2	($n=1$)		

splitting energies at Γ:

Δ_{cf}	-0.25 eV	$T = 77$ K	electroreflectance	75S1
Δ_{so}	0.31 eV			

Physical property	Numerical value	Experimental conditions	Experimental method remarks	Ref.
higher transition energies (in eV):				
		T [K]	Symmetry	
E_1	3.75	10	$\Gamma_4(\Gamma_{15})-\Gamma_1$	77S3
	3.74	77		
	3.65	300		
dE_1/dT	$4.1\cdot10^{-4}$ eV K^{-1}			
E_2	3.86	10	$\Gamma_3(\Gamma_{12})-\Gamma_1$ d-states	
	3.84	77		
	3.74	300		
dE_2/dT	$4.1\cdot10^{-4}$ eV K^{-1}			
E_3	3.96	10	$\Gamma_1(\Gamma_{12})-\Gamma_1$ d-states	
	3.92	77		
	3.74	300		
dE_3/dT	$4.1\cdot10^{-4}$ eV K^{-1}			
E_4	4.12	10	$\Gamma_5(\Gamma_{15})-\Gamma_1$	
	4.11	77		
	4.05	300		
dE_4/dT	$5.4\cdot10^{-4}$ eV K^{-1}			
E_5	4.19	10	$\Gamma_5(\Gamma_{15})-\Gamma_1$	
	4.17	77		
	4.05	300		
dE_5/dT	$5.4\cdot10^{-4}$ eV K^{-1}			
E_6	4.51	10	$E\parallel c$	
	4.49	77		
	4.39	300		
dE_6/dT	$4.5\cdot10^{-4}$ eV K^{-1}			
E_7	4.76	10	$E\parallel c$	
	4.75	77		
	4.77	300		
dE_7/dT	$1.4\cdot10^{-4}$ eV K^{-1}			
E_8	4.86	10	$E\parallel c$	
	4.85	77		
	4.77	300		
dE_8/dT	$1.4\cdot10^{-4}$ eV K^{-1}			
E_9	5.30	10	$E\perp c$	
	5.29	77		

Impurities and defects

photoluminescence spectrum: Fig. 29

unidentified states:

E	1.3 eV		activation energy of conductivity	75S1
	1.775 eV		stimulated emission	80M

Lattice properties

wavenumbers of infrared and Raman active phonons (in cm^{-1}, RT values):

		Symmetry I$\bar{4}$2d (F$\bar{4}$3m)			
$\bar{\nu}_{LO}/\bar{\nu}_{TO}$	274/248	Γ_5	(Γ_{15})	Γ_4, Γ_5: IR active, all Raman active	77K3,
	269/247	Γ_4	(Γ_{15})		76M
	213/208	Γ_5	(W_4)		76M
	244			not resolved in IR measurements	77K3
	238			Raman	76M

(continued)

Physical property	Numerical value	Experimental conditions	Experimental method remarks	Ref.
wavenumbers of infrared and Raman active phonons (continued)				
$\bar{\nu}_{LO}/\bar{\nu}_{TO}$	179	Γ_1　(W$_1$)		76M
	162/158	Γ_5　(X$_5$)		77K3,
	162/158	Γ_4　(W$_2$)		76M
	138/133	Γ_5　(W$_3$)		
	112		Raman	76M
	80/76	Γ_5 ?		77K3, 76M
	$-/48$	Γ_5 ?	only observed in Raman	76M
	12.5	Γ_3　(W$_2$)		76M

Transport properties (n-type samples)
see also Figs. 30, 31

resistivity, conductivity, Seebeck coefficient:

ϱ	$10^5\ \Omega$ cm	$T = 300$ K	different samples	75S1
σ	$8.8\ \Omega^{-1}$ cm^{-1}	300 K		
S	$-7 \cdot 10^{-2}$ V K^{-1}	300 K		

N.B. No anisotropy reported.
Ionic conduction has been reported [77T2, 77W2].

Optical properties
refractive indices:
coefficients in the formula: $n^2 = A + B/(1 - C/\lambda^2) + D/(1 - E/\lambda^2)$

	n_o	n_e		
A	4.6453	5.2912		76B3
B	2.2057	1.3970		
C	$0.1379 \cdot 10^{-12}$ m^2	$0.2845 \cdot 10^{-12}$ m^2		
D	1.8377	1.9282		
E	$1.600 \cdot 10^{-12}$ m^2	$1.600 \cdot 10^{-12}$ m^2		

valid for $0.725\ \mu$m $< \lambda < 13.5\ \mu$m

$n_e - n_o$	-0.033	$\lambda = 10\ \mu$m		77I
		λ [μm]		
n_o	2.8452	0.725		75S1
	2.7132	1.000		
	2.6376	2.000		
	2.6245	3.000		
	2.6189	4.000		
	2.6144	5.000		
	2.6113	6.000		
	2.6070	7.000		
	2.6032	8.000		
	2.5988	9.000		
	2.5939	10.000		
	2.5890	11.000		
	2.5837	12.000		
	2.5771	13.000		
	2.5731	13.500		

(continued)

Physical property	Numerical value	Experimental conditions	Experimental method remarks	Ref.
refractive indices (continued)				
		λ [µm]		
n_e	2.8932	0.725		75S1
	2.6934	1.000		
	2.6071	2.000		
	2.5925	3.000		
	2.5863	4.000		
	2.5819	5.000		
	2.5784	6.000		
	2.5743	7.000		
	2.5704	8.000		
	2.5659	9.000		
	2.5608	10.000		
	2.5505	11.000		
	2.5505	12.000		
	2.5439	13.000		
	2.5404	13.500		
dielectric constants:				
$\varepsilon(0)$	7.76	$T = 300$ K, $E \parallel c$	infrared	76M
	8.94	$E \perp c$		
	10.48	300 K, $E \parallel c$	infrared	77K3
	9.05	$E \perp c$		
$\varepsilon(\infty)$	6.11	300 K, $E \parallel c$	infrared	76M
	6.18	$E \perp c$		
	6.95	300 K, $E \parallel c$	infrared	77K3
	6.80	$E \perp c$		
two-photon absorption:				
β	$0.0014 \cdot 10^{-6}$ cm W^{-1}	$\lambda = 1.06$ µm		80M
nonlinear dielectric susceptibilities (in 10^{-12} m V^{-1}):				
d_{14}	33	$\lambda = 0.5 \cdots 12.5$ µm		76A
d_{36}	57.7	10.6 µm		76C
	65.4	2.12 µm		
Further properties				
melting point:				
T_m	1130 K			75S1

10.1.2.11 Silver gallium telluride (AgGaTe$_2$)

energy gap:				
E_g	1.32 eV	$T = 300$ K	electroreflectance	75S1
wavenumbers of IR active phonons (in cm^{-1}, RT values):				
		Symmetry		
$\bar{\nu}_{LO}/\bar{\nu}_{TO}$	221/204	Γ_5	infrared reflectivity	77K3
	217/198	Γ_4		
	203/206	Γ_5		
	137/134	Γ_5		
	137/136	Γ_4		
	117/115	Γ_5		
	52/53	Γ_4		

Physical property	Numerical value	Experimental conditions	Experimental method remarks	Ref.
conductivity and Seebeck coefficient:				
σ	$1.7 \cdot 10^{-4}\,\Omega^{-1}\,cm^{-1}$	$T = 300\,K$	(no anisotropy reported)	75S1
S	$7 \cdot 10^{-1}\,V\,K^{-1}$	300 K		
dielectric constants:				
$\varepsilon(0)$	14.5	$E \parallel c,\ T = 300\,K$		77K3
	15.0	$E \perp c$		
$\varepsilon(\infty)$	11.0	$E \parallel c,\quad 300\,K$		
	11.86	$E \perp c$		
Debye temperature:				
Θ_D	182.4 K	$T \to 0\,K$		77B3
melting point:				
T_m	950 K			76K5

10.1.2.12 Silver indium sulfide (AgInS$_2$)

Electronic and optical properties

a) chalcopyrite structure:

energy gap (in eV):					
$E_{g,\,dir}$	(A)	1.90	$T = 77\,K$	electroreflectance (slowly cooled sample)	74S
	(B, C)	2.06			
	(A)	1.87	300 K		
	(B, C)	2.02			
		1.91	300 K	photoconductivity (non polarized radiation; sample grown by Bridgman method)	81J
excitonic energy gap:					
E_{gx}	(A)	1.880 eV	$T = 77\,K$		74S
	(B)	2.045 eV			
splitting energy at Γ (in eV):					
Δ_{cf}		-0.165	$T = 77\,K$	electroreflectance	74S
		-0.15	300 K		
		-0.16	300 K	photoconductivity	81J

b) orthorhombic structure:

energy gap (in eV):					
$E_{g,\,dir}$	(A)	2.05	$T = 2\,K$	electroreflectance (sample quenched from $T > 1000\,K$: platelet with wurtzite c axis $\perp$ to major face)	74S
	(B)	2.12			
	(A)	2.081	300 K	electroreflectance (sample with ≈ 6 at% Li, bulk crystal)	
	(B)	2.144			
	(C)	2.168			
		2.06	300 K	photoconductivity (non polarized radiation; sample grown by chemical transport method)	81J

Physical property	Numerical value	Experimental conditions	Experimental method remarks	Ref.
excitonic energy gap:				
E_{gx} (A)	2.025 eV	$T=2$ K ($n=1$)	absorption (sample quenched from $T>1000$ K)	75S1
(B)	2.093 eV			
splitting energy at Γ (in eV):				
		T [K]		
$\varDelta_{cf}$	-0.069	2	electroreflectance (from orthorhombic crystal field; sample quenched from $T>1000$ K)	74S
	-0.063	300	(from orthorhombic crystal field; sample with ≈ 6 at% Li)	
	-0.024	300	(from wurtzite crystal field, sample with ≈ 6 at% Li)	
	-0.060	300	photoconductivity (from orthorhombic crystal field)	81J
	-0.030	300	(from wurtzite crystal field)	

d-like character of valence bands: 20% [75S1].

For a comparison of luminescence spectra of orthorhombic and chalcopyrite type AgInS$_2$, see [83R].

Transport and further properties

n-type samples (orthorhombic structure):

resistivity, mobility:

ϱ	$1\cdots30\ \Omega$ cm	$T=300$ K	(no anisotropy) thin films	80G3
μ_n	<1 cm^2 V^{-1} s^{-1}	300 K		
ϱ	$10^6\ \Omega$ cm	300 K	($\varrho = 3\cdot 10^2\ \Omega$ cm for chalcopyrite structure)	81J

melting point:

T_m	1150 (10) K			74S

10.1.2.13 Silver indium selenide (AgInSe$_2$)

Electronic properties

energy gap:

$E_{g, dir}$ (A)	1.24 eV	$T=300$ K	electroreflectance	75S1
(B)	1.33 eV			
(C)	1.60 eV			
	1.3 eV		photoemission	77R
dE_g/dT	0.0 eV K^{-1}	$80\cdots300$ K	reflectivity	77R
dE_g/dp	$2.7\cdot10^{-6}$ eV bar^{-1}	300 K	absorption	76S

d-like character of valence bands: 17% [75S1].

excitonic energy gap:

E_{gx} (A)	1.245 eV	$T=2$ K	absorption, reflectivity, photoluminescence	75S1

splitting energies:

$\varDelta_{cf}$	-0.12 eV	$T=300$ K		75S1
$\varDelta_{so}$	0.30 eV			

Physical property	Numerical value	Experimental conditions	Experimental method remarks	Ref.
higher transition energies (in eV):				
(from reflectivity spectra at $T = 80$ K on polycrystalline sample)				
E	3.28		$\Gamma - X$, pseudodirect ?	77R
	3.50			
	4.18			
	5.0		E_1-like	
	5.9			
	6.8			
	7.2		E_2-like	
	8.9			
	11.2			
	12.6			
	13.0			
	14.5			
	17.6		In 4d	
	18.5			
	19.4			
	20.2			
	20.9			
	21.8			

N.B. E_1, E_2-like denote that these probably correspond to the transitions denoted E_1, E_2 in zincblende compounds.

Physical property	Numerical value	Experimental conditions	Experimental method remarks	Ref.
conduction band, valence band and core state energies (in eV):				
(from photoemission spectra at 300 K; values relative to the valence band edge)				
$E(C_3)$	4.5			77R
$E(C_2)$	3.0			
$E(C_1)$	1.3			
$E(C_0)$	1.3		conduction band edge (energy gap)	
$E(V_0)$	0.0		valence band edge	
$E(V_1)$	-1.0			
$E(V_2)$	-4.3		Ag 4d	
$E(V_3)$	-4.3			
$E(V_4)$	$\cdots$			
$E(V_5)$	-12.4		Se 4s	
$E(d_{5/2})$	-16.7		In 4d (core states)	
$E(d_{3/2})$	-17.6			

Impurities and defects

Physical property	Numerical value	Experimental conditions	Experimental method remarks	Ref.
photoluminescence peaks:				
E	≈ 1.24 eV	$T = 2$ K	several lines (unidentified)	75S1
	1.19 eV	2 K	broad line (unidentified)	

Physical property	Numerical value	Experimental conditions	Experimental method remarks	Ref.
activation energy from impurity level:				
E_A	0.11 eV			75S1

Lattice properties

Physical property	Numerical value	Symmetry	Experimental method remarks	Ref.
wavenumbers of IR active phonons (in cm^{-1}, RT values):				
$\bar{\nu}_{LO}/\bar{\nu}_{TO}$	235/217	Γ_5	infrared reflectivity	78K2
	234/209	Γ_4		78P,
	213/206	Γ_5		78N3,
	164/155	Γ_5		78K2
				(continued)

Physical property	Numerical value	Experimental conditions	Experimental method remarks	Ref.
wavenumbers of IR active phonons (continued)				
$\bar{\nu}_{LO}/\bar{\nu}_{TO}$	161/148	Γ_4		78K2
	114/113	Γ_5		78P, 78N3
	62/66	Γ_5		78K2
	43/43	Γ_4		78K2
Transport properties				
resistivities, carrier concentrations, mobilities, Seebeck coefficients:				
p-type sample:				
ϱ	$<0.1\ \Omega\,\text{cm}$	$T = 300\ \text{K}$		78P
n-type samples (all data at $T = 300$ K):				
ϱ	$10^4\ \Omega\,\text{cm}$		annealed under maximal Se pressure	75S1
n	$8 \cdot 10^{11}\ \text{cm}^{-3}$			
μ_n	$750\ \text{cm}^2\,\text{V}^{-1}\,\text{s}^{-1}$			
ϱ	$0.02\ \Omega\,\text{cm}$		annealed under minimal Se pressure	75S1
n	$5 \cdot 10^{17}\ \text{cm}^{-3}$			
μ_n	$600\ \text{cm}^2\,\text{V}^{-1}\,\text{s}^{-1}$			
σ	$0.48\ \Omega^{-1}\,\text{cm}^{-1}$		unannealed	75S1
S	$-1.4 \cdot 10^{-1}\ \text{V}\,\text{K}^{-1}$			

For temperature and pressure dependence, see Figs. 32, 33.

    Steep rise of resistivity takes place slowly in time [78J]. Above $p = 4 \cdot 10^4$ bar low resistance, because of transition to NaCl like structure [78J].

    N.B. No anisotropy reported.

Optical properties

refractive indices:

		λ [μm]		
n_o	2.8265	1.05		75S1
	2.6761	2.00		
	2.6542	3.00		
	2.6463	4.00		
	2.6416	5.00		
	2.6381	6.00		
	2.6352	7.00		
	2.6318	8.00		
	2.6286	9.00		
	2.6251	10.00		
	2.6210	11.00		
	2.6167	12.00		
n_e	2.6838	2.00		75S1
	2.6592	3.00		
	2.6504	4.00		
	2.6451	5.00		
	2.6414	6.00		
	2.6379	7.00		
	2.6343	8.00		
	2.6310	9.00		
	2.6274	10.00		
	2.6229	11.00		
	2.6183	12.00		

Physical property	Numerical value	Experimental conditions	Experimental method remarks	Ref.
dielectric constant:				
$\varepsilon(0)$	10.73	$T = 300$ K, $E \parallel c$	infrared reflectivity	78K2
	11.94	$E \perp c$		
$\varepsilon(\infty)$	7.16	$T = 300$ K, $E \parallel c$		
	7.20	$E \perp c$		

For nonlinear dielectric susceptibility, see [72B].

Further properties
melting point:

T_m	1055 K			74S

10.1.2.14 Silver indium telluride (AgInTe$_2$)

energy gaps (in eV):

$E_{g, th}$	0.95	$T = 300 \cdots 350$ K	conductivity vs. T	75S1
$E_{g, dir}$	1.12	300 K	absorption	75M
$E_{g, ind}$	0.99	300 K		

wavenumbers of IR active phonons (in cm^{-1}):

		Symmetry		
$\bar{\nu}_{LO}/\bar{\nu}_{TO}$	$-/173$?	$E \perp [111]$	infrared reflectivity at 300 K,	78K2
	181/168	$E \parallel [111]$	unconventional polarization	78N3,
	181/166	$E \perp [111]$		78K2
	138/136	$E \perp [111]$		78K2
	136/131	$E \parallel [111]$		
	44/44	$E \parallel [111]$		
	43/42	$E \perp [111]$		

conductivity, Seebeck coefficient:

σ	0.78 Ω^{-1} cm^{-1}	$T = 300$ K	no anisotropy	75S1
S	$-7 \cdot 10^{-2}$ V K^{-1}			

dielectric constants:

$\varepsilon(0)$	7.68	$T = 300$ K, $E \parallel [111]$		78K2
	8.10	$E \perp [111]$		
$\varepsilon(\infty)$	6.38	300 K, $E \parallel [111]$		
	6.48	$E \perp [111]$		

For a study of thermal conduction in AgInTe$_2$, see [81B].

Debye temperature:

Θ_D	155.9 K	$T \to 0$ K		77B4

melting point:

T_m	960 (10) K			76K5

10.1.2.15 I-thallium-VI$_2$ compounds

CuTlS$_2$
Transport properties are semiconductor like [79G2].
The energy gap is estimated to be: $E_g = 1.39$ eV [83D].

CuTlSe$_2$
conductivity, Seebeck coefficient, melting point:

σ	$6 \cdot 10^3 \ \Omega^{-1}$ cm^{-1}	$T = 300$ K	no anisotropy	75S1
S	10^{-2} V K^{-1}	300 K		
T_m	680 K			

Physical property	Numerical value	Experimental conditions	Experimental method remarks	Ref.
CuTlTe$_2$				
conductivity, Seebeck coefficient, melting point:				
σ	$2.8 \cdot 10^3 \, \Omega^{-1} \, cm^{-1}$	$T = 300$ K	no anisotropy	75S1
S	$8 \cdot 10^{-2} \, V \, K^{-1}$	300 K		
T_m	650 K			
AgTlSe$_2$				
conductivity, Seebeck coefficient, melting point:				
σ	$10^{-5} \, \Omega^{-1} \, cm^{-1}$	$T = 300$ K	no anisotropy	75S1
S	$0.8 \, V \, K^{-1}$	300 K		
T_m	600 K			
AgTlTe$_2$				
conductivity, Seebeck coefficient, melting point:				
σ	$4.1 \cdot 10^2 \, \Omega^{-1} \, cm^{-1}$	$T = 300$ K	no anisotropy	75S1
S	$6 \cdot 10^{-2} \, V \, K^{-1}$	300 K		

For transport properties of AgTlSe$_2$, see also [82A].

T_m	560 K			

10.1.2.16 Chalcopyrite (CuFeS$_2$)

Normally Cu is in the Cu$^+$-state (EXAFS) and Fe in the Fe^{3+}-state. However changes in the stoichiometry can give Fe^{2+} in order to conserve charge neutrality [72A2, 78B1].

The electronic structure of CuFeS$_2$ has been discussed using results of X-ray emission and X-ray photo-electron spectra [82T2].

The highest energy occupied orbitals are found to be of Cu3d-S3p antibonding character. Theoretical results predict the lowest energy empty orbitals to be of Fe3d-S3p antibonding character with a small (<2 eV) separation from the highest occupied orbitals.

Other calculations (Xα-method [81H1]) for the antiferromagnetic phase show that the valence bands consist of many rather narrow bands constructed from S3p, Cu3d and majority spin Fe3d orbitals. The upper valence bands (top at the X point) consist mainly of Cu3d and Fe3d orbitals. The conduction bands (bottom at Γ) are composed mainly of 4s and 4p orbitals of Cu and Fe. Narrow d-bands (minority spin Fe3d orbitals) occur in the so-called fundamental gap (see Fig. 33α).

wavenumbers of IR and Raman active phonons (in cm^{-1}):

		T [K]	Symmetry $I\bar{4}2d$ (F$\bar{4}$3m)		
$\bar{\nu}_{LO}/\bar{\nu}_{TO}$	385/360	300	Γ_4	(Γ_{15})	all modes IR and	75K1
	371/357	300	Γ_5	(Γ_{15})	Raman active,
	330/322	300	Γ_5	(W$_4$)	except Γ_1
	296	4.5	Γ_1	(W$_1$)	
	272/262	300	Γ_4	(W$_2$)	
	267/263	300	Γ_5	(X$_5$)	
	$-/179$	4.5	Γ_5	(W$_3$)	
	$-/105$	4.5	Γ_5	(W$_4$)	
	$-/90$	4.5	Γ_4	(W$_2$)	
	$-/72$	4.5	Γ_5	(X$_5$)	

conductivity, mobility:

σ	$100 \, \Omega^{-1} \, cm^{-1}$	$T = 300$ K	p-type sample	75S1
μ_p	$30 \, cm^2 \, V^{-1} \, s^{-1}$	300 K	(no anisotropy)	

magnetic susceptibility:

χ_m	$10 \cdots 13$ $\cdot 10^{-4} \, cm^3 \, mol^{-1}$	$T = 300$ K	before anneal (χ_m in CGS-emu)	72A1
	$9.8 \cdot 10^{-4} \, cm^3 \, mol^{-1}$	300 K	after anneal	72A2

Physical property	Numerical value	Experimental conditions	Experimental method remarks	Ref.
Néel temperature:				
T_N	800 K		(antiferromagnetic)	72A2
melting point:				
T_m	1120 K or 1150 K		contradictory reports	75S1

10.1.2.17 Other I-iron-VI$_2$ compounds

CuFeSe$_2$

conductivity, mobility, melting point:

σ	700 Ω^{-1} cm^{-1}	$T = 300$ K	p-type	75S1
μ_p	20 cm^2 V^{-1} s^{-1}	300 K	(no anisotropy)	
T_m	850 K			

Fe is normally in the Fe^{3+}-state [76R].

CuFeTe$_2$

conductivity, mobility, melting point:

σ	400 Ω^{-1} cm^{-1}	$T = 300$ K	p-type	75S1
μ_p	50 cm^2 V^{-1} s^{-1}	300 K	(no anisotropy)	
T_m	1015 K			

AgFeSe$_2$

conductivity, mobility, melting point:

σ	1500 Ω^{-1} cm^{-1}	$T = 300$ K	n-type	75S1
μ_n	250 cm^2 V^{-1} s^{-1}	300 K	(no anisotropy)	
T_m	1009 K			

AgFeTe$_2$

conductivity, mobility, melting point:

σ	700 Ω^{-1} cm^{-1}	$T = 300$ K	n-type	75S1
μ_n	2000 cm^2 V^{-1} s^{-1}	300 K	(no anisotropy)	
T_m	953 K			

10.1.2.18 Solid solutions

(CuAlS$_2$)$_x$(CuAlSe$_2$)$_{1-x}$: exciton energies from phase-shift-difference spectroscopy: see Fig. 34.

(CuAlSe$_2$)$_x$(CuGaSe$_2$)$_{1-x}$: energy gap from absorption: see Fig. 34α.

(CuGaS$_2$)$_x$(CuGaSe$_2$)$_{1-x}$: band gap from absorption ($T = 77$ K): Fig. 35. CuGaS$_{1.5}$Se$_{0.5}$: $E_g = 2.14$ eV. CuGaSSe: 1.92 eV. CuGaS$_{0.5}$Se$_{1.5}$: 1.67 eV. (E_g-values from absorption measurements [73R]). Exciton energies from phase-shift-difference spectroscopy at $T = 20$ K: Fig. 36; photoluminescence spectrum: Fig. 37; phonon wavenumbers: Fig. 38.

(CuInS$_2$)$_x$(CuInSe$_2$)$_{1-x}$: energy gap and photoluminescence peaks: Fig. 39; cathodoluminescence peaks: Fig. 40; photoconductivity spectrum: Fig. 41.

(CuGaS$_2$)$_x$(CuInS$_2$)$_{1-x}$: phonon wavenumbers: Fig. 42.

(CuGaSe$_2$)$_x$(CuInSe$_2$)$_{1-x}$: energy gap: Fig. 43; CuGa$_{0.48}$In$_{0.52}$Se$_2$: E_g vs. T: Fig. 44; resistivity: Fig. 45.

(CuAlS$_2$)$_x$(CuFeS$_2$)$_{1-x}$: optical reflectivity: Fig. 46; activation energy of conductivity: Fig. 47(b); conductivity: Fig. 48; magnetic moment: Fig. 47(a).

For thermal expansion in CuGa$_{0.5}$In$_{0.5}$Se$_2$, see [82K].

(AgGaS$_2$)$_x$(AgGaSe$_2$)$_{1-x}$: phonon wavenumbers: Fig. 49.

For solid solutions of the type (Cu$_{1-x}$Ag$_x$)(In$_{1-y}$Ga$_y$)Te$_2$, see [81A2].

10.1.2.19 References for 10.1.2

53H   Hahn, H., Frank, G., Klingler, W., Meyer, A.-D., Störger, G.: Z. Anorg. Allgem. Chem. **271** (1953) 153.

58Z   Žuze, V.P., Sergeeva, V.M., Štrum, E.L.: Zh. Tekhn. Fiz. **28** (1958) 233; Soviet Phys.-Tech. Phys. (English Transl.) **3** (1958) 208.

64I   Imamov, R.M., Pinsker, Z.G.: Kristallografiya **9** (1964) 743.

65I   Imamov, R.M., Pinsker, Z.G.: Kristallografiya **10** (1965) 199; Soviet Phys.-Crystallogr. (English Transl.) **10** (1965) 148; C.A. **63** (1965) 2481e.

70M   Mal'sagov, A.U.: Fiz. Tekh. Poluprov. **4** (1970) 1417; Sov. Phys. Semicond. (English Transl.) **4** (1971) 1213.

71B   Boyd, G.D., Kasper, H., McFee, J.H.: IEEE J. Quant. Electron. QE-**7** (1971) 563.

72A1   Adams, R., Beaulieu, R., Vassiliades, M., Wold, A.: Mater. Res. Bull. **7** (1972) 87.

72A2   Adams, R., Russo, P., Arnoff, R., Wold, A.: Mater. Res. Bull. **7** (1972) 93.

72B   Boyd, G.D., Kasper, H., McFee, J.H., Storz, F.G.: IEEE J. Quant. Electron. QE-**8** (1972) 900.

73B   Bettini, M.: Solid State Commun. **13** (1973) 599.

73R   Robbins, M., Lambrecht, V.G.: Mater. Res. Bull. **8** (1973) 703.

74D   Donohue, P.C., Bierlein, J.D., Hanlon, J.E., Tarrett, H.S.: J. Electrochem. Soc. **121** (1974) 829.

74H   Holah, G.D., Webb, J.S., Montgomery, H.: J. Phys. C: Solid State Phys. **7** (1974) 3875.

74S   Shay, J.L., Tell, B., Shiavone, L.M., Kasper, H.M., Thiel, F.: Phys. Rev. B **9** (1974) 1719.

74T   Turner, E.H., Buehler, E., Kasper, H.: Phys. Rev. B **9** (1974) 558.

74Y   Yamamoto, N., Miyauchi, T.: Jpn. J. Appl. Phys. **13** (1974) 1919.

75A   Abrahams, S.C., Hsu, F.S.L.: J. Chem. Phys. **63** (1975) 1162.

75B1   Bridenbaugh, P., Tell, B.: Mater. Res. Bull. **10** (1975) 1127.

75B2   Badikov, V.V., Pivovarov, O.N., Skokov, Yu.V., Skrebneva, O.V., Trotsenko, N.K.: Kvant. Elektron. **2** (1975) 618; Sov. J. Quant. Electron. (English Transl.) **5** (1975) 350.

75D   Dirochka, A.I., Ivanova, G.S., Kurbatov, L.N., Sinitsyn, E.V., Kharakhorin, F.F., Kholina, E.N.: Fiz. Tekh. Poluprov. **9** (1975) 1128; Sov. Phys. Semicond. (English Transl.) **9** (1975) 742.

75G   Grimsditch, H.M., Holah, G.D.: Phys. Rev. B **12** (1975) 4377.

75H   Hengehold, R.L., Pedrotti, F.L.: J. Appl. Phys. **46** (1975) 5202.

75K1   Koschel, W.H., Bettini, M.: Phys. Status Solidi (b) **72** (1975) 729.

75K2   Kaufmann, U., Räuber, A., Schneider, J.: J. Phys. C: Solid State Phys. **8** (1975) L381.

75K3   Kaufmann, U.: Phys. Rev. B **11** (1975) 2478.

75K4   Kazmerski, L.L., Ayyagari, M.S., Sanborn, G.A.: J. Appl. Phys. **46** (1975) 4865.

75L   Lockwood, D.J., Montgomery, H.: J. Phys. C: Solid State Phys. **8** (1975) 3241.

75M   Mirgorodskii, V.M., Radautsan, S.I., Tsurkan, A.E.: Opt. Spektrosk. **39** (1975) 519; Opt. Spectrosc. USSR **39** (1975) 291.

75S1   Shay, J.L., Wernick, J.H.: Ternary Chalcopyrite Semiconductors: Growth, Electronics Properties and Applications. Pergamon, **1975**.

75S2   Suzuki, K., Kambara, T., Gondaira, K., Sato, K., Kondo, K., Teranishi, T.: J. Phys. Soc. Jpn. **39** (1975) 1310.

75T   Tell, B., Bridenbaugh, P.M.: Phys. Rev. B **12** (1975) 3330.

75Y   Yamamoto, N., Tohge, N., Miyauchi, T.: Jpn. J. Appl. Phys. **14** (1975) 192.

76A   Airoldi, G.: Riv. Nuovo Cimento **6** (1976) 295.

76B1   Braun, W., Cardona, M.: Phys. Status Solidi (b) **76** (1976) 251.

76B2   Bettini, M., Suga, S., Cho, K., Marshall, N.: Solid State Commun. **18** (1976) 17.

76B3   Bhar, G.C.: Appl. Opt. **15** (1976) 305.

76C   Choy, M.M., Beyer, R.L.: Phys. Rev. B **14** (1976) 1693.

76G1   Gan, J.N., Tauc, J., Lambrecht, V.G. Jr., Robbins, M.: Phys. Rev. B **13** (1976) 3610.

76G2   Gassmov, T.K., Gadjiev, V.A.: Izv. Akad. Nauk Az. SSSR **1976**, 108.

76J   Jayaraman, A., Narayanamurti, V., Kasper, H.M., Chin, M.A., Maines, R.G.: Phys. Rev. B **14** (1976) 3516.

76K1   Kaufmann, U.: Solid State Commun. **19** (1976) 213.

76K2   Kokta, M., Carruthers, J.R., Crasso, M., Kasper, H.M., Tell, B.: J. Electron. Mater. **5** (1976) 69.

76K3   Kazmerski, L.L., Ayyagari, M.S., Sanborn, G.A., White, F.R., Merrill, A.J.: Thin Solid Films **37** (1976) 323.

76K4   Kazmerski, L.L., Ayyagari, M.S., White, F.R., Sanborn, G.A.: J. Vac. Sci. Technol. **13** (1976) 139.

76K5	Kanellis, G., Kambas, C., Spyridelis, J.: Mater. Res. Bull. **11** (1976) 429.
76L	Look, D.C., Manthuruthil, J.C.: L. Phys. Chem. Solids **37** (1976) 173.
76M	Miller, A., Holah, G.D., Dunnett, W.D., Iseler, G.W.: Phys. Status Solidi (b) **78** (1976) 569.
76P1	Poplavnoi, A.S., Polygalov, Yu.İ, Ratner, A.M.: Izv. Vyssh. Uchebn. Zaved. Fiz. **6** (1976) 7.
76P2	Phil Won, Yu.: J. Appl. Phys. **47** (1976) 677.
76P3	Phil Won, Yu.: Solid State Commun. **18** (1976) 395.
76R	Reddy, K.V., Chetty, S.C.: Mater. Res. Bull. **11** (1976) 55.
76S	Sato, K., Kawakami, T., Teranishi, T., Kambara, T., Gondaira, K.: J. Phys. Soc. Jpn. **41** (1976) 937.
76T	Troeger, G.L., Rogers, R.N., Kasper, H.M.: J. Phys. C: Solid State Phys. **9** (1976) L73.
76Y1	Yamamoto, N.: Jpn. J. Appl. Phys. **15** (1976) 1909.
76Y2	Yu, P.W., Park, Y.S., Grant, J.T.: Appl. Phys. Lett. **28** (1976) 214.
77B1	Bodnat', I.V., Voroshilov, Yu.V., Lukomskii, A.I.: Izv. Akad. Nauk SSSR, Neorg. Mater. **13** (1977) 3935.
77B2	Bodnar, I.V., Karoza, A.G., Smirnova, G.F.: Phys. Status Solidi (b) **84** (1977) K65.
77B3	von Bardeleben, H.J., Goltzené, A., Schwab, C.: Inst. Phys. Conf. Ser. **35** (1977) 43.
77B4	Bachmann, K.J., Hsu, F.S.L., Thiel, F.A., Kasper, H.M.: J. Electron. Mater. **6** (1977) 431.
77G	Guseinov, G.D., Abdullaeva, S.G., Aksyanov, I.G., Kasumov, T.K., Nani, R.Kh.: Mater. Res. Bull. **12** (1977) 5157.
77H1	Haupt, H., Hess, K.: Inst. Phys. Conf. Ser. **35** (1977) 5.
77H2	Hörig, W., Neumann, H., Höbler, H.-J., Kühn, G.: Phys. Status Solidi (b) **80** (1977) K21.
77I	Iseler, G.W., Kildal, H., Menyuk, K.N.: Inst. Phys. Conf. Ser. **35** (1977) 73.
77J	Jayaraman, A., Dernier, P.D., Kasper, H.M., Maines, R.G.: High Temp.-High Press. **9** (1977) 97.
77K1	Kazmerski, L.L., Shieh, C.C.: Thin Solid Films **41** (1977) 35.
77K2	Kazmerski, L.L., Juang, Y.J.: J. Vac. Sci. Technol. **14** (1977) 769.
77K3	Kanellis, G., Kampas, K.: J. Phys. (Paris) **38** (1977) 833.
77L	Lockwood, D.J.: Inst. Phys. Conf. Ser. **35** (1977) 97.
77M1	Mandel, L., Tomlinson, R.D., Hampshire, M.J.: J. Appl. Crystallogr. **10** (1977) 130.
77M2	Mittleman, S.D., Singh, R.: Solid State Commun. **22** (1977) 659.
77M3	Mora, S., Romeo, N.: J. Appl. Phys. **48** (1977) 4826.
77P	Paorici, C., Romeo, N., Sberveglieri, G., Tarricone, L.: J. Lumin. **15** (1977) 101.
77R	Rife, J.C., Dexter, R.N., Bridenbaugh, P.M., Veal, B.W.: Phys. Rev. B **16** (1977) 4491.
77S1	Sermage, B., Voos, M.: Phys. Rev. B **15** (1977) 3935.
77S2	Sugai, S.: J. Phys. Soc. Jpn. **43** (1977) 592.
77S3	Sermage, B., Fishman, G.: Inst. Phys. Conf. Ser. **35** (1977) 139.
77T1	Tanaka, S., Kawani, S., Kobayashi, H., Sasakura, H.: J. Phys. Chem. Solids **38** (1977) 680.
77T2	Tell, B., Wagner, S., Kasper, H.M.: J. Electrochem. Soc. **124** (1977) 536.
77T3	Thwaites, M.J., Tomlinson, R.D., Hampshire, M.J.: Inst. Phys. Conf. Ser. **35** (1977) 237.
77T4	Thwaites, M.J., Tomlinson, R.D., Hampshire, M.J.: Solid State Commun. **23** (1977) 905.
77T5	Tyagunova, T.V., Kharakhorin, F.F., Kholina, E.N.: Izv. Akad. Nauk SSSR, Neorg. Mater. **13** (1977) 46; Inorg. Mater. (USSR) (English Transl.) **13** (1977) 36.
77T6	Teranishi, T., Sato, K., Saito, Y.: Inst. Phys. Conf. Ser. **35** (1977) 59.
77W1	Weinert, H., Neumann, H., Höbler, H.-J., Kühn, G., Nguyen Van Nam: Phys. Status Solidi (b) **81** (1977) K59.
77W2	Wagner, S.: Inst. Phys. Conf. Ser. **35** (1977) 205.
77Y	Yamamoto, N., Horinka, H., Okada, K., Miyauchi, T.: Jpn. J. Appl. Phys. **16** (1977) 1817.
77Z	Zielinger, J.P., Noguet, C., Tapiero, M.: Inst. Phys. Conf. Ser. **35** (1977) 145.
78B1	Ballal, M.M., Mande, C.: J. Phys. C: Solid State Phys. **11** (1978) 837.
78B2	Bodnar, I.V., Karoza, A.G., Smirnova, G.F.: Phys. Status Solidi (b) **86** (1978) K171.
78B3	von Bardeleben, H.J., Goltzené, A., Schwab, C., Feigelson, R.S.: Appl. Phys. Lett. **32** (1978) 741.
78D	Davis, J.G., Bridenbaugh, P.M., Wagner, S.: J. Electron. Mater. **7** (1978) 39.
78H1	Horinaka, H., Yamamoto, N., Miyauchi, T.: Jpn. J. Appl. Phys. **17** (1978) 521.
78H2	Hörig, W., Neumann, H., Schumann, B., Kühn, G.: Phys. Status Solidi (b) **85** (1978) K57.
78H3	Hwang, H.L., Sun, C.Y., Leu, C.Y., Cheng, C.L., Tu, C.C.: Rev. Phys. Appl. **13** (1978) 745.
78H4	Hörig, W., Neumann, H., Sobotta, H., Schumann, B., Kühn, G.: Thin Solid, Films **48** (1978) 67.

78J	Jayaraman, A., Tell, B., Maines, R.G.: Appl. Phys. Lett. **32** (1978) 21.
78K1	Kazmerski, L.L., Sprague, D.L., Cooper, R.B.: J. Vac. Sci. Technol. **15** (1978) 249.
78K2	Kanellis, G., Kampas, K.: Mater. Res. Bull. **13** (1978) 9.
78N1	Neumann, H., Hörig, W., Reccius, E., Möller, W., Kühn, G.: Solid State Commun. **27** (1978) 449.
78N2	Neumann, H., Nguyen Van Nam, Höbler, H.-J., Kühn, G.: Solid State Commun. **25** (1978) 899.
78N3	Nikolic, P.M., Stojilković, S.M., Petrović, Z., Dimitrijević, P.: Fizika **10** (1978) 98.
78P	Petrović, Z., Nikolić, P.M., Vujatović, S.S.: J. Eng. Phys. **17** (1978) 31.
78R	Riede, V., Sobotta, H., Neumann, H., Hoang Xuan Nguyen: Solid State Commun. **28** (1978) 449.
78S1	Sermage, B.: Solid State Electron. **21** (1978) 1361.
78S2	Sun, L.Y., Kazmerski, L.L., Clark, A.H., Ireland, P.J., Morton, D.W.: J. Vac. Sci. Technol. **15** (1978) 265.
78S3	Schumann, B., Georgi, C., Tempel, A., Kühn, G., Nguyen Van Nam, Neumann, H., Hörig, W.: Thin Solid Films **52** (1978) 45.
78T	Thwaites, M.J., Tomlinson, R.D., Hampshire, M.J.: Solid State Commun. **27** (1978) 727.
78V	Vecchi, M.P., Ramos, J., Giriat, W.: Solid State Electron. **21** (1978) 1609.
79B1	Bhar, G.C., Ghosh, G.: J. Opt. Soc. Am. **69** (1979) 730.
79B2	Bodnar', I.V., Voroshilov, Yu V., Karoza, A.G., Smirnova, G.F., Khudolii, V.A.: Izv. Akad. Nauk SSSR, Neorg. Mater. **15** (1979) 763; Inorg. Mater. (USSR) (English Transl.) **15** (1979) 567.
79B3	Boyarintsev, P.K., Ufimtsev, V.P., Khorakhoren, F.F., Kholina, E.N.: Fiz. Tekh. Poluprov. **13** (1979) 161; Sov. Phys. Semicond. (English Transl.) **13** (1979) 91.
79G1	Gorska, M., Beaulieu, R., Loferski, J.J., Roessler, B.: Solar Energy Mater. **1** (1979) 313.
79G2	Gardes, B., Brun, G., Raymond, A., Tedenac, J.C.: Mater. Res. Bull. **14** (1979) 943.
79H1	Hörig, W., Neumann, H., Reccius, E., Weinert, H., Kühn, G., Schumann, B.: Phys. Status Solidi (a) **51** (1979) 57.
79H2	Hörig, W., Möller, W., Neumann, H., Reccius, E., Kühn, G.: Phys. Status Solidi (b) **92** (1979) K1.
79I	Irie, T., Endo, S., Kimura, S.: Jpn. J. Appl. Phys. **18** (1979) 1303.
79J	Joshi, N.V.: J. Phys. Chem. Solids **40** (1979) 93.
79L	Li, P.W., Anderson, R.A., Plovnick, R.H.: J. Phys. Chem. Solids **40** (1979) 333.
79M1	Mandel, L., Tomlinson, R.D., Hampshire, M.S., Neumann, R.: Solid State Commun. **32** (1979) 201.
79M2	Mora, S., Romeo, N., Tarricone, L.: Solid State Commun. **29** (1979) 155.
79N1	Neumann, H., Peters, D., Schumann, B., Kühn, G.: Phys. Status Solidi (a) **52** (1979) 559.
79N2	Neumann, H., Hörig, W., Reccius, E., Sobotta, H., Schumann, B., Kühn, G.: Thin Solid Films **61** (1979) 13.
79N3	Neumann, H., Schumann, B., Peters, D., Tempel, A., Kühn, G.: Krist. Tech. **14** (1979) 379.
79N4	Neumann, H., Tomlinson, R.D., Nowak, E., Avgerinos, N.: Phys. Status Solidi (a) **56** (1979) K137.
79P	Paorici, C., Zanotti, L., Romeo, N., Sberveglieri, G., Tarricone, L.: Solar Energy Mater. **1** (1979) 3.
79R	Riede, V., Sobotta, H., Neumann, H., Hoang Xuan Nguyen, Möller, W., Kühn, G.: Phys. Status Solidi (b) **93** (1979) K93.
79S1	Stankiewicz, J., Giriat, W., Ramos, J., Vecchi, M.P.: Solar Energy Mater. **1** (1979) 169.
79S2	Schumann, B., Tempel, A., Kühn, G., Neumann, H., Peters, D., Hörig, W.: Krist. Tech. **14** (1979) 665.
79T1	Tempel, A., Schumann, B.: Thin Solid Films **59** (1979) 99.
79T2	Thwaites, M.J., Tomlinson, R.D., Hampshire, M.J.: Phys. Status Solidi (b) **94** (1979) 211.
80A1	Austinat, J., Nelkowski, H., Schrittenlacher, W.: Solid State Commun. **37** (1980) 285.
80A2	Abdinov, A.Sh., Mamedov, V.K.: Fiz. Tekh. Poluprov. **14** (1980) 892; Sov. Phys. Semicond. (English Transl.) **14** (1980) 526.
80B1	Bhar, G.C., Ghosh, G.C.: Appl. Opt. **19** (1980) 1029.
80B2	Brühl, H.G., Neumann, H.: Solid State Commun. **34** (1980) 225.
80B3	von Bardeleben, H.-J., Tomlinson, R.D.: J. Phys. C: Solid State Phys. **13** (1980) L1097.
80B4	von Bardeleben, H.-J., Schwab, C., Scharager, C., Muller, J.C., Siffert, P., Feigelson, R.S.: Phys. Status Solidi (a) **58** (1980) 143.

80C	Carlone, C., Olego, D., Jayaraman, A., Cardona, M.: Phys. Rev. B **22** (1980) 3877.
80D	Darný, R., Hill, A.E., Tomlinson, R.D.: Thin Solid Films **69** (1980) L11.
80G1	Grżeta-Plenković, B., Popović, S., Čelustka, B., Santić, B.: J. Appl. Crystallogr. **13** (1980) 311.
80G2	Gorska, M., Beaulieu, R., Loferski, J.J., Roessler, B., Beall, J.: Solar Energy Mater. **2** (1980) 343.
80G3	Górska, M., Beaulieu, R., Loferski, J.J., Roessler, B.: Thin Solid Films **67** (1980) 341.
80H1	Hörig, W., Neumann, H., Godmanis, I.: Solid State Commun. **36** (1980) 181.
80H2	Hörig, W., Neumann, H., Savalev, V., Lagzdonis, J.: Phys. Lett. **78A** (1980) 189.
80H3	Hwang, H.L., Tu, C.C., Maa, J.S., Sun, C.Y.: Solar Energy Mater. **2** (1980) 433.
80H4	Hwang, H.L., Cheng, C.L., Liu, L.M., Liu, Y.C., Sun, C.Y.: Thin Solid Films **67** (1980) 83.
80J	Jagadeesh, M.S., Seehra, M.S.: Solid State Commun. **36** (1980) 257.
80M	Miller, A., Ash, G.S.: Opt. Commun. **33** (1980) 297.
80P1	Poure, A., Aguero, G., Masse, G., Aicardi, J.P.: J. Phys. (Paris) **41** (1980) 707.
80P2	Piekoszewski, J., Loferski, J.J., Beaulieu, R., Beall, J., Roessler, B., Shewchun, J.: Solar Energy Mater. **2** (1980) 363.
80R	Riede, V., Neumann, H., Sobotta, H., Tomlinson, R.D., Elliott, E., Howarth, L.: Solid State Commun. **33** (1980) 557.
80S1	Sanchez-Porras, G.P., Wasim, S.M.: Phys. Status Solidi (a) **59** (1980) K175
80S2	Sobotta, H., Neumann, H., Riede, V., Kühn, G., Seltmann, J., Oppermann, D.: Phys. Status Solidi (a) **60** (1980) 531.
80T	Tomlinson, R.D., Elliott, E., Howarth, L., Hampshire, M.J.: J. Cryst. Growth **49** (1980) 115.
81A1	Abdinov, A.Sh., Gasanova, L.G., Mamedov, V.K.: Sov. Phys. Semicond. **15** (1981) 1302 (transl. from Fiz. Tekh. Poluprov. **15** (1981) 2245).
81A2	Avon, J.E., Woolley, J.C.: J. Appl. Phys. **52** (1981) 6423.
81B	Bellabarba, C., Wasim, S.M.: Phys. Status Solidi (a) **66** (1981) K105.
81H1	Hamajima, T., Kambara, T., Gondaira, K.I.: Phys. Rev. B **24** (1981) 3349.
81H2	Holah, G.D., Schenk, A.A., Perkovitz, S.: Phys. Rev. B **23** (1981) 6288.
81J	Joshi, N.V., Martinez, L., Echeverria, R.: J. Phys. Chem. Solids **42** (1981) 281.
81P	Poure, A., Legris, J.P., Aicarde, J.P.: J. Phys. C: Solid State Phys. **14** (1981) 521.
81S	Sobotta, H., Neumann, H., Kissinger, H., Kühn, G., Riede, V.: Phys. State Solidi (b) **103** (1981) K125.
81V	Vladimirov, V.E., Kopytov, A.V., Poplavnoi, A.S.: Kristallografiya **26** (1981) 619.
82A	Abou el Ela, A.H., Abdelmosen, N., Labib, H.H.A.: Phys. Status Solidi (a) **71** (1982) K219.
82B	Bohmhammel, K., Deus, P., Kühn, G., Möller, W.: Phys. Status Solidi (a) **71** (1982) 505.
82K	Kistaiah, P., Satyanarayana-Murthy, K., Krishna-Rao, K.V.: J. Phys. D **15** (1982) 1265.
82T1	Tyuterev, V.G., Skachkov, S.I., Brysneva, L.A.: Sov. Phys. Solid State **24** (1982) 1271 (transl. from Fiz. Tverd. Tela **24** (1982) 2236).
82T2	Tossell, J.A., Urch, D.S., Vaughan, D.J., Wiech, G.: J. Chem. Phys. **77** (1982) 77.
83B	Bodnar, I.V., Gil, N.L., Lukomskii, A.I.: Sov. Phys. Semicond. **17** (1983) 333 (transl. from Fiz. Tekh. Poluprov. **17** (1983) 530).
83D	Deshpande, A.P., Sapre, V.B., Mande, C.: J. Phys. C **16** (1983) L433.
83J	Jaffe, J.E., Zunger, A.: Phys. Rev. B **28** (1983) 5822.
83R	Rud, Yu.V., Parimbekov, Z.A.: Sov. Phys. Semicond. **17** (1983) 178 (transl. from Fiz. Tekh. Poluprov. **17** (1983) 281).

10.1.3 II–IV–V$_2$ compounds

10.1.3.0 Structure and energy gaps

Almost all the II–IV–V$_2$ compounds take the chalcopyrite structure. At high temperatures they tend to go into a disordered zincblende like state. Some of the compounds have also been prepared in an amorphous phase. The most important structural details are summarized in the following tables.

chalcopyrite structures (RT values):

Substance	a [Å]	c [Å]	c/a	x	d [g/cm^3]	Ref.
MgSiP$_2$	5.72 (2)	10.12 (25)	1.766 (3)			75S1
ZnSiP$_2$	5.400 (1)	10.438 (3)	1.933 (1)	0.2691		75S1
					3.35	69B
ZnSiAs$_2$	5.606 (5)	10.886 (5)	1.940 (3)	0.2658		75S1
					4.69	69B
ZnGeP$_2$	5.463 (3)	10.74 (3)	1.965 (5)	0.2582		75S1
					4.04	58P
ZnGeAs$_2$	5.671 (1)	11.153	1.966 (1)	0.2585*)		75S1
					5.26	58P
ZnSnP$_2$	5.652 (1)	11.305 (3)	2.000	0.239		75S1
ZnSnAs$_2$	5.8515 (5)	11.703 (1)	2.000			75S1
					5.53	69B
ZnSnSb$_2$	6.273	12.546	2.000			75S1
CdSiP$_2$	5.679 (1)	10.431	1.836 (1)	0.2967		75S1
					3.97	69B
CdSiAs$_2$	5.885 (1)	10.881 (1)	1.849	0.2893*)		75S1
CdGeP$_2$	5.740 (1)	10.776 (1)	1.878 (1)	0.2819*)		75S1
CdGeAs$_2$	5.943 (1)	11.220 (3)	1.888 (1)	0.2790*)		75S1
					5.6 (crystalline)	69B
					5.35 (amorphous)	
CdSnP$_2$	5.901 (1)	11.514 (4)	1.951 (1)	0.2621*)		75S1
CdSnAs$_2$	6.089 (5)	11.925 (10)	1.957 (2)	0.2612*)	5.71	75S1
						58P

*) Calculated.

ZnGeN$_2$ has a wurtzite-like monoclinic structure [70M]:

$a = c = 3.167$ Å
$b \quad = 5.194$ Å
$\beta \quad = 118°53'$

The compounds ZnSnP$_2$, ZnSnAs$_2$, (ZnSnSb$_2$?) tend to form domains with the tetragonal axis along the x, y or z direction of zincblende [70M].

high-temperature phases (at $p = 1$ bar, can be quenched to room temperature):

Substance	Transition	Structure	Ref.
MgGeP$_2$	0 K	disordered zincblende	75S1
ZnGeP$_2$	1223 (5) K	unknown	75S1
		existence disputed	78G3
ZnGeAs$_2$	1085 K	disordered zincblende	75S1
ZnSnP$_2$	990 K	disordered zincblende	75S1
ZnSnAs$_2$	900···920 K	disordered zincblende	75B3
ZnSnSb$_2$	unknown	disordered zincblende	75S1
CdGeAs$_2$	900 K	disordered zincblende	75S2, 75B3
CdSnAs$_2$	827···860 K	disordered zincblende	75B3, 75S1

solid solutions:

Substance	Structure	Lattice parameters
$ZnSiAs_{2x}P_{2(1-x)}$	chalcopyrite	Fig. 1
$Zn_xCd_{1-x}SiAs_2$	chalcopyrite	Fig. 2
$Zn_xCd_{1-x}SnAs_2$	chalcopyrite	Fig. 3
$CdSnP_{2x}As_{2(1-x)}$	chalcopyrite	Fig. 4
$CdGe_xSn_{1-x}P_2$	chalcopyrite	Fig. 5

amorphous phases:

Substance	Remarks	Ref.
$CdGeP_2$		75S1
$CdGeAs_2$	RDF: Fig. 6	
	crystallization temperature: 709 K	77S2
$CdSiAs_2$		
$CdSnAs_2$	RDF: Fig. 7	
$GdGeP_{2x}As_{2(1-x)}$		75S1
$CdGe_xSn_{1-x}As_2$		75S1
$CdSi_{0.2}Ge_{0.8}As_2$		77S2
$CdGe_{0.8}Sn_{0.2}As_2$		77S2
$CdGe_{0.8}Pb_{0.2}As_2$		77S2

energy gaps:

The room-temperature energy gaps of the II–IV–V_2 compounds are summarized in the following table. For further information on the electronic properties, see under each compound.

Substance	Energy gap [eV]	Type	Further data in section	Ref.
$ZnGeN_2$	2.67	?	10.1.3.4	74L
$CdSiP_2$	2.2···2.45	pseudodirect	10.1.3.10	79C, 75S1
	2.65	direct		77A3
$ZnSiP_2$	2.07 (1)	pseudodirect (indirect?)	10.1.3.2	79S
	2.96	direct		75S1
$ZnGeP_2$	2.05	pseudodirect	10.1.3.5	74V1
	2.365	direct		
$MgSiP_2$	2.03	pseudodirect	10.1.3.1	78A1
	2.26	indirect (?)		
	2.82	direct (?)		
$ZnSiAs_2$	1.74	pseudodirect	10.1.3.3	76R
	2.13 (1)	direct		76J
$CdGeP_2$	1.72	direct	10.1.3.12	75S1
$ZnSnP_2$	1.66	direct	10.1.3.7	75S1
$CdSiAs_2$	1.55	direct	10.1.3.11	76J
$CdSnP_2$	1.17	direct	10.1.3.14	70S
$ZnGeAs_2$	1.15	direct	10.1.3.6	75S1
$ZnSnAs_2$	0.745	direct	10.1.3.8	71K
$ZnSnSb_2$	0.7?	direct	10.1.3.9	73B
$CdGeAs_2$	0.57	direct	10.1.3.13	75S1
$CdSnAs_2$	0.26	direct	10.1.3.15	75S1

Physical property	Numerical value	Experimental conditions	Experimental method remarks	Ref.

10.1.3.1 Magnesium silicon phosphide (MgSiP$_2$)

energy gaps (in eV, at RT):

$E_{g,pseu}$	2.03	$E \perp c$	photoconductivity	78A1
	2.10	$E \perp c$		
$E_{g,ind}$	2.26	$E \perp c$	may be pseudodirect	78A1
	2.34	$E \perp c$		
	2.48	$E \perp c$		
	2.58	$E \perp (\parallel)c$		
	2.64	$E \perp c$		
$E_{g,dir}$	2.82	$E \parallel (\perp)c$	may be pseudodirect	78A1
	2.88	$E \perp c$		
	3.02	$E \perp c$		
$dE_g/dT = -4.5 \cdot 10^{-4}$ eV K^{-1}		$T = 77 \cdots 300$ K	main feature of spectrum	78A1

splitting energies:

Δ_{cf}	-0.08 eV	$T = 300$ K	experiment	78A1
	-0.72 eV		calculation	70G
Δ_{so}	0.16 eV	300 K		78A1

width of valence band:

ΔE	12.8 eV	$T = 300$ K	X-ray spectroscopy (ΔE: energy gap between emission ending and onset of absorption of corrected PK spectra)	80G1

higher interband transition energies:

E_1	3.22 eV	$T = 300$ K	photoconductivity, $E \perp c$	78A1
E_2	3.36 eV	300 K		

X-ray spectrum: Fig. 8.

resistivities, donor levels:

ϱ	$10^3 \cdots 10^4$ Ω cm	$T = 300$ K	n-type sample (no anisotropy reported)	75T, 78A1
E_d	$0.05 \cdots 0.08$ eV			78A1
ϱ	$10 \cdots 10^2$ Ω cm	$T = 300$ K	In doped sample	78A1
E_d	$0.02 \cdots 0.04$ eV			

10.1.3.2 Zinc silicon phosphide (ZnSiP$_2$)

Electronic properties

The electronic properties of ZnSiP$_2$ are not yet understood. The energy gap seems to be indirect or pseudodirect (lowest conduction band $\Gamma_3(X_3)$ [79S]) but the selection rules at the pseudodirect transition have not been explained.

energy gap (in eV) (values of the three peaks in the optical spectra):

		T [K]		
$E_{g,pseu}$	2.13 (1)	77; $E \perp c$	absorption, thermoabsorption, photoconductivity	65G, 75S1,
	weak			79S,
	2.23 (1)	$E \perp (\parallel)c$		75H
	2.27 (1)	$E \perp c$		

(continued)

Physical property	Numerical value	Experimental conditions	Experimental method remarks	Ref.
energy gap (continued)				
$E_{g, pseu}$	2.136 weak	124; $E\perp c$	absorption	75H
	2.252	$E\perp(\parallel)c$		
	2.294	$E\perp c$		
	2.07 (1) weak	300; $E\perp c$	absorption, thermoabsorption, photoconductivity	75H, 75R, 79S
	2.18 (3)	$E\perp(\parallel)c$		
	2.22 (3)	$E\perp c$		

The above transitions may be to a defect state below the conduction band.

$E_{g, dir}$	3.03	80; $\Gamma_{4v}(\Gamma_{15v})$ $-\Gamma_{1c}(\Gamma_{1c})$	thermoreflection	75R
	3.16	80; $\Gamma_{5v}(\Gamma_{15v})$ $-\Gamma_{1c}(\Gamma_{1c})$		
	2.96	300; $\Gamma_{4v}(\Gamma_{15v})$ $-\Gamma_{1c}(\Gamma_{1c})$	electroreflectance	75S1
	3.06	300; $\Gamma_{5v}(\Gamma_{15v})$ $-\Gamma_{1c}(\Gamma_{1c})$		
dE_g/dT	$-2.8\,(1)\cdot 10^{-4}$ eV K^{-1}	80$\cdots$300	absorption	75H, 75S
dE_g/dp	$-6.1\cdot 10^{-6}$ eV bar^{-1}	77$\cdots$300		78A2
splitting energies:				
Δ_{cf}	$-0.12\,(1)$ eV	$T=300$ K	thermoreflection, absorption (assuming the usual inter-	75H, 75S1
Δ_{so}	0.06 eV		pretation of the three bands)	
exciton binding energy:				
E_b	0.022 eV	$T=124$ K	absorption ($E\perp c$)	75H
intraconduction band transition:				
$E(\Gamma_3(X_3))-$ $E(\Gamma_2(X_1))$	0.76 eV	$T=160\cdots 377$ K	photomodulated absorption	76H
valence band width:				
ΔE	12.9 eV	$T=300$ K	X-ray spectroscopy	80G1

higher interband transition energies (in eV):

Differences of detail between different experiments but broad agreement on the main features:

		T [K]		
E_1	3.35$\cdots$3.6	80$\cdots$300	thermoreflection, reflection,	73S,
E_2	3.75$\cdots$4.0	80$\cdots$300	electroreflection	75R, 75S1, 79A1
E_3	4.5$\cdots$5.0	80$\cdots$300	reflection, features seen in	79A1
E_4	6.0$\cdots$7.0	300	all polarizations	
E_5	8$\cdots$9	300		

X-ray spectrum: Fig. 8

effective masses:				
m_p	$0.4\,(1)\,m_0$	$T=300$ K	Hall effect	76P1
m_n	$0.11\,(2)\,m_0$			

Physical property	Numerical value	Experimental conditions	Experimental method remarks	Ref.

Impurities and defects

There is some evidence that the pseudodirect gap (see above) is associated with defect states below the conduction band [79S].

intrinsic defect levels (unidentified) (in eV):

		T [K]		
E_d	0.08	120···400	photoconductivity, luminescence, thermally stimulated current	65G, 76G1
E_a	0.34	77···400	photoconductivity, luminescence, thermally stimulated current	65G, 76G1, 79K1
	0.48	77···300		79K1
	0.54	77···300		

extrinsic defect levels (in eV):

		T [K]		
E_d	0.041	Al; 77···1000	σ and R_H vs. T	76S
E_a	0.95	Cu; 77···1000	σ and R_H vs. T	76S
E_d	0.004···0.08	Ga; 77···1000	depends on concentration	76S, 76Z1, 78Z1
	0.0695	In; 77···1000	σ and R_H vs. T	76S, 76Z1

g-factor:

g^* (Mn^{2+})	2.004	$T = 20$ K	Mn^{2+} on Zn site; there are two magnetically inequivalent sites	76K1

diffusion coefficients: $D = D_0 \exp(-E/kT)$

		T [K]		
D_0	$1.6 \cdot 10^5$ cm^2 s^{-1}	Al; 1100···1200	no anisotropy reported	76Z1
E	1.1 eV			
D_0	$1.2 \cdot 10^{-3}$ cm^2 s^{-1}	Cu; 900···1100		
E	1.0 eV			
D_0	200 cm^2 s^{-1}	Ga; 1000···1200		
E	3.5 eV			
D_0	$2···6 \cdot 10^{-13}$ cm^2 s^{-1}	In; 1100		

Lattice properties

wavenumbers of Raman and infrared active phonons (in cm^{-1}, RT values):

		Symmetry I$\bar{4}$2d (F$\bar{4}$3m)			
$\bar{\nu}_{LO}/\bar{\nu}_{TO}$	518/494	Γ_4	(Γ_{15})	Raman and infrared spectra;	77H1,
	518/494	Γ_5	(Γ_{15})	Γ_4, Γ_5 IR-active,	79G,
	466	Γ_3	(W$_2$)	all Raman active	75P1
	464/461	Γ_5	(W$_4$)		
	359/343	Γ_4	(W$_2$)		
	337	Γ_1	(W$_1$)		
	335	Γ_3	(X$_3$)		
	327/321	Γ_5	(X$_5$)		
	264/246	Γ_5	(W$_3$)		
	187/185	Γ_5	(W$_4$)		
	145/145	Γ_4	(W$_2$)		
	130	Γ_3	(W$_2$)		
	102/102	Γ_5	(X$_5$)		

two-phonon Raman scattering: Fig. 9

Physical property	Numerical value	Experimental conditions	Experimental method remarks	Ref.
local phonon wavenumbers:				
$\bar{\nu}$	434 cm^{-1}	$T=300$ K; $E \parallel c$	Raman scattering, unexplained	77H1,
	445 cm^{-1}	$E \perp c$		79G
Transport properties				
carrier concentrations, mobilities, activation energies, resistivities:				
n-type samples:				
n	10^{17} cm^{-3}	$T=300$ K	conductivity	75S1,
μ_n	70···100 cm^2 V^{-1} s^{-1}	300 K	Hall effect; see also Figs. 10···14 for temperature dependence of σ, μ_n, R_H	65G, 76G1
μ_{imp}	0.02···0.35 cm^2 V^{-1} s^{-1}	$T=300$ K	in impurity band	76S
E_A	0.015···0.024 eV	77···1000 K	activation energy of hopping	
p-type samples:				
p	$6 \cdot 10^{13}$···$3 \cdot 10^{17}$ cm^{-3}	$T=300$ K		78G1
μ_p	4···11 cm^2 V^{-1} s^{-1}	300 K		
thin films:				
ϱ	1.5···3.0 Ω cm	$T=300$ K		75V

N.B. No anisotropy reported for transport properties.

Optical properties				
dielectric constants:				
$\varepsilon(0)$	11.7 (6)	$T=300$ K; $E \perp c$	infrared reflectivity	75P1
	11.15 (10)	$E \parallel c$		
$\varepsilon(\infty)$	9.26	300 K; $E \perp c$	infrared reflectivity	
	9.68	$E \parallel c$		
refractive index:				
n_0	3.31	$T=300$ K; $\lambda=600$ nm		75B8
	3.06	$\lambda=900$ nm		
birefringence: $\Delta n = n_e - n_0$: Fig. 15				
Δn	$7 \cdot 10^{-3}$	$T=300$ K	$\hbar\omega=0.6$ eV	78A2
	$6.5 \cdot 10^{-3}$	77 K		
$d\Delta n/dT$	$3.5 \cdot 10^{-6}$ K^{-1}	77···300 K		
$d\Delta n/dp$	$-0.8 \cdot 10^{-7}$ bar^{-1}			

luminescence and photoconductivity spectra:

photoluminescence: Fig. 16; cathodoluminescence: Figs. 17, 18; electroluminescence: Figs. 18, 19; photoconductivity: p-type: Fig. 20, n-type: Fig. 21

radiative recombination of a bound exciton with phonon side bands:

E	0.7 eV	$T=300$ K	(with Cu)	78G1
	1.2 eV		reported but not explained	78G2
	1.8 eV			77M1

For results on photoconductivity and photoluminescence, see also under "Impurities and defects".

Physical property	Numerical value	Experimental conditions	Experimental method remarks	Ref.
Further properties				
coefficients of thermal expansion (in 10^{-6} K^{-1}):				
α	7.8	$T = 300\cdots1400$ K	a axis	79K2,
	3.5		c axis	75M1
β	19.2	$300\cdots1400$ K	volume	
Debye temperature:				
Θ_{D}	445.2 (23) K	$T \to 0$ K	from heat capacity	81B
melting point:				
T_{m}	$1520\cdots1640$ K			75S1
Above $T = 900$ K oxidation [71L].				

10.1.3.3 Zinc silicon arsenide (ZnSiAs$_2$)

Electronic properties

   band structure: Fig. 21 α

energy gaps (in eV, at 300 K):

$E_{\mathrm{g, pseu}}$	1.74	$E \perp c$	photoconductivity	76R
	1.84	$E \perp (\parallel)c$	(see also [79Z])	
	1.93	$E \parallel (\perp)c$		
	2.03	$E \perp c$		
	2.09	$E \perp c$		
$E_{\mathrm{g, dir}}$ (A)	2.13 (1)		electroreflectance,	71S,
(B)	2.23 (1)		photoconductivity	76R,
(C)	2.69 (1)		(see also [79Z])	76J
$\mathrm{d}E_{\mathrm{g, dir}}/\mathrm{d}T$	$-3.5 \cdot 10^{-4}$ eV K^{-1}	$T = 77\cdots300$	birefringence vs. wavelength	78A2
$\mathrm{d}E_{\mathrm{g}}/\mathrm{d}p$	$1.6 \cdot 10^{-6}$ eV bar^{-1}	300	absorption	75S1

splitting energies:

Δ_{cf}	-0.13 eV	$T = 300$ K	electroreflectance	71S,
Δ_{so}	0.286 (6) eV			75T

higher transition energies (in eV):

		T [K]		
E	2.72 (2)	300	electroreflectance,	71S,
			photoconductivity	76R
	2.90	300	E_1-like	
	2.92	77	reflection	73S
	3.25	300	electroreflectance	71S
	3.00	77	reflection	73S
	3.50	300	electroreflectance	71S
	3.36	77	reflection	73S
	3.91	300	electroreflectance	71S
	3.80	77	reflection	73S
	4.04	300	electroreflectance	71S
	4.31	300		
	4.14	77	reflection	73S
	4.6	77		
	4.9	77	N.B. Measurements with unpolarized light.	

For further data and a comparison with the band structure of Fig. 21 α, see [81H].

Physical property	Numerical value	Experimental conditions	Experimental method remarks	Ref.
effective masses (in m_0):				
$m_{n,\perp}$	0.103		Γ_1-band, theoretical	75S1
$m_{n,\parallel}$	0.098			
$m_{n,ds}$	0.10		(m_{ds}: density of states mass,	
$m_{n,c}$	0.10		m_c: conductivity mass)	
$m_{p_1,\perp}$	(−6.1)			
$m_{p_1,\parallel}$	0.15			
$m_{p_1,ds}$	1.76			
$m_{p_1,c}$	0.42			
$m_{p_2,\perp}$	0.28			
$m_{p_3,\perp}$	0.43			
$m_{p_3,\parallel}$	2.29			
$m_{p_3,ds}$	0.75			
$m_{p_3,c}$	0.59			
$m_{n,\parallel}$	0.921		Γ_3-band	70G
$m_{n,\perp}$	0.124			

spin polarized photoelectrons: Fig. 22

Impurities and defects

defect levels (meV):

		T [K]		
E_a	0.19	77···300	transport	77M2
	0.27···0.3	77···300	photoluminescence, conductivity	77M2, 76I
	0.53	77		76I
	0.82	77		
	0.17	77	(sulfur)	
	0.05	77···300	transport; excess As	77M2
	0.043	77···300	transport; 2% Se	
E_d	>0.55	300	transport	76R

g-factor:

$g^*(\text{Mn}^{2+})$	2.006	$T = 20$ K	ESR (Mn^{2+} on Zn-site); there are magnetically non-equivalent sites	76K1

For influence of doping on transport and optical properties, see also under the corresponding subsection.

Lattice properties

wavenumbers of Raman active phonons (in cm^{-1}, RT values):

		Symmetry $\overline{14}2d$		
$\bar{\nu}_{LO}/\bar{\nu}_{TO}$	417/405	Γ_5		75M2
	401/388	Γ_4		
	269/–	Γ_4		
	262	Γ_3		
	236/–	Γ_5		
	210/–	Γ_5		
	203	Γ_1		
	162	Γ_1		
	133/–	Γ_5		
	108	Γ_3		
	77/–	Γ_5		

Physical property	Numerical value	Experimental conditions	Experimental method remarks	Ref.

Transport properties

carrier concentrations, mobilities, resistivities, Hall coefficients, Seebeck coefficient (RT values):

n-type (non-stoichiometric) sample:

n	10^9 cm^{-3}			76R
μ_n	40 cm^2 V^{-1} s^{-1}			

p-type samples (see also Figs. 23···26):

p [cm^{-3}]	ϱ [Ω cm]	R_H [cm^3 C^{-1}]	μ_p [cm^2 V^{-1} s^{-1}]	E_A [eV]	conditions of preparation:	
10^{16}	28.1	610	22	0.19	stoichiometric melt	77M2
$8.2 \cdot 10^{14}$	76.0	$7.6 \cdot 10^3$	100	0.27	transport, HgCl$_2$	
$8 \cdot 10^{16}$	2.4	78	32	–	transport, SiCl$_4$	
$2.6 \cdot 10^{16}$	3.26	237	73	0.022 and >0.05	excess As	
$4.3 \cdot 10^{17}$	0.72	14.6	20	0.043	2% Se (see also below)	
S	$1.1 \cdot 10^{-3}$ V K^{-1}		$T = 300$ K			77M2

The mobility at room temperature is determined by scattering from non-polar optical modes; $\mu \propto T^{-5/2}$ [76Z2].

For effect of electron irradiation on resistivity, see Fig. 27.

influence of doping (RT values):

Dopant	p [cm^{-3}]	μ_p [cm^2 V^{-1} s^{-1}]	Activity	
–	$4 \cdot 10^{14}$	50	–	73A
0.1% Li	$3 \cdot 10^{16}$	100	acceptor (A)	
0.1% Cu	$1 \cdot 10^{16}$	2	A	
0.1% Ag	$4 \cdot 10^{15}$	15	A	
1% Au	$8 \cdot 10^{14}$	70	inactive (I)	
0.1% Al	$2 \cdot 10^{18}$	100	A	
0.1% Ga	$1 \cdot 10^{19}$	50	A	
0.1% In	$2 \cdot 10^{15}$	10	I (A)	
0.1% S	$1 \cdot 10^{18}$	100	A	
0.1% Se	$3 \cdot 10^{18}$	30	A	
0.1% Te	$1 \cdot 10^{16}$	10	A	
0.1% Fe	$6 \cdot 10^{14}$	10	I	
0.1%Mn	$5 \cdot 10^{14}$	100···30	I	
0.1% V	$3 \cdot 10^{16}$	10	A	
0.1% Ti	$3 \cdot 10^{15}$	10	I (A)	
0.1% Sc	$3 \cdot 10^{15}$	60	I (A)	
0.1% Pt	$6 \cdot 10^{14}$	40	I	

N.B. No anisotropy reported for transport properties.

Optical properties

refractive indices: coefficients in the formula: $n^2 = A + B/(1 - C/\lambda^2) + D/(1 - E/\lambda^2)$ ($T = 300$ K) [76B3].

	n_o	n_e	
A	4.6066	4.9091	
B	5.6912	5.5565	
C	$0.1437 \cdot 10^{-12}$ m^2	$0.1578 \cdot 10^{-12}$ m^2	
D	1.316	1.287	
E	$7 \cdot 10^{-10}$ m^2	$7 \cdot 10^{-10}$ m^2	
		λ [µm]	
n_o	3.5579	0.70	75S1
	3.3551	1.00	
	3.2405	2.00	(continued)

Physical property	Numerical value	Experimental conditions	Experimental method remarks	Ref.
refractive indices (continued)				
		λ [μm]		
n_o	3.2210	3.00		75S1
	3.2133	4.00		
	3.2081	5.00		
	3.2025	6.00		
	3.1979	7.00		
	3.1930	8.00		
	3.1874	9.00		
	3.1810	10.00		
	3.1733	11.00		
	3.1626	12.00		
n_e	3.6201	0.70		75S1
	3.3928	1.00		
	3.2692	2.00		
	3.2481	3.00		
	3.2402	4.00		
	3.2345	5.00		
	3.2287	6.00		
	3.2241	7.00		
	3.2195	8.00		
	3.2138	9.00		
	3.2074	10.00		
	3.1996	11.00		
birefringence ($\Delta n = n_e - n_o$) (see also Fig. 21):				
Δn	0.026	$T = 300$ K	$\hbar\omega = 0.6$ eV	78A2
	0.025	77 K		
$d\Delta n/dT$	$7.9 \cdot 10^{-6}$ K^{-1}	77$\cdots$300 K		
$d\Delta n/dp$	$-2.3 \cdot 10^{-7}$ bar^{-1}			
luminescence peaks (in eV):				
E	2.26	$T = 77$ K	impurity band recombination	76I
	1.935		unexplained	
	1.48		from 0.82 eV level, activated by Au	
	1.32		unexplained	
For photoluminescence, see also under "Impurities and defects".				
nonlinear dielectric susceptibility:				
d_{36}	$109 \cdot 10^{-12}$ mV^{-1}	$\lambda = 10.6$ μm		72B
Further properties				
coefficients of thermal expansion (in 10^{-6} K^{-1}):				
α	7.6	$T = 300$ K	a axis	79K2
	4.0		c axis	
β	21.7	300$\cdots$800 K	volume	
thermal conductivity:				
κ	0.14 W cm^{-1} K^{-1}	$T = 300$ K		75S1
Debye temperature:				
Θ_D	346.6 (12) K	$T \rightarrow 0$ K	from heat capacity	81B
melting point:				
T_m	1370 K			75S1

Physical property	Numerical value	Experimental conditions	Experimental method remarks	Ref.

10.1.3.4 Zinc germanium nitride (ZnGeN$_2$)

energy gap:

E_g	2.67 eV	$T = 300$ K	absorption (unpolarized)	74L

carrier concentration, resistivity, mobility:

		T [K]		
n	$10^{18} \cdots 10^{19}$ cm^{-3}	300	n-type sample	74L
ϱ	$0.3 \cdots 0.4$ Ω cm	300	(no anisotropy reported)	
μ_n	$0.5 \cdots 5$ cm^2 V^{-1} s^{-1}	$100 \cdots 300$		

10.1.3.5 Zinc germanium phosphide (ZnGeP$_2$)

Electronic properties:

band structure (and Brillouin zone): Fig. 28 (for another calculation, see also [81H])

energy gaps and other band-band transitions (in eV):

(a)	reflectivity ($T = 5$ K)		74V1
(b)	reflectivity (80K)		77R1
(c)	thermoreflectance (120 K)		74V1
(d)	electroreflectance (300 K)		74V1
(e)	wavelength modulated adsorption (77 K)		74V1

	(a)	(b)	(c)	(d)	(e)	Symmetry
$E_{g,\,pseu}$				2.05	2.08 2.145 $\big\}$	$\Gamma_{4v} - \Gamma_{3c}(\Gamma_{15v} - X_{3c})$
	2.14					$\Gamma_{5v} - \Gamma_{3c}(\Gamma_{15v} - X_{3c})$
				2.11	2.21	
	2.29					
					2.3	$\Gamma_{5v} - \Gamma_{2c}(\Gamma_{15v} - X_{1c})$
	2.31					
$E_{g,\,dir}$	2.51		2.46	2.365		$\Gamma_{4v} - \Gamma_{1c}(\Gamma_{15v} - \Gamma_{1c})$
	2.63		2.53	2.43 $\big\}$		$\Gamma_{5v} - \Gamma_{1c}(\Gamma_{15v} - \Gamma_{1c})$
	2.67		2.59	2.50		
E	3.02	3.03	3.02	2.92		$X_{1v} - X_{1c}$, $N_{1v} - N_{1c}(E_1)$
	3.08		3.15	3.07		$N_{2v} - N_{1c}$
	3.20	3.22	3.22	3.22		$N_{2v} - N_{1c}$
	3.41	3.43	3.48	3.52		$\Sigma_{2v} - \Sigma_{1c}$
	3.73	3.70	3.75	3.77		$X_{1v} - X_{1c}(E_2)$
	4.17	4.19				$\Lambda; \Sigma$
	4.3	4.44				Δ
	4.46					$\Gamma_v - \Gamma_c(E_2)$
	4.73	4.86				
	4.79					$X_v - \Sigma_c$
	4.92	5.2				Δ
		5.6				
		6.2				$N(E_1')$
		6.6				$N(E_1 + \Delta_1')$
		7.7				
		7.9				
		9.6				(E_1'')
		11.7 $\big\}$				
		13.1				trans. originating at Zn3d
		14.2				
		16.0				trans. originating at P3s

N.B. E_1, E_1' etc. is the nomenclature of the corresponding peaks of the III–V analog GaP.
Notation as in Fig. 28.

(continued)

Physical property	Numerical value	Experimental conditions	Experimental method remarks	Ref.
energy gaps (continued)				
$dE_{g,dir}/dT$	$-(2.5\cdots3.5)$ $\cdot 10^{-4}$ eV K^{-1}	$T = 77\cdots300$ K		78A2, 80B2
$dE_{g,dir}/dp$	$5.6 \cdot 10^{-6}$ eV bar^{-1}		refractive index vs. wavelength	78A2
	$1.2 \cdot 10^{-6}$ eV bar^{-1}		absorption	75S1
splitting energies:				
Δ_{cf}	-0.08 eV	$T = 300$ K	electroreflectance	75S1
Δ_{so}	0.09 eV			
conduction band, valence band and core states (in eV):				
relative to upper valence band edge (from photoemission (XPS) at $T = 80$ K):				
$E(C_3)$	4.6			74V2,
$E(C_2)$	3.5			77R1
$E(C_1)$	2.1		Γ (X)	
$E(C_0)$	2.4		Γ_1 (Γ_1)	
$E(V_0)$	0.0		$\Gamma_{4,5}$ (Γ_{15}) Zn$-$P bonds	
$E(V_1)$	-0.9		Zn$-$P bonds	
$E(V_2)$	-3.0		Ge$-$P bonds	
$E(V_3)$	-7.3		Ge	
$E(V_4)$	-12.8		P 3s	
$E(d_{5/2})$	-9.3		Zn 3d, core state	
$E(d_{3/2})$	-9.8		Zn 3d, core state	
effective masses (in m_o):				
$m_{n,\perp}$	0.108		theory	75S1
$m_{n,\parallel}$	0.105			
$m_{n,ds}$	0.11		(m_{ds}: density of states mass,	
$m_{n,c}$	0.11		m_c: conductivity mass)	
$m_{p_1,\perp}$	(-1.8)			
$m_{p_1,\parallel}$	0.15			
$m_{p_1,ds}$	0.79			
$m_{p_1,c}$	0.39			
$m_{p_2,\perp}$	0.32			
$m_{p_3,\perp}$	0.38			
$m_{p_3,\parallel}$	(-6.2)			
$m_{p_3,ds}$	0.97			
$m_{p_3,c}$	0.56			
m_p	0.5 (1)	$T = 260\cdots430$ K	transport	72S
Impurities and defects				
data from optical spectra:				
luminescence peaks (meV):				
		T [K]		
E	2.5	300	band gap (direct), observed with most dopants	76G2
	2.0	300	pseudodirect gap or impurity band; stronger in non-stoichiometric samples	76G2
	1.6	77	acceptor at 0.55 ev (see below)	77A2
	$1.32\cdots1.35$	77	P vacancy; strongly for $E\parallel c$	77R2
	1.745	77	associated with isoelectronic defects: Cd, Si	77A2

(continued)

Physical property	Numerical value	Experimental conditions	Experimental method remarks	Ref.

data from optical spectra (continued)
photoconductivity: In—ZnGeP$_2$ junction:

E	0.9···1.8 eV	$T = 300···400$ K		75G3
	1.93 eV		minimum (band gap?)	
	2.06 eV		minimum (band gap?)	

Dopant	p [cm^{-3}]	Photoconductivity maximum [eV] [75G3; 76G2]	Photoluminescence maximum [eV] [77A2]	
		($T = 300$ K)	($T = 300$ K)	($T = 77$ K)
–	$0.9···40 \cdot 10^{10}$	2.0; 2.5	1.39···1.44; 1.50···1.625	
0.1% Cu	$10^{10}···10^{11}$	2.0; 2.4	1.46; 1.53	
1% Cu	10^{14}	2.14; 2.43	1.425 (230)	
0.1% Ga	10^{17}		1.49	
0.1% In	$10^{11}···10^{16}$	2.0; 2.5	1.425 (230)	
1% In	$4 \cdot 10^{16}$		1.435 (230)	
5% In	$4 \cdot 10^{16}$	2.0		
0.1% Se	$2 \cdot 10^{13}···2 \cdot 10^{14}$	2.0; 2.5	1.32 (21)	
1% Fe	$6 \cdot 10^{11}$		1.335 (200)	
0.1% Cd	$3 \cdot 10^{10}$		1.42 (25); 1.765 (90)	
1% Si	$8 \cdot 10^{10}$	2.0	1.335 (200); 1.745 (110)	

acceptor ionization energy:

		T [K]		
E_a	0.35 eV	260···430	transport	72S
	0.55 eV	77	luminescence (see above)	75G3, 76G2, 77A2
	0.50 eV	300···425	transport in large single crystals	78G3

g-factors:

g^* (Mn^{2+})	2.002 (1)	$T = 300$ K	ESR	75B2
g^* (P antisite)	2.007 (4)	20 K	ESR	76K2

Lattice properties

wavenumbers of infrared and Raman active phonons (in cm^{-1}, $T = 78$ K):

		Symmetry I$\bar{4}$2d (F$\bar{4}$3m)			
$\bar{\nu}_{LO}/\bar{\nu}_{TO}$	408/399	Γ_4	(Γ_{15})	Γ_4, Γ_5 IR active, all Raman active	75B1,
	404/387	Γ_5	(Γ_{15})		75G2,
	389	Γ_3	(W$_2$)		74B,
	376/369	Γ_5	(W$_4$)		75G4
	364	Γ_2	(X$_1$)	calculated [75B1]	
	359/344	Γ_4	(W$_2$)		
	331/329	Γ_5	(X$_5$)		
	328	Γ_1	(W$_1$)		
	247	Γ_3	(X$_3$)		
	204/202	Γ_5	(W$_3$)		
	142	Γ_5	(W$_4$)		
	120	Γ_3	(W$_2$)		
	96	Γ_5	(X$_5$)		

two-phonon infrared spectrum: Fig. 30, two-phonon Raman spectrum: Fig. 9
    For dielectric constants see "Optical properties".

Physical property	Numerical value	Experimental conditions	Experimental method remarks	Ref.

Transport properties

For temperature dependence of σ, R_H, μ, see also Fig. 31.

carrier concentrations, mobilities, Seebeck coefficients, conductivities (at $T = 300$ K):

p-type samples

p	10^{13} cm^{-3}			72S
μ_p	18 cm^2 V^{-1} s^{-1}			72S
	25$\cdots$60 cm^2 V^{-1} s^{-1}		In—ZnGeP$_2$ junction	75G3
S	$1.2 \cdot 10^{-3}$ V K^{-1}			75S1
p	$5 \cdot 10^{10}$ cm^{-3}		large single crystals	78G3
σ	$1.4 \cdot 10^{-7}$ Ω^{-1} cm^{-1}		(high-pressure furnace)	
μ_p	20 cm^2 V^{-1} s^{-1}			
p	$2.2 \cdot 10^{11}$ cm^{-3}			78G3
σ	$6.6 \cdot 10^{-7}$ Ω^{-1} cm^{-1}			
μ_p	20 cm^2 V^{-1} s^{-1}			

For effect of irradiation with 2 MeV electrons on resistivity, see Fig.29.

N.B. No anisotropy reported.

Optical properties

refractive indices: coefficients in the formula $n^2 = A + B/(1 - C/\lambda^2) + D/(1 - E/\lambda^2)$

	A	B	C [10^{-12} m^2]	D	E [10^{-12} m^2]	T [K]	
n_o	4.6171	5.4709	0.1495	1.4912	662.55	670	80B3
n_e	4.6791	5.6826	0.1567	1.4577	662.55		
n_o	4.5654	5.3382	0.1419	1.4913	662.55	470	
n_e	4.6732	5.4842	0.1499	1.4581	662.55		
n_o	4.5209	5.2917	0.1376	1.4911	662.55	370	
n_e	4.6559	5.4001	0.1460	1.4580	662.55		
n_o	4.4492	5.3338	0.1353	1.4736	662.55	320	79B1
n_e	4.5717	5.4513	0.1435	1.4262	662.55		
n_o	4.4733	5.2658	0.1338	1.4909	662.55	270	
n_e	4.6332	5.3422	0.1426	1.4580	662.55		
n_o	4.3761	5.2540	0.1275	1.4903	662.55	170	80B3
n_e	4.5801	5.2747	0.1368	1.4576	662.55		
n_o	4.2889	5.2688	0.1228	1.4898	662.55	90	
n_e	4.5285	5.2463	0.1326	1.4572	662.55		

		λ [μm]	
n_o	3.5052	0.64	75S1
	3.2478	1.00	
	3.1490	2.00	
	3.1304	3.00	
	3.1223	4.00	
	3.1149	5.00	
	3.1101	6.00	
	3.1040	7.00	
	3.0961	8.00	
	3.0880	9.00	
	3.0788	10.00	
	3.0689	11.00	
	3.0552	12.00	

(continued)

Physical property	Numerical value	Experimental conditions	Experimental method remarks	Ref.
refractive indices (continued)				
		λ [μm]		
n_e	3.5802	0.64		75S1
	3.2954	1.00		
	3.1889	2.00		
	3.1693	3.00		
	3.1608	4.00		
	3.1533	5.00		
	3.1480	6.00		
	3.1420	7.00		
	3.1350	8.00		
	3.1272	9.00		
	3.1183	10.00		
	3.1087	11.00		
	3.0949	12.00		
birefringence ($\Delta n = n_e - n_o$):				
Δn	0.037	$T = 300$ K	$\hbar\omega = 0.6$ eV	78A2
$\mathrm{d}\Delta n/\mathrm{d}T$	$11 \cdot 10^{-6}$ K^{-1}	77$\cdots$300 K		
$\mathrm{d}\Delta n/\mathrm{d}p$	$-2.4 \cdot 10^{-7}$ bar^{-1}	300 K	see also Fig. 15	
dielectric constants:				
ε_{11}	15		$f = 1\cdots7\cdot10^7$ Hz	74T2
ε_{33}	12			
linear electrooptic coefficients:				
r_{41}	$1.6 \cdot 10^{-12}$ mV^{-1}	RT,	at constant stress or strain	74T2
r_{63}	$-0.8 \cdot 10^{-12}$ mV^{-1}	$\lambda = 3390$ nm	at constant stress	
nonlinear dielectric susceptibility:				
d_{14}	$75 \cdot 10^{-12}$ mV^{-1}	$\lambda = 1.06$ μm		76A
	$111 \cdot 10^{-12}$ mV^{-1}	10.6 μm		71B2, 72M

For photoconductivity and photoluminescence data, see "Impurities and defects".

Further properties
coefficients of thermal expansion (in 10^{-6} K^{-1}):

α	1.8	$T = 300\cdots1300$ K	a axis	75M1
	5.0		c axis	
β	20	300$\cdots$1300 K	volume	
thermal conductivity:				
κ	0.18 W cm^{-1} K^{-1}			75S1
Debye temperature:				
Θ_D	428 K	$T \to 0$ K		75A3
melting point:				
T_m	1300 (3) K			75S1

Physical property	Numerical value	Experimental conditions	Experimental method remarks	Ref.

10.1.3.6 Zinc germanium arsenide (ZnGeAs$_2$)

Electronic and optical properties

energy gaps (in eV):

$E_{g,dir}$	(A) 1.15	$T = 300$ K	electroreflectance	75S1
	(B) 1.19			
	(C) 1.48			
$E_{g,th}$	1.16	0 K	conductivity	75A1
$dE_{g,dir}/dp$	$9.7 \cdot 10^{-6}$ eV bar^{-1}			75S1

splitting energies:

Δ_{cf}	-0.06 eV	$T = 300$ K		75S1
Δ_{so}	0.31 eV			

higher transition energies (in eV) (from reflectivity spectra at 300 K):

E	2.46		E_1-like	77R1
	2.70			
	3.24			
	3.46			
	3.8			
	4.50		E_2-like	
	4.92 ⎫		transitions near X	
	5.06 ⎭		(from zone boundary)	
	5.36 ⎫		transitions near L and Σ	
	5.70 ⎭			
	6.0		E_1-like	
	6.7		$E_1 + \Delta_1$-like	
	7.6			
	8.0		strong feature, unexplained	
	9.6		E_1''-like	
	11.7 ⎫			
	13.0 ⎬		Zn 3d	
	14.2 ⎭			
	16.0		As 3s	

N.B. E_1, E_2-like etc. denote that these probably correspond to the transitions denoted E_1, E_2 in the zincblende analog GaAs.

conduction band, valence band and core states (in eV):

relative to upper valence band edge (from photoemission data at $T = 80$ K): 77R1

$E(C_4)$	4.8			
$E(C_3)$	3.7			
$E(C_2)$	2.5			
$E(C_1)$	–			
$E(C_0)$	1.2		conduction band minimum	
$E(V_0)$	0.0		valence band maximum	
$E(V_1)$	-0.3			
$E(V_2)$	-0.9			
$E(V_3)$	-2.9			
$E(V_4)$	-7.1		s-like cation states	
$E(V_5)$	-13.0			
$E(d_{5/2})$	-9.2		Zn 3d, core state	
$E(d_{3/2})$	-9.7		Zn 3d, core state	

photoemission spectra in the amorphous phase: Fig. 32, spin polarized photoelectron spectra: Fig. 22

Physical property	Numerical value	Experimental conditions	Experimental method remarks	Ref.
effective masses (in m_o):				
$m_{n,\perp}$	0.060		theoretical	75S1
$m_{n,\parallel}$	0.058			
$m_{n,ds}$	0.059		(m_{ds}: density of states mass;	
$m_{n,c}$	0.059		m_c: conductivity mass)	
$m_{p_1,\perp}$	0.82			
$m_{p_1,\parallel}$	0.084			
$m_{p_1,ds}$	0.38			
$m_{p_1,c}$	0.21			
$m_{p_2,\perp}$	0.14			
$m_{p_3,\perp}$	0.25			
$m_{p_3,\parallel}$	0.41			
$m_{p_3,ds}$	0.29			
$m_{p_3,c}$	0.29			
luminescence peaks:				
E	1.13 (5) eV	$T = 300$ K	probably band gap	74S
	1.14 (2) eV		probably band gap	
Transport properties				
carrier concentration, resistivity, mobility, Seebeck coefficient (at $T = 300$ K):				
p-type sample:				
p	$7.2 \cdot 10^{18} \cdots$ $1.2 \cdot 10^{19}$ cm^{-3}			74S
ϱ	$2.7 \cdots 5.8 \cdot 10^{-2}$ Ω cm		(no anisotropy reported for transport properties)	
μ_p	$19.6 \cdots 24$ cm^2 V^{-1} s^{-1}			
S	$2.5 \cdot 10^{-4}$ V K^{-1}			75S1
Further properties				
thermal conductivity:				
κ	0.11 W cm^{-1} K^{-1}	$T = 300$ K		75S1
melting point:				
T_m	$1120 \cdots 1145$ K			75S1

10.1.3.7 Zinc tin phosphide (ZnSnP$_2$)

ZnSnP$_2$ has a c/a ratio of 2 and grows with a domain structure in which all three possible directions for the tetragonal axis are present [70G]. No anisotropic properties have been reported.

Electronic properties

energy gap:

$E_{g,dir}$	(A, B)	1.66 eV	$T = 300$ K	electroreflectance	75S1
	(C)	1.75 eV			
dE_g/dp		$7.5 \cdot 10^{-6}$ eV bar^{-1}	300 K	absorption	75S1

splitting energies:

Δ_{cf}	0	$T = 300$ K	($c/a = 2$)	75S1
Δ_{so}	0.09 eV			

Physical property	Numerical value	Experimental conditions	Experimental method remarks	Ref.
higher transition energies (in eV):				
E	2.80	$T = 300$ K	electroreflectance	75K1
	2.96		(unpolarized light)	
	3.11			
	3.6			
	4.65		reflectance (unpolarized light)	
	5.7			
	7.7			

Lattice properties

wavenumbers of infrared active phonons (in cm^{-1}):

		Symmetry			
$\bar{\nu}_{LO}$	327$\cdots$330	Γ_4	$T = 300$ K	infrared reflectivity	75S1, 75B1
	368	Γ_5	300 K		75B1
$\bar{\nu}$	322	–	300 K	disordered structure	75S1

Theoretical values of other phonons in [75B1].

Transport properties:

carrier concentration, mobility, resistivity, activation energy (at 300 K):

p-type sample (see also Fig. 33):

p	$(3\cdots10)\cdot10^{16}$ cm^{-3}			77R3
μ_p	$10\cdots60$ cm^2 V^{-1} s^{-1}			
ϱ	$4.1\cdot10^2\ \Omega$ cm			78B3
	$1.1\cdot10^4\ \Omega$ cm		after irradiation with $5\cdot10^{17}$ electrons with $E = 2$ MeV	
E_A	0.11 eV		activation energy of conductivity	75S1

For effect of irradiation on resistivity, see Fig. 29.

Optical properties

dielectric constants:

$\varepsilon(0)$	10.0	$T = 300$ K	chalcopyrite	75S1
$\varepsilon(\infty)$	8.08			
$\varepsilon(0)$	10.8	300 K	disordered zincblende	
$\varepsilon(\infty)$	8.3			

photoluminescence peaks:

E	1.28 eV	$T = 77$ K	disordered form or low excitation intensity	75S1, 77R3
	1.52 eV		higher excitation intensity	

Further properties

melting point:

T_{perit}	1200 K		peritectic	75S1

Physical property	Numerical value	Experimental conditions	Experimental method remarks	Ref.

10.1.3.8 Zinc tin arsenide (ZnSnAs$_2$)

ZnSnAs$_2$ has a c/a ratio of 2. It is thus difficult to grow single crystals. Generally there is a domain structure with the tetragonal axis parallel to either x, y or z in each domain [70G, 75S1]. No anisotropic properties have been reported.

Electronic properties

energy gap (in eV):

$E_{g,dir}$ (A, B)	0.745	$T = 300$ K	electroreflectance	71K,
(C)	1.185			75S1
	0.63 eV	300 K	epitaxial layers; absorption	78B4
dE_g/dT	$2.7 \cdot 10^{-4}$ eV K^{-1}			71S
dE_g/dp	$8.2 \cdot 10^{-6}$ eV bar^{-1}		absorption	75S1

splitting energies:

Δ_{cf}	≈ 0 eV	$T = 300$ K	($c/a = 2$)	75S1
Δ_{so}	0.34 eV			

higher transition energies (in eV):

		T [K]		
E	2.24 ⎫	295	electroreflectance	71K
	2.35 ⎬		$N_{1v}(L_{3v}) - N_{1c}(L_{1c})$	
	2.56 ⎭			
	2.79		$\Gamma_{5v} - \Gamma_{2c}$; $T_{3v} + T_{4v} - T_{1c} + T_{2c}$	
	3.06 ⎫		Γ; T ($X_{5v} - X_{1c}$)	
	3.68 ⎭			
	3.89 ⎫			
	4.06 ⎪		$\Gamma_{15v} - \Gamma_{15c}$	
	4.32 ⎬			
	4.75 ⎭			
	2.34	77	electroreflectance	75S1
	2.46		(unpolarized light)	
	2.63			
	2.67			
	2.31	300	reflectance	75K1
	2.56		(unpolarized light)	
	3.06			
	4.54			
	6.5 ⎫		$N_{1v}(L_{3v}) - N_{1c}(L_{3c}, L_{1c})$	
	7.9 ⎭			

effective mass:

m_p	0.35 m_o	$T = 5 \cdots 200$ K	transport	80D1

Impurities and defects

defect levels:

E	0.589 (1) eV	$T = 55 \cdots 181$ K	photoluminescence peak	80D1
E_d	$0.184 \cdots 0.201$ eV	$55 \cdots 181$ K		
E_a	0.020 eV	$55 \cdots 181$ K		

Probable donor acceptor pair formation with Coulomb binding energy $E_b = 0.003 \cdots 0.020$ eV [80D1].

Physical property	Numerical value	Experimental conditions	Experimental method remarks	Ref.
Transport properties				
All attempts at doping unsuccessful.				
carrier concentration, mobility, Hall coefficient (at $T = 300$ K):				
p-type (intrinsic) sample (see also Fig. 35):				
p	$5 \cdot 10^{17} \dots 10^{20}$ cm^{-3}			75K2
μ_p	$80 \dots 90$ cm^2 V^{-1} s^{-1}		normal sample	75K2
	190 cm^2 V^{-1} s^{-1}		relatively pure sample	75S1
R_H	2.2 cm^3 C^{-1}			75K2
Impurity band conduction important after irradiation [76K3].				
For effects of irradiation on R_H and μ, see Fig. 34; see also [75K2, 76B4].				
For change from p-type to n-type during irradiation, see [75K2, 76B4].				
mobilities in irradiated samples (in cm^2 V^{-1} s^{-1}, at 300 K):				
μ_p	66		valence band; after $2 \cdot 10^{17}$ e cm^{-2}	76K3
μ_{imp}	11		impurity band; after $2 \cdot 10^{17}$ e cm^{-2}	
μ_p	12		valence band; after $2 \cdot 10^{18}$ e cm^{-2}	
μ_{imp}	1.5		impurity band; after $2 \cdot 10^{18}$ e cm^{-2}	
data for epitaxial layers (at $T = 300$ K) (on GaAs; InAs):				
p	10^{19} cm^{-3}			78B4
ϱ	$0.012 \dots 0.015$ Ω cm			
μ_p	$25 \dots 34$ cm^2 V^{-1} s^{-1}			
Further properties				
dielectric constant:				
$\varepsilon(0)$	15.6	$T = 300$ K		80D1
Debye temperature:				
Θ_D	271.1 (27) K	$T \rightarrow 0$ K		65A
	266.4 (31) K	$T \rightarrow 0$ K	from heat capacity	81B
melting point:				
T_m	1048 (3) K			75B3

10.1.3.9 Zinc tin antimonide (ZnSnSb$_2$)

energy gap:				
E_g	0.4 eV	$T = 77$ K	absorption	73B
	0.7 eV	300 K	(unpolarized light)	
effective masses:				
m_n	$0.025\,m_o$	$T = 77$ K, 300 K		73B
m_{p2}	$0.031\,m_o$			
m_{p3}	$0.25\,m_o$			
carrier concentration, mobility, Seebeck coefficient (at 300 K):				
p-type sample:				
p	10^{20} cm^{-3}			75S1
μ_p	70 cm^2 V^{-1} s^{-1}			
S	36 μV K^{-1}?			

Physical property	Numerical value	Experimental conditions	Experimental method remarks	Ref.

10.1.3.10 Cadmium silicon phosphide (CdSiP$_2$)

Electronic properties

CdSiP$_2$ is the only chalcopyrite material with a positive crystal field splitting (i.e. $E(\Gamma_5) > E(\Gamma_4)$). This is related to the large anionic displacement. The energy gap is generally said to be pseudodirect. The size of the pseudodirect gap has however proved difficult to measure accurately.

band structure: Fig. 35α

energy gaps (in eV):

		T [K]		
$E_{g, pseu}$	2.08	90	thermoabsorption	79C
	2.2	300	absorption	70G, 79C
	2.45	300	electroreflectance	75S1
$E_{g, dir}$ (B)	2.71	90	thermoreflectance; $\Gamma_{5v}(\Gamma_{15v})$ $-\Gamma_{1c}(\Gamma_{1c})$	77A3
(C)	2.75			
(A)	2.945		$\Gamma_{4v} - \Gamma_{1c}$	
(B, C)	2.65	293	thermoreflectance; $\Gamma_{5v}(\Gamma_{15v})$ $-\Gamma_{1c}(\Gamma_{1c})$	77A3
(A)	2.86		$\Gamma_{4v} - \Gamma_{1c}$	
(A, B, C)	2.75	300	electroreflectance; $\Gamma_{5v}(\Gamma_{15v})$ $-\Gamma_{1c}(\Gamma_{1c})$	75S1
$dE_{g, dir}/dT$	$-3.5 \cdot 10^{-4}$ eV K^{-1}		refractive index	78A2
$dE_{g, dir}/dp$	$-0.7 \cdot 10^{-6}$ eV bar^{-1}	77···300 K	absorption (pseudodirect)	75S1
	$3.2 \cdot 10^{-6}$ eV bar^{-1}		refractive index	78A2

excitonic energy gap:

E_{gx}	2.085 eV	$n = 1$; $T = 80$ K	absorption	80G2

splitting energies:

Δ_{cf}	0.20 eV	$T = 90$ K	thermoabsorption	77A3
Δ_{so}	0.07 eV			

higher transition energies (in eV):

(a)	thermoreflectance	($T = $ 90 K)	77A3
(b)	thermoreflectance	(293 K)	77A3
(c)	reflection	(300 K)	79A1
(d)	electroreflectance	(300 K)	75S1

(a)	(b)	(c)	(d)	Symmetry
		3.16 ⎱ 3.24 ⎰	3.3	N (L$_{3v}$ − L$_{1c}$)
3.34	3.27			
3.41	3.35	3.30 ⎱		
3.71	3.64	3.55 ⎰	3.7	T; Γ (X$_{5v}$ − X$_{1c}$)
3.82	3.74	3.67		
4.37	4.30	4.15		Γ; T (X$_{5v}$ − X$_{3c}$)
4.63	4.53	4.42		N (Σ)
		5.3		Γ (Γ_{15v} − Γ_{15c})
		6.4		
		7.4 ⎱ 8.0 ⎰		N(L$_{3v}$ − L$_{3c}$)
		8.8 ⎱ 10.7 ⎰		N (L$_{3v}$ − L$_{1c}$)

Physical property	Numerical value	Experimental conditions	Experimental method remarks	Ref.
effective masses:				
$m_{n\parallel}$	1.068 m_o			70G
$m_{n\perp}$	0.124 m_o			
valence band width:				
ΔE	13.4 eV	$T = 300$ K	X-ray spectroscopy	80G1
X-ray spectrum: Fig. 8				
Impurities and defects				
defect levels (in eV):				
from photoconductivity:				
E_d	0.9	$T = 173\cdots300$ K	K centre	77B
	1.2		R centre	
E_a	0.8	$173\cdots300$ K	R centre	
			(band gap: 2.08 eV assumed)	
	0.4		M centre	
from cathodoluminescence:				
E	1.15$\cdots$1.40	$T = 80$ K	deep impurities	80G2
	1.80$\cdots$2.00		recombination at Cd vacancies	
	2.00$\cdots$2.20		N impurity	
	2.080		resonance with exciton (2.085 eV)	
	2.20$\cdots$2.75		surface effects	
g-factors:				
g^* (P antisite)	2.005 (3)	$T = 20$ K	ESR	76K2
g^* (Mn^{2+})	2.004 (1)	20 K	ESR, Mn^{2+} on Zn site; there are magnetically non-equivalent Zn sites	76K1

Amorphisation by irradiation with $5 \cdot 10^{17}$ deuterons/cm^2 of energy $E = 13.5$ MeV; amorphous regions around In^{111} nuclei [78U].

Lattice properties

wavenumbers of Raman active phonons (in cm^{-1}, RT values):

		Symmetry I$\bar{4}$2d (F$\bar{4}$3m)		
$\bar{\nu}_{LO}/\bar{\nu}_{TO}$	512/489	Γ_5	(Γ_{15})	76G3,
	510/487	Γ_4	(Γ_{15})	75B1
	456/454	Γ_5	(W_4)	
	326	Γ_1	(W_1)	
	318/304	Γ_4	(W_2)	
	316	Γ_3	(W_2, X_5)	
	301	Γ_3	(W_2, X_5)	
	290/285	Γ_5	(X_5)	
	256/252	Γ_5	(W_3)	
	160/156	Γ_5	(W_4)	
	109	Γ_4	(W_2)	
	88	Γ_3	(W_2)	
	68	Γ_5	(X_5)	

two-phonon Raman scattering: Fig. 9

Physical property	Numerical value	Experimental conditions	Experimental method remarks	Ref.
Transport properties				
carrier concentrations, mobilities, resistivities (at 300 K):				
n-type samples:				
n	$10^{14}\cdots10^{15}\,\mathrm{cm^{-3}}$		(no anisotropy reported)	75S1
μ_n	$80\cdots150\,\mathrm{cm^2\,V^{-1}\,s^{-1}}$			
ϱ	$\approx 10^6\,\Omega\,\mathrm{cm}$			
n	$10^{10}\cdots10^{12}\,\mathrm{cm^{-3}}$			77B
μ_n	$90\,(5)\,\mathrm{cm^2\,V^{-1}\,s^{-1}}$			
E_A	$0.48\,\mathrm{eV}$		activation energy of σ in the presence of Sn	75S1
Optical properties				
refractive indices and birefringence (see also Fig. 15):				
n_o	3.414	$\lambda = 0.6328\,\mu\mathrm{m}$, $T = 300\,\mathrm{K}$		78I1
n_e	3.379	$\lambda = 0.6328\,\mu\mathrm{m}$, $T = 300\,\mathrm{K}$		
Δn	-0.035	$\lambda = 0.6328\,\mu\mathrm{m}$, $T = 300\,\mathrm{K}$		
	0	$\hbar\omega = 2.41\,\mathrm{eV}$, $T = 300\,\mathrm{K}$ $\hbar\omega = 2.48\,\mathrm{eV}$, $T = 77\,\mathrm{K}$ $\hbar\omega = 2.50\,\mathrm{eV}$, $T = 4.2\,\mathrm{K}$		78A3
	-0.051	$\hbar\omega = 0.6\,\mathrm{eV}$, $T = 300\,\mathrm{K}$		78A2
$\mathrm{d}\Delta n/\mathrm{d}T$	$-1.15\cdot10^{-6}\,\mathrm{K^{-1}}$	$T = 77\cdots300\,\mathrm{K}$		
$\mathrm{d}\Delta n/\mathrm{d}p$	$-2.1\cdot10^{-7}\,\mathrm{bar^{-1}}$			
optical activity:				
θ	$620°\,\mathrm{mm^{-1}}$			78A3
g_{11}	$7.1\cdot10^{-3}$		gyration tensor component	78A3

For photoconductivity and luminescence data, see "Impurities and defects".

Further properties				
coefficient of linear thermal expansion:				
α	$11\cdot10^{-6}\,\mathrm{K^{-1}}$	$T = 300\cdots570\,\mathrm{K}$	a axis	78N
	0		c axis	

Oxidation (formation of CdO and P_2O_3 in air) at $T = 850\,\mathrm{K}$ [71L].

melting point:				
T_m	$1390\,\mathrm{K}$			75S1

Physical property	Numerical value	Experimental conditions	Experimental method remarks	Ref.

10.1.3.11 Cadmium silicon arsenide (CdSiAs$_2$)

Electronic properties

energy gap (in eV):

$E_{g, dir}$	(A) 1.55	$T = 300$ K	electroreflectance	75S1, 75T
	(B) 1.74			
	(C) 1.99			
	1.635	1.7 K	photoluminescence	75S1, 76M1
dE_g/dT	$-2.3 \cdot 10^{-4}$ eV K^{-1}		photoconductivity	76L
dE_g/dp	10^{-5} eV bar^{-1}	300 K	absorption	75S1

splitting energies:

Δ_{cf}	-0.24 eV	$T = 300$ K	photoluminescence	75S1, 76M1
Δ_{so}	0.297 eV			

higher transition energies (in eV):

E	2.50	$T = 300$ K	electroreflectance, weakly polarized	75S1
	2.57			
	2.99			
	3.10			

effective masses (in m_o):

$m_{n, \perp}$	0.084		calculated	75S1
$m_{n, \parallel}$	0.074		(m_{ds}: density of states mass;	
$m_{n, ds}$	0.080		m_c: conductivity mass)	
$m_{n, c}$	0.080			
$m_{p_1, \perp}$	(-3.6)			
$m_{p_1, \parallel}$	0.090			
$m_{p_1, ds}$	1.07			
$m_{p_1, c}$	0.27			
$m_{p_2, \perp}$	0.21			
$m_{p_3, \perp}$	0.29			
$m_{p_3, \parallel}$	7.8			
$m_{p_3, ds}$	0.86			
$m_{p_3, c}$	0.47			

Impurities and defects

defect levels (in eV):

E_a	0.20···0.21	$T = 1.7$ K	photoluminescence, photo-conductivity, Hall effect	75S1, 75D1, 76M1, 76D1
	0.030	80 K	cathodoluminescence (stimulated emission)	75S1
E_d	0.003	1.7 K	photoluminescence	75S1

For further data, see also under "Optical properties".

Physical property	Numerical value	Experimental conditions	Experimental method remarks	Ref.
Transport properties				
carrier concentrations, mobilities:				
p-type samples:				
p	$6 \cdot 10^{15}$ cm^{-3}	$T = 300$ K		75S1,
μ_p	$300 \cdots 500$ cm^2 V^{-1} s^{-1}	300 K		75D1
n-type samples (formed by diffusion of indium):				
n	10^{17} cm^{-3}			75D1

The mobility in the valence bands is limited by scattering by non polar optical phonons; $\mu_p \propto T^{-5/2}$ [76Z2].
  For effect of irradiation on resistivity, see Fig. 27.
  N.B. No anisotropy reported for transport properties.

Physical property	Numerical value	Experimental conditions	Experimental method remarks	Ref.
Optical properties				
photoluminescence peaks (in eV):				
E	1.635	$T = 4.2$ K, $E \parallel c$	$\Gamma_{1c} - \Gamma_{4v}$: band-band transition	76M1
	1.615		Γ_{1c}-acceptor: $E_a = 0.02$ eV	
	1.598		deep level $\rightarrow \Gamma_{4v}$	
	1.593		deep level $\rightarrow \Gamma_{4v}$	
	1.635	1.7 K	as above (different samples)	75S1
	1.632			
	1.614			
	1.611			
photoconductivity peaks (in eV):				
E	1.47	$T = 300$ K, $E \parallel c$	n-type, crystal field split level,	76L,
	1.5	$E \perp c$	associated with In	76D1
	1.52		p-type, release from $E_a = 0.02$ eV level	

For data on photoconductivity and photoluminescence, see also under "Impurities and defect".

Physical property	Numerical value	Experimental conditions	Experimental method remarks	Ref.
Further properties				
melting point:				
T_m	> 1120 K			75S1

10.1.3.12 Cadmium germanium phosphide (CdGeP$_2$)

Electronic properties

energy gap (in eV):

			T [K]		
$E_{g, dir}$	(A)	1.72	300	electroreflectance	75S1
	(B)	1.90			
	(C)	1.99			
		1.803	77	photoluminescence ($E \parallel c$)	77R4
dE_g/dT		$-(2.5\cdots3) \cdot 10^{-4}$ eV K^{-1}		photoconductivity	75B4
dE_g/dp		$9.7 \cdot 10^{-6}$ eV bar^{-1}	300	absorption	75S1

Physical property	Numerical value	Experimental conditions	Experimental method remarks	Ref.
excitonic energy gap:				
E_{gx}	1.686 eV	$T = 300$ K, $E \parallel c$	photoconductivity	75B4
	1.712 eV	$E \perp c$		
splitting energies:				
Δ_{cf}	-0.2 eV	$T = 300$ K	electroreflectance	75S1
Δ_{so}	0.11 eV			
intraconduction band energy:				
$E(\Gamma_3) - E(\Gamma_1)$	0.2 eV	$T = 300$ K		70G
higher transition energies (in eV):				
E	2.62	$T = 300$ K	reflection, electroreflectance (unpolarized light)	73S
	3.50		reflection	
	4.25			
	4.9			
effective masses (in m_o):				
$m_{n,\perp}$	0.088		calculated	75S1
$m_{n,\parallel}$	0.088		(m_{ds}: density of states mass;	
$m_{n,ds}$	0.085		m_c: conductivity mass)	
$m_{n,c}$	0.085			
$m_{p_1,\perp}$	(-1.3)			
$m_{p_1,\parallel}$	0.099			
$m_{p_1,ds}$	0.56			
$m_{p_1,c}$	0.28			
$m_{p_2,\perp}$	0.23			
$m_{p_3,\perp}$	0.26			
$m_{p_3,\parallel}$	(1.7)			
$m_{p_3,ds}$	0.49			
$m_{p_3,c}$	0.36			
Impurities and defects				
defect levels (in eV):		T [K]		
E_d	0.32		P vacancy? (luminescence)	77R4
	0.25 (2)	2.4···300	In (transport)	75B5, 75B6
E_a	0.5	300	? (photoconductivity)	75B4
	0.42···0.49	2.4···300	In (photoconductivity)	75B5, 75B6
	0.37···0.46	77, 300	Cu (transport, luminescence)	76B5

For the influence of doping on transport and optical properties, see the corresponding subsections.

Physical property	Numerical value	Experimental conditions	Experimental method remarks	Ref.

Lattice properties

wavenumbers of infrared and Raman active phonons (in cm^{-1}, RT values):

		Symmetry I$\bar{4}$2d (F$\bar{4}$3m)		
$\bar{\nu}_{LO}/\bar{\nu}_{TO}$	400/384	Γ_5 (Γ_{15})	Γ_4, Γ_5 IR active, all Raman active	75B1,
	399/389	Γ_4 (Γ_{15})	(for conflicting IR results,	74B
	373	Γ_3 (W$_2$)	see [75G5])	
	364/355	Γ_5 (W$_4$)		
	321	Γ_2 (W$_1$)		
	317/297	Γ_4 (W$_2$)		
	293/289	Γ_5 (X$_5$)		
	225	Γ_3 (X$_3$)		
	181/179	Γ_5 (W$_3$)		
	122	Γ_5 (W$_4$)		
	88	Γ_4 (W$_2$)		
	85	Γ_3 (W$_2$)		
	63	Γ_5 (X$_5$)		

two-phonon absorption: Fig. 36

Transport properties

carrier concentrations, mobilities, conductivities, resistivities, Seebeck coefficients (at 300 K):

n-type samples (In or Sb doped or excess P):

n	$5 \cdot 10^{11} \dots 10^{15}$ cm^{-3}			75B4, 76M3, 75S1
μ_n	100 cm^2 V^{-1} s^{-1}			75S1
σ	$5 \cdot 10^{-5}$ Ω^{-1} cm^{-1}			76M3
S	$-1.2 \cdot 10^{-3}$ V K^{-1}		undoped	75S1

p-type samples (Cu or Bi doped or excess Cd):

p	$3 \dots 4 \cdot 10^9$ cm^{-3}		undoped	75B4, 76B5
ϱ	$1 \dots 2 \cdot 10^7$ Ω cm			76B5
μ_p	$30 \dots 60$ cm^2 V^{-1} s^{-1}			
p	$5 \cdot 10^{11} \dots 4 \cdot 10^{13}$ cm^{-3}		Cu doped	
ϱ	$2.5 \cdot 10^5 \dots 4.3 \cdot 10^3$ Ωcm			
μ_p	$60 \dots 50$ cm^2 V^{-1} s^{-1}			

The mobility obeys $\mu \propto T^{3/2}$ [76B5, 76M3].

p	10^{15} cm^{-3}		Bi doped	75S1
μ_p	25 cm^2 V^{-1} s^{-1}			

N.B. No anisotropy reported.
For effects of irradiation on the resistivity, see Fig. 29.

Physical property	Numerical value	Experimental conditions	Experimental method remarks	Ref.

Optical properties

refractive indices: coefficients in the formula $n^2 = A + B/(1 - C/\lambda^2) + D/(1 - E/\lambda^2)$:

	A	B	C [10^{-12} m^2]	D	E [10^{-12} m^2]	T [K]	
n_o	5.9677	4.2286	0.2021	1.6351	671.33	293	79B1
n_e	6.1573	4.0970	0.2330	1.4925	671.33	293	
n_o	6.3737	3.9281	0.2069	1.6686	671.33	391	
n_e	6.9280	3.4442	0.2803	1.5515	617.33	391	

		λ [μm]		
n_o	3.4833	0.80		75S1
	3.3560	1.00		
	3.2255	2.00		
	3.2065	3.00		
	3.1963	4.00		
	3.1887	5.00		
	3.1800	6.00		
	3.1735	7.00		
	3.1669	8.00		
	3.1585	9.00		
	3.1508	10.00		
	3.1372	11.00		
	3.1241	12.00		
	3.1165	12.50		
n_e	3.3902	1.00		75S1
	3.2406	2.00		
	3.2175	3.00		
	3.2063	4.00		
	3.1991	5.00		
	3.1906	6.00		
	3.1846	7.00		
	3.1785	8.00		
	3.1707	9.00		
	3.1636	10.00		
	3.1517	11.00		
	3.1402	12.00		
	3.1332	12.50		

nonlinear dielectric susceptibility:

d_{36}	$162 \cdot 10^{-12}$ mV^{-1}	$\lambda = 10.6$ μm		72B

electrooptic coefficients:

r_{41}	$3.0 \cdot 10^{-12}$ mV^{-1}	$T = 300$ K,		79V
r_{63}	$3.2 \cdot 10^{-12}$ mV^{-1}	$\lambda = 3.39$ μm		

photoluminescence peaks (in eV):

undoped samples:

		T [K]		
E	1.310	300	P vacancy?	76B5
	1.765	4.2, $E \parallel c$	$\Gamma_{1c} \rightarrow$ local level	77R4
	1.767	77, $E \parallel c$	$\Gamma_{1c} \rightarrow$ local level	
	1.803	77, $E \parallel c$	$\Gamma_{1c} \rightarrow \Gamma_{4v}$ (band-band transition)	

Cu doped sample:

E	1.305	300		76B5

(continued)

Physical property	Numerical value	Experimental conditions	Experimental method remarks	Ref.
photoluminescence peaks (continued) In doped samples:				
E	1.32···1.44	77	$E_c(\Gamma_1) \rightarrow E_a(In_{Ge})$, In on Ge site or $E_d(In_{Cd}) \rightarrow E_a(In_{Ge})$, In on neighboring Cd and Ge sites	76M2, 78M1
	1.53···1.585	77	$E_d(In_{Cd}) \rightarrow E_v(\Gamma_4)$, In on Cd site or $E_c(\Gamma_1) \rightarrow E_a(In_{Cd}\text{-}V_{Cd})$, In on Cd site with Cd vacancy	76M2, 78M1
	1.34	4.2	compare above for possible explanations	76M3
	1.610			
	1.665			
	1.715			
	1.727			
	1.765			
	1.785			
	1.805		band-band	

All above luminescence lines polarized $E \| c$ [77R2].
The luminescence is very sensitive to annealing, etc. [78M1].

photoconductivity peaks (in eV):

		T [K]		
E	1.686	300, $E \| c$	also photo EMF from homojunction	75B4
	1.712	300, $E \perp c$	from exciton ($n=1$), crystal field split	
	1.74···1.76	77		75B5
	1.45	300	In doped (also photoelectric effect; not due to $E_d = 0.25$ eV level [75B6])	

For photoconductivity and luminescence data, see also under "Impurities and defects".

Further properties
coefficients of thermal expansion (in 10^{-6} K^{-1}):

α	8.9	$T = 300···1030$ K	a axis	75M1
	0.37		c axis	
β	1.8	$300···1030$ K	volume	

thermal conductivity:

κ	0.11 W cm^{-1} K^{-1}	$T = 300$ K		75S1

Debye temperature:

Θ_D	340 K	$T \rightarrow 0$ K		75A3

melting point:

T_m	1073 K			75B3

Physical property	Numerical value	Experimental conditions	Experimental method remarks	Ref.

10.1.3.13 Cadmium germanium arsenide (CdGeAs$_2$)

Electronic properties

energy gaps (in eV):

			T [K]		
$E_{g,dir}$	(A)	0.57	300	electroreflectance	75S1
	(B)	0.73			
	(C)	1.02			
		0.65	77	photoconductivity ($\Gamma_{4v} - \Gamma_{1c}$)	76B6
		0.53	300	absorption	70G, 77I
		0.61	115		70G
$E_{g,th}$		0.673	0	Hall effect	78I2
$dE_{g,th}/dT$		$-3.5 \cdot 10^{-4}$ eV K^{-1}	0...300		
dE_g/dp		$9.3 \cdot 10^{-6}$ eV bar^{-1}	300	absorption	75S1

splitting energies:

Δ_{cf}		-0.21 eV	$T = 300$ K	electroreflectance	75S1
Δ_{so}		0.33 eV			

higher transition energies (in eV):

E		2.0	$T = 300$ K	electroreflectance, weakly polarized	75S1
		2.09			
		2.44			
		2.58			
				disagrees with [73S]	

other structure:

λ		5.5 μm	$T = 300$ K	absorption; p-type material only; lower to highest valence band; disappears on heat treatment of starting materials, i.e. due to oxygen? [80B4]	78I2

effective masses (in m_o):

m_n		0.26		Hall effect	78I2
m_p		0.035			
		0.027		thermoelectric power; Nernst-Ettingshausen effect	70G
$m_{n,\perp}$		0.039		calculated	75S1
$m_{n,\parallel}$		0.030		(m_{ds}: density of states mass;	
$m_{n,ds}$		0.036		m_c: conductivity mass)	
$m_{n,c}$		0.035			
$m_{p1,\perp}$		0.67			
$m_{p1,\parallel}$		0.034			
$m_{p1,ds}$		0.25			
$m_{p1,c}$		0.094			
$m_{p2,\perp}$		0.079			
$m_{p3,\perp}$		0.13			
$m_{p3,\parallel}$		0.56			
$m_{p3,ds}$		0.22			
$m_{p3,c}$		0.18			

Physical property	Numerical value	Experimental conditions	Experimental method remarks	Ref.
Impurities and defects				
defect levels (in eV):				
		T [K]		
E_d	0.04	77, 300	crystal field split	76B6
	0.07			
E_a	0.15	77, 300	start of photoresponse	76B6
			transport	76B7
	0.30	77, 300, 450	transport	78I2
	0.38	77$\cdots$300		77I

For photoconductivity and photoluminescence, see also under "Optical properties".

amorphous samples: absorption (with different dopants): Fig. 37

Lattice properties

wavenumbers of infrared active phonons (in cm^{-1}, RT values):

		Symmetry $I\bar{4}2d$ ($F\bar{4}3m$)		
$\bar{\nu}_{LO}/\bar{\nu}_{TO}$	280/272	Γ_5	(Γ_{15})	77H2
	278/270	Γ_4	(Γ_{15})	
	258/255	Γ_5	(W_4)	
	210/203	Γ_4	(W_2)	
	206/200	Γ_5	(X_5)	
	101/159	Γ_5	(W_3)	
	98/95	Γ_5	(W_4)	

elastic moduli (in 10^{10} Nm^{-2}):

c_{11}	9.45	$T = 290$ K	from ultrasonic wave velocity	82H
c_{12}	5.96			
c_{13}	5.97			
c_{33}	8.34			
c_{44}	4.21			
c_{66}	4.08			

For pressure and temperature dependence and further elastic properties, see [82H].

Transport properties

a) chalcopyrite structure:

carrier concentrations, mobilities, Seebeck coefficients:

p-type samples:

		T [K]		
p	$(0.7\cdots2)\cdot10^{16}$ cm^{-3}	300		78I2,
μ_p	$140\cdots400$ cm^2 V^{-1} s^{-1}	300		76B7

n-type samples (by vacuum anneal, or In, Al, Te doping) [76B7]:

n	$4\cdot10^{16}\cdots10^{18}$ cm^{-3}	$100\cdots500$		75S1,
μ_n	$1000\cdots$	$100\cdots500$		78I2
	4000 cm^2 V^{-1} s^{-1}			
S	$-1.9\cdot10^{-4}$ V K^{-1}	300		75S1

Effect of irradiation on conductivity and Hall coefficient: Fig. 38; see also [78B5].
Pressure dependence of resistivity: Fig. 39.
N.B. No anisotropy reported.

Physical property	Numerical value	Experimental conditions	Experimental method remarks	Ref.

b) glasses:

resistivity vs. doping concentration: Fig. 40; conductivity and Hall coefficient in CdGe$_x$As$_2$ glasses: [76M4]

mobility:

μ_{dr}	$0.1 \cdots 1$ cm^2 V^{-1} s^{-1}	$T = 300$ K	carrier type unclear	75S2

Optical properties

refractive indices: coefficients in the formula $n^2 = A + B/(1 - C/\lambda^2) + D/(1 - E/\lambda^2)$

	A	B	C [10^{-12} m^2]	D	E [10^{-12} m^2]	T [K]	
n_o	10.1064	2.2988	1.0872	1.6247	1370	300	76B3
n_e	11.8018	1.2152	1.6971	1.6922	1370	300	
$d(n_e - n_o)/dT$	$2.3 \cdot 10^{-6}$ K^{-1}		$\lambda = 10.6$ µm				76B3

		λ [µm]		
n_o	3.6076	2.3		75S1
	3.5645	3.0		
	3.5448	4.0		
	3.5336	5.0		
	3.5251	6.0		
	3.5200	7.0		
	3.5157	8.0		
	3.5120	9.0		
	3.5078	10.0		
	3.5031	11.0		
	3.4977	12.0		
	3.4950	12.5		
n_e	3.6775	3.0		75S1
	3.6402	4.0		
	3.6249	5.0		
	3.6134	6.0		
	3.6073	7.0		
	3.6030	8.0		
	3.5988	9.0		
	3.5942	10.0		
	3.5896	11.0		

dielectric constant:

$\varepsilon(\infty)$	15.4	$E \parallel c$, $T = 300$ K	infrared reflectivity	77H2
	15.2	$E \perp c$		

non-linear dielectric susceptibilities (at $T = 300$ K):

		λ [µm]		
d_{14}	$2.36 \cdot 10^{-10}$ mV^{-1}	1.06	second harmonic generation	76A
	$6.1 \cdot 10^{-10}$ mV^{-1}	$10.25 \cdots 2.54$		79K3
d_{36}	$3.51 \cdot 10^{-10}$ mV^{-1}	10.6		72B
$d^{(3)}$	$7 \cdot 10^{-18}$ m^2 V^{-2}	$10.25 \cdots 2.54$	third harmonic generation	79K3
$d^{(4)}$	$7.9 \cdot 10^{-27}$ m^3 V^{-3}	10.24	fourth harmonic generation $d^{(3)}$ and $d^{(4)}$ contain significant (dominant?) contributions from cascade processes [79K3]	

Physical property	Numerical value	Experimental conditions	Experimental method remarks	Ref.
photoconductivity and photo EMF peaks (in eV):				
		T [K]		
E	0.65	77, $E \parallel c$, $E \perp c$	$\Gamma_{4v} - \Gamma_{1c}$; band-band transition	76B6
	0.61	77, $E \parallel c$	$E_v(\Gamma_4) - E_d$; crystal field split	
	0.58	77, $E \perp c$		
	0.53	77, $E \perp c$	p-type only	
	0.51	77, $E \parallel c$	p-type only	
	0.50	300, $E \parallel c$	p-type only	
	0.65	300, $E \perp c$	p-type only	
photoluminescence peak:				
E	0.607 eV	$T = 1.7$ K	possibly stimulated emission	75S1
Further properties				
coefficient of linear thermal expansion:				
α	$1 \cdot 10^{-6}$ K^{-1}	$T = 370 \cdots 570$ K	c axis	78I2
	$8 \cdots 9 \cdot 10^{-6}$ K^{-1}		a axis (large anisotropy)	
magnetic susceptibility:				
χ_m	$-103 \cdot 10^{-6}$ cm^3 mol^{-1}	$T = 300$ K	(χ_m in CGS-emu), measured; powder sample	77S2
	$31.4 \cdot 10^{-6}$ cm^3 mol^{-1}		van Vleck contribution	
thermal conductivity:				
κ	0.42 W cm^{-1} K^{-1}	$T = 300$ K		75S1
Debye temperature:				
Θ_D	240.9 (14) K	$T \to 0$ K	from heat capacity	81B
	257 (2) K	$T = 4.2$ K	from elastic moduli	82H
melting point:				
T_m	943 K			77S2, 75B3

10.1.3.14 Cadmium tin phosphide (CdSnP$_2$)

Electronic properties
band structure: Fig. 41
energy gap (in eV):

			T [K]		
$E_{g,dir}$	(A)	1.17	300	electroreflectance	70S
	(B)	1.25			
	(C)	1.33			
	(A)	1.23	77	photoluminescence	76M6
	(B)	1.32			
	(C)	1.40			
		1.237	4.2	photoluminescence	78M2
		1.24	1.7	absorption	75S1
dE_g/dT		$-2.8 \cdot 10^{-4}$ eV K^{-1}			80Z, 78M4
dE_g/dp		$7.3 \cdot 10^{-6}$ eV bar^{-1}	300	absorption	75S1

Physical property	Numerical value		Experimental conditions	Experimental method remarks	Ref.
excitonic energy gap (in eV):					
E_{gx}	(A)	1.2343	$T = 1.7$ K	photoluminescence	75S1
	(B)	1.2353			
	(C)	1.2360			
splitting energies:					
Δ_{cf}		-0.10 eV	$T = 300$ K	electroreflectance	70S
Δ_{so}		0.10 eV			
higher transition energies (in eV, at 300 K):					
E		2.56	E, R	electroreflectance (E), reflection (R)	73S,
		2.65	E, R	(unpolarized light)	75K1,
		2.92	E, R		70S
		3.00	E		
		3.26	E		
		3.53	E, R		
		4.33	E, R		
		5.5	R		
		7.3	R		
effective masses (in m_o):					
$m_{n, \perp}$		0.060		calculated	75S1
$m_{n, \parallel}$		0.056		calculated	75S1
$m_{n, ds}$		0.035		(m_{ds}: density of states mass)	80Z
$m_{n, c}$		0.059		calculated	75S1
				(m_c: conductivity mass)	
$m_{p_1, \perp}$		(-6.6)		calculated	75S1
$m_{p_1, \parallel}$		0.69		calculated	75S1
$m_{p_1, ds}$		0.2			80Z
$m_{p_1, c}$		0.20		calculated	75S1
$m_{p_2, \perp}$		0.14		calculated	75S1
$m_{p_3, \perp}$		0.17		calculated	75S1
$m_{p_3, \parallel}$		2.6		calculated	75S1
$m_{p_3, ds}$		0.43		calculated	75S1
$m_{p_3, c}$		0.25		calculated	75S1
Impurities and defects					
defect levels (in eV):					
			T [K]		
E_d		0.32	$250 \cdots 300$	in n-type Cu doped samples, activation energy of conductivity, current oscillations, relaxation time	75P2
		0.004	$24 \cdots 900$	transport, photoconductivity	78M4, 80Z
		0.012	$24 \cdots 900$	transport	80Z
E_a		$0.13 \cdots 0.14$	77	Cu doped; photoluminescence	71B, 76M5
		$0.07, 0.13 \cdots 0.14$	77, 300	Cu doped; photoconductivity	78M4
			77, 300	Cu doped; transport	78M3
		0.37	300	Cu doped; photovoltage	76M5

For further data, see also under "Transport properties" and "Optical properties".

Physical property	Numerical value	Experimental conditions	Experimental method remarks	Ref.

Lattice properties

wavenumbers of infrared active phonons (in cm^{-1}, RT values):

		Symmetry I$\bar{4}$2d (F$\bar{4}$3m)		
$\bar{\nu}_{LO}/\bar{\nu}_{TO}$	353/–	Γ_4 (Γ_{15})		78B1
	–/389	Γ_5 (Γ_{15})		
	327/318	Γ_5 (W$_4$)		
	–/295	Γ_4 (W$_2$)		
	285/279	Γ_5 (X$_5$)		

Transport properties

carrier concentrations, mobilities, conductivities (at 300 K):

n-type (undoped) samples:

n	$10^{15}\cdots10^{18}$ cm^{-3}		for temperature dependence of	75S1,
μ_n	$400\cdots2000$ cm^2 V^{-1} s^{-1}		n, μ_n, σ and $R_{H,n}$, see Figs. 42, 43	71B,
σ	$2\cdots10$ Ω^{-1} cm^{-1}			76P3,
				80Z

Cu doped (i.e. heavily compensated) samples:

n	$5\cdot10^{14}$ cm^{-3}			76P3
μ_n	800 cm^2 V^{-1} s^{-1}		$\mu_n \propto T^{3/2}$	76P3
ϱ	$10^4\cdots10^6$ Ω cm			75P2,
				76M5

p-type (Cu or Ag doped) samples:

p	10^{14} cm^{-3}			78M3
μ_p	$90\cdots150$ cm^{2}V^{-1} s^{-1}			78D,
				78M3
ϱ	$10\cdots10^5$ Ω cm			71B,
				78M4

For effect of irradiation on resistivity, see Fig. 29.
   N.B. No anisotropy reported.

Optical properties

dielectric constants:

$\varepsilon(0)$	11.8	$T = 300$ K	infrared, unpolarized	75S1
$\varepsilon(\infty)$	10.0	300 K		

photoluminescence peaks:

Ag-doped:

		T [K]		
λ	1.004 µm	1.7	band-band	75S1
	1.028 µm	1.7	Ag complexes	
	1.072 µm	1.7	Ag complexes	
	1.124 µm	1.7	weak	
			quenched by Cu	71B

(continued)

Physical property	Numerical value	Experimental conditions	Experimental method remarks	Ref.
photoluminescence peaks (continued)				
Cu-doped:				
E	1.10 eV	77, $E \parallel c$		76M5
		$E \perp c$	$\approx 2 \cdots 2.5$ meV higher	76M6
	1.17 eV	77	?	76M2
	1.16 eV	4.2	crystal field split	78M2
	1.185 eV	4.2	crystal field split	
	1.237 eV	4.2	$\Gamma_{4v} - \Gamma_{1c}$, band-band transition	
photoconductivity and photovoltage data:				
E	1.15 eV	$T = 300$ K, $E \parallel c$	n-type; $\Gamma_{4v} - \Gamma_{1c}$; band-band transition	76M5
	1.18 eV	300 K, $E \perp c$	n-type; ?	
	1.10 eV	300 K, $E \perp c$	p-type; Cu level, $E_a = 0.07$ eV	78M4

Stimulated emission is observed at $\lambda \approx 1.010 \cdots 1.015 \,\mu$m at RT [75S1].

For data on photoluminescence and photoconductivity, see also under "Impurities and defects".

Further properties

magnetic susceptibilities:

χ_g	$-0.295 \cdot 10^{-6}$ cm^3 g^{-1}	$T = 300$ K	(χ in CGS-emu)	74T1
χ_m	$-86.4 \cdot 10^{-6}$ cm^3 mol^{-1}	300 K	total molar susceptibility	
	$+57.9 \cdot 10^{-6}$ cm^3 mol^{-1}		van Vleck paramagnetic part	
	$-144.3 \cdot 10^{-6}$ cm^3 mol$^-$		diamagnetic part	

N.B. No anisotropy reported

melting point:

T_m	840 K			75S1

10.1.3.15 Cadmium tin arsenide (CdSnAs$_2$)

Electronic properties

band structure: see Fig. 44

energy gap (in eV):

			T [K]		
$E_{g, \text{dir}}$	(A)	0.26	300	electroreflectance	75S1
	(B)	0.30			
	(C)	0.79			
		0.26	300	photoconductivity	81D
		0.278	77	absorption	75D2
		0.80	77	electroreflectance	72K
		0.28	0 (?)	absorption	75D2
dE_g/dT		$-2.2 \cdot 10^{-4}$ eV K^{-1}	$0 \cdots 300$	transport	75D2
		$-2.3 \cdot 10^{-4}$ eV K^{-1}	300	photoconductivity	81D
dE_g/dp		$16 \cdot 10^{-6}$ eV bar^{-1}	300		75A2

splitting energies:

Δ_{cf}	-0.06 eV	$T = 300$ K	electroreflectance	75S1
	-0.049 eV			80D3
Δ_{so}	0.48 eV			75S1
$d\Delta_{cf}/dp$	$-1.5 \cdot 10^{-6}$ eV bar^{-1}	300 K		80D2

Physical property	Numerical value	Experimental conditions	Experimental method remarks	Ref.
other gaps:				
ΔE	0.17$\cdots$0.18 eV	$T = 77$ K	between 2 conduction bands	75D2
higher transition energies (in eV):				
		T [K]		
E	2.02$\cdots$2.05	300	electroreflectance, reflection	73S,
	2.12$\cdots$2.13		(unpolarized light)	75K1
	2.33$\cdots$2.43			
	2.62	300	electroreflectance, reflection	73S
	3.00$\cdots$3.04			
	3.6$\cdots$3.8			
	4.1$\cdots$4.6	300	electroreflectance, reflection	73S,
				75K1
	5.3$\cdots$5.6			
	7.2			
	2.13	77, $E \parallel c$	electroreflectance, reflection	73S,
				72K
	2.25	77	electroreflectance, reflection	72K
	2.48$\cdots$2.51	77, $E \perp c$	electroreflectance, reflection	73S,
				72K
	2.60	77	electroreflectance, reflection	72K
	3.6$\cdots$3.76	77	electroreflectance, reflection	73S
	4.2$\cdots$4.6	77		
photoemission (XPS) data (in eV):				
E	-2 (2)	$T = 300$ K	(E relative to upper valence	74V2
	-7 (1)		band edge)	
	-10.5 (1)		Cd 4d	
effective masses (in m_o):				
$m_{n, \perp}$	0.05		Faraday effect	78K,
$m_{n, \parallel}$	0.048			75D2
$m_{n, ds}$	0.014		(m_{ds}: density of states mass)	
$m_{n, c}$	0.017		calculated	
			(m_c: conductivity mass)	
$m_{p_1, \perp}$	0.10		calculated	75D2,
$m_{p_1, \parallel}$	0.018		calculated	80D2
$m_{p_1, ds}$	0.5		transport	
$m_{p_1, c}$	0.041		calculated	
$m_{p_2, \perp}$	0.031		calculated	80D2
$m_{p_2, ds}$	0.027		transport	
$m_{p_3, \perp}$	0.12		calculated	75S1
$m_{p_3, \parallel}$	0.16			
$m_{p_3, ds}$	0.13			
$m_{p_3, c}$	0.13			

Physical property	Numerical value	Experimental conditions	Experimental method remarks	Ref.
Transport properties				
carrier concentrations, resistivities, Hall coefficients, mobilities, Seebeck coefficients, magnetoresistance:				
n-type samples (for T- and p-dependence, see also Figs. 45$\cdots$49):				
		T [K]		
n	10^{18} cm^{-3}	300		76S,
ϱ	$4.41 \cdot 10^{-4} \cdots$	77$\cdots$300		75A1,
	$1.35 \cdot 10^{-3}\,\Omega$ cm			75D2,
R_{H}	$-2.5 \cdots -13$ cm^3 C^{-1}	77$\cdots$300		76D2
μ_{n}	$1.1 \cdots 1.5$	300		
	$\cdot 10^4$ cm^2 V^{-1} s^{-1}			
	$2.5 \cdot 10^4$ cm^2 V^{-1} s^{-1}	500		75S1
S	$-8.5 \cdot 10^{-5}$ V K^{-1}	300		
p-type samples (values for $T = 300$ K; see also Figs. 50$\cdots$52):				
p	$1.9 \cdot 10^{17} \cdots$			75S1,
	$1.8 \cdot 10^{19}$ cm^{-3}			75A2,
				76D2,
				80D2,
				80D3
σ	$35 \cdots 116\,\Omega^{-1}$ cm^{-1}			80D2,
R_{H}	$1.4 \cdots 0.595$ cm^3 C^{-1}			80D3
μ_{p}	$42.6 \cdots 36$ cm^2 V^{-1} s^{-1}		heavy holes	80D3
	$546 \cdots 510$ cm^2 V^{-1} s^{-1}		light holes	
$\Delta\varrho/\varrho$	$0.49 \cdots 0.54 \cdot 10^{-3}$		$B = 1.8$ T	
S	$2.25 \cdots 2.32 \cdot 10^{-4}$ V K^{-1}			
N.B. No anisotropy reported.				
Optical properties				
dielectric constant:				
$\varepsilon(0)$	12.1	$T = 300$ K	unpolarized	80P
luminescence and photoconductivity: maximum at $\lambda \approx 1.3\,\mu$m (at 300 K) ($CdSnAs_2$—InP junction) [80P]				
For photoelectric properties, see also [81D].				
Further properties				
linear thermal expansion coefficient:				
α	$3.2 \cdot 10^{-6}$ K^{-1}	$T = 300$ K	a axis	80P
	$4.5 \cdot 10^{-6}$ K^{-1}		c axis	
magnetic susceptibilities:				
χ_{g}	$-0.265 \cdot 10^{-6}$ cm^3 g^{-1}	$T = 300$ K	(χ in CGS-emu)	74T1
χ_{m}	$-101 \cdot 10^{-6}$ cm^3 mol^{-1}	300 K	total molar susceptibility	
	$-161.6 \cdot 10^{-6}$ cm^3 mol^{-1}		diamagnetic part	
	$60.6 \cdot 10^{-6}$ cm^3 mol^{-1}		paramagnetic part	
N.B. No anisotropy reported.				
Debye temperature:				
Θ_{D}	234.4 (53) K	$T \to 0$ K		65A
melting point:				
T_{m}	868 (2) K			75B3,
				75S1

Physical property	Numerical value	Experimental conditions	Experimental method remarks	Ref.

10.1.3.16 Solid solutions

$(ZnSiAs_2)_{1-x}(CdSiAs_2)_x$:
electroreflectance: Fig. 53

$(CdSnAs_2)_{1-x}(CdSnP_2)_x$:
energy gaps:

$E_{g,dir}$	0.42 eV	x = 0.1	$T = 300$ K	77G
	0.69 eV	0.2		
	0.56 eV	0.3		

magnetic susceptibility: Fig. 54

electron concentrations (in 10^{18} cm^{-3}) **and mobilities** (in 10^3 cm^2 V^{-1} s^{-1}):

$n = 2.1$	$\mu_n = 4.5$	x = 0.1	$T = 300$ K	77G
1.2	5.0	0.2		
0.9	2.1	0.3		
1.6	3.9	0.4		
≈ 1	1.5	0.8		

$(CdGeP_2)_{1-x}(CdSnP_2)_x$:
energy gaps:

$E_{g,dir}$	1.33 eV	x = 0.68	$T = 300$ K	77G
	1.23 eV	0.87		

For absorption and luminescence spectra and a determination of further band parameters, see [81R].

electron concentrations and mobilities:

$n = \approx 10^{17}$ cm^{-3}	$\mu_n = 1.5 \cdot 10^3$ cm^2 V^{-1} s^{-1}	x = 0.68	$T = 300$ K	77G
$8 \cdot 10^{16}$ cm^{-3}	$1.6 \cdot 10^3$ cm^2 V^{-1} s^{-1}	0.87		

$(CdSi_{0.2}Ge_{0.8}As_2)$:
magnetic susceptibility (χ_m in CGS–emu):

χ_m	$-97.5 \cdot 10^{-6}$ cm^3 mol^{-1}	$T = 300$ K	total (powder sample)	77S2
	$34.6 \cdot 10^{-6}$ cm^3 mol^{-1}		paramagnetic part	

$(CdGe_{0.8}Sn_{0.2}As_2)$:
magnetic susceptibility (χ_m in CGS–emu):

χ_m	$-103 \cdot 10^{-6}$ cm^3 mol^{-1}	$T = 300$ K	total (powder sample)	77S2
	$35.7 \cdot 10^{-6}$ cm^3 mol^{-1}		paramagnetic part	

$(CdGe_{0.8}Pb_{0.2}As_2)$:
magnetic susceptibility (χ_m in CGS–emu):

χ_m	$-40.3 \cdot 10^{-6}$ cm^3 mol^{-1}	$T = 300$ K	total (powder sample)	77S2
	$103 \cdot 10^{-6}$ cm^3 mol^{-1}		paramagnetic part	

10.1.3.17 References for 10.1.3

58P   Pfister, H.: Acta Cryst. **11** (1958) 221.

63G   Goryunova, N.A., Kesamanly, F.P., Osmanov, E.O.: Fiz. Tverd. Tela **5** (1963) 2031; Soviet Phys. Solid State (English Transl.) **5** (1963) 1484; Vajpolin, A.A., Gašimzade, F.M., Goryunova, N.A., Kesamanly, F.P., Naselov, D.N., Osmanov, E.O., Rud', Ju.V.: Izv. Akad. Nauk. SSSR, Ser. Fiz. **28** (1964) 1085.

65A   Leroux-Hugon, P., Veyssie, J.J.: Phys. Status Solidi **8** (1965) 561.

65G   Goryunova, N.A., Kesamanly, F.P., Nasledov, D.N., Negreskul, V.V., Rud'Yu, V., Slobodchikov, S.V.: Fiz. Tverd. Tela **7** (1965) 1312; Sov. Phys. Solid State (English Transl.) **7** (1965) 1060.

69B   Berger, L.T., Prochokhan, V.D.: "Ternary Diamond Like Semiconductors", New York, London: Consultants Bureau, **1969**.

70G   Goryunova, N.A., Poplavnoi, A.S., Polygalov, Yu.I., Chaldyshev, V.A.: Phys. Status Solidi **39** (1970) 9.

70M   Maunaye, M., Lang, T.: Mater. Res. Bull. **5** (1970) 793.

70S   Shay, T.L., Buehler, E., Wernick, T.H.: Phys. Rev. **B2** (1970) 4104.

71B1   Buehler, E., Wernick, T.H., Shay, T.L.: Mater. Res. Bull. **6** (1971) 303.

71B2   Boyd, G., Kasper, H., McFee, J.M.: IEEE J. Quant. Electron. QE **7** (1971) 563.

71K   Kriviate, G.Z., Korneev, E.F., Shileika, A.Yu.: Fiz. Tekh. Poluprovodn. **13** (1971) 2242; Sov. Phys. Semicond. (English Transl.) **5** (1972) 1961.

71L   Lyalikova, R.Yu., Mirovich, L.V.: Izv. Akad. Nauk SSSR, Neorg. Mater. **7** (1971) 1453; Inorg. Mater. (USSR) (English Transl.) **7** (1971) 1292.

71S   Shay, T.L., Buehler, E., Wernick, T.H.: Phys. Rev. **B3** (1971) 2004.

72B   Boyd, G.D., Buehler, E., Storz, F.G., Wernick, J.H.: IEEE J. Quant. Electron. QE-**8** (1972) 419.

72K   Karavaev, F.G., Kriviate, G.Z., Polygalov, Yu.I., Chaldyshev, V.A., Shileika, A.Yu.: Fiz. Tekh. Poluprovodn. **6** (1972) 2211; Sov. Phys. Semicond, (English Transl.) **6** (1973) 1863.

72M   Miller, R.C., Nordland, W.A.: Phys. Rev. **B5** (1972) 4931.

72S   Somogyi, K., Bertóti, I.: Jpn. J. Appl. Phys. **11** (1972) 103.

73A   Averkieva, G.K., Prochukhan, V.D., Tashtanova, M.: Izv. Akad. Nauk SSSR, Neorg. Mater. **9** (1973) 487; Inorg. Mater. (USSR) (English Transl.) **9** (1973) 435.

73B   Berger, L.I., Kradinova, L.V., Petrov, V.M., Prochukhan, V.D.: Izv. Akad. Nauk SSSR, Neorg. Mater. **9** (1973) 1258; Inorg. Mater. (USSR) (English Transl.) **9** (1973) 1118.

73S   Sobolev, V.V.: Izv. Akad. Nauk SSSR, Neorg. Mater. **9** (1973) 1060; Inorg. Mater. (USSR) (English Transl.) **9** (1973) 947.

73V   Varea de Alvarez, C., Cohen, M.L.: Phys. Rev. Lett. **30** (1973) 979.

74B   Bettini, M., Miller, A.: Phys. Status Solidi (b) **66** (1974) 579.

74L   Larson, W.L., Maruska, H.P., Stevenson, D.A.: J. Electrochem. Soc. **121** (1974) 1674.

74S   Scholl, F.W., Cory, E.S.: Mater. Res. Bull. **9** (1974) 1511.

74T1   Trifosova, E.P., Baidakov, L.A.: Izv. Akad. Nauk SSSR, Neorg. Mater. **10** (1974) 2125; Inorg. Mater. (USSR) (English Transl.) **10** (1974) 1824.

74T2   Turner, E.H., Buehler, E., Kasper, H.: Phys. Rev. **B39** (1974) 558.

74V1   Varea de Alvarez, C., Cohen, M.L., Kohn, S.E., Petroff, Y., Shen, Y.R.: Phys. Rev. **B10** (1974) 5175.

74V2   Varea de Alvarez, C., Cohen, M.L., Lev, L., Kowalczyk, S.P., McFeely, F.R., Shirley, D.A., Grant, R.W.: Phys. Rev. **B10** (1974) 596.

75A1   Averkieva, G.K., Prochukhan, V.D., Rud', Yu.V., Tashtanova, M.: Izv. Akad. Nauk SSSR, Neorg. Mater. **11** (1975) 607.

75A2   Amirkhanov, Kh.I., Daunov, M.I., Magomedov, A.B., Magomedov, Ya.B., Emirov, S.N.: High Temp. High Pressure **7** (1975) 690.

75A3   Abrahams, S.C., Hsu, F.S.L.: J. Chem. Phys. **63** (1975) 1162.

75B1   Bettini, M.: Phys. Status Solidi (b) **69** (1975) 201.

75B2   Baran, N.P., Tychina, I.I., Tregub, I.G., Tkachuk, I.Yu., Chermenko, L.I., Sherbyna, I.P.: Fiz. Tekh. Poluprovodn. **9** (1975) 2366; Sov. Phys. Semicond. (English Transl.) **9** (1975) 1527.

75B3   Borshchevskii, A.S., Kotsyuruba, E.S.: Fiz. Tekh. Poluprovodn. **9** (1975) 2346; Sov. Phys. Semicond. (English Transl.) **9** (1976) 1513.

75B4   Borshchevskii, A.S., Lebedev, A.A., Mal'tseva, I.A., Ovezovk, K., Rud', Yu.V., Undalov, Yu.K.: Fiz. Tekh. Poluprovodn. **9** (1975) 1949; Sov. Phys. Semicond. (English Transl.) **9** (1976) 1278.

75B5   Borshchevskii, A.S., Kunaev, A.M., Kusainov, S.G., Lebedov, A.A., Rud', Yu.V., Undalov, Yu.K.:
        Fiz. Tekh. Poluprovodn. **9** (1975) 1021; Sov. Phys. Semicond, (English Transl.) **9** (1976) 673.
75B6   Baitenev, N.A., Borshchevskii, A.S., Kusainov, S.G., Rud', Yu.V., Undalov, Yu.K.: Fiz. Tekh.
        Poluprovodn. **9** (1975) 401; Sov. Phys. Semicond, (English Transl.) **9** (1976) 267.
75B7   Borshchevskii, A.S., Shantsovoi, T.M.: Izv. Akad. Nauk, Neorg. Mater. **11** (1975) 2158; Inorg.
        Mater (USSR) (English Transl.) **11** (1975) 1853.
75B8   Bondriot, H., Foeller, B., Schneider, H.A.: Phys. Status Solidi (a) **30** (1975) K121.
75D1   Dovletmuradov, Ch., Ovezov, K., Prochukhan, V.D., Rud', Yu.N., Serginov, M.: Pišma Zh. Eksp.
        Teor. Fiz. **1** (1975) 878.
75D2   Daunov, M.I., Magdiev, B.N., Magomedov, A.B.: Fiz. Tekh. Poluprovodn. **9** (1975) 1747; Sov.
        Phys. Semicond. (English Transl.) **9** (1976) 1147.
75G1   Gorban', I.S., Gorynya, V.A., Lugovoi, V.I., Tychina, I.I.: Ukr. Fiz. Zh. **20** (1975) 1428.
75G2   Grigor'eva, V.S., Markov, Yu.F., Rybakova, T.V.: Fiz. Tverd. Tela **17** (1975) 1993.
75G3   Grigor'eva, V.S., Lebedov, A.A., Ovezov, K., Prochukhan, V.D., Rud', Yu.V., Yakovenko, A.A.:
        Fiz. Tekh. Poluprovodn. **9** (1975) 1605; Sov. Phys. Semicond. (English Transl.) **9** (1976) 1058.
75G4   Gorban', I.S., Gorynya, V.A., Lugovoi, V.I., Tychina, I.I.: Fiz. Tverd. Tela **17** (1975) 2631; Sov.
        Phys. Solid State (English Transl.) **17** (1976) 1749.
75G5   Gorban', I.I., Gorynya, V.A., Seryi, V.I., Tychina, I.I., Il'in, M.A: Fiz. Tverd. Tela **17** (1975)
        44.
75H   Humphreys, R.G., Pamplin, B.R.: J. Phys. (Paris) **36** (1975) C3–155.
75K1   Kavaliauskas, T., Krivaite, G., Sileika, A.: Litov. Fiz. Sb. **15** (1975) 605.
75K2   Krivov, M.A., Melev, V.G., Klimov, V.N., Khlystova, A.S.: Fiz. Tekh. Poluprovodn. **9** (1975)
        1211; Sov. Phys. Semicond. (English Transl.) **9** (1975) 807.
75K3   Kończewicz, L., Porowski, S., Polushina, I.K.: High Temp. High Pressure **7** (1975) 716.
75M1   Miller, A., Humphreys, R.G., Chapman, B.: J. Phys. (Paris) **36** (1975) C3–31.
75M2   Markov, Yu.F., Gromova, T.M., Rud', Yu.B., Tashtanova, M.: Fiz. Tverd. Tela **17** (1975) 1226;
        Sov. Phys. Solid State (English Transl.) **17** (1975) 796.
75P1   Poplavnoi, A.S., Tyuterev, V.G.: Fiz. Tverd. Tela **17** (1975) 1055; Sov. Phys. Solid State (English
        Transl.) **17** (1975) 672.
75P2   Podol'skii, V.V., Karpovich, I.A., Zvonkov, B.N.: Fiz. Tekh. Poluprovodn. **9** (1975) 2188; Sov.
        Phys. Semicond. (English Transl.) **9** (1976) 1422.
75R   Raudonis, A.V., Shileika, A.Yu.: Fiz. Tekh. Poluprovodn. **9** (1975) 1539; Sov. Phys. Semicond.
        (English Transl.) **9** (1975) 1014.
75S1   Shay, T.L., Wernick, T.H.: "Ternary Chalcopyrite Semiconductors" Growth, Electronic Properties
        and Applications. Pergamon, **1975**.
75S2   Škácha, T.: Czech. J. Phys. **B25** (1975) 1397.
75T   Trykozko, R.T.: Mater. Res. Bull. **10** (1975) 489.
75V   Vlasenko, V.V., Lozovskii, V.N., Popov, V.P.: Kristallografiya **20** (1975) 1082; Sov. Phys. Crystal-
        logr. (English Transl.) **20** (1976) 662.
76A   Airoldi, G.: Riv. Nuovo Cimento **6** (1976) 295.
76B1   Becherer, H., Kirsten, P.: Phys. Status Solidi (b) **75** (1976) K19.
76B2   Braun, W., Cardona, M.: Phys. Status Solidi (b) **76** (1976) 251.
76B3   Bhar, G.C.: Appl. Opt. **15** (1976) 305.
76B4   Brudnyi, V.N., Budnitskii, D.L., Krivov, M.A., Melov, V.G.: Phys. Status Solidi (a) **35** (1976)
        425.
76B5   Borshchevskii, A.S., Mal'tseva, I.A., Rud', Yu.V., Undalov, Yu.K.: Fiz. Tekh. Poluprovodn. **10**
        (1976) 1101; Sov. Phys. Semicond. (English Transl.) **10** (1976) 655.
76B6   Borshchevskii, A.S., Dagina, N.E., Lebedov, A.A., Ovezov, K., Polushina, I.K., Rud', Yu.V.:
        Fiz. Tekh. Poluprovodn. **10** (1976) 1905; Sov. Phys. Semicond. (English Transl.) **10** (1976)
        1136.
76B7   Borshchevskii, A.S., Dagina, N.E., Lebedov, A.A., Ovezov, K., Polushina, I.K., Rud', Yu.V.:
        Fiz. Tekh. Poluprovodn. **10** (1976) 1571; Sov. Phys. Semicond. (English Transl.) **10** (1976)
        934.
76D1   Dovletmuradov, Ch., Ovezov, K., Prochukhan, V.D., Rud', Yu.V., Serginov, M.: Fiz. Tekh. Polu-
        provodn. **10** (1976) 1659; Sov. Phys. Semicond. (English Transl.) **10** (1976) 986.
76D2   Daunov, M.I., Magomedov, A.B.: Fiz. Tekh. Poluprovodn. **10** (1976) 641; Sov. Phys. Semicond.
        (English Transl.) **10** (1976) 383.
76G1   Grishchenko, G.A., Ljubchenko, A.V., Tychina, I.I., Ukr. Fiz. Zh. **21** (1976) 303.

76G2	Grigoreva, V.S., Lebedov, A.A., Prochukhan, V.D., Rud', Yu.V., Yakovenko, A.A.: Phys. Status Solidi (a) **36** (1976) K51.
76G3	Gorban', I.S., Garynya, V.A., Gugovoĭ, V.I., Tychina, I.I., Tkachuk, I.Yu.: Fiz. Tverd Tela **18** (1976) 1777; Sov. Phys. Solid State (English Transl.) **18** (1976) 1036.
76H	Humphreys, R.G.: J. Phys. C **9** (1976) 4491.
76I	Imenkov, A.N., Prochukhan, V.D., Negreskul, V.V., Rud', Yu.V., Tashtanova, M.: Izv. Akad. Nauk SSSR, Neorg. Mater. **12** (1976) 112; Inorg. Mater. (USSR) (English Transl.) **12** (1976) 93.
76J	Joullie, A.M., Alibert, C., Gallay, J., Deschanvres, A.: Solid State Commun. **19** (1976) 369.
76K1	Kaufmann, U., Räuber, A., Schneider, J.: Phys. Status Solidi (b) **74** (1976) 169.
76K2	Kaufmann, U., Schneider, J., Räuber, A.: Appl. Phys. Lett. **29** (1976) 312.
76K3	Krivov, M.A., Melev, V.G.: Izv. Vyssh. Uchebn. Zaved. Fiz. **4** (1976) 134.
76L	Lebedov, A.A., Ovezov, K., Prochukhan, V.D., Rud', Yu.V., Serginov, M.: Pis'ma Zh. Eksp. Teor. Fiz. **2** (1976) 385.
76M1	Mal'tseva, I.A., Prochukhan, V.D., Rud', Yu.V., Serginov, M.: Fiz. Tekh. Poluprovodn. **10** (1976) 1222; Sov. Phys. Semicond. (English Transl.) **10** (1976) 727.
76M2	Mal'tseva, I.A., Rud', Yu.V., Undalov, Yu.K.: Fiz. Tekh. Proluprovodn. **10** (1976) 400; Sov. Phys. Semicond. (English Transl.) **10** (1976) 240.
76M3	Mal'tseva, I.A., Rud', Yu.V., Undalov, Yu.K.: Ukr. Fiz. Zh. **21** (1976) 1503.
76M4	Meimaris, D., Katsika, V., Roiles, M.: J. Appl. Phys. **47** (1976) 685.
76M5	Medvedkin, G.A., Ovezov, K., Rud', Yu.V., Sokolova, V.I.: Fiz. Tekh. Poluprovodn. **10** (1976) 2081; Sov. Phys. Semicond. (English Transl.) **10** (1976) 1239.
76M6	Mal'tseva, I.A., Rud', Yu.V.: Pis'ma Zh. Eksp. Teor. Fiz. **2** (1976) 266.
76P1	Pasemann, L., Cordts, W., Heinrich, A., Monecke, J.: Phys. Status Solidi (b) **77** (1976) 527.
76P2	Poltavtsev, Yu.G.: Izv. Akad. Nauk SSSR, Neorg. Mater. **12** (1976) 2127; Inorg. Mater. (USSR) (English Transl.) **12** (1976) 1740.
76P3	Podol'skii, V.V., Karpovich, I.A., Zvonkov, B.N.: Fiz. Tekh. Poluprovodn. **10** (1976) 1004; Sov. Phys. Semicond. (English Transl.) **10** (1976) 594.
76R	Rud', Yu.V., Ovezov, K.: Fiz. Tekh. Poluprovodn. **10** (1976) 951; Sov. Phys. Semicond. (English Transl.) **10** (1976) 561.
76S	Siegel, W., Ziegler, E.: Exp. Tech. Phys. **24** (1976) 141.
76Z1	Ziegler, E., Siegel, W., Kühnel, G., Cobet, U.: Exp. Tech. Phys. **24** (1976) 249.
76Z2	Ziegler, E., Siegel, W., Heinrich, A.: Phys. Status Solidi (a) **36** (1976) 491.
77A1	Ambrazyavichyus, G.A., Babonas, G.A., Shileika, A.Yu.: Litov. Fiz. Sb. **17** (1977) 205.
77A2	Averkieva, G.K., Grigoreva, V.S., Mal'tseva, I.A., Prochukhan, V.D., Rud', Yu.V.: Phys. Status Solidi (a) **39** (1977) 453.
77A3	Ambrazevičius, G.A., Babonas, G.A., Šileika, A.Yu.: Phys. Status Solidi (b) **82** (1977) K45.
77B	Baltramiejunas, R., Vaitkus, J., Veleckas, D., Tychina, I., Tkachyuk, I.: Litov. Fiz. Sb. **17** (1977) 621.
77G	Golikova, O.A., Orlov, V.M.: Izv. Akad. Nauk SSSR, Neorg. Mater. **13** (1977) 970; Inorg. Mater (USSR) (English Transl.) **13** (1977) 792.
77H1	Humphreys, R.G.: Inst. Phys. Conf. Ser. **35** (1977) 105.
77H2	Holah, G.D., Miller, A., Dunnett, W.D., Iseler, G.W.: Solid State Commun. **23** (1977) 75.
77I	Isomura, S., Takahashi, S., Masumoto, K.: Jpn. J. Appl. Phys. **16** (1977) 1723.
77M1	Mal'tseva, I.A., Mamedov, A., Medvedkin, G.A., Rud', Yu.V.: Fiz. Tekh. Poluprovodn. **11** (1977) 2153; Sov. Phys. Semicond. (English Transl.) **11** (1977) 1264.
77M2	Mercey, B., Chippaux, D., Deschanvres, A.: Mater. Res. Bull. **12** (1977) 613.
77R1	Rife, J.L., Dexter, R.N., Bridenbaugh, P.M., Veal, B.W.: Phys. Rev. **B16** (1977) 4491.
77R2	Rud', Yu.V., Mal'tseva, I.A.: Fiz. Tverd. Tela **19** (1977) 870; Sov. Phys. Solid State (English Transl.) **19** (1977) 505.
77R3	Rud', Yu.V., Mal'tseva, I.A.: Fiz. Tekh. Poluprovodn. **11** (1977) 1033; Sov. Phys. Semicond. (English Transl.) **11** (1977) 612.
77R4	Rud', Yu.V., Mal'tseva, I.A.: Fiz. Tekh. Poluprovodn. **11** (1977) 201; Sov. Phys. Semicond. (English Transl.) **11** (1977) 116.
77S1	Siegel, W., Becherer, H., Kühnel, G., Ziegler, E.: Phys. Status Solidi (a) **41** (1977) 75.
77S2	Satow, T. Uemura, O., Watanabe, S.: Phys. Status Solidi (a) **44** (1977) 731.
78A1	Averkieva, G.K., Mamedov, A., Prochukhan, V.D., Rud', Yu.V.: Fiz. Tekh. Poluprovodn. **12** (1987) 1732; Sov. Phys. Semicond. (English Transl.) **12** (1978) 1025.

78A2	Ambrazevičius, G.A., Babonas, G.: Litov. Fiz. Sb. **18** (1978) 765.
78A3	Ambrazyavichyus, G.A., Babonas, G., Karpus, V.: Fiz. Tekh. Poluprovodn. **12** (1978) 2034; Sov. Phys. Semicond. (English Transl.) **12** (1978) 2110.
78B1	Brudnyi, V.N. Krivov, M.A., Potapov, A.I., Prochukhan, V.D., Rud', Yu.V.: Fiz. Tekh. Poluprovodn. **12** (1978) 1109; Sov. Phys. Semicond. (English Transl.) **12** (1978) 659.
78B2	Brudnyi, V.N., Budnitskii, D.L., Krivov, M.A., Masagutova, R.V., Prochukhan, V.D., Rud', Yu.V.: Phys. Status Solidi (a) **50** (1978) 379.
78B3	Brudnyi, V.N., Krivov, M.A., Popatov, A.I., Masagutova, R.V., Prochukhan, V.D., Rud', Yu.V.: Pis'ma Zh. Eksp. Teor. Fiz. **4** (1978) 41.
78B4	Bedair, S.M., Littlejohn, M.A.: J. Electrochem. Soc. **125** (1978) 952.
78B5	Brudnyi, V.N., Krivov, M.A., Potapov, A.I., Polushina, I.K., Prochukhan, V.D., Rud', Yu.V.: Phys. Status Solidi (a) **49** (1978) 761.
78D	Danilov, V.I., Zvonkov, B.N., Karpovich, I.A., Kunstevich, T.S., Podol'skii, V.V.: Kristallografiya **23** (1978) 892; Sov. Phys. Crystallogr. (English Transl.) **23** (1978) 505.
78G1	Grishchenko, G.A., Ljubchenko, A.V., Tychina, I.I.: Ukr. Fiz. Zh. **23** (1978) 958.
78G2	Grishchenko, G.A., Ljubchenko, A.V., Puzin, I.B., Tychina, I.I.: Ukr. Fiz. Zh. **23** (1978) 1539.
78G3	Girault, B., Gouskov, A., Bougnot, J.: Mater. Res. Bull. **13** (1978) 457.
78I1	Itoh, N., Fujinaga, T., Nakau, T.: Jpn. J. Appl. Phys. **17** (1978) 951.
78I2	Iseler, G.W., Kildal, H., Menyuk, N.: J. Electron. Mater. **7** (1978) 737.
78K	Karavaev, G.F., Barisenko, S.I.: Izv. Vyssh. Uchebn. Zaved. Fiz. **6** (1978) 28.
78M1	Mal'tseva, I.A., Mamedov, A., Rud', Yu.V., Undalov, Yu.K.: Phys. Status Solidi (a) **50** (1978) 139.
78M2	Mal'tseva, I.A., Rud', Yu.V., Sokolova, V.I., Smirnova, A.D.: Ukr. Fiz. Zh. **23** (1978) 46.
78M3	Mal'tseva, I.A., Mamedov, A., Rud', Yu.V., Sokolova, V.I.: Ukr. Fiz. Zh. **23** (1978) 1355.
78M4	Medvedkin, G.A., Rud', Yu.V., Valov, Yu.A., Sokolova, V.I.: Phys. Status Solidi (a) **45** (1978) K95.
78N	Nakau, T., Nimura, H., Ozaki, Y., Kamada, S., Hisamatsu, T.: Jpn. J. Appl. Phys. **17** (1978) 1677.
78U	Unterrucker, A., Hausbrand, J.: Phys. Status Solidi (a) **46** (1978) 125.
78Z1	Ziegler, E., Siegel, W., Kühnel, G., Buhrig, E.: Phys. Status Solidi (a) **48** (1978) K63.
78Z2	Ziegler, E., Siegel, W., Kühnel, G.: Phys. Status Solidi (a) **49** (1978) K205.
79A1	Ambrazevičius, G.A., Babonas, G., Šileika, A.: Phys. Status Solidi (b) **95** (1979) 643.
79A2	Alward, J.F., Fong, C.Y., Wooton, F.: Phys. Rev. **B19** (1979) 6337.
79B1	Bhar, G.C., Ghosh, G.C.: J. Opt. Soc. Am. **69** (1979) 730.
79B2	Barnes, N.P., Eckhardt, R.C., Gettemy, D.J., Edgett, L.B.: IEEE J. Quant. Elec. **15** (1979) 1074.
79C	Cordts, W., Heinrich, A., Monecke, J.: Phys. Status Solidi (b) **96** (1979) 201.
79G	Gorban, I.S., Gorynya, V.A., Lugovoi, V.I., Krasnolob, N.P., Salivon, G.I., Tychina, I.I.: Phys. Status Solidi (b) **93** (1979) 531.
79K1	Kühnel, G., Siegel, W., Ziegler, E.: Phys. Status Solidi (a) **54** (1979) 315.
79K2	Kozkina, I.I.: Vestn. Leningr. Univ. Fiz. Khim. **13** (1979) 83.
79K3	Kildal, H., Iseler, G.W.: Phys. Rev. **B19** (1979) 5218.
79O	Okunev, V.D., Yurov, A.G.: Pis'ma Zh. Eksp. Teor. Fiz. **5** (1979) 161.
79R	Risbug, S.H.: Metall. Trans. **10A** (1979) 1953.
79S	Shirakawa, T., Okamura, K., Nakai, J.: Phys. Lett. **73A** (1979) 442.
79V	Vohl, P.: J. Electron. Mater. **8** (1979) 517.
79Z	Zürcher, P., Meier, F.: J. Appl. Phys. **50** (1979) 3687.
80B1	Brudnyi, V.N., Krivov, M.A., Mamedov, A., Potapov, A.I., Prochukhan, V.D., Rud', Yu.V.: Phys. Status Solidi (a) **60** (1979) K57.
80B2	Bhar, G.C., Ghosh, G.C.: Appl. Opt. **19** (1980) 1029.
80B3	Bhar, G.C., Ghosh, G.C.: IEEE J. Quant Elec. **16** (1980) 838.
80B4	Borshchevsky, A.S., Route, R.K., Feigelson, R.S.: Mater. Res. Bull. **15** (1980) 409.
80D1	Duncan, W.M., Schreiner, A.F., Bedair, S.M., Littlejohn, M.A.: J. Lumin. **21** (1980) 137.
80D2	Daunov, M.I., Magomedov, A.B.: Fiz. Tekh. Poluprovodn. **14** (1980) 341; Sov. Phys. Semicond. (English Transl.) **14** (1980) 199.
80D3	Daunov, M.I., Magomedov, A.B., Ramazanova, A.E.: Phys. Status Solidi (a) **60** (1980) 651.
80G1	Gusatinskii, A.N., Bunin, M.A., Blokhin, M.A., Minin, V.I., Prochukhan, V.D., Averkieva, G.K.: Phys. Status Solidi (b) **100** (1980) 739.
80G2	Gorban, I.S., Krys'kov, Ts.A., Tennakun, M., Tychina, I.I., Chukichev, M.V.: Fiz. Tekh. Poluprovodn **14** (1980) 975; Sov. Phys. Semicond. (English Transl.) **14** (1980) 577.

80P	Popov, A.S., Trifonova, E.P.: Phys. Status Solidi (a) **58** (1980) 679.
80Z	Ziegler, E., Siegel, W., Kühnel, G.: Phys. Status Solidi (a) **65** (1980) 625.
81B	Bohmhammel, K., Deus, P., Schneider, H.A.: Phys. Status Solidi (a) **65** (1981) 563.
81D	Dovletmuradov, Ch., Lebedev, A.A., Rud', Yu.V., Serginaov, M., Skoryukin, V.E.: Sov. Phys. Semicond. **15** (1981) 1369 (transl. from Fiz. Tekh. Poluprovodn. **15** (1981) 2357).
81H	Heinrich, A., Cordts, W., Monecke, J.: Phys. Status Solidi (b) **107** (1981) 319.
81R	Rud', Yu.V., Vaipolin, A.A., Kalevich, E.S., Medvekin, G.A., Parimbekov, Z.A., Sokolova, V.I.: Sov. Phys. Semicond. **15** (1981) 1374 (transl. from Fiz. Tekh. Poluprovodn. **15** (1981) 2366).
82H	Hailing, T., Saunders, G.A., Lambson, W.A., Feigelson, R.S.: J.Phys. C **15** (1982) 1399.

10.1.4 I_2–IV–VI_3 compounds

10.1.4.0 Structure

The I_2–IV–VI_3 compounds have received very little attention. Their structures are not known in detail. They generally adopt a disordered zincblende-like form with some tendency to superstructure formation, usually tetragonal, chalcopyrite like. In this case there is some confusion about the indexing of the c axis. In the table below a chalcopyrite-like convention will be adopted (i.e. $c \approx 2a$). (N.B. Anisotropy of transport properties has never been reported for tetragonal chalcopyrite structure.)

Substance	Bravais lattice	Lattice parameters a, b, c in Å	Remarks	Ref.
ternary compounds:				
Cu_2SiS_3	tetragonal	$a = 5.290$ $c = 10.015$	$T < 1030$ K $d = 3.63$ g cm^{-3}	69B
	hexagonal (wurtzite like)	$a = 3.684$ $c = 6.004$	$T > 1030$ K; can be quenched to room temperature $d = 3.81$ g cm^{-3}	
Cu_2SiSe_3	no reports			
Cu_2SiTe_3	cubic (tetragonal?)	$a = 5.93$	$d = 5.47$ g cm^{-3}	69B
Cu_2GeS_3	tetragonal	$a = 5.317$ $c = 10.438$	$T < 940$ K possible superstructure as below (monoclinic); $d = 4.45$ g cm^{-3}	69B, 74K
	monoclinic; C_s^3—Bm, C_2^3—B2 or C_{2h}^3—B2/m	$a = 7.464 (5)$ $b = 22.38 (1)$ $c = 10.640 (1)$ $\gamma = 91°52'$		74K
	cubic (disordered)	$a = 5.317$	$T > 940$ K	69B
Cu_2GeSe_3	tetragonal	$a = 5.5913$ $c = 10.977 (2)$	$T = 303$ K; $d = 5.57$ g cm^{-3}	69B, 70S
		$a = 5.6022 (6)$ $c = 10.984 (2)$	$T = 393$ K	70S
		$a = 5.6103 (6)$ $c = 10.987 (3)$	$T = 473$ K	
		$a = 5.6193 (10)$ $c = 10.990 (3)$	$T = 573$ K	
		$a = 5.6193 (10)$ $c = 10.994 (5)$	$T = 673$ K	

(continued)

Substance	Bravais lattice	Lattice parameters a, b, c in Å	Remarks	Ref.
tenary compounds (continued)				
Cu$_2$GeSe$_2$	monoclinic	$a=$ 5.512 $b=$ 5.598 $c=$ 5.486 $\beta=$ 98.7°	$T=300$ K; Ge-deficient	75S
	cubic (disordered)	$a=$ 5.568	$T=300$ K; Ge-excess	75S
Cu$_2$GeTe$_3$	tetragonal	$a=$ 5.959 $c=$ 11.858	$T=300$ K; probably 2 phases	77S
			$d=5.95$ g cm^{-3}	69B
Cu$_2$SnS$_3$	cubic (disordered)	$a=$ 5.445	$T=300$ K; possible super-structure, see below (monoclinic)	69B
	monoclinic	$a=23.10$ (1) $b=c=6.25\cdot 3$n $\alpha=101°$	$T=300$ K; $d=5.02$ g cm^{-3}	
Cu$_2$SnSe$_3$	cubic (disordered)	$a=$ 5.6877 (2)	$T=300$ K $d=5.94$ g cm^{-3}	72S, 77S 69B
Cu$_2$SnTe$_3$	cubic (disordered)	$a=$ 6.049	$T=300$ K; probably 3 phases $d=6.51$ g cm^{-3}	77S 69B
Ag$_2$SnS$_3$	not zincblende			69B
Ag$_2$SnSe$_3$	like Ag$_2$SnS$_3$			69B
Ag$_2$SnTe$_3$	zincblende like			69B
solid solutions:				
(Cu$_2$GeSe$_3$)$_x$(Cu$_2$SnSe$_3$)$_{1-x}$				
$x=1.0$	tetragonal	$a=$ 5.58 $c=$ 10.96		69B
0.75	tetragonal	$a=$ 5.60 $c=$ 11.08		
0.66	cubic	$a=$ 5.60		
0.50	cubic	$a=$ 5.61		
0.33	cubic	$a=$ 5.63		
0.25	cubic	$a=$ 5.65		

Physical property	Numerical value	Experimental conditions	Experimental method remarks	Ref.

10.1.4.1 Copper silicon-VI$_3$ compounds

Cu$_2$SiS$_3$:

thermal conductivity:

κ	$2.3\cdot 10^{-2}$ W cm^{-1} K^{-1}	$T=300$ K	with excess Si	69B

melting point:

T_m	1150 K		phase transition at 1030 K	69B

Cu$_2$SiTe$_3$:

transport: metallic — 69B

Mössbauer effect: Te^{125}, 35.48 keV, 130 K

δ	1.8 (5) $\cdot 10^{-4}$ m s^{-1}		shift (relative to ZnTe)	73D
$\varDelta$	4.10 (8) $\cdot 10^{-3}$ m s^{-1}		quadrupole splitting	

Physical property	Numerical value	Experimental conditions	Experimental method remarks	Ref.

10.1.4.2 Copper germanium sulfide (Cu_2GeS_3)

energy gap:

$E_{g,th}$	0.3 eV		transport	69B, 64K
	0.53 eV		high-temperature form	70A1

conductivities, Seebeck coefficients, Hall coefficients, carrier concentrations, mobilities (at 300 K except where otherwise stated):

For temperature dependence of σ, S, ϱ, R_H, see Figs. 1$\cdots$4.

p-type samples:

σ	$1.9\ \Omega^{-1}\,cm^{-1}$			62P
S	$1\cdots3\cdot10^{-4}\ V\,K^{-1}$			
σ	$17.3\ \Omega^{-1}\,cm^{-1}$			69B
R_H	$28.5\ cm^3\,C^{-1}$			
p	$3\cdot10^{17}\ cm^{-3}$			
μ_p	$360\ cm^2\,V^{-1}\,s^{-1}$			

n-type samples:

ϱ	$3.2\cdot10^2\ \Omega\,cm$			74K
	$2.7\cdots8\cdot10^4\ \Omega\,cm$	$T=77\ K$		
R_H	$-1.1\cdot10^3\ cm^3\,C^{-1}$			
μ_n	$3\ cm^2\,V^{-1}\,s^{-1}$			

linear thermal expansion coefficient:

α	$7.2\cdot10^{-6}\ K^{-1}$	$T=300\ K$		64B

thermal conductivity:

κ	$12\cdot10^{-3}\ W\,cm^{-1}\,K^{-1}$	$T=300\ K$	with excess Ge	69B
	$7.7\cdot10^{-3}\ W\,cm^{-1}\,K^{-1}$	300 K		64B

Debye temperature:

Θ_D	254 K	$T=300\ K$		69B

heat capacity:

c	$0.51\ J\,g^{-1}\,K^{-1}$	$T=300\ K$		69B

Young's modulus:

E	$1.4\cdot10^{11}\ dyn\,cm^{-2}$	$T=300\ K$		69B

melting point:

T_m	1220 (10) K		phase transition at 940 K	69B

10.1.4.3 Copper germanium selenide (Cu_2GeSe_3)

All values are for the tetragonal form except where otherwise stated.

energy gaps (in eV):

E_g	0.94 (5)	$T=293\ K$	photoconductivity	64K
	0.79 (5)	77 K		
$E_{g,th}$	0.25		transport, contradictory results	64K
	0.6 (1)			
	1.1			

defect levels:

E_a	0.016 eV	$T=77\cdots500\ K$	Se vacancies (transport)	71E
	0.010 eV			

Physical property	Numerical value	Experimental conditions	Experimental method remarks	Ref.
conductivities, Seebeck coefficients, Hall coefficients, carrier concentrations, mobilities (at 300 K):				
p-type samples:				
σ	$50\ \Omega^{-1}\,cm^{-1}$		for temperature dependence of	62P
S	$7\cdots 10\cdot 10^{-5}\,V\,K^{-1}$		σ, $S\ R_H$ and ϱ, see Figs. 1, 2, 5, 6	
ϱ	$4\cdot 10^{-2}\,\Omega\,cm$			71E
R_H	$1.2\cdot 10^{-2}\,cm^3\,C^{-1}$			
p	$6\cdot 10^{20}\,cm^{-3}$			
σ	$5.71\ \Omega^{-1}\,cm^{-1}$		cubic form	69B
R_H	$57.0\,cm^3\,C^{-1}$			
p	$1.5\cdot 10^{17}\,cm^{-3}$			
μ_p	$283\,cm^2\,V^{-1}\,s^{-1}$		$\mu \propto T^{-3/2}$ after annealing [71E]	
linear thermal expansion coefficient				
α	$8.4\cdot 10^{-6}\,K^{-1}$	$T=300\,K$		64B
thermal conductivity:				
κ	$2.4\cdot 10^{-2}\,W\,cm^{-1}\,K^{-1}$	$T=300\,K$	with excess Ge	69B
Debye temperature:				
Θ_D	$168\,K$	$T=300\,K$		69B
heat capacity:				
c	$0.34\,J\,g^{-1}\,K^{-1}$	$T=300\,K$		69B
Young's modulus:				
E	$9.1\cdot 10^{10}\,dyn\,cm^{-2}$	$T=300\,K$		69B
melting point:				
T_m	$2050\,(10)\,K$			69B

10.1.4.4 Copper germanium telluride (Cu_2GeTe_3)

transport: metallic

conductivity, Seebeck coefficient:				
σ	$1.4\cdots$	$T=300\,K$	temperature dependence: Figs. 1, 2	62P,
	$3.9\cdot 10^3\,\Omega^{-1}\,cm^{-1}$		(no anisotropy reported)	69B
S	$10^{-5}\,V\,K^{-1}$	$300\,K$		62P
thermal conductivity:				
κ	$0.13\,W\,cm^{-1}\,K^{-1}$	$T=300\,K$	with excess Ge; see Fig. 7	69B
Mössbauer effect: Te^{125}, 35.48 keV, 130 K				
δ	$1.1\,(3)\cdot 10^{-4}\,ms^{-1}$		shift (relative to ZnTe)	73D
Δ	$5.75\,(10)\cdot 10^{-3}\,ms^{-1}$		quadrupole splitting	
melting point:				
T_m	$800\,(50)\,K$			69B

10.1.4.5 Copper tin sulfide (Cu_2SnS_3)

energy gaps:				
E_g	$0.91\,(1)\,eV$	$T=293\,K$	photoconductivity	64K
	$0.93\,(1)\,eV$	$77\,K$		
$E_{g,th}$	$0.59\,eV$		transport	70A1

Physical property	Numerical value	Experimental conditions	Experimental method remarks	Ref.
conductivities, Seebeck coefficients, Hall coefficients, carrier concentrations, mobilities (at 300 K except where otherwise stated):				
p-type samples:				
σ	$0.49\,\Omega^{-1}\,cm^{-1}$		for temperature dependence	62P
S	$1\cdots6\cdot10^{-4}\,V\,K^{-1}$		of σ and S, see Figs. 1, 2	
σ	$39.6\,\Omega^{-1}\,cm^{-1}$			64B
R_H	$10.2\,cm^3\,C^{-1}$			
p	$6.1\cdot10^{17}\,cm^{-3}$			
μ_p	$605\,cm^2\,V^{-1}\,s^{-1}$			
n-type samples (monoclinic form):				
ϱ	$1.3\cdots3.6\cdot10^{-2}\,\Omega\,cm$		for temperature dependence	74K
	$1.9\cdots5.5\cdot10^{-2}\,\Omega\,cm$	$T=77\,K$	of ϱ and R_H, see Figs. 8, 9	
R_H	$-0.06\,cm^3\,C^{-1}$			
n	$1.2\cdot10^{20}\,cm^{-3}$			
μ_n	$0.50\,cm^2\,V^{-1}\,s^{-1}$			
linear thermal expansion coefficient:				
α	$7.8\cdot10^{-6}\,K^{-1}$	$T=300\,K$		64B
thermal conductivity:				
κ	$2.8\cdot10^{-2}\,W\,cm^{-1}\,K^{-1}$	$T=300\,K$		69B
Debye temperature:				
Θ_D	$168\,K$	$T=300\,K$		69B
heat capacity:				
c	$0.34\,J\,g^{-1}\,K^{-1}$	$T=300\,K$		69B
Young's modulus:				
E	$1.16\cdot10^{11}\,dyn\,cm^{-2}$	$T=300\,K$		69B
melting point:				
T_m	$1120\,(10)\,K$			69B

10.1.4.6 Copper tin selenide (Cu_2SnSe_3)

Physical property	Numerical value	Experimental conditions	Experimental method remarks	Ref.
energy gaps:				
E_g	$0.96\,(5)\,eV$	$T=293\,K$	photoconductivity	64K
	$0.77\,(5)\,eV$	$77\,K$		
$E_{g,th}$	$0.6\cdots0.83\,eV$		transport	62P, 64K, 70A1
conductivity, Seebeck coefficient, Hall coefficient, carrier concentration, mobility (at 300 K):				
p-type samples:				
σ	$71\cdots91\,\Omega^{-1}\,cm^{-1}$		for temperature dependence	69B, 62P
			of σ and S, see Figs. 1, 2, 10, 11	
S	$2.5\cdot10^{-4}\,V\,K^{-1}$			62P
p	$1.4\cdot10^{18}\,cm^{-3}$			
R_H	$12.3\,cm^3\,C^{-1}$			69B
p	$5.1\cdot10^{17}\,cm^{-3}$			
μ_p	$870\,cm^2\,V^{-1}\,s^{-1}$			

Physical property	Numerical value	Experimental conditions	Experimental method remarks	Ref.
linear thermal expansion coefficient:				
α	$8.9 \cdot 10^{-6}\,K^{-1}$	$T = 300\,K$		64B
	$20.4 \cdot 10^{-6}\,K^{-1}$	$300\,K$	(decreases at higher T)	72S
thermal conductivity:				
κ	$3.5 \cdot 10^{-2}\,W\,cm^{-1}\,K^{-1}$	$T = 300\,K$		69B
Debye temperature:				
Θ_D	$148\,K$	$T = 300\,K$		69B
heat capacity:				
c	$0.31\,J\,g^{-1}\,K^{-1}$	$T = 300\,K$		69B
Young's modulus:				
E	$7.5 \cdot 10^{10}\,dyn\,cm^{-2}$	$T = 300\,K$		69B
melting point:				
T_m	$970\,(5)\,K$			69B

10.1.4.7 Copper tin telluride (Cu_2SnTe_3)

conductivity, Seebeck coefficient:

p-type sample:

σ	$1.4 \cdot 10^4\,\Omega^{-1}\,cm^{-1}$	$T = 300\,K$	for temperature dependence,	62P
S	$3 \cdot 10^{-5}\,V\,K^{-1}$	$300\,K$	see Figs. 1, 2, 12, 13	

thermal conductivity:

κ	$0.144\,W\,cm^{-1}\,K^{-1}$	$T = 300\,K$	with excess Sn	69B
	$70\,W\,cm^{-1}\,K^{-1}$	$300\,K$	for temperature dependence, see Fig. 14	70A2

Mössbauer effect: Te^{125}, 35.48 keV, 130 K

δ	$2.4\,(4)\,10^{-4}\,ms^{-1}$		shift (relative to ZnTe)	73D
Δ	$4.47\,(6)\,10^{-3}\,ms^{-1}$		quadrupole splitting	

melting point:

T_m	$683\,K$			69B

10.1.4.8 Silver–IV–VI$_3$ compounds

Ag_2GeSe_3:

energy gaps:

E_g	$0.91\,(5)\,eV$	$T = 77\,K$	photoconductivity	64K
$E_{g,th}$	$0.9\,(1)\,eV$		transport	

photoconductivity: Fig. 15

conductivity, carrier concentration, mobility (at 300 K):

σ	$27\,\Omega^{-1}\,cm^{-1}$			69B
p	$2 \cdot 10^{17}\,cm^{-3}$			
μ_p	$850\,cm^2\,V^{-1}\,s^{-1}$			

melting point:

T_m	$810\,K$			69B

Physical property	Numerical value	Experimental conditions	Experimental method remarks	Ref.
Ag$_2$GeTe$_3$:				
energy gap:				
$E_{g,th}$	0.25 eV		transport	69B
conductivity, carrier concentration, mobility (at 300 K):				
σ	92 Ω^{-1} cm^{-1}			69B
p	$8 \cdot 10^{17}$ cm^{-3}			
μ_p	720 cm^2 V^{-1} s^{-1}			69B
melting point:				
T_m	600 K			69B
Ag$_2$SnS$_3$:				
energy gap:				
$E_{g,th}$	0.5 (1) eV		transport	64K
conductivity: Fig. 16				
Ag$_2$SnSe$_3$:				
energy gaps:				
E_g	0.81 (5) eV	$T = 293$ K	photoconductivity	64K
	0.84 (5) eV	77 K		
$E_{g,th}$	0.7 (1) eV		transport	
photoconductivity: Fig. 15				
conductivity, carrier concentration, mobility (at 300 K):				
σ	146 Ω^{-1} cm^{-1}			69B
p	10^{18} cm^{-3}			
μ_p	910 cm^2 V^{-1} s^{-1}			
melting point:				
T_m	760 K			69B
Ag$_2$SnTe$_3$:				
energy gap:				
$E_{g,th}$	0.08 eV		transport	69B
conductivity, carrier concentration, mobility (at 300 K):				
σ	48 Ω^{-1} cm^{-1}			69B
p	$5 \cdot 10^{17}$ cm^{-3}			
μ_p	600 cm^2 V^{-1} s^{-1}			
melting point:				
T_m	590 K			69B

10.1.4.9 References for 10.1.4

62P   Palatnik, L.S., Koshkin, V.M., Gal'chinetskii, L.P., Kolesnikov, V.I., Komnik, Yu.F.: Fiz. Tverd.
           Tela **4** (1962) 1430.
64B   Berger, L.I., Balanevskaya, A.E.: Fiz. Tverd. Tela **6** (1964) 1311.
64K   Kharakhorin, F.F., Petrov, V.M.: Fiz. Tverd. Tela **6** (1964) 2867.
69B   Berger, L.I., Prochukhan, V.D.: Ternary Diamond-like Semiconductors. New York: Consultants
           Bureau, **1969**.
70A1   Aliev, S.N., Magomedov, Ya.B., Shchegol'kova, Ya.B.: Fiz. Tekh. Poluprovodn. **4** (1970) 2306.
70A2   Aliev, S.N., Magomedov, Ya.B., Shchegol'kova, Ya.B.: Teplofiz. Vys. Temp. **8** (1970) 672.
70S   Sharma, B.B.: Phys. Status Solidi (a) **2** (1970) K13.
71E   Endo, S., Sudo, I., Irie, T.: Jpn. J. Appl. Phys. **10** (1971) 218.
72S   Sharma, B.B., Chavada, F.R.: Phys. Status Solidi (a) **14** (1972) 639.
73D   Dragunas, A.K., Makaryunas, K.V., Adominaite, M.L.: Fiz. Tekh. Poluprovodn. **7** (1973) 1599.
74K   Khanafer, M., Gorochov, O., Rivet, J.: Mater. Res. Bull. **9** (1974) 1543.
75S   Sharma, B.B., Ayyar, R., Singh, H.: Phys. Status Solidi (a) **29** (1975) K17.
77S   Sharma, B.B., Ayyar, R., Singh, H.: Phys. Status Solidi (a) **40** (1977) 691.

10.1.5 I_3–V–VI_4 compounds

The I_3–V–VI_4 compounds have received very little attention from semiconductor physicists, in spite of the existence of three naturally occurring minerals, enargite, luzonite and famatinite, and the prediction of useful non-linear optical properties [77S].

There are no reports of such compounds containing silver.

Although their structures are not cubic, there are no reports of anisotropic properties in these compounds.

10.1.5.0 Structure

Besides the usual disordered zincblende or wurtzite-like phases the I_3–V–VI_4 compounds adopt either the wurtzite-like enargite structure, space group C_{2v}^7—$Pmn2_1$ (see Fig. 1) or the zincblende-like famatinite structure, space group D_{2d}^{11}—$I\bar{4}2m$ (see Fig. 2).

Substance	Structure	Lattice parameters [Å]	Remarks	Ref.
Cu_3PS_4	enargite	$a=$ 7.296 (2)	$T=300$ K, phase transition $T=370$ K	72G,
		$b=$ 6.319 (2)	(transport)	75H
		$c=$ 6.072 (2)		
Cu_3PSe_4	enargite	$a=$ 7.697 (2)	$T=300$ K	72G
		$b=$ 6.661 (2)		
		$c=$ 6.381 (2)		
Cu_3AsS_4	enargite	$a=$ 7.407 (1)	$T>580$ K [69T]?	70A1
		$b=$ 6.436 (1)	see Fig. 3	
		$c=$ 6.154 (1)		
	famatinite	$a=$ 5.290	$T<580$ K [69T]?	57G
		$c=$ 10.465	usually called "luzonite"	
Cu_3AsSe_4	famatinite	$a=$ 5.570 (3)	$T<713$ K [69A]	69B
		$c=$ 10.957 (5)		
	disordered zincblende	$a=$ 5.5	$T>713$ K [69A], $d=7.02$ g/cm^3 [67A]	69B

(continued)

Substance	Structure	Lattice parameters [Å]	Remarks	Ref.
table (continued)				
Cu_3SbS_4	famatinite	$a= 5.385\,(1)$ $c = 10.754\,(2)$	$T = 300$ K, $d = 4.635$ g/cm³ [57G]	72G
	cubic: O_h^5—Fm3m	$a = 10.74$	thin films	69B
	disordered zincblende	$a= 5.28$	high temperature, $d = 4.71$ g/cm³ [61A]	69B
Cu_3SbSe_4	famatinite	$a= 5.645\,(1)$ $c = 12.275\,(2)$	$T < 688$ K [69A]	72G
	disordered zincblende	$a= 5.6$	$T > 688$ K [69A], $d = 5.94$ g/cm³ [67A]	69B

solid solution:

Cu_3AsS_4–Cu_3SbS_4 (enargite-luzonite-famatinite): phase diagram, see Fig. 3.

Physical property	Numerical value	Experimental conditions	Experimental method remarks	Ref.

10.1.5.1 Copper thiophosphate (Cu₃PS₄)

energy gap:

$E_{g,\,th}$	2 eV		transport	75H

wavenumbers of infrared and Raman active phonons (in cm⁻¹, RT values):

		Symmetry		
$\bar{v}$	60	external modes?		78T
	65			
	68			
	80			
	86			
	92			
	106			
	112			
	117			
	138			
	171			
	192			
	203			
	210			
	244			
	285	A′	Raman	75H
	298	A″	Raman	
	300	A′	Raman	
	307	A′	IR + Raman	
	320	A″	Raman	
	392	A′	Raman; PS₄ symmetric stretch	
	512	A′	IR + Raman	
	523	A′	IR	
	539	A′	IR + Raman	

resonant Raman scattering: overtones, see Fig. 6; excitation profiles, see Fig. 7.

conductivity, Seebeck coefficient:

σ	$2 \cdot 10^{-2} \ldots$ $3 \cdot 10^{-4}\ \Omega^{-1}\,cm^{-1}$	$T = 300$ K	for temperature dependence see Figs. 4, 5	75H
S	$8.5 \cdot 10^{-4}$ V K⁻¹	260 K		

Physical property	Numerical value	Experimental conditions	Experimental method remarks	Ref.

10.1.5.2 Enargite-luzonite (Cu_3AsS_4)

Naturally occurring mineral, usually with considerable addition of Sb impurities.
Enargite: wurtzite-like structure.
Luzonite: zincblende-like structure.

All data are for enargite.

energy gaps (in eV):

E_g	1.48	$T = 77$ K	photoconductivity	69B
	1.24	300 K	absorption	
$E_{g, th}$	0.8		transport	69B

conductivity, Hall coefficient, carrier concentration, mobility, Seebeck coefficient (at 300 K):

n-type sample:

σ	200 Ω^{-1} cm^{-1}		unexplained difference between [69B] and [70A2]	70A2
	0.095 Ω^{-1} cm^{-1}		temperature dependence of σ	69B
R_H	-0.093 cm^3 C^{-1}		and S: Figs. 8, 9	
n	$7.8 \cdot 10^{19}$ cm^{-3}			
μ_n	0.008 cm^2 V^{-1} s^{-1}			
S	$-1.3 \cdot 10^{-4}$ V K^{-1}			

p-type conduction is achieved by doping with Cl, I, Mn [69B].

linear thermal expansion coefficient:

α	$3.2 \cdot 10^{-6}$ K^{-1}	$T = 300$ K		69B

thermal conductivity:

κ	$3.2 \cdot 10^{-2}$ W cm^{-1} K^{-1}	$T = 300$ K		69B

magnetic susceptibilities:

χ_g	0.28 m^3 g^{-1}	$T = 300$ K	SI-units	69B
χ_m	110 m^3 mol^{-1}	300 K		

melting point:

T_m	931 K		phase transition at 580 K?	69T, 70A2

10.1.5.3 Copper arsenic selenide (Cu_3AsSe_4)

energy gaps (in eV):

E_g	0.65	$T = 77$ K	photoconductivity	69B
	0.88	300 K	absorption	
$E_{g, th}$	0.76		transport	69B
	0.26?			69A

conductivity, Hall coefficient, carrier concentration, mobility, Seebeck coefficient (at 300 K):

p-type samples:

σ	215 Ω^{-1} cm^{-1}		temperature dependence of σ	69B
R_H	2.77 cm^3 C^{-1}		and S: see Figs. 8, 9	
n	$2.7 \cdot 10^{18}$ cm^{-3}			
μ_n	505 cm^2 V^{-1} s^{-1}			
S	$1.2 \cdot 10^{-4}$ V K^{-1}			
	$\approx 10^{-5}$ V K^{-1}		unexplained difference between [69B] and [70A2]	70A2

Physical property	Numerical value	Experimental conditions	Experimental method remarks	Ref.
linear thermal expansion coefficient:				
α	$9.5 \cdot 10^{-6}\,K^{-1}$	$T = 300\,K$		69B
thermal conductivity:				
κ	$1.9 \cdots 2.7$ $\cdot 10^{-2}\,W\,cm^{-1}\,K^{-1}$	$T = 300\,K$		69A, 69B
Debye temperature:				
Θ_D	$169\,K$	$T = 300\,K$		69B
magnetic susceptibilities:				
χ_g	$0.15\,m^3\,g^{-1}$	$T = 300\,K$	SI-units	69B
χ_m	$87\,m^3\,mol^{-1}$	$300\,K$		
melting point:				
T_m	$733\,K$			70A2

10.1.5.4 Famatinite (Cu_3SbS_4)

Naturally occurring mineral (from Sierra de Famatina, Argentina). Usually with considerable addition of As impurities.

All data probably for tetragonal form.

energy gap:				
$E_{g,th}$	$0.46\,eV$		transport	78A

n-type conduction: especially with Zn, p-type conduction: with Cl, I, Mn [69B].

thermal conductivity:				
κ	$2.7 \cdot 10^{-2}\,W\,cm^{-1}\,K^{-1}$	$T = 300\,K$		78A
magnetic susceptibilities:				
χ_g	$0.12\,m^3\,g^{-1}$	$T = 300\,K$	SI-units	69B
χ_m	$53\,m^3\,mol^{-1}$	$300\,K$		
melting point:				
T_m	$830\,K$			69B

10.1.5.5 Copper antimony selenide (Cu_3SbSe_4)

Electronic properties, impurities and defects

There is a considerable confusion about the size of the electronic energy gap, probably due to a large concentration of defects.

energy gap (in eV):				
E_g	0.31	$T = 300\,K$	absorption	69B
	0.808	$90\,K$	cathodoluminescence	76D
	0.11	$300\,K$	transmission	69N
$E_{g,th}$	0.42		transport	69B
	0.13			69N
	0.76			76D
dE_g/dT	$1.27 \cdot 10^{-5}\,K^{-1}$		cathodoluminescence	76D

Physical property	Numerical value	Experimental conditions	Experimental method remarks	Ref.
effective mass:				
m_p	$0.73\ m_o$	$T = 320$ K	Seebeck effect	69N
defect states:				
E	0.795 eV	$T = 90$ K	cathodoluminescence; gap $= 0.808$ eV	76D
E_a	0.013 eV	90 K		

Transport properties

hole concentrations, mobilities, Seebeck coefficients and Nernst coefficient (at 300 K):

p-type samples (see Figs. 8⋯12 for temperature dependence of transport parameters):

p	$7 \cdot 10^{18}$ cm^{-3}		unannealed	76D
	$6 \cdots 8 \cdot 10^{17}$ cm^{-3}		annealed	
μ_p	$60 \cdots 40$ cm^2 V^{-1} s^{-1}			
S	$4 \cdot 10^{-4}$ V K^{-1}			
p	10^{19} cm^{-3}			69N
μ_p	$50 \cdots 60$ cm^2 V^{-1} s^{-1}			
S	$2 \cdot 10^{-4}$ V K^{-1}			
$B_\perp$	$2 \cdots 3$ $\cdot 10^{-3}$ cm^2 K^{-1} s^{-1}	$T = 100 \cdots 350$ K	transverse Nernst-Ettingshausen effect	

Further properties

linear thermal expansion coefficient:

α	$1.24 \cdot 10^{-5}$ K^{-1}	$T = 300$ K		69B

thermal conductivity:

κ	1.46 $\cdot 10^{-2}$ W cm^{-1} K^{-1}	$T = 300$ K		69B
	2.2 $\cdot 10^{-2}$ W cm^{-1} K^{-1}	300 K		69A

Debye temperature:

Θ_D	131 K	$T = 300$ K		69B

Young's modulus:

E	$25 \cdot 10^{12}$ dyn cm^{-2}	$T = 300$ K		76D

magnetic susceptibilities:

χ_g	0.20 m^3 g^{-1}	$T = 300$ K	SI-units	69B
χ_m	130 m^3 mol^{-1}	300 K		

melting point:

T_m	700 K		phase transition at 688 K	69A

Physical property	Numerical value	Experimental conditions	Experimental method remarks	Ref.

10.1.5.6 Other I_3–V–VI_4 compounds

Cu_3AsTe_4

σ, S vs. T: Fig. 13

thermal conductivity:

κ	$5.9 \cdot 10^{-2}$ W cm^{-1} K^{-1}	$T = 300$ K		77G

melting point:

T_m	≈ 600 K			77G

Cu_3SbTe_4

σ, S vs. T: Fig. 13

thermal conductivity:

κ	$5 \cdot 10^{-2}$ W cm^{-1} K^{-1}	$T = 300$ K		77G

melting point:

T_m	≈ 600 K			77G

10.1.5.7 References for 10.1.5

57G   Gaines, R.V.: Am. Mineralogist **42** (1957) 766.

61A   Alieva, A.G., Pinsker, Z.G.: Kristallografiya **6** (1961) 204; Soviet Phys. Crystallogr. (English Transl.) **6** (1961) 161.

67A   Annamamedov, R., Berger, L.T., Petrov, V.M., Slobodchikov, S.V.: Izv. Akad. Nauk SSSR, Neorg. Mater. **3** (1967) 1370.

69A   Aliev, S.N., Gadzhiev, G.G., Magomedov, Ya.B.: Fiz. Tekh. Poluprovodn. **3** (1969) 1709; Sov. Phys. Semicond. (English Transl.) **3** (1970) 1437.

69B   Berger, L.I., Prochukhan, V.D.: "Ternary Diamond-like Semiconductors". New York, London: Consultants Bureau **1969**.

69N   Nakanishi, H., Endo, S., Irie, T.: Jpn. J. Appl. Phys. **8** (1969) 443.

69T   Tanelli, G.: Rend. Accad. Lincei Sci. Fis. Mat. Nat. **46** (1969) 196.

70A1   Adiwidjaja, G., Löhn, J.: Acta Crystallogr. **B26** (1970) 2878.

70A2   Aliev, S.N., Magomedov, Ya.B., Shchegol'kova, N.V.: Fiz. Tekh. Poluprovodn. **4** (1970) 2306.

70P   Poplavnoi, A.S., Tyuterev, V.G.: Kristallografiya **15** (1970) 230; Sov. Phys. Crystallogr. (English Transl.) **15** (1970) 193.

72G   Garin, J., Parthé, E.: Acta Crystallogr. **B28** (1972) 3672.

75H   Hindia, T.A., Valov, Yu.A.: Phys. Status Solidi (a) **30** (1975) K41.

76D   Dirochka, A.I., Ivanova, G.S., Kurbatov, L.N., Sinitsyn, E.V., Kharakhorin, F.F., Kholina, E.I.: Izv. Akad. Nauk SSSR, Neorg. Mater. **12** (1976) 339; Inorg. Mater. (USSR) (English Transl.) **12** (1976) 290.

77G   Gadzhiev, G.G., Magdiev, B.N.: Teplofiz. Vys. Temp. **15** (1977) 425.

77S   Samanta, L.K., Bhar, G.C.: Phys. Status Solidi (a) **41** (1977) 331.

78A   Amirkhanov, Kh.I., Gadzhiev, G.G., Magomedov, Ya.B.: Teplofiz. Vys. Temp. **16** (1978) 1232.

78T   Temperini, M.L.A., Sala, O., Bernstein, H.J.: Chem. Phys. Lett. **59** (1978) 10.

10.1.6 II–III$_2$–VI$_4$ compounds

10.1.6.0 Structure

The II–III$_2$–VI$_4$ compounds almost all have one of the two structures, defect stannite (space group D$_{2d}^{11}$–I$\bar{4}$2m) or defect chalcopyrite (space group S$_4^2$–I$\bar{4}$) (see introduction to this chapter). The structures of the others will be discussed under these compounds. (Data in the following tables are RT values).

Substance	Space group	a Å	c Å	c/a	x	y	z	d g cm^{-3}	Ref.
ZnAl$_2$S$_4$ (β')	C$_{6v}^4$–P6$_3$mc	3.764	6.142	–	–	–	–	2.63	55H
(α)	O$_h^7$–Fd3m	9.988	–	–	–	–	–	3.30	55H
ZnAl$_2$Se$_4$	S$_4^2$–I$\bar{4}$	5.49	10.8	1.98	0.26	0.25	0.13	4.37	55H
ZnAl$_2$Te$_4$	S$_4^2$–I$\bar{4}$ or D$_{2d}^{11}$–I$\bar{4}$2m	5.90	12.0	2.03	0.25	0.25	0.125	4.91	55H
ZnGa$_2$S$_4$	S$_4^2$–I$\bar{4}$ or D$_{2d}^{11}$–I$\bar{4}$2m	5.26	10.4	1.97	0.25	0.25	0.125	3.7	55H
ZnGa$_2$Se$_4$	S$_4^2$–I$\bar{4}$ or D$_{2d}^{11}$–I$\bar{4}$2m	5.48	10.9	2.00	0.25	0.25	0.125	5.13	55H
ZnGa$_2$Te$_4$	S$_4^2$–I$\bar{4}$ or D$_{2d}^{11}$–I$\bar{4}$2m	5.92	11.8	2.00	0.25	0.25	0.125	5.57	55H
ZnIn$_2$S$_4$	see below								
ZnIn$_2$Se$_4$	S$_4^2$–I$\bar{4}$	5.69	11.4	2.00	0.26	0.22	0.13	5.36	55H
ZnIn$_2$Te$_4$	S$_4^2$–I$\bar{4}$	6.11	12.2	2.00	0.25	0.25	0.125		55H
CdAl$_2$S$_4$	S$_4^2$–I$\bar{4}$	5.55	10.3	1.82	0.26	0.25	0.13	3.04	55H
CdAl$_2$Se$_4$	S$_4^2$–I$\bar{4}$	5.73	10.6	1.85	0.27	0.27	0.135	4.50	55H
CdAl$_2$Te$_4$	S$_4^2$–I$\bar{4}$ or D$_{2d}^{11}$–I$\bar{4}$2m	5.99	12.1	2.03	0.25	0.25	0.125	5.10	55H
CdGa$_2$S$_4$	S$_4^2$–I$\bar{4}$	5.56	10.0	1.80	0.27	0.26	0.14	3.97	55H
CdGa$_2$Se$_4$	S$_4^2$–I$\bar{4}$	5.73	10.7	1.87	0.25	0.26	0.13	6.28	55H
CdGa$_2$Te$_4$	S$_4^2$–I$\bar{4}$	6.08	11.7	1.93	0.27	0.26	0.135	5.63	55H
CdIn$_2$S$_4$	see below								
CdIn$_2$Se$_4$	see below								
CdIn$_2$Te$_4$	S$_4^2$–I$\bar{4}$	6.19	12.3	2.00	0.26	0.24	0.13	5.88	55H
CdTl$_2$Se$_4$	hexagonal	4.28	6.67	–	–	–	–	–	76S1
HgAl$_2$S$_4$	S$_4^2$–I$\bar{4}$	5.47	10.2	1.87	0.28	0.26	0.137	4.08	55H
HgAl$_2$Se$_4$	S$_4^2$–I$\bar{4}$	5.69	10.7	1.88	0.275	0.265	0.138	5.02	55H
HgAl$_2$Te$_4$	S$_4^2$–I$\bar{4}$ or D$_{2d}^{11}$–I$\bar{4}$2m	5.99	12.0	2.01	0.25	0.25	0.125	5.79	55H
HgGa$_2$S$_4$	S$_4^2$–I$\bar{4}$	5.49	10.2	1.86	0.275	0.265	0.139	4.95	55H
HgGa$_2$Se$_4$	S$_4^2$–I$\bar{4}$	5.70	10.7	1.88	0.25	0.25	0.125	6.10	55H
HgGa$_2$Te$_4$	S$_4^2$–I$\bar{4}$ or D$_{2d}^{11}$–I$\bar{4}$2m	6.00	12.0	2.00	0.25	0.25	0.125		55H
								6.42	60W
HgIn$_2$S$_4$	O$_h^7$–Fd3m	10.834	–	–	–	–	–	5.79	50H
HgIn$_2$Se$_4$	S$_4^2$–I$\bar{4}$	5.75	11.7	2.04	0.27	0.23	0.137	6.26	55H
HgIn$_2$Te$_4$	S$_4^2$–I$\bar{4}$	6.17	12.3	2.00	0.27	0.23	0.135	6.34	55H
	D$_{2d}^{11}$–I$\bar{4}$2m	6.17	12.3	2.00	–	–	–	–	76M2

pecularities for ZnIn$_2$S$_4$, CdIn$_2$S$_4$ and CdIn$_2$Se$_4$:

a) ZnIn$_2$S$_4$:

ZnIn$_2$S$_4$ has a large number of polytypes. All polytypes however are comprised of hexagonal layers with a = 3.85 (2) Å (the differences being in the packing of the layers) and c = $N \cdot$ 3.086 (3) Å where $N = 4Z$ and $Z = 1, 2, 3 \cdots$ is the number of formula units in the unit cell.

The reported polytypes are [70D]:

Polytype	Z	a [Å]	c [Å]
$ZnIn_2S_4$ (I)	1	3.85	12.34
$ZnIn_2S_4$ (II) a	2	3.85	24.68
$ZnIn_2S_4$ (II) b	2	3.85	24.68
$ZnIn_2S_4$ (III) a	3	3.85	37.02
$ZnIn_2S_4$ (III) b	3	3.85	37.02
$ZnIn_2S_4$ (III) c	3	3.85	37.02
$ZnIn_2S_4$ (IV)	4	3.85	49.36
$ZnIn_2S_4$ (V)	5	3.85	61.70
$ZnIn_2S_4$ (VI) a	6	3.85	74.04
$ZnIn_2S_4$ (VI) b	6	3.85	74.04
$ZnIn_2S_4$ (VI) c	6	3.85	74.04
$ZnIn_2S_4$ (XII) a	12	3.85	148.08
$ZnIn_2S_4$ (XII) b	12	3.85	148.08
$ZnIn_2S_4$ (XIV)	14	3.85	172.76
$ZnIn_2S_4$ (XXIV) a	24	3.85	296.16
$ZnIn_2S_4$ (XXIV) b	24	3.85	296.16

(In the literature polytype notation α, β, and γ is also used with α: space group C_{3v}^5, $a=3.85$ Å, $c=37.06$ Å; β: space group C_{3v}^1, $a=3.85$ Å, $c=12.34$ Å; γ: space group D_{3d}^3, $a=3.85$ Å, $c=24.68$ Å; see e.g. [79A2].)

$ZnIn_2S_4$ (I): space group C_{3v}^1—P3m1, lattice parameters (at 300 K): $a=3.85$ (2) Å, $c=12.34$ Å [70D].

atomic coordinates [70D]:

Atom	Position	z
In_0	$(\frac{2}{3} \frac{1}{3} z)$	0.375
In_t	$(\frac{1}{3} \frac{1}{3} z)$	0.062
Zn	$(0\ 0\ z)$	0.685
S_1	$(0\ 0\ z)$	0.000
S_2	$(\frac{1}{3} \frac{2}{3} z)$	0.262
S_3	$(0\ 0\ z)$	0.488
S_4	$(\frac{1}{3} \frac{2}{3} z)$	0.747

The S atoms are therefore arranged in an ABAB pattern. In_t and Zn in tetrahedral positions and In_0 in octahedral positions between S_2 and S_3. This gives a layer structure as follows: – $S_1In_tS_2In_0S_3ZnS_4$ –.

$ZnIn_2S_4$ (II) a: space group D_{3d}^3—P$\bar{3}$m1, lattice parameters (at 300 K): $a=3.85$ (2) Å, $c=24.68$ (4) Å [71D1]. S atoms arranged in ABABABAB pattern.

atomic coordinates [71D1]:

Atom	Position	$2z$
In_o	$(\frac{2}{3} \frac{1}{3} z)$	0.375 (10)
In_t	$(\frac{1}{3} \frac{2}{3} z)$	0.062 (2)
Zn	$(0\ 0\ z)$	0.686
S_1	$(0\ 0\ z)$	0.0
S_2	$(\frac{1}{3} \frac{2}{3} z)$	0.260 (16)
S_3	$(0\ 0\ z)$	0.488 (10)
S_4	$(\frac{1}{3} \frac{2}{3} z)$	0.748 (10)

This gives the larger structure: – $S_1In_tS_2In_0S_3ZnS_4$ $S_4'Zn'S_3'In_0'S_2'In_t'S_1'$ – where the S' etc. are related to S by inversion about an octahedral position between S_4–S_4' or S_1–S_1'.

ZnIn$_2$S$_4$ (II) b: space group C_{6v}^4—P6$_3$mc, lattice parameters (at 300 K): $a = 3.85$ (2) Å, $c = 24.68$ (4) Å [72D].
S atoms arranged in ABAC–ACAB pattern.

atomic coordinates [72D]:

Atom	Position	$2z$
In$_0$	$(\frac{2}{3}\,\frac{1}{3}\,z)$	0.376
In$_t$	$(\frac{1}{3}\,\frac{2}{3}\,z)$	0.064
Zn	$(0\,0\,z)$	0.688
S$_1$	$(0\,0\,z)$	0.0
S$_2$	$(\frac{1}{3}\,\frac{2}{3}\,z)$	0.264
S$_3$	$(0\,0\,z)$	0.488
S$_4$	$(\frac{2}{3}\,\frac{1}{3}\,z)$	0.752

layer structure: $\square$-S$_1$In$_t$S$_2$In$_0$S$_3$ZnS$_4$ – twice/unit cell.

ZnIn$_2$S$_4$ (III) a: space group C_{3v}^5—R3m, lattice parameters (at 300 K): $a = 3.85$ (2) Å, $c = 37.02$ (4) Å [71D1];
density $d = 4.38$ g/cm^3 [62L], $= 4.42$ g/cm^3 [69R].
S atoms arranged in -ABCA-CABC-BCAB- pattern.

atomic coordinates [71D1]:

Atom	Position	$3z$
In$_0$	$(0\,0\,z)$	0.378 (15)
In$_t$	$(\frac{1}{3}\,\frac{2}{3}\,z)$	0.068 (21)
Zn	$(\frac{2}{3}\,\frac{1}{3}\,z)$	0.691 (27)
S$_1$	$(0\,0\,z)$	0.0
S$_2$	$(\frac{1}{3}\,\frac{2}{3}\,z)$	0.266 (9)
S$_3$	$(\frac{2}{3}\,\frac{1}{3}\,z)$	0.490 (18)
S$_4$	$(0\,0\,z)$	0.756 (24)

layer structure: $\square$-S$_2$In$_t$S$_2$In$_0$S$_3$ZnS$_4$ – three times/unit cell; diagram, see Fig. 1.

spinel modification: lattice parameter: $a = 10.59$ Å, density $d = 4.74$ g/cm^3. Prepared at 40 kbar and 670 K and quenched to room temperature [69R].

b) CdIn$_2$S$_4$:

Spinel type, space group T_d^2—F$\bar{4}$3m; lattice parameters (at 300 K): $a = 10.797$ Å [55H]; $a = 10.818$ [76G3]; density $d = 4.93$ g/cm^3 [50H], $= 5.0$ g/cm^3 [70C].

There is some confusion regarding the exact type of spinel structure. Whether it is normal or inverse spinel, In on tetrahedral sites and Cd and In randomly distributed on the octahedral sites, or only partially inverted is in doubt [75S]. There is evidence of a structural phase transition, possibly an ordering of the cations at $T = 403$ K [70C].

c) CdIn$_2$Se$_4$:

$\square$-CdIn$_2$Se$_4$ has an ordered vacancy structure which is unique to this compound (see Fig. 2).
α-phase: space group D_{2d}^1—P$\bar{4}$2m, lattice parameters (at 300 K): $a = 5.81$ Å, $c = a$ [55H, 79P3, 80M2]; density $d = 5.54$ g/cm^3 [55H].
This phase is sometimes referred to as pseudocubic since $c = a$ although the space group is tetragonal.
β-phase: space group S_4^2—I$\bar{4}$, lattice parameters (at 300 K): $a = 5.81$ Å, $c = 2a$ [79P3, 80M2].
γ-phase: space group S_4^2—I$\bar{4}$, lattice parameters (at 300 K): $a = 5.81$ Å, $c = 4a$ [79P3, 80M2].
The β and γ phases have not been identified in detail.

solid solutions:

ZnAl$_2$S$_4$—ZnIn$_2$S$_4$: phase diagram, see Fig. 3.
CdGa$_2$Se$_4$—CdIn$_2$Se$_4$: phase diagram, see Fig. 4.

other phases:

CdGa$_2$Te$_4$: disordered zincblende, $T > 590$ K, $a = 6.09$ Å [77P]; amorphous: crystallizes at $T = 560$ K [77P]; body centred tetragonal reported, $a = 8.56$ Å, $c = 48.24$ Å; high T; no more details [77P].

CdIn$_2$S$_4$: phase transition at $T = 403$ K; ordering of cations? [70C].

Physical property	Numerical value	Experimental conditions	Experimental method remarks	Ref.

10.1.6.1 □-ZnGa₂–VI₄ compounds

□-ZnGa₂S₄:

There are no experimental determinations of the electronic band gap of □-ZnGa₂S₄. A pseudopotential calculation predicts a gap of 3.6 eV [79P1].

wavenumbers of Raman active phonons (in cm^{-1}):

		Symmetry		
$\bar{v}$	230	A	Raman spectroscopy (see Fig. 4 α)	83L1
	320			
	367			
	108	E (TO)		
	137			
	260			
	357			
	371			
	392	E (LO)		
	170	B (TO)		
	278			
	380			
	399	B (LO)		

For respective results on ZnGa₂Se₄, see [83L2].

photoconductivity data:

ϱ_d	$4 \cdot 10^{11}\,\Omega$ cm	$T = 300$ K	dark resistivity	69B1
ϱ_i	$1.5 \cdot 10^{10}\,\Omega$ cm	300 K	under illumination of 10^{14} photons/s at $\lambda = 0.39\,\mu$m	
E	3.8 eV		maximum response	61B

N.B. No anisotropy reported.

smoothed values of the heat capacity of several ZnGa₂VI₄ compounds (in J/mol K) [75M]:

	ZnGa₂S₄	ZnGa₂Se₄	ZnGa₂Te₄	T [K]	ZnGa₂S₄	ZnGa₂Se₄	ZnGa₂Te₄	T [K]
C_p	2.92	4,98	8.62	13	100.5	128.4	142.2	120
	4.93	6.50	15.48	15	106.6	133.9	146.2	130
	8.78	12.97	26.78	20	113.0	138.5	149.7	140
	13.85	22.02	36.27	25	119.0	142.7	152.8	150
	21.38	29.82	45.14	30	124.2	146.9	155.4	160
	28.70	36.89	52.76	35	128.6	149.8	157.8	170
	35.06	43.93	61.92	40	133.0	153.0	159.8	180
	40.63	51.04	68.99	45	136.8	155.6	161.6	190
	45.40	57.32	75.94	50	139.0	157.7	163.1	200
	49.79	68.60	82.59	55	143.3	160.1	164.5	210
	53.97	69.87	88.41	60	145.3	161.0	165.8	220
	57.03	76.57	94.77	65	148.7	163.2	157.9	230
	62.55	82.05	101.3	70	151.7	164.1	169.7	240
	66.94	88.94	108.0	75	154.1	164.8	170.4	250
	69.96	94.34	113.3	80	156.3	165.5	171.2	260
	74.01	99.36	118.2	85	158.4	165.8	172.5	270
	77.95	104.3	122.8	90	160.5	166.4	173.5	280
	81.84	108.7	126.9	95	162.5	166.5	174.8	290
	85.69	113.0	130.7	100	164.6	166.6	174.8	300
	93.22	121.5	137.4	110				

Physical property	Numerical value	Experimental conditions	Experimental method remarks	Ref.
melting point:				
T_m	> 1620 K			71C
$\square$-ZnGa$_2$Se$_4$:				
heat capacity: see table under $\square$-ZnGa$_2$S$_4$.				
magnetic susceptibility:				
χ_m	$-151\,(2) \cdot 10^{-6}$ cm^3/mol	$T = 300$ K	χ_m in CGS-emu; powder sample	71M
$\square$-ZnGa$_2$Te$_4$:				
heat capacity: see table under $\square$-ZnGa$_2$S$_4$.				
magnetic susceptibility:				
χ_m	$-184\,(3) \cdot 10^{-6}$ cm^3/mol	$T = 300$ K	χ_m in CGS-emu; powder sample	71M

10.1.6.2 Zinc thioindiate ($\square$-ZnIn$_2$S$_4$)

All data are for the layered structure unless otherwise stated. Usually the polytype is ZnIn$_2$S$_4$ (III) a, but it is not always given.

Electronic properties, impurities and defects

band structure: Fig. 5

energy gap:				
E_g	2.96 eV	$T = 77$ K, $E \perp c$	reflectivity	79A1
	2.87 (1) eV	300 K; $E \perp c$	absorption	75R1

higher interband transition energies (in eV):

(A, B, C are the three uppermost valence bands, 1, 2 the two lowest conduction bands of Fig. 5.)

$E_0\,(\Gamma_{3v} - \Gamma_{1c}) = E_g$		reflectivity at 77 and 300 K ($E \perp c$)	79A1
$E_0'\,(\Gamma_{3v} - \Gamma_{1c})$	3.56		
$E_1\,(Z_A - Z_1)$	4		
$E_1'\,(Z_B - Z_1)$	4.7		
$E_2\,(A-1)$	5.1		
$E_3\,(B-1)$	5.7		
$E_4\,(B-2)$	7.4		
$E_5\,(C-1)$	8.6		

photoemission peak energies (in eV):

(critical points in density of states, all energies measured from valence band maximum, RT values)

E_{vb}	-18.61 (3)	In 4d$_{3/2}$	80M1
	-17.77 (3)	In 4d$_{5/2}$	
	-11.4 (2)		
	-8.55 (5)	Zn 3d	
	-5.90 (5)		
	-4.30 (5)		
	-3.10 (5)		
	-1.6 (1)		
	-0.90 (5)		78C
E_{cb}	9.1 (2)	constant initial state	80M1
	11.9 (2)	spectroscopy	
	12.9 (2)		
	15.1 (4)		
	17.3 (2)		

Physical property	Numerical value	Experimental conditions	Experimental method remarks	Ref.

trap states in the energy gap:

ZnIn$_2$S$_4$ displays a long exponential tail in the optical absorption (Fig. 6). This together with luminescence intensity, phosphorescence decay, thermoluminescence glow [75G] and photoconductivity relaxation time measurements gives an exponential distribution of traps (Fig. 7) $D(E) = Ae^{-\alpha E}$ where α is less than 36 eV^{-1} [75G].

A level at 1.77 eV above the valence band is responsible for the capture of hot electrons and gives rise to n-type negative resistance [72R1].

spinel modification:

energy gap:

E_g	2.4 eV	$T = 80$ K	photoluminescence and resonant Raman scattering	78U2

photoluminescence peak energies:

E	1.52 eV	$T = 80$ K	tetrahedral cation vacancies	78U2
	1.29 eV	80 K	In on tetrahedral sites	

Lattice properties

For structure, polytypes, lattice parameters etc., see section 10.1.6.0.

frequencies of IR and Raman active phonons (in 10^{12} s^{-1}, at 300 K):

ZnIn$_2$S$_4$ (III) a:

	Infrared transmission [77H]	Raman scattering [72B, 76A]	Symmetry [76A]
ν	20.55		
	19.41		
	16.23		
	13.5		
	11.52		
	11.19	11.19 (9)	A$_1$
	10.83	10.84 (10)	A$_1$
	10.5	10.42 (76)	
	10.17		
	9.96		
	9.66	9.57 (2)	A$_1$
	9.33		
	8.97	9.12 (2)	
	8.76		
	8.19	8.22 (2)	
	8.01		
	7.8		
	7.62	7.59 (6)	A$_1$
	6.99	6.96 (6)	
	6.48		
	5.76	5.7 (6)	
	2.49	3.09 (6)	A$_1$
	1.23	2.22 (6)	A$_1$

Physical property	Numerical value	Experimental conditions	Experimental method remarks	Ref.

ZnIn$_2$S$_4$ (spinel modification) [78U1]:

	Infrared transmission $T = 300$ K	Raman scattering $T = 80$ K	$T = 300$ K	Symmetry	
v		2.22	2.16	F_{2g}	
		5.61	5.52	E_g	
	6.75	6.72	6.63	F_{1u}	
		7.68	7.59	F_{2g}	
	8.1	8.25	8.16	F_{1u}	
	9.3	9.45	9.36	F_{1u}	
	10.35	10.35	10.26	F_{1u}	
		11.28	11.16	A_{1g}	
		18.6		F_{1u}	

Transport properties

conductivity and Hall mobility: Fig. 8

activation energies:

E_A	0.045 eV	$T = 100 \cdots 300$ K	conductivity	76G1
	0.02 eV	$100 \cdots 300$ K	Hall effect	

The ac conductivity shows a dependence $\sigma \propto \omega^n$ where n is a power between $0.6 \cdots 0.8$ [76G1]. Various S and N type negative resistance effects have been reported in [79Z, 70M, 77K1].

Optical properties (see also under "Electronic properties")

refractive index: Fig. 9

dielectric constant:

$\varepsilon(\infty)$	6.0	$T = 300$ K, $E \perp c$	infrared	76A

Further properties

magnetic susceptibility:

χ_m	$-144\,(1) \cdot 10^{-6}$ cm^3/mol	$T = 300$ K	χ_m in CGS-emu; powder sample	71M

10.1.6.3 Zinc indium selenide ($\square$-ZnIn$_2$Se$_4$)

energy gap:

$E_{g,dir}$	2.05 eV	$T = 85$ K	absorption, reflectivity (unpolarized)	73M1
	2.0 eV	300 K	photoconductivity	
dE_g/dT	$-3.1 \cdot 10^{-9}$ eV K^{-1}	$88 \cdots 300$ K	photoconductivity (unpolarized)	74M

at 4.2 K and 50 kG no Landau levels were observed [74F2].

photoconductivity peak energy:

E	2.35 eV	$T = 300$ K		74F2

defect levels:

E	1.29 eV	$T = 77$ K	luminescence band	76G2
dE/dT	0.9 meV K^{-1}			
E	2.4 eV	$55 \cdots 200$ K	luminescence excitation	
	2.05 eV	< 170 K	luminescence excitation (band gap)	

There is an exponential distribution of trap states at the conduction band edge which reaches to about 0.4 eV below the edge [76M1].

Physical property	Numerical value	Experimental conditions	Experimental method remarks	Ref.
resistivities, Seebeck coefficient:				
ϱ_d	$4.0\,(5)\cdot10^8\,\Omega\,cm$	$T=300\,K$	n-type; dark resistivity	74M
S	$-7.5\,(7)\cdot10^{-4}\,V\,K^{-1}$	300 K		
ϱ_i	$1.8\cdot10^4\,\Omega\,cm$	300 K	under illumination of 10^{14} photons/s at $\lambda=0.58\,\mu m$	69B1

N.B. No anisotropy reported.

magnetic susceptibility:				
χ_m	$-193\,(1)\cdot10^{-6}\,cm^3/mol$	$T=300\,K$	χ_m in CGS-emu; powder sample	71M

10.1.6.4 Zinc indium telluride (□-ZnIn$_2$Te$_4$)

energy gaps (in eV):

		$T\,[K]$		
$E_{g,ind}$	1.41	85	unpolarized reflection	73M1,
	1.35	300		74M
$E_{g,dir}$	1.90	85		
	1.87	300		
dE_g/dT	$-6\cdot10^{-4}\,eV\,K^{-1}$			

defect levels:

There is an exponential distribution of traps down to 0.4 eV below the conduction band.

E_d	0.01 eV	$T=85,\,300\,K$	photoconductivity	73M1

resistivities, Seebeck coefficient:

		$T\,[K]$		
ϱ_d	$5.0\,(5)\cdot10^6\,\Omega\,cm$	300	n-type	74M
	$6\cdot10^8\,\Omega\,cm$	300	p-type	69B2
S	$-4.5\,(7)\cdot10^{-4}\,V\,K^{-1}$	300		74M
ϱ_i	$1.7\cdot10^4\,\Omega\,cm$	300	under illumination of 10^{14} photons/s at $\lambda=0.58\,\mu m$	69B1

For photoconductivity, see also [69B2].
N.B.: No anisotropy reported.

magnetic susceptibility:				
χ_m	$-238\,(3)\cdot10^{-6}\,cm^3/mol$	$T=300\,K$	χ_m in CGS-emu; powder sample	71M

10.1.6.5 Cadmium thiogallate (□-CdGa$_2$S$_4$)

Electronic properties

There is some confusion about the nature of the energy gap. At room temperature absorption studies give a direct gap of 3.44 eV [69B1] or 3.25 eV [77K3] or an indirect gap of 3.05 eV [77R]. Reflectivity gives a first peak at 3.58 eV [77G, 72A1, 73A1] and photoconductivity at 3.35 eV [77K3] or 3.65 eV [75R2]. Absorption in thin films gives 3.50⋯3.63 eV [71K2]. On the other hand the thermal gap is 2.92 eV [75R2].

reflectivity peak energies (in eV):

		$T\,[K]$		
E	3.735	4.2; $E\parallel c$	sharp exciton-like peaks	79A4
	3.721	77; $E\parallel c$		
	3.747	4.2; $E\perp c$		
	3.733	77; $E\perp c$		

Physical property	Numerical value	Experimental conditions	Experimental method remarks	Ref.
photoconductivity peak energies (unpolarized radiation):				
E	3.65 eV	$T = 300$ K	shifts to lower energy at higher electric fields	75R2
	3.80 eV	300 K		
	2.9...3.15 eV	300 K	see "Impurities and defects"	80R

Impurities and defects
defect levels (in eV):

		T [K]		
E	1.508	500...600	activation energy of conductivity	77K3
	2.9...3.15	300	maximum in photoconductance; disappears at high electric fields	80R
	1.15 (8)	293...500	activation energies of photoconductance	75R2
	0.40 (6)	293...500		
	2.7 (1)	4.2...300	maximum of photoluminescence	78K1
E_d	0.55 (1)	4.2...300	photoluminescence and photoconductivity	78K1
	0.65 (1)	4.2...300		
E_a	0.60	4.2...300	thermal activation energy, photoluminescence	78K1

concentration of native defects: 10^{21} cm^{-3} [78K1].
Ag doping (0.16 (2) at % Ag): trap at 0.65 eV vanishes [78K1].
In doping (3.4 (2) at % In): new levels at 0.17 eV, 0.48 eV, no thermally stimulated conductivity [78K1].

Lattice properties

For structure; lattice parameters etc., see section 10.1.6.0.

wavenumbers of Raman and infrared active phonons (in cm^{-1}, at 300 K):

	Raman scattering [78S1, 83L1]		IR transmission [77S2] and reflectivity [79K1]	Symmetry
	TO	LO		
$\bar{\nu}$	371	389...394	364...368	B (Γ_{15})
	364, 365	382...387	364	E (Γ_{15})
		359...361	–	A (W$_1$)
		349 ([78S1] only)	326	B (W$_2$)
	320...323	354 ([78S1] only)	332, 333	E (W$_4$)
		308...311	–	A (W$_1$)
	291 ([83L1] only)			B
	253...257	263 ([78S1] only)	256...262	B (W$_2$)
		218, 219	–	A (X$_1$)
	–		242 ([79K1] only)	B
	237...241	248 ([78S1] only	240...244	E (X$_5$)
	161...165	168 ([78S1] only)	166...168	B (X$_3$)
	136	141 ([78S1] only)	136 ([77S2] only)	E (W$_3$)
	84...86	89 ([78S1] only)	82...88	E (W$_4$)

For Raman spectra, see Fig. 4α. For the effect of ordered vacancies on the vibrational spectra, see also [83N].

dielectric constants:

$\varepsilon (0)$	9.6	$E \parallel c$; $T = 300$ K		79K1
	12.3	$E \perp c$		
$\varepsilon (\infty)$	6.2	$E \parallel c$;	300 K	
	8.3	$E \perp c$		

Considerable structure in the reflectivity above $\bar{\nu} = 400$ cm^{-1} is due to two-phonon processes [77R].

Physical property	Numerical value	Experimental conditions	Experimental method remarks	Ref.
Transport properties				
conductance and thermoelectric power: Fig. 10				
resistivity, mobility:				
ϱ	$0.8 \cdots 6 \cdot 10^{13}\,\Omega\,\text{cm}$	$T = 293\,\text{K}$	n-type	75R2
μ_n	$10\left(\dfrac{T}{120}\right)^{1.35}\,\text{cm}^2\,\text{V}^{-1}\,\text{s}^{-1}$	$120 \cdots 300\,\text{K}$	decreases in the presence of thermally stimulated current	76K

N.B. No anisotropy reported.

Ag and In doping cause little change in conductivity, but see "Impurities and defects" for photoconductivity and luminescence [78K1].

Optical and further properties
See also under "Electronic properties".

refractive index:				
n_o	2.3	$\lambda = 500\,\text{nm}$		71K3
linear electrooptical coefficient:				
r_{13}	$0.37 \cdot 10^{-12}\,\text{m/V}$	$\lambda = 500\,\text{nm}$		71K3
r_{63}	$3.5 \cdot 10^{-12}\,\text{m/V}$			
nonlinear dielectric susceptibility:				
d_{36}	$40.2 \cdot 10^{-12}\,\text{m/V}$	$\lambda = 1.064\,\mu\text{m}$		74L

heat capacity and Debye temperature [72M]:

C_p J/mol K	Θ_D K	T K	C_p J/mol K	Θ_D K	T K
2.720	173	10	107.1	414	120
5.456	180	15	112.9	430	130
12.31	202	20	119.5	434	140
22.26	201	25	125.4	443	150
33.39	210	30	130.5	450	160
41.18	231	35	134.6	463	170
45.27	244	40	137.9	483	180
48.53	264	45	140.5	494	190
51.04	285	50	142.8	510	200
53.05	306	55	145.3	526	210
54.31	330	60	147.8	550	220
60.17	342	65	150.2	571	230
64.02	350	70	152.5	578	240
68.03	367	75	154.9	600	250
72.05	373	80	156.2	625	260
76.23	384	85	156.6	635	270
80.50	387	90	159.0	660	280
84.85	390	95	160.5	682	290
89.33	394	100	162.0	716	300
98.32	396	110			

Physical property	Numerical value	Experimental conditions	Experimental method remarks	Ref.

10.1.6.6 Cadmium gallium selenide (□-CdGa$_2$Se$_4$)

Electronic properties

The size and nature of the electronic energy gap is disputed, ranging from an indirect gap of 1.97 eV [74R] to a direct gap of 2.57 eV [79B2]. The higher value has been observed in reflection [73A1, 77G], absorption and thermoreflectance [79B2] and photoconductivity [77K3]. This confusion is common to all such compounds. In the case of □-CdGa$_2$Se$_4$, however, the lower values may well be due to the effects of impurities or disorder. There is some consistency in the results of different groups who find the higher values [73A1, 77G, 77K2, 79B2].

energy gap (in eV):

$E_{g, dir}$	2.57	$T = 300$ K; $E \perp c$	absorption and thermoreflectance (spectra are of similar	79B2
	2.63	$E \parallel (\perp) c$	structure to ternary	
	3.02	$E \parallel (\perp) c$	chalcopyrite semiconductors but the crystal field splitting is larger and positive [77M2])	
	2.27	$T = 300$ K; $E \perp (\parallel) c$	wavelength modulated reflection (see Fig. 10α)	81K
	2.43	$E \perp c$		
	2.56	$E \parallel (\perp) c$		
	2.68	$T = 90$ K; $E \perp c$		79B2
	2.75	$E \parallel (\perp) c$		
	3.15	$E \parallel (\perp) c$		
dE_g/dT	$5 \cdot 10^{-4}$ eV K^{-1}	$T = 77 \cdots 300$ K	photoconductivity	79K2

splitting energies (in eV):

		T [K]		
Δ_{so}	0.41	300		79B2
	0.42	90		
Δ_{cf}	0.10	300		
	0.13	90		

higher transition energies (in eV):

E	4.04	$T = 90$ K; $E \parallel (\perp) c$	absorption and thermoreflectance	79B2
	4.16	$E \parallel (\perp) c$		
	4.53	$E \perp (\parallel) c$		
	4.86	$E \perp c$		
	5.12	$E \parallel (\perp) c$		
	3.08	$T = 300$ K; $E \parallel (\perp) c$	wavelength modulated reflection	81K
	3.14	$E \parallel (\perp) c$		
	3.81	$E \perp (\parallel) c$		
	3.88	$E \perp (\parallel) c$		
	4.17	$E \perp (\parallel) c$		
	4.55	$E \parallel c$		
	4.99	$E \perp (\parallel) c$		
	5.12	$E \parallel (\perp) c$		
	5.21	$E \parallel c$		
	5.4	$E(\perp) \parallel c$		

Physical property	Numerical value	Experimental conditions	Experimental method remarks	Ref.
Impurities and defects				
defect levels (in eV):				
		T [K]		
E	2.25	300	peaks in photoconductivity,	71R
	2.43	300	unpolarized	
	2.81	300		
	2.94	300		
	2.09	300		
E_t	0.33	105···300	electron traps in photo- conductivity; thermally stimulated current	72A2
E_r	0.67 (9)	105···300	recombination centres	72A2
	0.77	105···300		
data on Au doped samples:				
E_t	0.26	293	thermally stimulated current	73A2
	1.53	105	photoconductivity, unpolarized	
	1.17			
	1.46	77	photoluminescence	
	1.14			
	0.05···0.3	293	quasi continuous trap dis- tribution; thermally stimulated current	78S2
E_r	0.51	105···300	thermally stimulated current	73A2
	0.49			
E_A	0.56 (1)	77···500	activation energy of impurity conduction	78S2

Lattice properties

For structure, lattice parameters etc., see section 10.1.6.0.

wavenumbers of infrared and Raman active phonons (in cm^{-1}, at 300 K):

	TO	LO	Symmetry		
$\bar{\nu}$	254···264	275···280	B	B, E are IR active,	79B3,
	250···262	274···278	E	all are Raman active	79K3,
	241···250	242···249	E	*) not mentioned in [82B, 83L1]	79M,
	220···222	232···237	B		82B,
		210 ([83L1] only)	A		83L1
	194···196	198···202	B		
		188	A		
	176*)	178*)	B		
	174···178	180···183	E		
		141	A		
	124···125	127···128	B		
	105	105	E		
	76	76	B		
	68	70	E		
		53*)	E		
		47*)	E		

For Raman spectra, see Fig. 4 α, For a discussion of the influence of ordered vacancies on the lattice vibration spectrum, see [82T].

Physical property	Numerical value		Experimental conditions	Experimental method remarks	Ref.
two phonon Raman mode:					
$\bar{\nu}_R$	277 cm^{-1}	2A	$T = 300$ K(?)		79M
dielectric constants:					
$\varepsilon(0)$	9.7		$T = 300$ K(?); $E \perp c$		79K3
	8.2		$E \parallel c$		
$\varepsilon(\infty)$	6.7		$T = 300$ K(?); $E \perp c$		
	6.2		$E \parallel c$		

Transport properties

For conductance and thermoelectric power, see Fig. 11.

resistivities (in Ω cm):

		T [K]		
ϱ	10^{13}	300	n-type	71R
ϱ_d/ϱ_i	3.10^4	293	illumination 10^3 lux; ϱ_d: dark	73A2
	2.10^6	200	resistivity; ϱ_i: under illumination	

data on Au doped samples:

ϱ	$4 \cdot 10^{10}$	293	n-type	73A2
ϱ_d/ϱ_i	$6 \cdot 10^5$	293	illumination 10^3 lux	
	$7 \cdot 10^7$	200		

mobility:

μ_n	17 cm^2 V^{-1} s^{-1}	$T = 293$ K	(under 10^3 lux)	73A2

N.B. No anisotropy reported.

Further properties

linear thermal expansion coefficient (in 10^{-6} K^{-1}):

α	18.7	$T = 300\cdots873$ K	polycrystal	78K2
	4.9	$300\cdots873$ K	c axis	
	14.0		a axis	

magnetic susceptibility:

χ_m	$-169\,(2) \cdot 10^{-6}$ cm^3/mol	$T = 300$ K	χ_m in CGS-emu; powder sample	71M

melting point:

T_m	1250 K		78K2

10.1.6.7 Cadmium gallium telluride ($\square$-CdGa$_2$Te$_4$)

energy gap:

$E_{g,\,dir}$	1.5 eV	$T = 300$ K	unpolarized; reflectivity ($\bar{4}$ or bct structure)	77P
	1.4 eV	300 K	(disordered zincblende)	

Physical property	Numerical value	Experimental conditions	Experimental method remarks	Ref.

10.1.6.8 Cadmium thioindiate (CdIn$_2$S$_4$)

(See also Landolt-Börnstein, NS, Vol. III/12b)

Electronic properties

band structure: Fig. 12

There is general agreement that the band gap is indirect and that the valence band maximum is not at Γ but probably along the Σ direction [110] [77B1]. The values of the indirect and direct gaps are however somewhat inconsistent.

energy gaps (in eV):

		T [K]		
$E_{g,ind}$	2.44	0	absorption	80N
	2.20	0	photovoltaic effect	79A3
	2.18	77		
	2.17	100		
	2.11	300		
	2.21	300	reflectivity	74F1
	2.28	300	absorption	80N
$E_{g,dir}$	2.75	0	absorption	80N
	2.71	0	photovoltaic effect	79A3
	2.65	77		
	2.63	100		
	2.47	300		
	2.5	300	reflectivity	74F1
	2.62	300	absorption	80N
$E_{g,th}$	2.2		resistivity vs. T	76E

structure at higher energies: Fig. 13

photoemission peak energies (in eV) (all energies relative to valence band maximum at 300 K):

E	1.6		S(3p)	78I
	2.15		Cd(5s) + In(5p)	80C
	3.9		Cd(5s) + In(5p)	78I
	5.9		In(5s)	
	10.0		Cd(4d)	
	11.8		S(3s)	

g-value:

g_c	1.616	T not given	stoichiometric	76E
	2.001		with excess sulfur	

effective masses:

m_n	0.19 m_0	$T = 70 \cdots 500$ K	Hall effect and Seebeck effect	76E
	0.17 m_0		Faraday effect	74F1
m_{ds}	0.3 m_0		Seebeck effect	72A3

Impurities and defects

Some authors report considerable fine structure at the absorption edge [73R1, 73R2], others a fairly sharp edge [78A]. This probably reflects the confusion about the structure, since a partially inverted spinel is equivalent to a normal or an inverted structure with a large density of antisite defects.

Physical property	Numerical value	Experimental conditions	Experimental method remarks	Ref.
a) native defects:				
defect levels (in eV):				
		T [K]		
E	1.65	4.2···300	luminescence maximum, In vacancy (tetrahedral)	71D2
	1.42···1.44	4.2···300	luminescence maximum, In vacancy (octahedral)	
	0.7	77	above valence band edge; unidentified	77E2
	1.2	77	above valence band edge; unidentified	
E_a	0.25	77, 300	probably Cd on In site, present only with excess S, photoconductivity	78A
E_d	0.2	70···500	probably S vacancies, Hall effect	76E
	0.5	70···500		

An exponential trap distribution between 0 and 0.1 eV under E_c with a density of $\approx 9 \cdot 10^{19}$ cm^{-3} is observed in photoconductivity. It is probably due to Cd-In antisite defects [78A].

photoluminescence peak energies:

E	1.46 eV	$T = 300$ K	A band, Gaussian	78G2
	1.62 eV	78 K	A band, Gaussian	
	1.83 eV	78 K	B band, Gaussian	

(For A, B bands, see Fig. 12.)

photoconductivity peak energies (in eV, at 293 K):

E	2.08		$E_{g, ind}$	72R2,
	2.27		$E_{g, dir}$	79A3
	2.5			
	2.7		See also [69B2].	

excitation spectrum energies:

E	2.2 eV	$T = 2···300$ K	A band, ionization of centre 0.3 eV from E_v; A, B bands: see Fig. 12	78G2, 80G
	2.35 eV		A, B bands, release of minority carriers	
	2.65 eV		A, B bands, interband transitions	

activation energies for quenching of luminescence emission:

ΔE_1	0.11 (4) eV	$T = 77···300$ K	A band; A, B bands: see Fig. 12	78G2
ΔE_2	0.32 (2) eV		A band	
ΔE	0.10 (5) eV		B band	

b) chromium doped samples:

defect levels:

E	0.7 eV	$T = 77$ K	above valence band, photoconductivity and photoluminescence	77E2
	1.2 eV	300 K	photoconductivity	72Y
	1.8 eV	300 K	photoconductivity	

There is considerable fine structure in the photoluminescence, probably due to scattering of phonons [77S1, 79U].

Physical property	Numerical value	Experimental conditions	Experimental method remarks	Ref.
c) cobalt doped samples:				
defect levels (in eV):				
		T [K]		
E	1.46	2	Co^{2+} on tetrahedral site: luminescence	79G
	1.14	2	Co^{2+} on octahedral site: luminescence	
	0.62	4.2···290	absorption	78U2
	1.61			

There is considerable fine structure [78U2].

d) iron doped samples:				
defect levels (in eV):				
E	0.265	$T = 6···300$ K	absorption; Fe^{2+} on tetrahedral site Jahn-Teller splitting	73W
	0.282			
	0.305			
	0.336			
	1.611		Fe^{2+} on tetrahedral site (but see Co, above)	
	0.992		Fe^{2+} on octahedral site	

e) manganese doped samples:

Mn^{2+} can be found on tetrahedral or octahedral sites, but preferentially on octahedral: EPR [71K1].

Lattice properties

For structure, lattice parameters etc, see section 10.1.6.0. For lattice properties, see also section 10.2.1, p. 162.

phonon wavenumbers (in cm^{-1}):
IR active phonons:

		T [K]		
$\bar{v}$	307	300	T_{1u}	75S, 73Y
	215			
	171			
	68			
	293	440		
	206			
	169			
	67			

Raman active phonons:

$\bar{v}$	366	300	A_{1g}	75S, 73Y
	312		T_{2g}	
	247		T_{2g} (very weak)	
	185		E_g	
	93		T_{2g}	
	374	76	A_{1g}	
	316		T_{2g}	
	192		E_g	
	96		T_{2g}	

N.B. Possible order-disorder transition at 403 K [70C].

Physical property	Numerical value	Experimental conditions	Experimental method remarks	Ref.
phonon energies (in meV):				
E	22.9	$T = 4.2$ K	E_g	74S
	31.6		T_{2g} from tunneling	
	37.9		T_{1u} spectroscopy	
	46.4		A_{1g}	

For phonon wavenumbers, see also [76W]. Infrared absorption at $340\,cm^{-1}$ due to Fe doping [74B, 78U1].

Resonant Raman scattering from mode at $184\,cm^{-1}$ [75K].

elastic moduli (in $10^{10}\,Nm^{-2}$):				
c_{11}	12.15 (10)	$T = 300$ K	Brillouin scattering	80Y
c_{12}	2.46 (46)			
c_{44}	2.57 (5)			
elastooptic constants:				
p_{11}	0.027 (5)	$T = 300$ K	Brillouin scattering	80Y
p_{12}	0.093 (9)			
p_{44}	-0.033 (2)			
dielectric constants:				
$\varepsilon(0)$	17		from effective charge	72A3
$\varepsilon(\infty)$	10		estimated from optical measurements; see also Fig. 13	

Transport properties

resistivity: Fig. 14; Hall coefficient: Fig. 15; Seebeck coefficient: Fig. 16; Nernst coefficient: Fig. 17. See also Figs. 6···9 of section 10.2.1, p. 383.

carrier concentrations, mobilities, Hall coefficient, conductivity, resistivity:

n-type samples:

		T [K]		
n	$1.3 \cdots 1.6 \cdot 10^{19}\,cm^{-3}$	125	stoichiometric samples	76E
	$2.7 \cdot 10^{13}$	125	samples with excess sulfur	76E
	$\cdots 4.8 \cdot 10^{17}\,cm^{-3}$			
	$7.2 \cdot 10^{15}\,cm^{-3}$	300		73A3
	$10^{13} \cdots 1.4 \cdot 10^{19}\,cm^{-3}$	300		72A3
μ_H	$320 \cdots 400\,cm^2\,V^{-1}\,s^{-1}$	125	stoichiometric samples	76E
	$25 \cdots 100\,cm^2\,V^{-1}\,s^{-1}$	125	samples with excess sulfur	76E
	$200\,cm^2\,V^{-1}\,s^{-1}$	300		74F1
	$35\,cm^2\,V^{-1}\,s^{-1}$	300		73A3
	$10 \cdots 250\,cm^2\,V^{-1}\,s^{-1}$	300		72A3
R_H	$-1021\,cm^3\,C^{-1}$	300		73A3
σ	$3.48 \cdot 10^{-2}\,\Omega^{-1}\,cm^{-1}$	300		73A3
ϱ	$5 \cdot 10^7\,\Omega\,cm$	300	n-type	69B2

magnetoresistance: At $T = 77$ K the magnetoresistance is negative and not proportional to H^2 [74I]. At $T = 300$ K, however, the magnetoresistance is positive with a normal H^2 dependence (Fig. 18) [76E, 73A3].

other transport properties: oscillations of the photoconductivity and non-linear $I-U$ characteristics together with a memory effect have been reported in [72A1, 74U].

Optical and further properties

refractive index:

n	2.55	$\lambda = ?$		74F1

For further optical properties, see other subsections and section 10.2.1, p. 162.
thermal conductivity: Fig. 19

Physical property	Numerical value	Experimental conditions	Experimental method remarks	Ref.
magnetic susceptibility:				
χ_m	$-140\,(2)\,cm^3/mol$	$T = 300\,K$	χ_m in CGS-emu; powder sample	71M

heat capacity and Debye temperature:

C_p J/mol K	Θ_D K	T K	C_p J/mol K	Θ_D K	T K	
2.594	181	10	106.8	371	110	72M
6.615	192	15	113.3	392	120	
12.64	205	20	119.2	403	130	
19.38	216	25	124.8	415	140	
26.24	228	30	130.1	422	150	
36.18	237	35	134.8	433	160	
46.02	240	40	139.1	442	170	
54.06	258	45	143.2	454	180	
59.91	270	50	146.8	466	190	
64.85	280	55	150.0	480	200	
68.78	288	60	153.6	486	210	
73.39	300	65	156.3	503	220	
77.57	308	70	158.6	512	230	
81.59	315	75	160.5	532	240	
85.52	320	80	162.3	551	250	
89.33	334	85	164.0	574	260	
92.88	342	90	165.5	596	270	
96.48	349	95	166.3	622	280	
99.99	354	100	166.7	645	290	
			166.8	690	300	

melting point

T_m	1378 K			70C

10.1.6.9　Cadmium indium selenide (□-CdIn$_2$Se$_4$)

α-phase:

energy gaps (in eV):

		T [K]		
$E_{g, ind}$	1.51 (3)	300	absorption, unpolarized	72K2, 77T
	1.61 (1)	83	photoconductivity, unpolarized	77T
	1.6		calculated	76M1, 77B2
$E_{g, dir}$	1.65···1.73	300	absorption, photoconductivity, thermoreflectance, unpolarized	72K2, 77T
	1.77···1.84	83		77T
	1.8		calculated	76M1, 77B2

Conduction band minimum at Γ, valence band maximum disputed.

higher interband transition energies (in eV):

E	1.95	$T = 300\,K$	reflectivity, unpolarized	77K2
	2.50			
	3.55			
	4.15			

Physical property	Numerical value	Experimental conditions	Experimental method remarks	Ref.
photoemission peak energies (in eV) (critical points in valence band density of states, at 300 K):				
E	1.05 (1)		below E_V	78P
	1.75 (1)			
	6.05 (1)			
	10.1 (1)		Cd 4d states	
effective mass:				
m_n	0.15 (1) m_0	$T = 300$ K	thermoelectric power	72K1
β-phase:				
energy gaps:				
$E_{g,\,ind}$	1.30 eV	$T = 300$ K	absorption, unpolarized	72K2
$E_{g,\,dir}$	1.49 eV	300 K	absorption, unpolarized	72K2
higher interband transition energies (in eV):				
E	1.85	$T = 300$ K	reflectivity, unpolarized	72K2
	2.75			
	3.75			
	4.25			
carrier concentration, Hall coefficient, Seebeck coefficient, resistivity (at 300 K, phase unclear):				
n-type samples:				
n	10^{17} cm^{-3}		intrinsic	72K3
R_H	-0.5 cm^3 C^{-1}		not very T-dependent	72K1
S	-100 µV K^{-1}			
ϱ	$8 \cdot 10^5$ Ω cm			69B1
p-type samples:				
S	70 µV K^{-1}			69B1
N.B. No anisotropy reported.				

10.1.6.10 Cadmium indium telluride ($\square$-CdIn$_2$Te$_4$)

energy gaps (in eV):				
$E_{g,\,ind}$	1.15	$T = 293$ K	absorption, unpolarized	72K3
	1.1		calculated	76M1
$E_{g,\,dir}$	1.25	$T = 293$ K	absorption, unpolarized	72K3
	1.2		calculated	76M1
conductivity, Seebeck coefficient, mobility, carrier concentration:				
n-type samples:		T [K]		
σ	10^{-7} Ω^{-1} cm^{-1}	293	no anisotropy reported	73P
S	-300 µV K^{-1}	293		
μ_H	72 cm^2 V^{-1} s^{-1}	394		
n	10^{14} cm^{-3}	300		69B1
μ_n	4000 cm^2 V^{-1} s^{-1}	300		
photoconductivity peak energy:				
E	1.2 eV	$T = 293$ K	n-type sample	73P
magnetic susceptibility:				
χ_m	-258 (3) $\cdot 10^{-6}$ cm^3/mol	$T = 300$ K	χ_m in CGS-emu; powder sample	71M

Physical property	Numerical value	Experimental conditions	Experimental method remarks	Ref.

10.1.6.11 Cadmium thallium selenide (CdTl$_2$Se$_4$)

energy gap:

E_g	0.8 eV		calculated	76S1

defects:

can be doped with Cu, Ag, Au [79S].

conductivity:

σ	$10^{-1} \cdots 10^{-5} \, \Omega^{-1} \, cm^{-1}$	$T = 77$ K	no anisotropy	76S1

melting point:

T_m	1200 (50) K			76S1

10.1.6.12 Mercury thioaluminate (□-HgAl$_2$S$_4$)

Electronic properties

There are no reported experiments on the electronic properties of HgAl$_2$S$_4$. A band structure calculation predicts a direct gap of 2.54 eV and a crystal field splitting of 0.14 eV [80P1].

10.1.6.13 Mercury thiogallate (□-HgGa$_2$S$_4$)

energy gaps:

$E_{g,\,ind}$	2.79 eV	$T = 300$ K	these are alternative inter-	61B
$E_{g,\,dir}$	2.84 eV	300 K	pretations of the same data (unpolarized optical measurements)	61B
	2.53 eV		calculated	79P2

resistivities:

ϱ_d	$10^{10} \, \Omega$ cm	$T = 300$ K	dark resistivity	69B1
ϱ_i	$7 \cdot 10^4 \, \Omega$ cm	300 K	under illumination of 10^4 photons/s at 0.49 μm	69B1

N.B. No anisotropy reported for transport properties.

photoconductivity peak energy:

E	2.53 eV	$T = 300$ K	maximum of photoconductivity, unpolarized	61B

refractive indices (at RT):

n_o	n_e	$n_o - n_e$	λ [μm]		
2.6592	2.5979	0.0613	0.5495	see also [80B] and [76L]	79B1
2.6334	2.5748	0.0586	0.5747		
2.6112	2.5549	0.0563	0.6009		
2.5890	2.5349	0.0541	0.6328		
2.5796	2.5264	0.0532	0.6500		
2.477	2.482	0.045	1.6760		
2.472	2.458	0.014	1.7500		
2.444	2.403	0.041	2.6500		
2.439	2.398	0.041	3.5400		
2.414	2.372	0.042	7.1500		
2.400	2.358	0.042	8.7300		
2.380	2.337	0.043	10.4000		
2.369	2.329	0.040	11.0000		

Physical property	Numerical value	Experimental conditions	Experimental method remarks	Ref.
nonlinear dielectric susceptibility:				
d_{36}	$4.0 \cdot 10^{-11}$ m V^{-1}	$\lambda = 1.06$ μm	see also [80B]	76L

10.1.6.14 Mercury gallium selenide (□-HgGa$_2$Se$_4$)

energy gap:				
E_g	1.95 eV	$T = 300$ K	absorption, unpolarized	61B
	1.66 eV		calculated	79P2
dE_g/dT	$-(7\cdots8) \cdot 10^{-4}$ eV K^{-1}	$77\cdots300$ K	photoconductivity	77L
defect levels:				
E	1.4 eV	$T = 300$ K	start of photoconduction	77L
	1.6 eV	300 K	maximum photoconduction	
	$1.95\cdots2.1$ eV	300 K	maximum photoconduction (band gap)	
resistivities:				
ϱ_d	$10^8\cdots10^{10}$ Ω cm	$T = 300$ K		69B1
ϱ_i	$2.7 \cdot 10^4$ Ω cm	300 K	under illumination of 10^4 photons/s at 0.62 μm	

N.B. No anisotropy reported.

10.1.6.15 Mercury indium telluride (□-HgIn$_2$Te$_4$)

All experimental work discussed below was done on the same crystals with the defect stannite structure (D_{2d}^{11}—I$\bar{4}$2m). There is some evidence that the Hg and In atoms are at least partially disordered [76M2, 77M2].

energy gap (in eV):

		T [K]		
$E_{g, dir}$	0.96	4, $E \perp c$	electroreflectance, absorption	77M2
	0.9	300, $E \perp c$		
	1.09	4, $E \parallel (\perp) c$		
	0.94	300, $E \parallel (\perp) c$		
dE_g/dT	$2.8 \cdot 10^{-4}$ eV K^{-1}	$4\cdots300$	electroreflectance, see Fig. 20	77M2

wavenumbers of infrared and Raman active phonons (in cm^{-1}, at 300 K) ($k = 0$):

		Symmetry		
$\bar{\nu}$	42	B_2 (LO)	Raman active	76M2
	50	B_1	Raman active	
	61	B_2 (LO)	Raman active	
	74	B_2 (LO)	Raman active	
	100	A_1	Raman active (very strong)	
	114	A_2	calculated (inactive)	
	118 (122)	E (TO)	infrared (Raman) active	
	127	E (LO)	infrared (Raman) active	
	132	A_1	Raman active	
	247	B_2 (TO)	infrared active	
	153	B_2 (LO)	infrared (Raman) active	
	155	B_1	Raman active	
	160	E (TO)	infrared (Raman) active	
	169	E (LO)	infrared (Raman) active	
	180 (184)	B_2 (TO)	infrared (Raman) active	
	184 (189)	B_2 (LO)	infrared (Raman) active	
	181	E (TO)	Raman active	
	188	E (LO)	Raman active	

Physical property	Numerical value	Experimental conditions	Experimental method remarks	Ref.
dielectric constants:				
$\varepsilon(0)$	9.64	$T = 300\,\mathrm{K}$; $E \parallel c$	infrared reflectivity	76M2
	11.06	$E \perp c$		
$\varepsilon(\infty)$	8.53	$300\,\mathrm{K}$; $E \parallel c$		
	8.57	$E \perp c$		
elastic moduli (in $10^{10}\,\mathrm{Nm^{-2}}$):				
c_{11}	4.31	$T = 77\,\mathrm{K}$	ultrasonic wave attenuation	76S2
c_{12}	2.54		For pressure dependence and further	
c_{13}	2.18		elastic properties, see [82H].	
c_{33}	4.47			
c_{44}	2.14			
c_{66}	2.41			
bulk modulus:				
B	$2.99 \cdot 10^{10}\,\mathrm{Nm^{-2}}$	$T = 77\,\mathrm{K}$		76S2
carrier concentration, mobility:				
n	$3.5 \cdot 10^{15}\,\mathrm{cm^{-3}}$	$T = 300\,\mathrm{K}$	n-type samples, no anisotropy	69B1
μ_{n}	$200\,\mathrm{cm^2\,V^{-1}\,s^{-1}}$			

10.1.6.16 Solid solutions

$CdGa_2S_{4x}Se_{4(1-x)}$:
energy gap:

$E_{\mathrm{g,dir}}$	2.8 eV	$T = 300\,\mathrm{K}$	$CdGa_2SSe_3$ absorption, unpolarized	77K3
	2.67 eV	$300\,\mathrm{K}$	$CdGa_2S_2Se_2$	
	2.37 eV	$300\,\mathrm{K}$	$CdGa_2S_3Se$	

For band structure calculations, see [80P2].

higher transition energies:

E	3.27 eV	$T = 300\,\mathrm{K}$	$CdGa_2S_2Se_2$	72K1
			photoconductivity, unpolarized	
	3.18 eV	$300\,\mathrm{K}$	$CdGa_2S_3Se$	

photoconductivity peak energies:

E	2.53 eV	$T = 300\,\mathrm{K}$	$CdGa_2SSe_3$	77K3
	2.48 eV	$300\,\mathrm{K}$	$CdGa_2S_2Se_2$ (broad peaks)	
	2.07 eV	$300\,\mathrm{K}$	$CdGa_2S_3Se$	

conductance and thermoelectric power: Figs. 21, 22, 23

$CdIn_2Se_{4x}Te_{4(1-x)}$:
energy gap: Fig. 24

effective mass:

m_{n}	$0.15\,(1)\,m_{\mathrm{o}}$	$T = 300\,\mathrm{K}$	thermoelectric power ($x = 0 \cdots 0.6$)	72K1

electron concentration (in $10^{17}\,\mathrm{cm^{-3}}$, at $300\,\mathrm{K}$):

n	151		$x = 1.0$ (two different samples	72K1
	79		1.0 for each x-value)	
	57		0.8	
	25		0.8	
	33		0.6	
	9.1		0.6	
	13.4		0.4	
	5.9		0.4	

Physical property	Numerical value	Experimental conditions	Experimental method remarks	Ref.
Hall coefficient:				
R_H	$-5 \cdot 10^{-1}$ cm^3 C^{-1}	$T = 300$ K, x = 1.0	n-type samples, no anisotropy	72K1
	-10 cm^3 C^{-1}	300 K, x = 0.4		
Seebeck coefficient:				
S	-100 µV K^{-1}	$T = 300$ K, x = 1.0	n-type samples, no anisotropy	72K1
	-300 µV K^{-1}	300 K, x = 0.4		

N.B. x = 1 phase not given.
For some results on the $Co_xZn_{1-x}In_2S_4$ system, see [83F].

10.1.6.17 References for 10.1.6

50H Hahn, H., Klingler, W.: Z. Anorg. Allg. Chem. **263** (1950) 177.
55H Hahn, H., Frank, G., Klinger, W., Störger, A.D., Störger, G.: Z. Anorg. Allg. Chem. **279** (1955) 241.
59C Callaway, J.: Phys. Rev. **113** (1959) 1046.
60W Woolley, J.C., Ray, B.: J. Phys. Chem. Solids **16** (1960) 102.
61B Beun, J.A., Nitsche, R., Lichtensteiger, M.: Physica **27** (1961) 448.
62L Lappe, F., Niggli, A., Nitsche, R., White, J.G.: Z. Kristallogr. **117** (1962) 146.
64P Parthé, E.: "Crystal Chemistry of Tetrahedral Structures". New York: Gordon and Breach, **1964**.
69B1 Berger, L.T., Prochukhan, V.D.: "Ternary Diamond like Semiconductors". New York, London: Consultants Bureau, **1969**.
69B2 Boltivets, N.S., Drobyazko, V.P., Mityurev, V.K.: Sov. Phys. Semicond. **2** (1969) 867.
69R Range, K.J., Becker, W., Weiss, A.: Z. Naturforsch. **24b** (1969) 811.
70C Czajy, W.: Phys. Kondens. Mater. **10** (1970) 299.
70D Donika, F.G., Radautsan, S.I., Semiletov, S.A., Donika, T.V., Mustya, I.G., Zhitar, V.F.: Kristallografiya **15** (1970) 813; Sov. Phys. Crystallogr. (English Transl.) **15** (1971) 695.
70M Moro, S., Paorici, C., Romeo, N.: J. Appl. Phys. **42** (1970) 2061.
71A1 Allakhverdiev, K.P., Antonov, V.B., Nani, R.Kh., Salaev, E.Yu.: Fiz. Tekh. Poluprovodn. **5** (1971) 2370; Sov. Phys. Semicond. (English Transl.) **5** (1971) 190.
71A2 Abdullaev, G.B., Agaev, V.G., Antonov, V.B., Nani, R.Kh., Salaev, E.Yu.: Fiz. Tekh. Poluprovodn. **5** (1971) 2132; Sov. Phys. Semicond. (English Transl.) **5** (1971) 1854.
71C Chedzey, H.A., Marshall, D.J., Parfitt, H.T., Robertson, D.S.: J. Phys. D: Appl. Phys. **4** (1971) 1320.
71D1 Donika, F.G., Radautsan, S.I., Kiosse, G.A., Semiletov, S.A., Donika, T.V., Mustya, I.G.: Kristallografiya **16** (1971) 235; Sov. Phys. Crystallogr. (English Transl.) **16** (1971) 190.
71D2 Damaskin, I.A., Pyshkin, S.L., Radautsan, S.I., Tazlavan, V.E.: Phys. Status Solidi **(a) 6** (1971) 425.
71K1 Kerr, R.K., Schwerdtfeger, C.F.: J. Phys. Chem. Solids **32** (1971) 2007.
71K2 Korpel, A.: Appl. Solid State Sci. Vol. **3**, p. 72; Wolfe, R. (ed.), New York: Academic Press, 1972.
71K3 Kaminow, J.P., Turner, E.H.: Handbook of Lasers, p. 447; Pressley, R.J. (ed.), Cleveland: Chemical Rubber Co. **1971**.
71M Manca, P., Muntoni, C., Raga, F., Spiga, A.: Phys. Status Solidi **(b) 44** (1971) 51.
71R Radautsan, S.I., Zhitar, V.F., Kosnichan, I.G., Shmiglyuk, M.I.: Fiz. Tekh. Poluprovodn. **5** (1971) 2240; Sov. Phys. Semicond. (English Transl.) **5** (1971) 1959.
72A1 Abdullaev, G.B., Guseinova, D.A., Kerimova, T.G., Nani, R.Kh.: Phys. Status Solidi **(b) 54** (1972) K115.

72A2	Abdullaev, G.B., Agaev, V.G., Antonov, V.B., Nani, R.Kh., Salaev, E.Yu.: Fiz. Tekh. Poluprovodn. **6** (1972) 1729; Sov. Phys. Semicond. (English Transl.) **6** (1973) 1492.
72A3	Allakhverdiev, K.P., Antonov, V.B., Nani, R.Kh., Salaev, E.Yu.: Sov. Phys. Semicond. **5** (1972) 2080.
72B	Baldini, G., Aggarwall, R.L., Lax, B., Shin, S.H., Tsang, J.C.: Lett. Nuovo Cimento **5** (1972) 1062.
72D	Donika, F.G., Radautsan, S.I., Semiletov, S.A., Kiosse, G.A., Mustya, I.G.: Kristallografiya **17** (1972) 663; Sov. Phys. Crystallogr. (English Transl.) **17** (1972) 575.
72K1	Koval, L.S., Arushanov, E.K., Radautsan, S.I.: Phys. Status Solidi **(a) 9** (1972) K73.
72K2	Koval, L.S., Markus, M.M., Radautsan, S.I.: Phys. Status Solidi **(a) 9** (1972) K69.
72K3	Koval, L.S., Radautsan, S.I., Sobolev, V.V.: Izv. Akad. Nauk SSSR, Neorg. Mater **8** (1972) 2021; Inorg. Mater. (USSR) (English Transl.) **8** (1973) 1776.
72M	Mamedov, K.K., Aliev, M.M., Kerimov, I.G., Nani, R.Kh: Phys. Status Solidi **(a) 9** (1972) K149.
72R1	Romeo, N., Vigil, O.: Phys. Status Solidi **(a) 10** (1972) 447.
72R2	Radautsan, S.I., Molodyan, I.P., Syrba, N.N., Tezlevan, V.E., Shipilka, M.A.: Phys. Status Solidi **(b) 49** (1972) K175.
72Y	Yokoyama, Y., Tsukahara, S., Satoh, T., Tsushima, T.: J. Phys. Soc. Jpn. **32** (1972) 1169.
73A1	Abdullaev, G.B., Guseinova, D.A., Kerimova, T.G., Nani, R.Kh.: Fiz. Tekh. Poluprovodn. **7** (1973) 840; Sov. Phys. Semicond. (English Transl.) **7** (1973) 575.
73A2	Abdullaev, G.B., Agaev, V.G., Antonov, V.B., Mamedov, A.A., Nani, R.Kh., Salaev, E.Yu.: Fiz. Tekh. Poluprovodn. **7** (1973) 1051; Sov. Phys. Semicond. (English Transl.) **7** (1973) 717.
73A3	Aresti, A., Congiu, A.: Phys. Status Solidi **(a) 16** (1973) K55.
73B	Bosacchi, A., Bosacchi, B., Franchi, S., Hernandez, L.: Solid State Commun. **13** (1973) 1805.
73M1	Manca, P., Raga, F., Spiga, A.: Phys. Status Solidi **(a) 16** (1973) K105.
73M2	Makaryumas, K.V., Dragunas, A.K., Adominaite, M.L.: Zh. Eksp. Teor. Fiz. **64** (1973) 985; Sov. Phys. JETP (English Transl.) **37** (1973) 449.
73P	Paulavicius, A., Jasutis, V., Vesione, T., Tolutis, V.: Liet. Fiz. Rinkinys **13** (1973) 561.
73R1	Radautsan, S.I., Syrba, N.N., Tezlevan, V.E., Sherban, K.F., Baran, N.P.: Phys. Status Solidi **(b) 57** (1973) K93.
73R2	Radautsan, S.I., Syrba, N.N., Tezlevan, V.E., Sherban, K.F., Strumban, E.E.: Phys. Status Solidi **(a) 15** (1973) 295.
73S	Slagsvold, B.J.: J. Phys. Chem. Solids **34** (1973) 1752.
73W	Wittekoek, S., van Stapele, R.P., Wijma, A.W.J.: Phys. Rev. **B 7** (1973) 1667.
73Y	Yamamoto, K., Murakawa, T., Ohbayashi, Y., Hiroyasu, S., Abe, K.: J. Phys. Soc. Jpn. **35** (1973) 1258.
74A1	Anedda, A.: Appl. Opt. **13** (1974) 1595.
74A2	Abdullaev, G.B., Guseinova, D.A., Kerimova, T.G., Nani, R.Kh.: Fiz. Tekh. Poluprovodn. **8** (1974) 1210; Sov. Phys. Semicond. (English Transl.) **8** (1974) 785.
74B	van den Boom, H., Haanstra, J.H.: J. Raman Spectrosc. **2** (1974) 265.
74E	Endo, S., Irie, T.: J. Phys. Soc. Jpn. **36** (1974) 1210.
74F1	Fiejeta, H., Okada, Y.: Jpn. J. App. Phys. 13 (1974) 1823.
74F2	Fortin, E., Raga, F.: Solid State Commun. 14 (1974) 847.
74I	Iwai, T., Endo, S., Irie, T.: J. Phys. Soc. Jpn. **36** (1974) 307.
74L	Levine, B.F., Bettea, C.G., Kasper, H.M.: IEEE J. Quant, Electron. **10** (1974) 904.
74M	Manca, P., Raga, F., Spiga, A.: Nuovo Cimento **19 B** (1974) 15.
74R	Radautsan, S.I., Syrba, N.N., Tezlevan, V.E., Baran, N.P., Blas, V.D., Titov, V.A.: "Properties of some Semiconductor Materials and Devices" Shtiintsa, Kishinev **1974**, 3.
74S	Suzuki, M., Komorita, K.: Phys. Lett. **49 A** (1974) 139.
74U	Ueno, M., Endo, S., Irie, T.: Jpn. J. App. Phys. **13** (1974) 580.
75G	Guzzi, M., Baldini, G.: J. Lumin. **9** (1975) 514.
75K	Koshinzuka, N., Yokoyama, Y., Hiruma, H., Tsushima, T.: Solid State Commun. **16** (1975) 1011.
75M	Mamedov, K.K., Kerimov, I.G., Veliev, R.K., Mekhtiev, M.I.: Izv. Akad. SSSR, Neorg. Mater. **11** (1975) 2062; Inorg. Mater. (USSR) (English Transl.) **11** (1975) 1767.
75R1	Radautsan, S.I., Zhitar, V.F., Railyon, V.Ya.: Fiz. Tekh. Poluprovodn. **9** (1975) 2278; Sov. Phys. Semicond. (English Transl.) **9** (1975) 1476.
75R2	Radautsan,.S.I., Zhitar, V.F., Dona, V.S.: Fiz. Tekh. Poluprovodn. **9** (1975) 1018; Sov. Phys. Semicond. (English Transl.) **9** (1975) 670.
75S	Shimizu, H., Ohbayashi, Y., Yamamoto, K., Abe, K.: J. Phys. Soc. Jpn. **38** (1975) 750.

76A	Arama, E.D., Vinogradov, E.A., Zhizhin, G.N., Zhitar, V.F., Mel'nik, N.N., Radautsan, S.I.: Dokl. Akad. Nauk SSSR **231** (1976) 1343.
76E	Endo, S., Irie, T.: J. Phys. Chem. Solids **37** (1976) 201.
76G1	Gombia, E., Romeo, N., Sberveglieri, G., Paorici, C.: Phys. Status Solidi **(a) 36** (1976) 631.
76G2	Grilli, E., Guzzi, M., Molteni, R.: Phys. Status Solidi **(a) 37** (1976) 399.
76G3	Glidewell, C.: Inorg. Chim. Acta **19** (1976) L45.
76H	Hernandez, L., Vigil, O., Gonzalez, F.: Phys. Status Solidi **(a) 36** (1976) 33.
76K	Kivits, P.: J. Phys. C: Solid State **9** (1976) 605.
76L	Levine, B.F., Bettea, C.G., Kasper, H.M., Thiel, F.A.: IEEE, J. Quant. Electron. QE **12** (1976) 367.
76M1	Meloni, F., Aymerich, F., Mula, G., Baldereschi, A.: Helv. Phys. Acta **49** (1976) 687.
76M2	Miller, A., Lockwood, D.J., MacKinnon, A., Weaire, D.: J. Phys. C: Solid State **9** (1976) 2997.
76S1	Sultanov, S., Maylonov, Sh., Karimov, S., Kurbanov, Kh.M., Madazimov, A.: Izv. Akad. Nauk SSSR, Neorg. Mater **12** (1976) 115; Inorg. Mater. (USSR) (English Transl.) **12** (1976) 96.
76S2	Saunders, G.A., Seddon, T.: J. Phys. Chem. Solids **37** (1976) 873.
76W	Wakamura, K., Arai, T., Kudo, K.: J. Phys. Soc. Jpn. **41** (1976) 130.
77B1	Baldereschi, A., Meloni, F., Aymerich, F., Mula, G.: Inst. Phys. Conf. Ser. **35** (1977) 193.
77B2	Baldereschi, A., Meloni, F., Aymerich, F., Mula, G.: Solid State Commun. **21** (1977) 113.
77E1	Endo, S.: J. Phys. Soc. Jpn. **43** (1977) 132.
77E2	Endo, S.: Jpn. J. App. Phys. **16** (1977) 2055.
77G	Guseinova, D.A., Kerimova, T.G., Nani, R.Kh.: Fiz. Tekh. Poluprovodn. **11** (1977) 1135; Sov. Phys. Semicond. (English Transl.) **11** (1977) 670.
77H	Hermann, H.: Phys. Status Solidi **(b) 82** (1977) 513.
77K1	Kobzarenko, V.N., Donika, F.G., Zhitar, V.F., Radautsan, S.I.: Dokl. Akad, Nauk. SSSR **235** (1977) 445; Sov. Phys. Dokl. (English Transl.) **22** (1977) 445.
77K2	Kshirsagar, S.T.: Thin Solid Films **45** (1977) L5.
77K3	Kshirsagar, S.T., Sinha, A.P.B.: J. Mater. Sci. **12** (1977) 1614.
77K4	Kisiel, A., Turowski, M., Czeppe, T., Giriat, W.: Inst. Phys. Conf. Ser. **35** (1977) 259.
77L	Lebedov, A.A., Metlinskii, P.N., Rud, Yu.V., Tyrziu, V.G.: Fiz. Tekh. Poluprovodn. **11** (1977) 1038; Sov. Phys. Semicond. (English Transl.) **11** (1977) 615.
77M1	Mocharnyuk, G.F., Babyuk, T.I., Derid, O.P., Lazarenko, L.S., Markus, M.M., Radautsan, S.I.: Dokl. Akad. Nauk SSSR **237** (1977) 821; Sov. Phys. Dokl. (English Transl.) **22** (1977) 749.
77M2	MacKinnon, A., Miller, A., Ross, G.: Inst. Phys. Conf. Ser. **35** (1977) 171.
77P	Paulavichyus, A.B., Yasutis, V.V., Paukshte, Yu.A., Burneika, I.P.: Litov. Fiz. Sb. **17** (1977) 787.
77R	Radautsan, S.I., Syrba, N.N., Nebola, I.I., Tyrziu, V.G., Bercha, D.M.: Fiz. Tekh. Poluprovodn. **11** (1977) 69; Sov. Phys. Semicond, (English Transl.) **11** (1977) 38.
77S1	Sato, K., Yokoyama, Y., Tsushima, T.: J. Phys. Soc. Jpn. **42** (1977) 559.
77S2	Slivka, V.Yu., Peresh, E.Yu.: Ukr. Fiz. Zh. **22** (1977) 1951.
77T	Trykozko, R.: Inst. Phys. Conf. Ser. **35** (1977) 249.
78A	Anedda, A., Garbato, L., Raga, F., Serpi, A.: Phys. Status Solidi **(a) 50** (1978) 643.
78B	Bornsall, S.B., Hummel, F.A.: J. Solid State Chem. **25** (1978) 379.
78C	Cerrina, F., Abbati, I., Braicovich, L., Levy, F., Margaritondo, G.: Solid State Commun. **26** (1978) 99.
78G1	Grilli, E., Guzzi, M., Anedda, A., Raga, F., Serpi, A.: Solid State Commun. **27** (1978) 105.
78G2	Grilli, E., Cappelletti, P., Guzzi, M.: Phys. Status Solidi **(a) 50** (1978) K93.
78I	Ihara, H., Abe, H., Endo, S., Irie, T.: Solid State Commun. **28** (1978) 563.
78K1	Kivits, P., Wijnakker, M., Claasen, J., Geerts, J.: J. Phys. C: Solid State Phys. **11** (1978) 2361.
78K2	Khuseinov, B., Mavlonov, Sh., Umarov, B.S.: Izv. Akad. Nauk SSSR, Neorg. Mater. **14** (1978) 863; Inorg. Mater. (USSR) (English Transl.) **14** (1978) 675.
78P	Picco, P., Abbati, I., Braicovich, L., Cerrina, F., Levy, F., Margaritondo, G.: Phys. Lett. **65A** (1978) 467.
78S1	Suslikov, L.M., Nebola, I.I., Peresh, E.Yu., Voroshilov, Yu.V., Bercha, D.M., Slivka, V.Yu.: Fiz. Tverd. Tela **20** (1978) 3186; Sov. Phys.-Solid State (English Transl.) **20** (1978) 1840.
78S2	Symorov, V.F., Bezryadin, N.N., Roviskii, A.P., Sysoev, B.I.: Izv. Vyssh. Uchebn. Zaved. Fiz. **4** (1978) 127.
78U1	Unger, W.K., Farnworth, B., Irwin, J.C., Pink, H.: Solid State Commun. **25** (1978) 913.
78U2	Unger, W.K., Meuth, H., Irwin, J.C., Pink, H.: Phys. Status Solidi **(a) 46** (1978) 81.

78U3   Ueno, M., Nakanishi, H., Irie, T.: J. Phys. Soc. Jpn. **44** (1978) 2013.
79A1   Anedda, A., Cugusi, L., Grilli, E., Guzzi, M., Raga, F., Spiga, A.: Solid State Commun. **29** (1979) 829.
79A2   Aymerich, F., Meloni, F., Mula, G.: Solid State Commun. **29** (1979) 235.
79A3   Anedda, A., Fortin, E.: J. Phys. Chem. Solids **60** (1978) 653.
79A4   Areshkin, A.G., Zhitar, V.F., Radautsan, S.I., Rallyan, V.Yu., Suslina, L.G.: Fiz. Tekh. Poluprovodn. **13** (1979) 337; Sov. Phys. Semicond. (English Transl.) **13** (1979) 194.
79B1   Badikov, V.V., Matveev, I.N., Panyatin, V.L., Pshenichnikov, S.M., Repyakhova, T.M., Rychok, O.V., Rozenson, A.E., Trotsenko, N.K., Ustinov, N.D.: Kvant. Elektron **6** (1979) 1807; Sov. J. Quant. Electron. (English Transl.) **9** (1979) 1068.
79B2   Bacewicz, R., Trykozko, R., Borghesi, A., Cambiaghi, M., Reguzzoni, E.: Phys. Lett. **75A** (1979) 121.
79B3   Bacewicz, R., Lottici, P.P., Razzetti, C.: J. Phys. C: Solid State **12** (1979) 3603.
79G   Graber, N., Wagner, H.J., Schwerdtfeger, C.F.: J. Phys. Soc. Jpn. **46** (1979) 1953.
79H   Horiba, R., Nakanishi, N., Endo, S., Irie, T.: Surface Sci. **86** (1979) 498.
79K1   Kerimova, T.G., Nani, R.Kh., Salaev, E.Yu., Shteinshraiber, V.Yu.: Fiz. Tverd. Tela **21** (1979) 2791; Sov. Phys.-Solid State (English Transl.) **21** (1979) 1605.
79K2   Kerimova, T.G., Mamedov, Sh.S., Mekhtiev, N.M., Nani, R.Kh., Salaev, E.Yu.: Fiz. Tekh. Poluprovodn. **13** (1979) 494; Sov. Phys. Semicond. (English Transl.) **13** (1979) 291.
79K3   Kerimova, T.G., Nani, R.Kh., Salaev, E.Yu., Shteinshraiber, V.Yu., Aliev, A.A.: Fiz. Tverd. Tela **21** (1979) 1899; Sov. Phys.-Solid State (English Transl.) **21** (1979) 1093.
79M   MacKinnon, A.: J. Phys. C **12** (1979) L655.
79P1   Panyutin, V.L., Ponedel'nikov, B.E., Rozenson, A.E., Chizhkov, V.I.: Izv. Vyssh. Uchebn. Zaved. Fiz. **8** (1979) 57; Sov. Phys. J. (English Transl.) **22** (1979) 857.
79P2   Panyutin, V.L., Ponedel'nikov, B.E., Rozenson, A.E., Chizhkov, V.I.: Fiz. Tekh. Poluprovodn. **13** (1979) 1211; Sov. Phys. Semicond. (English Transl.) **13** (1979) 711.
79P3   Przedmojski, J., Patosz, B.: Phys. Status Solidi (a) **51** (1979) K1.
79S   Sultanov, S., Karimov, S.K.: Izv. Akad. Nauk SSSR, Neorg. Mater. **15** (1979) 1290; Inorg. Mater. (USSR) (English Transl.) **15** (1979) 1009.
79U   Ueno, M.: J. Phys. Soc. Jpn. **46** (1979) 1877.
79Z   Zhitar, V.F., Moldovyan, N.A., Radautsan, S.I.: Fiz. Tekh. Poluprovodn. **13** (1979) 1886; Sov. Phys. Semicond. (English Transl.) **13** (1979) 1100.
80B   Badikov, V.V., Matveev, I.N., Panyutin, V.L., Pshenichnikov, S.M., Rozenson, A.I., Skrebneva, S.V., Trotsenko, N.K., Ustinov, N.D.: Sov. J. Quantum Electron. **10** (1980) 1302.
80C   Cerrina, F., Quaresima, C., Abbati, I., Braicovich, L., Picco, P., Margaritondo, G.: Solid State Commun. **33** (1980) 429.
80G   Grilli, E., Guzzi, M., Cappelletti, P., Moskalonov, A.V.: Phys. Status Solidi (a) **59** (1980) 755.
80M1   Margaritondo, G., Stoffel, N.O., Levy, F.: J. Phys. C **13** (1980) 277.
80M2   Manolikas, C., Bartzokas, D., van Tendeloo, G., van Landuyt, T., Amelinckx, S.: Phys. Status Solidi (a) **59** (1980) 425.
80N   Nakanishi, H.: Jpn. J. Appl. Phys. **19** (1980) 103.
80P1   Panyutin, V.L., Ponedel'nikov, B.E., Rozenson, A.E., Chizhikov, V.I.: Fiz. Tekh. Poluprovodn. **14** (1980) 1230; Sov. Phys. Semicond. (English Transl.) **14** (1980) 728.
80P2   Panyutin, V.L., Ponedel'nikov, B.E., Rozenson, A.E., Chizhikov, V.I.: J. Phys. (Paris) **41** (1980) 1025.
80R   Radautsan, S.I., Dopa, V.S., Zhitar, V.F., Strumban, E.E.: Phys. Status Solidi (a) **57** (1980) K79.
80Y   Yamada, M., Shirai, T., Yamamoto, K., Abe, K.: J. Phys. Soc. Jpn. **48** (1980) 874.
81K   Kerimova, T.G., Mamedov, Sh.S., Nani, R.Kh.: Fiz. Tekh. Poluprovodn. **15** (1981) 138; Sov. Phys. Semicond. (English Transl.) **15** (1981) 81.
82B   Razzetti, C., Lottici, P.P., Bacewicz, R.: J. Phys. **C15** (1982) 5657.
82H   Hailing, T., Saunders, G.A., Lambson, W.A.: Phys. Rev. **B26** (1982) 5786.
82T   Tyuterev, V.G., Skachkov, S.I., Bysneva, L.A.: Fiz. Tverd. Tela **24** (1982) 2236; Sov. Phys. Solid State (English Transl.) **24** (1982) 1271.
83F   Fiorani, D., Gastaldi, L., Viticoli, S.: Solid State Commun. **48** (1983) 865.
83L1   Lottici, P.P., Razzetti, C.: J. Phys. **C16** (1983) 3449.
83L2   Lottici, P.P., Razzetti, C.: Solid State Commun. **16** (1983) 681.
83N   Nebola, I.I., Suslikov, L.M., Dordyai, V.S., Vysochanskii, Yu.M., Kharkalis, N.R., Slivka, V.Yu.: Fiz. Tverd. Tela **24** (1983) 3631; Sov. Phys. Solid State (English Transl.) **24** (1983) 2069.

10.1.7 Other ordered vacancy compounds

The compounds considered in this section are all members of the series II_3VI_3–III_2VI_3. Only those combinations where some ordering of the cations occurs will be considered.

10.1.7.0 Structure

The Zn–In–Te compounds have layered structures, like $ZnIn_2Te_4$, with Zn in tetrahedral sites and In in either tetrahedral or octahedral sites. The Hg–(Ga, In)–Te compounds have ordered vacancy or defect tetrahedral structures. There is some evidence that the vacancies are ordered and the other cations disordered on the remaining sites.

Substance	Structure	Lattice parameters*) [Å]	Remarks	Ref.
$Zn_2In_2S_5$	layered hexagonal	$a = 3.85(2)$ $c = 3.086(3) \times 5n$	general formula for all polytypes	70D
IIa	space group C_{6v}^4—$P6_3mc$ 2 layers/unit cell S-packing: ABABA/B etc. 1-layer: $SZnSIn_0SZnSIn_tS$	$a = 3.85(2)$ $c = 30.85(4)$	$In_0 = $ In on octahedral site $In_t = $ In on tetrahedral site	72D
IIIa	space group C_{3v}^5—R3m 3 layers/unit cell S-packing: ABCBC/BCACA/CABAB/ 1-layer: $SZnSIn_0SZnSIn_tS$	$a = 3.85(2)$ $c = 46.27(4)$		70D
$Zn_3In_2S_6$	space group D_{3d}^3—$P\bar{3}m1$ S-packing: ABAB etc. 1-layer: S(Zn, In)$SZnSIn_0S$	$a = 3.85(2)$ $c = 18.5(3)$	(Zn, In) = Zn and In randomly on tetrahedral site, see Fig. 1	67D
$Hg_3Ga_2Te_6$	space group T_d^1—$P\bar{4}3m$ $3 \times 3 \times 3$ zincblende cubes	$a = 18.648(3)$	$T > 773$ K disordered zincblende-like	76S
$Hg_5Ga_2Te_8$	space group T_d^2—$F\bar{4}3m$	$a = 12.472(4)$	ordered; Hg, Ga random	71S
$Hg_3In_2Te_6$	space group T_d^1—$P\bar{4}3m$ $3 \times 3 \times 3$ zincblende cubes disordered zincblende	$a = 18.870(3)$ $a = 6.289$	$T < 595$ K $T > 595$ K	70M, 77M
$Hg_5In_2Te_8$	space group T_d^2—$F\bar{4}3m$	$a = 12.661(2)$ $a = 12.655(3)$	annealed unannealed ordered; Hg, Ga random	71M
	rocksalt		$p > 42(2)$ kbar	72P

*) For $T = 300$ K if not otherwise stated.

10.1.7.1 Zinc indium sulfide ($Zn_2In_2S_5$)

energy gap and further transition energies (in eV): (IIIa phase)

		T [K]		
$E_{g, dir}$	2.65	77; $E \perp c$	reflection, transmission	69R
	2.50	300; $E \perp c$		
	2.7	300; $E \perp c$	absorption	71R
	2.62	300; $E \perp c$	reflection	71R
E	2.96	300; $E \perp c$	reflection	71R
	4.60	300; $E \perp c$		

Physical property	Numerical value	Experimental conditions	Experimental method remarks	Ref.
photoconductivity maxima (at 300 K): (IIIa phase)				
E	2.44 eV	$E \perp c$		71R
	2.68 eV	unpolarized		
	3.0 eV	$E \parallel c$		
conductivity:				
σ	$10^{-12}\,\Omega^{-1}\,cm^{-1}$	$T = 300$ K; $E \perp c$	IIIa phase	71R

10.1.7.2 Zinc indium sulfide ($Zn_3In_2S_6$)

energy gap:				
E_g	2.8 eV	$T = 300$ K; $E \perp c$	absorption	66Z
	2.82 eV	77 K; $E \perp c$	photocapacitance	
	2.76 eV	295 K; $E \perp c$		
dE_g/dT	$-2.8 \cdot 10^{-4}\,eV\,K^{-1}$		photocapacitance	66Z
photoluminescence maxima:				
E	1.83 eV	$T = 300$ K; $E \perp c$		66Z
	2.06 eV	77 K; $E \perp c$		
defect level:				
E	2.05 eV		from band edge; optical quenching	66Z
resistivity:				
ϱ	$> 10^{10}\,\Omega\,cm$	$T = 300$ K; $E \perp c$		66Z

10.1.7.3 Mercury gallium telluride ($Hg_3Ga_2Te_6$)

elastic moduli (in 10^{11} dyn cm^{-2}):				
c_{11}	4.55	$T = 77$ K		76S
c_{12}	2.49			
c_{44}	2.14			
For pressure dependence and further elastic properties, see [82H].				
melting point:				
T_m	980 K		phase transition at 673···773 K	70R, 76S

10.1.7.4 Mercury gallium telluride ($Hg_5Ga_2Te_8$)

elastic moduli (in 10^{11} dyn cm^{-2}):				
c_{11}	4.97	$T = 77$ K		71S
c_{12}	3.14			
c_{44}	2.18			
For pressure dependence and further elastic properties, see [82H].				
conductivity, Seebeck coefficient, Hall coefficient, mobility (at 300 K):				
σ	$6 \cdot 10^{-3}\,\Omega^{-1}\,cm^{-1}$		p-type sample	62S
S	$8.1 \cdot 10^{-4}\,V\,K^{-1}$			
R_H	$1.5 \cdot 10^4\,cm^3\,C^{-1}$			
μ_p	90 cm^2 V^{-1} s^{-1}			
Debye temperature:				
Θ_D	151 (6)	$T = 77$ K		71S
melting point:				
T_m	970 K			70R

Physical property	Numerical value	Experimental conditions	Experimental method remarks	Ref.

10.1.7.5 Mercury indium telluride ($Hg_3In_2Te_6$)

wavenumbers of infrared active phonons (in cm^{-1}, at 300 K):

$\bar{\nu}$	118	A	broad peaks	77M
	158	B	broad peaks	
	238	2A	two-phonon peak	
	304	2B	two-phonon peak	
	395	2A+B	three-phonon peak	

Raman scattering: Fig. 2.

elastic moduli (in 10^{11} dyn cm^{-2}):

c_{11}	4.33	$T = 77$ K		70S
c_{12}	2.64			
c_{44}	1.99			

For pressure dependence and further elastic properties, see [82H].

phase transition at 595 (5) K: [70M]
resistivity: Fig. 3

10.1.7.6 Mercury indium telluride ($Hg_5In_2Te_8$)

energy gap:

$E_{g,th}$	0.71 eV		transport	62S

elastic moduli (in 10^{11} dyn cm^{-2}):

c_{11}	5.02	$T = 77$ K		71S
c_{12}	3.33			
c_{44}	2.08			

For pressure dependence and further elastic properties, see [82H].

conductivities, carrier concentrations, Hall coefficient, mobility, Seebeck coefficient (at 300 K):
p-type sample:

σ	$2.3 \cdot 10^{-4} \, \Omega^{-1} \, cm^{-1}$			62S
p	$7 \cdot 10^{12} \, cm^{-3}$			
R_H	$9 \cdot 10^5 \, cm^3 \, C^{-1}$			
μ_p	$210 \, cm^2 \, V^{-1} \, s^{-1}$			
S	$2.3 \cdot 10^{-4} \, V \, K^{-1}$			

n-type sample: high-pressure modification ($p = 55$ kbar)

ϱ	$10^{-4} \, \Omega \, cm$			72P
n	$3 \cdot 10^{22} \, cm^{-3}$			

For pressure dependence of transport parameters, see Fig. 4.

Debye temperature:

Θ_D	142 (6) K	$T = 77$ K		71S

phase transition at 720 K: [70S]

10.1.7.7 References for 10.1.7

62S   Spencer, P.M., Pamplin, B.R., Wright, D.A.: "International Conference on the Physics of Semiconductors", Exeter (Inst. of Phys.) **1962**, p.244.

66Z   Zhitar, V.F., Oksman, Ya., Radautsan, S.I., Smirnov, V.: Phys. Status Solidi **15** (1966) K105.

67D   Donika, F.G., Kiosse, G.A., Radautsan, S.I., Semiletov, S.A., Zhitar, V.F.: Kristallografiya **12** (1967) 854; Sov. Phys. Crystallogr. (English Transl.) **12** (1968) 745.

69R   Radautsan, S.I., Donika, F.G., Kiosse, G.A., Mustya, I.G., Zhitar, V.F.: Phys. Status Solidi **34** (1969) K129.

70D   Donika, F.G., Radautsan, S.I., Kiosse, G.A., Semiletov, S.A., Donika, T.V., Mustya, I.G.: Kristallografiya **15** (1970) 816; Sov. Phys. Crystallogr. (English Transl.) **15** (1971) 698.

70M   Maynell, C.A., Saunders, G.A., Seddon, T.: Phys. Lett. **31A** (1970) 338.

70R   Ray, B., Spencer, P.M., Younger, P.A.: J. Phys. D **3** (1970) 37.

70S   Saunders, G.A., Seddon, T.: J. Phys. Chem. Solids **31** (1970) 2495.

71M   McCartney, H.: Ph.D. Thesis, University College, London **1971**.

71R   Radautsan, S.I., Andriesh, A.M., Mustya, I.G., Donika, F.G.: Fiz. Tekh. Poluprovodn. **5** (1971) 578; Sov. Phys. Semicond. (English Transl.) **5** (1971) 512.

71S   Saunders, G.A., Seddon, T.: Phys. Lett. **34A** (1971) 443.

72D   Donika, F.G., Radautsan, S.I., Semiletov, S.A., Donika, T.V., Mustya, I.G.: Kristallografiya **15** (1972) 666.

72P   Pitt, G.D., McCartney, J.H., Lees, J., Wright, D.A.: J. Phys. D **5** (1972) 1330.

76S   Saunders, G.A., Seddon, T.: J. Phys. Chem. Solids **37** (1976) 873.

77M   MacKinnon, A., Miller, A., Lockwood, D.J., Ross, G., Holah, G.D.: Inst. Phys. Conf. Ser. **35** (1977) 119.

82H   Hailing, T., Saunders, G.A., Lambson, W.A.: Phys. Rev. **B26** (1982) 5786.

10.1.8 Quaternary compounds

Only those compounds will be considered in this section which have an ordered superstructure which is different from that of the series of mixed crystals of which they are a special case. Thus $Cu_2ZnGeSe_4$ is considered but $CuZn_2GaSe_4$ is not since the latter is not different from the general phase $CuGaSe_2-ZnSe$.

In 10.1.8 data for all substances are arranged according to the headings: "Structure" (0), "magnetic properties" (1) and "other properties" (2).

10.1.8.0 Structure

The $I_2-II-IV-VI_4$ compounds generally adopt one of the two structures stannite (Fig. 1a) or wurtz-stannite (Fig. 1b).

The former (stannite) has space group $D_{2d}^{11}-I\bar{4}2m$ and is a superlattice of zincblende, like chalcopyrite, famatinite, thiogallate.

The latter (wurtz-stannite) is the corresponding wurtzite based structure with space group $C_{2v}^7-Pmn2_1$. The lattice vectors of this orthorhombic structure are related to those of wurtzite by:

$$a = 2a_w; \quad b = 3 \cdot a_w; \quad c = c_w$$

In addition, there are a few, not fully explained zincblende like structures. Some of these may be similar to the pseudocubic, $D_{2d}^1-P\bar{4}2m$, structure of $\square$-$CdIn_2Se_4$ (sect. 10.1.6.9).

A large number of I–III–IV–Se_4 compounds have been investigated in [81G]. It is found that these compounds have the chalcopyrite structure with lattice vacancies and group I atoms arranged at random on the cation sublattice and the group III and group IV atoms at random on the other, i.e. $(I\square)$ $(III IV)Se_4$.

Substance	Structure	Lattice parameters*) [in Å, angles in °]			Remarks	Ref.
Cu_2MnSiS_4	wurtz-stannite	$a = 7.533$	$b = 6.435$	$c = 5.179$		74S
$Cu_2MnSiSe_4$	wurtz-stannite	7.912	6.774	6.501		74S
Cu_2MnGeS_4	wurtz-stannite	7.608	6.511	6.236		74S
$Cu_2MnGeSe_4$	wurtz-stannite	7.979	6.865	6.557		74S
Cu_2MnSnS_4	stannite	$a = 5.49$		$c = 10.72$		74S
$Cu_2MnSnSe_4$	stannite	5.744		11.423		74S
Cu_2FeSiS_4	wurtz-stannite	$a = 7.404$	$b = 6.411$	$c = 6.14$		74S
$Cu_2FeSiSe_4$	stannite	$a = 5.549$		$c = 10.951$		74S
Cu_2FeGeS_4	stannite	5.330		10.528	briartite	74S
$Cu_2FeGeSe_4$	stannite	5.590		11.072		74S
Cu_2FeSnS_4	stannite	5.46		10.725	stannite	74S
$Cu_2FeSnSe_4$	stannite	5.664		11.33		74S
Cu_2CoSiS_4	stannite	5.270		10.327		74S
Cu_2CoGeS_4	stannite	5.30		10.48		74S
$Cu_2CoGeSe_4$	zincblende-like	$a = 5.601$	$b = 5.561$	$c = 5.500$	orthorhombic	74S
Cu_2CoSnS_4	stannite	$a = 5.402$		$c = 10.805$		74S
Cu_2NiSiS_4	zincblende-like	$a = 5.143$	$b = 5.311$	$c = 5.179$	monoclinic	74S
				$\beta = 89.60°$		
Cu_2NiGeS_4	zincblende-like	$a = 5.332$	$b = 5.263$	$c = 5.227$	orthorhombic	74S
$Cu_2NiGeSe_4$	zincblende-like	$a = 5.581$		$c = 5.500$	pseudocubic, $D_{2d}^1 - P\bar{4}2m$	74S
Cu_2NiSnS_4	zincblende-like	$a = 5.425$			pseudocubic?	74S
Cu_2SnSe_4	zincblende-like	5.705			$(c/a = 1)$	
Cu_2ZnSiS_4	wurtz-stannite	$a = 7.435$	$b = 6.396$	$c = 6.135$		74S
$Cu_2ZnSiSe_4$	wurtz-stannite	7.823	6.720	6.440		74S
Cu_2ZnGeS_4	wurtz-stannite	7.504	6.474	6.185		74S
$Cu_2ZnGeSe_4$	stannite	$a = 5.622$		$c = 11.06$		74S
Cu_2ZnSnS_4	stannite	5.427		10.848	kesterite	74S
$Cu_2ZnSnSe_4$	stannite	5.681		11.34		74S
Cu_2CdSiS_4	wurtz-stannite	$a = 7.598\,(8)$	$b = 6.486\,(6)$	$c = 6.258\,(11)$		72C
$Cu_2CdSiSe_4$	wurtz-stannite	7.990	6.824	6.264		74S
Cu_2CdGeS_4	wurtz-stannite	7.692\,(2)	6.555\,(2)	6.299\,(2)		69P
$Cu_2CdGeSe_4$	stannite	$a = 5.657\,(5)$		$c = 10.988\,(10)$		69P
Cu_2CdSnS_4	stannite	5.586		10.834		74S
$Cu_2CdSnSe_4$	stannite	5.814		11.47		74S
Cu_2HgSiS_4	wurtz-stannite	$a = 7.592$	$b = 6.484$	$c = 6.269$		74S
$Cu_2HgSiSe_4$	wurtz-stannite	7.962	6.823	6.569		74S
Cu_2HgGeS_4	stannite	$a = 5.490$		$c = 10.550$		74S
$Cu_2HgGeSe_4$	stannite	5.694		11.02		74S
Cu_2HgSnS_4	stannite	5.566		10.88		74S
$Cu_2HgSnSe_4$	stannite	5.818		11.48		74S
$Ag_2ZnGeSe_4$	zincblende-like	$a = 4.269\,(5)$		$c = 5.659\,(5)$	pseudocubic	69P
Ag_2CdGeS_4	wurtz-stannite	$a = 8.044\,(8)$	$b = 6.849\,(5)$	$c = 6.593$		69P
Ag_2CdSnS_4	wurtzite-like	4.111\,(5)	7.038\,(5)	6.685\,(5)	space group $C_{2v}^{12} - Cmc2_1$	69P
$Ag_2CdSnSe_4$	wurtzite-like	4.262\,(5)	7.324\,(5)	6.979\,(5)	space group $C_{2v}^{12} - Cmc2_1$	69P

*) at $T = 300$ K

Physical property	Numerical value	Experimental conditions	Experimental method remarks	Ref.

10.1.8.1 Magnetic properties

Cu_2FeGeS_4 (briartite):
Néel temperature:

T_N	12.3 (3) K			73I

antiferromagnetic: $k = (\frac{1}{2} \, 0 \, \frac{1}{2})$, $S \parallel c$ (Mössbauer) [73I]

Cu_2FeSnS_4 (stannite):
Néel temperature:

T_N	7 K			72G

antiferromagnetic: $S \perp c$, magnetic group I_p 222 or $I_p$mm2 [72G]; see Fig. 2

$Cu_2FeSnSe_4$:

no transition observed				70E

susceptibility:

χ_g	$2 \cdot 10^{-2}$ m^3 kg^{-1}	$T = 300$ K	SI-units	70E
	$5 \cdot 10^{-2}$ m^3 kg^{-1}	100 K		

10.1.8.2 Other properties

Cu_2ZnSiS_4:
energy gap:

E_g	3.25 eV	$T = 300$ K	absorption, unpolarized	77S

conduction: insulator [77S]

$Cu_2ZnSiSe_4$:
energy gaps:

E_g	2.33 eV	$T = 300$ K	absorption, unpolarized	77S
$E_{g,\,th}$	0.3 eV		transport	77S

resistivity:

ϱ	$2 \cdot 10^3$ Ω cm	$T = 300$ K	p-type conduction, no anisotropy	77S
	$5 \cdot 10^6$ Ω cm	77 K		

Cu_2ZnGeS_4:
energy gaps:

E_g	2.1 eV	$T = 300$ K	absorption, unpolarized	77S
$E_{g,\,th}$	0.03 eV		transport	77S

resistivity:

ϱ	1 Ω cm	$T = 300$ K	no anisotropy	77S
	20 Ω cm	77 K		

$Cu_2ZnGeSe_4$:
energy gaps:

E_g	1.29 eV	$T = 300$ K	absorption, unpolarized	77S
$E_{g,\,th}$	0.01 eV		transport	77S

resistivity:

ϱ	$2 \cdot 10^{-3}$ Ω cm	$T = 300$ K	no anisotropy	77S
	$4 \cdot 10^{-3}$ Ω cm	77 K		

Physical property	Numerical value	Experimental conditions	Experimental method remarks	Ref.
Cu_2FeGeS_2 (briartite):				
splitting of Fe d-bands:				
$\bar{v}(\Gamma_3 - \Gamma_5)$	$1430\ cm^{-1}$	$T = 4.2\ K$		73I
$Cu_2FeSnSe_4$:				
defect level:				
$E_a = 0.012\ eV$		$T = 100 \cdots 300\ K$		70E
resistivity, Hall coefficient, mobility:				
		$T\ [K]$		
ϱ	$2 \cdot 10^{-2}\ \Omega\ cm$	$100 \cdots 300$	p-type conduction	70E
R_H	$3.8 \cdot 10^{-2}\ cm^3\ C^{-1}$	300		
	$1.5 \cdot 10^{-1}\ cm^3\ C^{-1}$	100		
μ_H	$1.5\ cm^2\ V^{-1}\ s^{-1}$	300		
	$5\ cm^2\ V^{-1}\ s^{-1}$	100		

N.B.: no anisotropy reported.

optical energy gaps for further compounds [82G]:

Compounds	$E_g\ [eV]$		
$AgAlGeSe_4$	2.02	all values at RT	
$AgAlSnSe_4$	1.85		
$AgGaGeSe_4$	1.85		
$AgGaSnSe_4$	1.70		
$AgInGeSe_4$	1.58		
$AgInSnSe_4$	0.94		
$CuAlGeSe_4$	2.25		
$CuAlSnSe_4$	1.90		
$CuGaGeSe_4$	1.87		
$CuGaSnSe_4$	1.42		
$CuInGeSe_4$	1.26		
$CuInSnSe_4$	0.71		

10.1.8.3 References for 10.1.8

69P   Parthé, E., Yvon, K., Deitch, R.H.: Acta Crystallogr. **B25** (1969) 1164.
70E   Endo, S., Irie, T.: J. Phys. Soc. Jpn. **29** (1970) 1393.
72C   Chapuis, G., Niggli, A.: Acta Crystallogr. **B28** (1972) 1626.
72G   Ganal, U., Herman, E., Shtrikman, S.: J. Phys. Chem. Solids **33** (1972) 1873.
73I   Imbert, P., Varret, F., Wintenberger, M.: J. Phys. Chem. Solids **34** (1973) 1675.
74S   Schäfer, W., Nitsche, R.: Mater. Res. Bull. **9** (1974) 945.
77S   Schleich, D.M., Wold, A.: Mater. Res. Bull. **12** (1977) 111.
81G   Goodchild, R.G., Hughes, O.H., Woolley, J.C.: Phys. Status Solidi **(a) 68** (1981) 239.
82G   Goodchild, R.G., Hughes, O.H., Lopez-Rivera, S.A., Woolley, J.C.: Can. J. Phys. **60** (1982) 1096.

Physical property	Numerical value	Experimental conditions	Experimental method remarks	Ref.

10.2 Ternary transition-metal compounds

In this chapter, numerous compounds with unknown semiconducting properties have been included, mainly for the sake of completeness. Data of magnetic and related properties of most of the ternary transition-metal compounds (and of their solid solutions) have been omitted in this volume, as they are extensively presented in other Landolt-Börnstein volumes (Vol. III/4b, 12b and 12c). With semiconducting ferroelectric substances, we restrict the data to those which are of prime importance for a semiconductor physicist. For further information, the reader is referred to Landolt-Börnstein, Vol. III/16a, b (Ferroelectrics).

If not stated otherwise, all data in the tables below are measured at room temperature.

10.2.1 Chalcogenides MeM$_2$S$_4$ (Me = Mn, Fe, Co, Ni, Cd; M = In, Ga, Sb, Sn, Rh)
(See also Landolt-Börnstein, Vol. III/12b, section 4.3)

In a ternary spinel AB$_2$X$_4$ two extreme distributions of the cations on the available sites are possible: the normal distribution A(B$_2$)X$_4$, and the inverse distribution B(AB)X$_4$ where in each case the ions written in parentheses occupy octahedral sites. Intermediate distributions are also known. The unit cell contains eight formula units of AB$_2$X$_4$ and is generally referred to the space group O$_h^7$—Fd3m. If not stated otherwise all compounds listed below crystallize in the spinel structure.

MnIn$_2$S$_4$
lattice parameters:

a	10.72 (1) Å			72S
d(In−In)	3.78 Å		Debye-Scherrer	

density:

d	4.44 g/cm^3		theoretical	72S
	4.39 (5) g/cm^3		experimental	

resistivity:

ϱ	10^5···10^8 Ω cm			72S

IR absorption wavenumbers:

$\bar{\nu}$	326 cm^{-1}			71L
	238 cm^{-1}			

FeIn$_2$S$_4$
lattice parameters (in Å):

a	10.61 (1)		see also [76G]	72S
	10.62			73K
	10.618	(u = 0.25907)		78H
d(In−In)	3.75		Debye-Scherrer	72S

density:

d	4.59 g/cm^3		theoretical	72S
	4.49 (5) g/cm^3		experimental	

IR absorption wavenumbers (in cm^{-1}):

$\bar{\nu}$	328			71L
	242			
	192		shoulder	
	92			
	74			

Physical property	Numerical value	Experimental conditions	Experimental method remarks	Ref.
CoIn$_2$S$_4$				
lattice parameters:				
a	10.58 (1) Å		see also [76G]	72S
u	0.384			76G
d(In−In)	3.75 Å			72S
density:				
d	4.67 g/cm^3		theoretical	72S
	4.62 (5) g/cm^3		experimental	
IR absorption wavenumbers (in cm^{-1}):				
$\bar{v}$	336			71L
	246			
	190		shoulder?	
	94			
	78			
NiIn$_2$S$_4$				
lattice parameters:				
a	10.50 (1) Å		see also [76G]	72S
u	0.384			76G
d(In−In)	3.71 Å			72S
density:				
d	4.78 g/cm^3		theoretical	72S
	4.71 (5) g/cm^3		experimental	
IR absorption wavenumbers (in cm^{-1}):				
$\bar{v}$	329			71L
	257			
	99			
	79			
Solid solutions based on MIn$_2$S$_4$(M = Mn, Fe, Ni)				
MnIn$_2$S$_{4-x}$Se$_x$:				
lattice parameters (in Å):				
a	10.746 (3)	$x = 0.2$	slow scan diffraction and	72S
	10.787 (3)	0.5	Guinier films	
	10.830 (3)	1.0		
d(In−In)	3.79	$x = 0.2$		
	3.81	0.5		
	3.83	1.0		
densities (in g/cm^3):				
d	4.49 (5) (4.51)	$x = 0.2$	experimental (theoretical)	72S
	4.55 (5) (4.61)	0.5		
	4.73 (5) (4.79)	1.0		
FeIn$_2$S$_{4-x}$Se$_x$:				
lattice parameters (in Å):				
a	10.643 (3)	$x = 0.2$	slow scan diffraction and	72S
	10.670 (3)	0.5	Guinier films	
d(In−In)	3.76	$x = 0.2$		
	3.77	0.5		

Böhm, Scharmann

Physical property	Numerical value	Experimental conditions	Experimental method remarks	Ref.
densities (in g/cm^3):				
d	4.64 (5) (4.65)	x = 0.2	experimental (theoretical)	72S
	4.70 (5) (4.78)	0.5		

NiIn$_2$S$_{3.5}$Se$_{0.5}$:

lattice parameters:

a	10.574 (3) Å			72S
d(In−In)	3.73 Å			

density:

d	4.93 (5) (4.94) g/cm^3		experimental (theoretical)	72S

For In spinels with cation substitutions, see Landolt-Börnstein, Vol. III/12b, Sect. 4.3.7.

MnGa$_2$S$_4$

structure: monoclinic, space group $C_{2h}^6 - C2/c$ (low temperature α-phase) [81R].

There are six phases in the pressure range 20···100 kbar. For the phase diagram, see Fig. 1. Phase IV is hexagonal with a trimolecular unit cell ($a = 3.695$ (3) Å, $c = 36.35$ (5) Å, $V = 429.8$ (13) Å^3) [71Y].

lattice parameters:

a	12.746 Å	α-phase		81R
b	22.609 Å			
c	6.394 Å			
β	108.78°			

density:

d	3.29 g/cm^3			77R

energy gap:

$E_{g, th}$	1.2 eV		from conductivity (600 K < T < 750 K)	77R

conductivity:

σ	$3.98 \cdot 10^{-4}\,\Omega^{-1}\,cm^{-1}$		polycrystalline material	77R

thermal conductivity:

κ	$4.1 \cdot 10^{-1}\,W\,cm^{-1}\,K^{-1}$		polycrystalline material	71Y

For data on **CdGa$_2$S$_4$**, see sections 10.1.6.0 and 10.1.6.6.

MnSb$_2$S$_4$

lattice parameters:

a	9.664 Å			77R
b	6.49 Å			
c	7.14 Å			
α	30°48′			
β	133°5′			
γ	82°18′			

density:

d	3.53 g/cm^3			77R

energy gap:

$E_{g, th}$	1 eV		from conductivity (600 K < T < 750 K)	77R

Physical property	Numerical value	Experimental conditions	Experimental method remarks	Ref.
conductivity:				
σ	$2.51 \cdot 10^{-4}\,\Omega^{-1}\,cm^{-1}$		polycrystalline material	77R
thermal conductivity:				
κ	$2.2 \cdot 10^{-1}\,W\,cm^{-1}\,K^{-1}$		polycrystalline material	77R

CoRh$_2$S$_4$

structure: cubic, space group O_h^7—Fd3m [64B]

lattice parameter:				
a	9.74 Å			65B
	9.80 Å			76K
Néel temperature:				
T_N	400 K			65B
resistivity:				
ϱ	10 Ω cm	RT	see also Fig. 2	65B
	1000 Ω cm	$T = 0$ K		

parameters of solid solutions:

Co$_{1-x}$Cu$_x$Rh$_2$S$_4$: Seebeck coefficient and resistivity at 80 K: Fig. 2.

Co$_{1-x}$Fe$_x$Rh$_2$S$_4$: Temperature dependence of resistivity: Fig. 3. Lattice constant for a sample with x = 0.5: $a = 9.85$ Å [76K], $E_{g,th} = 0.02$ eV (from conductivity) [76K].

Fe(FeRh)S$_4$

This compound is a ferrimagnetic p-type semiconductor [73H].

lattice parameters of Fe(FeRh)S$_4$ and related compounds:

a [Å]	T_f [°C]			
9.87 ($u = 0.391$)	850···920	Fe(FeRh)S$_4$	samples prepared by mixing powder	73H
9.67 ($u = 0.388$)	850···920	Co(CoRh)S$_4$	of metallic elements, metal	
9.60 ($u = 0.386$)	850···920	Ni(NiRh)S$_4$	sulfides and sulfur, heated at	
			250···700 °C for three days	
			and then refired for about a	
			week at the *firing temperature*	
			T_f given in the third column	
Cu$_{0.5}$Fe$_{0.5}$(Fe$_x$Rh$_{2-x}$)S$_4$:				
9.85	900	x = 0	p-type semiconductor for x = 0 with	73H
9.83	900	0.2	$\varrho = 1.1\,\Omega$ cm (293K) and 15.1 Ω cm	
9.82	870	0.4	(88K) and $S = 155\,\mu V\,K^{-1}$	
9.82	870	0.5	[70P]	
9.81	870	0.6		
9.80	800	0.8		
9.80	800	1.0		

further data on Fe(FeRh)S$_4$:

energy gap:				
$E_{g,th}$	0.05 eV		from conductivity	73H
Seebeck coefficient:				
S	20 $\mu V\,K^{-1}$		for further data on S and ϱ, see [70P]	73H

Physical property	Numerical value	Experimental conditions	Experimental method remarks	Ref.
resistivity:				
ϱ	$2.2 \cdot 10^{-3}\,\Omega\,cm$	RT		73H
	$1.7 \cdot 10^{-2}\,\Omega\,cm$	$T = 80\,K$		
Curie temperature:				
T_C	480 K			73H

For further Rh spinels and Rh spinels with substitutions, see Landolt-Börnstein, Vol. III/12b, Sect. 4.3.3.

$Cd_{1-x}Fe_x(FeSn)S_4$

lattice parameters, Curie temperatures, and firing temperatures:

x	a [Å]	T_C [K]	T_f [°C]	
0	10.61	145	580⋯650	73H
0.2	10.56		580⋯650	
0.3	10.54	123	580⋯650	
0.4	10.51		580⋯650	
0.5	10.48	153	580⋯650	
0.6	10.46	170	530	
0.7	10.42	202	530	
0.8	10.39	245	510	
0.85	10.37	285	480	
0.9	10.35	338	480	

lattice parameters (in Å) of related compounds:

		T_f [°C]	compound	
a	10.40	580⋯600	$Li_{0.5}Fe_{0.5}(FeSn)S_4$	73H
	10.33	750	$Cu_{0.5}Fe_{0.5}(FeSn)S_4$	
	10.53	700	$Ag_{0.5}Fe_{0.5}(FeSn)S_4$	
	10.34	800	$Zn_{0.5}Fe_{0.5}(FeSn)S_4$	
	10.56	650	$Hg_{0.5}Fe_{0.5}(FeSn)S_4$	
	10.44	750	$Cu_{0.5}Mn_{0.5}(MnSn)S_4$	
	10.21	750	$Cu_{0.5}Co_{0.5}(CoSn)S_4$	
	10.30	750	$Cu_{0.5}Ni_{0.5}(NiSn)S_4$	
	10.43	800	$Cu_{0.5}Mn_{0.5}(Mn_{0.75}Sn_{1.25})S_4$	
	10.35	800	$Cu_{0.5}Fe_{0.5}(Fe_{0.75}Sn_{1.25})S_4$	
	10.21	800	$Cu_{0.5}Co_{0.5}(Co_{0.75}Sn_{1.25})S_4$	
	10.31	800	$Cu_{0.5}Ni_{0.5}(Ni_{0.75}Sn_{1.25})S_4$	

further data on $Cd_{1-x}Fe_x(FeSn)S_4$:

The compound is a p-type semiconductor [73H]; for temperature dependence of electrical resistivity, see Fig. 4, of magnetoresistance, see Fig. 5.

Seebeck coefficients (in $\mu V\,K^{-1}$):

S	100	x = 0.5		73H
	90	0.7		
	55	0.8		
	50	0.85		
	30	0.9		

activation energies of resistivity (in eV):

E_A	0.13	x = 0.5	$85\,K < T < 330\,K$	73H
	0.09	0.7		
	0.07	0.8		
	0.04	0.85		
	0.02	0.9		

Physical property	Numerical value	Experimental conditions	Experimental method remarks	Ref.

$CdIn_2S_4$ (See also sections 10.1.6.0 and 10.1.6.8.)

lattice parameters:

a	10.82 Å			76G
u	0.386			

density:

d	$5.0\,\mathrm{g\,cm^{-3}}$		experimental	70C

conductivity:

σ	$2\cdot 10^{-8}\,\Omega^{-1}\,\mathrm{cm^{-1}}$		see also Figs. 6, 7 (conflicting results)	69B

activation energy for conductivity: Fig. 8, see also [73A]; Seebeck coefficient: Fig. 9

IR absorption wavenumbers (in $\mathrm{cm^{-1}}$):

$\bar{\nu}$	311		see also [76W]	71L
	231			
	170		shoulder	
	68			

For further optical properties on pure and doped material, see [63S, 69C, 70B, 74F, 78G].

References for 10.2.1

63S   Springford, M.: Proc. Phys. Soc. (London) **82** (1963) 1029.
64B   Blasse, G., Schipper, D.J.: J. Inorg. Nucl. Chem. **26** (1964) 1467.
65B   Blasse, G.: Phys. Lett. **19** (2) (1965) 110.
69B   Brokhovetskii, B.G., Balakirev, V.F., Dobrovinskii, R.Yu., Men, A.N., Popov, G.P., Chufarov, G.I.: Dokl. Akad. Nauk SSSR **187** (1969) 839; Dokl. Phys. Chem. (English Transl.) **187** (1969) 513.
69C   Czaja, W., Krausbauer, L.: Phys. Status Solidi **33** (1969) 191.
69L   Lotgering, F.K.: J. Phys. Chem. Solids **30** (1969) 1429.
69M   Van Maaren, M.H., Harland, H.B.: Phys. Lett. **30A** (1969) 204.
70B   Brown, M.R., Roots, K.G., Shand, W.A.: J. Phys. **C3** (1970) 1323.
70C   Czaja, W.: Phys. Kondens. Mater. **10** (1970) 299.
70L   Locher, P.R., van Stapele, R.P.: J. Phys. Chem. Solids **31** (1970) 2643.
70P   Plumier, R., Lotgering, F.K.: Solid State Commun. **8** (1970) 477.
70S   Van Stapele, R.P., Lotgering, F.K.: J. Phys. Chem. Solids **31** (1970) 1547.
71L   Lutz, H.D., Fehér, M.: Spectrochim. Acta **A 27** (1971) 357.
71Y   Yokota, M., Syono, Y., Minomura, S.: J. Solid State Chem. **3** (1971) 520.
72S   Schlein, W.S., Wold, A.: J. Solid State Chem. **4** (1972) 286.
73A   Aresti, A., Congiu, A.: Phys. Status Solidi (**a**) **16** (1973) K55.
73B   Barz, H.: Mater. Res. Bull. **8** (1973) 983.
73H   Harada, S.: Mater. Res. Bull. **8** (1973) 1361.
73K   Kanomata, T., Ido, H.: J. Phys. Soc. Jpn. **34** (1973) 554.
73N   Nakanishi, H., Endo, S., Irie, T.: Jpn. J. Appl. Phys. **12** (1973) 1646.
74F   Fujita, H., Okada, Y.: Jpn. J. Appl. Phys. **13** (1974) 1823.
75B   Brasen, D., Vandenberg, J.M., Robbins, M., Willens, R.H., Reed, W.A., Sherwood, R.C., Pinder, X.J.: J. Solid State Chem. **13** (1975) 298.
76E   Endo, S., Irie, T.: J. Phys. Chem. Solids **37** (1976) 201.
76G   Glidewell, C.: Inorg. Chim. Acta **19** (1976) L45.
76K   Kondo, H.: J. Phys. Soc. Jpn. **41** (1976) 1247.
76W   Wakamura, K., Arai, T., Kudo, K.: J. Phys. Soc. Jpn. **41** (1976) 130.
77R   Rustamov, P.G., Guseinova, M.A., Alidzhanov, M.A., Safarov, M.G.: Inorg. Mater. (USSR) **12** (1976) 1218.
78G   Grilli, E., Cappelletti, P., Guzzi, M.: Phys. Status Solidi (**a**) **50** (1978) K93.
78H   Hill, R.J., Craig, J.R., Gibbs, G.V.: J. Phys. Chem. Solids **39** (1978) 1105.
81R   Rimet, R., Buder, R., Schlenker, C., Roques, R., Zanchetta, J.V.: Solid State Commun. **37** (1981) 693.

Physical property	Numerical value	Experimental conditions	Experimental method remarks	Ref.

10.2.2 Chalcogenides MCr$_2$S$_4$ (M = Cd, Fe, Co, Cu, Hg, Zn, Mn, V, Ba)

All these compounds have been treated in Vol. III/12b. We therefore concentrate here on properties which are important to the semiconductor physicist. For magnetic properties and further information and references we refer to Vol. III/12b, section 4.2.

CdCr$_2$S$_4$

CdCr$_2$S$_4$ was discovered in 1965 as a ferromagnetic and semiconducting material [65B].

Crystal structure

structure: spinel (Fig. 1), space group O$_h^7$—Fd3m.

Growth by chemical transport or vapor liquid transport [71P] (see also [74G, 75M]).

coordinates of the atomic sites (the numbers refer to Fig. 1):

Number	Cd	Cr	S	80K
1	(0, 0, 0)	(5/8, 5/8, 5/8)	(u, u, u)	
2	(1/4, 1/4, 1/4)	(5/8, 7/8, 7/8)	$(u, \bar{u}, \bar{u})$	
3		(7/8, 5/8, 7/8)	$(\bar{u}, u, \bar{u})$	
4		(7/8, 7/8, 5/8)	$(\bar{u}, \bar{u}, u)$	
5			$(1/4-u, 1/4-u, 1/4-u)$	
6			$(1/4-u, 1/4+u, 1/4+u)$	
7			$(1/4+u, 1/4-u, 1/4+u)$	
8			$(1/4+u, 1/4+u, 1/4-u)$	

lattice parameters:

a	10.242 (1) Å		temperature dependence of the	77K1
u	0.3901 (3)		lattice parameter: see Fig. 2;	
a	10.244 Å		for interatomic distances,	66H1
a	10.2375 (5) Å		see [72B]; for CdCr$_2$ (S$_{1-x}$Se$_x$)$_4$,	74K
u	0.391		see Fig. 90 of Sect. 10.2.3;	
a	10.207 Å		for Cd$_{1-x}$Co$_x$Cr$_2$S$_4$, see Fig. 53;	80K,
u	0.375		further lattice data: see below	76G1

Electronic properties

energy gap (in eV):

E_g	1.57		optical measurements	66H1
	2.48		photoconductivity	72L
			($\lambda = 520$ nm, single crystal)	
	2.1		magnetoreflectance	71A2
	2.3		polarized modulated	73P1
			magnetoreflectance	
			for energy levels and band structure, see Figs. 3···5 (see also [72N, 71W])	
			For characteristic optical transitions at $T = 1.5$K, see Fig. 3; for temperature dependence of the activation energy of resistivity and optical transitions, see Fig. 6. See also Figs. 17, 29, 32, 33.	

Physical property	Numerical value	Experimental conditions	Experimental method remarks	Ref.

Lattice properties

    lattice vibrations: a group theoretical analysis [70S] shows that there are 42 phonon modes at $q=0$: 12 Raman active modes: $1\Gamma_1^+(1)$, $1\Gamma_{12}^+(2)$, $3\Gamma_{25}^+(3)$, 15 infrared active modes: $5\Gamma_{15}(3)$ and 15 inactive modes: $2\Gamma_2^-(1)$, $2\Gamma_{12}^-(2)$, $1\Gamma_{15}^+(3)$ $2\Gamma_{25}^-(3)$.

wavenumbers of Raman active modes (in cm^{-1}):

		line	T [K]	relative intensity at 40 K	half width cm^{-1}	assign- ment	
$\bar{\nu}$	101 (2)	A	300	–	5	mixed	70S
	105 (2)		40	12	5		
	256 (2)	C	300	–	5	Γ_{12}^+	
	257 (2)		40	44	5		
	280.0 (15)	D	300	–	5	Γ_{25}^+	
	281.0 (15)		40	49	5		
	351 (2)	E	300	–	8	mixed	
	353 (2)		40	74	9		
	394 (2)	F	300	–	7	Γ_1^+	
	396 (2)		40	112	8	mixed	
	460	G	300	6			
	506	H	300	5			
	600	I	300	6			

For Raman spectra, see also under "Optical properties".

wavenumbers of infrared active modes (in cm^{-1}):

		T [K]		
$\bar{\nu}_{TO}$	379.3 (3)	79	reflectance	71L1
	326.6 (4)			
	376.9 (2)	300		
	321.6 (3)			
$\bar{\nu}_{LO}$	392.5 (3)	79		
	350.4 (4)			
	389.9 (3)	300		
	347.2 (3)			

For IR spectra, see also [76P].

linear expansion coefficient (in 10^{-6} K^{-1}):

		T [K]		
α	5.10	179		77K1
	5.60	223		
	6.10	273		
	6.60	323		
	7.05	373		
	7.50	423		
	7.90	473		
	8.25	523		
	8.46	573		
	8.50	623		
	8.50	673		

Grüneisen constant:

γ	1.69 (5)	$T = 373 \cdots 673$ K	from linear expansion and heat capacity	77T

compressibility:

κ	$5.63 \cdot 10^{-12}$Pa^{-1}			77T

Physical property	Numerical value	Experimental conditions	Experimental method remarks	Ref.

Transport properties

Most samples show n-type conductivity. The temperature dependence of conductivity (resistivity) for several pure and doped samples are shown in Figs. 7···11; for magnetoresistance data of Ga doped sample, see Fig. 12, for thin films, see Fig. 13. For resistivity and magnetoresistance on Cd$_{1-x}$Fe$_x$Cr$_2$Se$_4$, see Figs. 49···51.

activation energy of conductivity: (Fig. 6)

E_A	0.15 eV	$T = 120 \cdots 140$ K	see also [72L]	78R
	0.17···0.35 eV	200 K		77G1

influence of annealing treatment on the resistivity of hot-pressed samples [73L]:

Sample number	ϱ(300 K) [Ω cm]	E_A [eV] $T = 270 \cdots 370$ K	Annealing treatment, remarks
1	10^7	0.47	no treatment, undoped sample
2	10^9	0.47	60 h at 600°C with $p_S = 665$ mbar; slow cooling (10 deg/h), undoped sample (p_S: sulfur pressure)
3	10^5	0.09	60 h at 600°C with $p_S = 66.5 \cdot 10^{-6}$ bar; slow cooling (10 deg/h), undoped sample
4	10^{11}	0.70	sample (2 at% Ga doped) obtained by annealing a hot-pressed disk together with powder of the same composition 24 h at 700°C in a sealed quartz-capsule evacuated at RT
5	10^6	0.43	starting material: sample 4, 24 h at 550°C with $p_S = 6.65 \cdot 10^{-7}$ bar, quenched to RT
6	240	0.16	starting material: sample 4, 24 h at 550°C with $p_S = 6.65 \cdot 10^{-9}$ bar, quenched to RT
7	40	0.12	2% Ga doped, no treatment
8	3	0.063	starting material: sample 7, 60 h at 600°C with $p_{Cd} = 93.1$ mbar, quenched to RT, second phase present
9	10^{11}	0.54	starting material: sample 7, 24 h at 700°C with $p_S = 1.013$ bar, quenched to RT

Physical property	Numerical value	Experimental conditions	Experimental method remarks	Ref.

Optical properties

The absorption edge is found to exhibit a shift to higher energies (blue shift) below the Curie temperature [69B, 70W].

absorption coefficient as a function of stoichiometry (in cm^{-1}):

(the three K-values given refer to $\lambda = 9$, 10 and 11 μm, respectively; the stoichiometry is given by the ratio $Cr^{3+} : Cd^{2+}$ in the third column)

K	44.3, 34.6, 30.2	1.953	five different hot-pressed samples	74P
	4.7, 4.6, 5.3	1.982		
	3.0, 2.7, 3.2	2.002		
	1.9, 2.1, 2.6	2.030		
	3.0, 2.7, 3.2	2.040		

absorption coefficient as a function of hot pressing temperature and pressure (in cm^{-1}):

		T_p [°C]	d [%]		
K	68,43	750	99.69	the two K-values given refer to	73P2
	16,16	800	99.82	$\lambda = 1.3$ μm and $\lambda = 2.0$ μm,	
	13,15	850	99.87	respectively; T_p: pressing	
	5,8	900	99.99	temperature; d: density, based on 4.2590 g cm^{-3} as theoretical density of CdCr$_2$S$_4$	
		p [bar]			
	55, 9	87.90		the two K-values given refer to	
	29, 4	155.13		$\lambda = 3.0$ μm and $\lambda = 10.0$ μm,	
	22, 4	262.00		respectively; p: applied	
	17, 3	351.60		pressure of hot pressing	
	10, 5	1161.77			
	8, 4	2323.53			

index of refraction:

		λ [μm]		
n	2.84	15	ellipsometric technique, RT	71M
	2.84	10		
	2.84	5	for index of refraction for longer	
	2.86	2.5	wavelengths, see [73P2]	
	2.89	2.0		
	2.97	1.5		
	3.13	1.2		
	3.37	1.0		
	3.46	0.95		
	3.58	0.90		
	3.75	0.85		
	3.86	0.80		
	3.57	0.633		

dielectric constants:

		T [K]		
$\varepsilon(0)$	9.66 (23)	79		71L1
	9.79 (23)	300		
$\varepsilon(\infty)$	7.84 (20)	79		
	7.84 (20)	300		

Physical property	Numerical value	Experimental conditions	Experimental method remarks	Ref.

figures and further references:

absorption: Figs. 14⋯22 (for dependence on magnetic field, see [69B]; for transmission on samples with different stoichiometry, see [74P]); Faraday rotation: Fig. 23 (for dependence on temperature, magnetic field, sample type, see also [71C, 71M, 71W, 74J, 74T, 75J, 76J, 78K]); Raman spectrum: Figs. 24, 25 (see also [70S, 77K2]); reflectivity: Figs. 26, 27; IR reflectance: Figs. 40, 41 of Sect. 10.2.3 (see also Fig. 66); magnetoreflectance: Figs. 28, 29 (for Kerr effect, see [70W, 71W]); photoconductivity: Figs. 30⋯32; luminescence: Fig. 33

$Cd_{1-x}Co_xCr_2S_4$: absorption and Faraday rotation: Figs. 22, 64, 67

Further properties

density:

d	4.26 g/cm^3		X-ray	77K1
	4.23 (2) g/cm^3		pycnometric	75B

Curie temperature:

T_C	84.5 K			66H1, 77G2

Debye temperature:

Θ_D	415 (5) K	172 K < T < 673 K	heat capacity	77K1

heat capacity (in cal/mol K):

(the two values given refer to C_p and C_v, respectively)

		T [K]		
C	34.15, 33.97	200	for temperature dependence, see Fig. 34	77K1
	35.59, 35.32	223		
	36.80, 36.48	248		
	37.71, 37.34	273		
	38.07, 37.96	298.15		
	38.78, 38.50	323		
	39.70, 39.28	373		
	40.50, 39.95	423		
	40.90, 40.24	473		
	41.27, 40.48	523		
	41.57, 40.67	573		
	41.84, 40.84	623		
	42.04, 40.95	673		

$FeCr_2S_4$

Crystal structure

structure: spinel (Fig. 1), space group O_h^7—Fd3m.

Growth by vapor transport or vapor liquid transport [71P].

lattice parameters (in Å):

a	9.995			75V
	9.9893			69S
	9.995		(u = 0.3850)	66R, 73R1
$d(Cr-S)$	2.403			66R
	2.413			64S

Transport properties

carrier concentration:

p	$1 \cdot 10^{19}$ cm^{-3}	$T = 77$ K	p-type conduction according to sign of Hall and Seebeck coefficient	78W
	$2 \cdot 10^{19}$ cm^{-3}	200 K		

Physical property	Numerical value	Experimental conditions	Experimental method remarks	Ref.
hole mobility:				
μ_p	0.2 cm^2/Vs	$T = 77$ K		78W
	0.3 cm^2/Vs	200 K		
activation energy for conductivity:				
E_A	0.02 eV	$T > T_C$	(see also Figs. 41 and 69)	71G, 68G
	0.22 eV	"for higher temperatures"		77E

In $Fe_{1-x}Cr_{2-x}S_4$ the Seebeck coefficient shows p-type conduction at 300 K between x = 0 and x = 0.4 decreasing from 80 μV/K to 3 μV/K [70R].

figures and further references:

electrical conductivity: Figs. 35···37 (for time dependence, see [73N, 74N], for temperature dependence, see also [76G2]); resistivity: Figs. 38···41; magnetoresistance: Figs. 42···45 (for temperature dependence, see also [71G, 72G], for anisotropy of magnetoresistance, see [73G1, 74L2, 76G2]); Hall effect: Figs. 46, 47; Seebeck coefficient: Fig. 48 (see also Fig. 37); $FeCr_2Se_{4-x}S_x$: resistivity and Seebeck coefficient: Fig. 104 of section 10.2.3; $Cd_{1-x}Fe_xCr_2S_4$: resistivity: Figs. 49, 50, magnetoresistance: Fig. 51; $Co_{1-x}Fe_xCr_2S_4$: resistivity: Figs. 58, 59, magnetoresistance: Fig. 59, activation energy of conductivity: Fig. 60; $Cu_xFe_{1-x}Cr_2S_4$: see below

Optical and further properties

energy level scheme: see [71S1]; IR absorption: see [71L2]

density:

d_{exp}	3.83 g/cm^3			51H

Curie temperature:

T_C	190 K			67H
	185 K			74R

heat capacity: Fig. 52

$CoCr_2S_4$

Crystal structure

structure: spinel (Fig. 1), space group O_h^7—Fd3m.

Growth by vapor transport or vapor-liquid transport [71P], [74G].

lattice parameters (in Å):

a	9.9158		For the dependence of the lattice	69S
	9.9213		parameter in $Cd_{1-x}Co_xCr_2S_4$,	72C
	9.936	($u = 0.383$)	see Fig. 53; phase diagram of	74R
$d(Cr-S)$	2.412		$CoCr_2S_4$—$CoCr_2Se_4$: Fig. 54.	66R

Transport properties

p-type conduction found in [65B]. For temperature dependence of resistivity and magnetoresistance, see Figs. 55···58 and [74L]; resistivity and thermal activation energy of $Co_{1-x}Fe_xCr_2S_4$: Figs. 59, 60.

Optical properties

absorption coefficient as a function of stoichiometry (in cm^{-1}):

(the three K-values refer to $\lambda = 9$, 10 and 11 μm, respectively, the stoichiometry is given by the ratio $Cr^{3+} : Co^{2+}$ for pressed samples in the third column, for powder in the fourth column)

K					
	47.7, 45.0, 47.6	1.951	1.950		74P
	43.0, 41.7, 41.3	1.977	1.975		
	17.0, 16.1, 16.2	1.993	1.993		
	9.0, 8.3, 8.7	2.008	2.005		
	11.3, 9.8, 10.3	2.034	2.034		

Physical property	Numerical value	Experimental conditions	Experimental method remarks	Ref.

figures and further references:

imaginary part of the dielectric constant: Fig. 61; absorption: Figs. 62, 63; transmission: Figs. 65, 66; Faraday effect: Fig. 67. For energy level diagram, see [73A1]; for Kerr effect, see [73A2, 74A]. For reflectance and reflectance circular dichroism of CoCr$_2$S$_4$ and Cd$_x$Co$_{1-x}$Cr$_2$S$_4$, see [73A1, 74A, 75A]; for absorption and Faraday rotation on Cd$_x$Co$_{1-x}$Cr$_2$S$_4$, see Figs. 22, 64, 67.

Further properties
density:

d	3.959 g cm^{-3}			72C

Curie temperature:

T_C	221 K		Faraday rotation	73C
	223 K		differential scanning calorimetry	
	235 K			74R

Debye temperature:

Θ_D	405 K	173 K < T < 673 K	heat capacity	77S

CuCr$_2$S$_4$
structure: spinel (Fig. 1), space group O$_h^7$—Fd3m.

This material exhibits metallic conduction and is ferromagnetic with a moment of about 5 μ$_B$ per molecule [64L, 65B]. For physical properties, see Landolt-Börnstein, Vol. III/12b, section 4.2.2.2. For data on semiconducting CuCr$_2$S$_{4-x}$Se$_x$, see section 10.2.3.

Fe$_{1-x}$Cu$_x$Cr$_2$S$_4$
structure: spinel (Fig. 1), space group O$_h^7$—Fd3m.

The transition from semiconducting to metallic behavior occurs near x = 0.85 [79A] (Fig. 68). Thermal activation energy of conductivity: Fig. 69.

For compositional dependence of Curie temperature, peak value of magnetoresistance, and of the spontaneous Hall coefficient, see [79A].

resistivity, carrier concentration, Hall mobility for a sample with x = 0.42 [79A]:

		T [K]		
ϱ	1.94 · 10^{-2} Ω cm	16	T_C = 315 K	
	0.64 · 10^{-2} Ω cm	78		
	1.87 · 10^{-2} Ω cm	300		
n	1.3 · 10^{20} cm^{-3}	16	For data on Seebeck coefficient,	
	2.3 · 10^{20} cm^{-3}	78	resistivity and activation	
	2.5 · 10^{20} cm^{-3}	300	energy, see also [75O2] and	
$\mu_{H,n}$	2.6 cm^2/Vs	16	[69L].	
	4.2 cm^2/Vs	78		
	1.3 cm^2/Vs	300		

XPS spectra: Figs. 70···72. For semiconducting Cu$_{0.5}$Fe$_{1.5}$Cr$_2$Se$_{1.5}$S$_{2.5}$, see [73G3].

HgCr$_2$S$_4$
structure: spinel (Fig. 1), space group O$_h^7$—Fd3m.

Growth by vapor transport [71P]; ferromagnetic semiconductor.

lattice parameters (in Å):

a	10.237			77G2
	10.2350 (6)		temperature dependence: Fig. 2	73K, 74K
	10.2006	(u = 0.392)		76G4
d(Cr−S)	2.415			65B

Physical property	Numerical value	Experimental conditions	Experimental method remarks	Ref.
density:				
d_{exp}	5.32 g/cm^3			51H
Curie temperature:				
T_C	36 K			65B
energy gap:				
E_g	1.42 eV		temperature dependence: Figs. 73, 74; magnetic field dependence: [70L]	70H
vibrational wavenumbers (in cm^{-1}):				
$\bar{v}$	392			75L2
	344			
	243			
	112			
activation energy of conductivity (in eV):				
		T [°C]		
E_A	0.36	20···100	for temperature dependence of conductivity, see Fig. 75	75O1
	1.2	100···400		
	2.0	400···600		
optical transmission: see Fig. 76				

$ZnCr_2S_4$
structure: spinel (Fig. 1), space group O_h^7—Fd3m.
     Growth by vapor transport [71P].

lattice parameters (in Å):				
a	9.986	($u = 0.384$)	See also Fig. 90 of section 10.2.3 for lattice parameters on $ZnCr_2(S_{1-x}Se_x)_4$.	68M, 76G4
	9.983			73R1
$d(Cr-Cr)$	3.53			70M
$d(Cr-S)$	2.400			65B
$d(Zn-S)$	2.322			76K1
vibrational wavenumbers (in cm^{-1}):				
$\bar{v}$	392		For IR spectra on $Zn_{1-x}Hg_xCr_2S_4$ and $ZnCr_2(S_{1-x}Se_x)_4$, see Figs. 98 and 99 of section 10.2.3.	71L2, 75L2
	344			
	246			
	115			

Néel temperature:

T_N	<20 K			77G2

For resistivity of pressed powders, see [71S2].

$MnCr_2S_4$
structure: spinel (Fig. 1), space group O_h^7—Fd3m.
     Growth by vapor transport [71P].

lattice parameters (in Å):				
a	10.110	($u = 0.386$)		76G4
	10.107			68T
$d(Cr-Cr)$	3.25			71L2
$d(Cr-S)$	2.420			66R
$d(Mn-S)$	2.39			71L2

Physical property	Numerical value	Experimental conditions	Experimental method remarks	Ref.
density				
d_{exp}	3.44 g/cm^3			31P
vibrational wavenumbers (in cm^{-1}):				71L2
$\bar{v}$	381			
	323			
	260			
	120			
Curie temperature:				
T_C	74···95 K			73G2

$V_xCr_{3-x}S_4$

structure: monoclinic, space group C_{2h}^3—I2/m [57J, 66H1].

lattice parameters: Fig. 77

Curie (Néel) temperatures (in K):				
T_C	5 (1)	x = 0.2	determined from $M - H$ curve	79T
	7.0 (5)	0.4		
	7.0 (5)	0.6		
	7 (1)	0.8		
	5.5 (5)	1.0		
	–	0		
T_N	5.0	x = 0	determined from resistivity	
T_C	6.0	0.2	anomaly	
	6.5	0.4		
	7.5	0.6		
	7.8	0.8		
	5.5	1.0		
activation energies of conductivity (in meV) (RT values):				
E_A	6.6	x = 0.6	for temperature dependence of	79T
	9.9	0.8	resistivity, see Figs. 78, 79	
	3.0	1.0		

$BaCr_2S_4$

structure: hexagonal with nine formula units in the unit cell [71O].

lattice parameters:				
a	21.969 (3) Å			71O
c	3.429 (2) Å			
energy gap:				
E_g	1.0···1.5 eV		optical absorption edge	71O
Néel temperature:				
T_N	68 K			71O

For temperature dependence of the resistivity for compounds MCr_2S_4 (M = Ba, Eu, Sr) and activation energies, see Fig. 80.

References for 10.2.2

31P   Passerini, L., Baccaredda, M.: Atti Accad. Naz. Lincei, Cl. Sci. Fis. Mat. Nat. Rend. **14** (1931) 33.

51H   Hahn, H.: Z. Anorg. Allgem. Chem. **264** (1951) 184.

57J   Jellinek, F.: Acta Crystallogr. **10** (1957) 620.

64L   Proc. Int. Conf. on Magnetism, Nottingham (London IPPS) **1964**, 533.

64S   Shirane, G., Cox, D.E., Pickart, S.J.: J. Appl. Phys. **35** (1964) 954.

65B   Baltzer, P.K., Lehmann, H.W., Robbins, M.: Phys. Rev. Lett. **15** (1965) 493.

66H1   Holt, S.L., Bouchard, R.J., Wold, A.: J. Phys. Chem. Solids **27** (1966) 755.

66H2   Haacke, G., Beegle, L.C.: Phys. Rev. Lett. **17** (1966) 427.

66R   Raccah, P.M., Bouchard, R.J., Wold, A.: J. Appl. Phys. **37** (1966) 1436.

67H   Haacke, G., Beegle, L.C.: J. Phys. Chem. Solids **28** (1967) 1699.

68G   Goldstein, L., Gibart, P.: C.R. Acad. Sci. Ser. **B 269** (1968) 471.

68H   Haacke, G., Beegle, L.C.: J. Appl. Phys. **39** (1968) 656.

68M   Miyatani, K., Wada, Y., Okamoto, F.: J. Phys. Soc. Jpn. **25** (1968) 369.

68T   Tressler, R.E., Hummel, F.A., Stubican, V.S.: J. Am. Ceram. Soc. **51** (1968) 391.

69B   Berger, S.B., Ekstrom, L.: Phys. Rev. Lett. **23** (1969) 1499.

69L   Lotgering, F.K., van Stepele, R.P., van der Steen, G.H.A.M, van Wieringen, J.S.: J. Phys. Chem. Solids **30** (1969) 799.

69P   Pellerin, Y., Gibart, P.: C.R. Acad. Sci. Ser. **B 269** 615.

69S   Shick, L.K., von Neida, A.R.: J. Cryst. Growth **5** (1969) 313.

70H   Harbeke, G., Lehmann, H.W.: Solid State Commun. **8** (1970) 1281.

70L   Lehmann, H.W., Harbeke, G.: Phys. Rev. **B1** (1) (1970) 319.

70M   Motida, K., Niyahara, S.: J. Phys. Soc. Jpn. **29** (1970) 516.

70R   Robbins, M., Wolff, R., Kurtzig, A.J., Sherwood, R.C., Miksovsky, M.A.: J. Appl. Phys. **41** (1970) 1086.

70S   Steigmeier, E.F., Harbeke, G.: Phys. Kondens. Mater. **12** (1970) 1.

70W   Wittekoek, S., Bongers, P.F.: IBM J. Res. Dev. **14** (1970) 312.

71A1   Ahrenkiel, R.K., Moser, F., Carnall, E., Martin, T., Pearlman, D., Lyu, S.L., Coburn, T., Lee, T.H.: Appl. Phys. Lett. **18** (1971) 171.

71A2   Ahrenkiel, R.K., Moser, F., Lyu, S., Pidgeom, C.R.: J. Appl. Phys. **42** (1971) 1452.

71C   Coborn, T.J., Moser, F., Ahrenkiel, R.K., Teegarden, K.-J.: IEEE Trans. Magn. **7** (1971) 392.

71F   Fujita, H., Okada, Y., Okamoto, F.: J. Phys. Soc. Jpn. **31** (1971) 610.

71G   Goldstein, L., Gibart, P.: AIP Conf. Proc. **5** (1971) 883.

71L1   Dietz, R.E., Lee, T.H.: J. Appl. Phys. **42** (1971) 1141.

71L2   Lutz, H.D., Fehèr, M.: Spectrochim. Acta **27A** (1971) 357.

71L3   Lugscheider, W., Pink, H., Weber, K., Zinn, W.: Z. Angew. Phys. **32** (1971) 80.

71M   Moser, F., Ahrenkiel, R.K., Carnall, E., Coburn, T., Lyu, S.L., Lee, T.H., Martin, T., Pearlman, D.: J. Appl. Phys. **42** (1971) 1449.

71O   Omloo, W.P.F.A.M., Bommerson, J.C., Heikens, H.H., Risselada, H., Vellinga, M.B., Van Bruggen, C.F., Haas, C., Jellinek, F.: Phys. Status Solidi **5** (1971) 349.

71P   von Philipsborn, H.: J. Cryst. Growth **9** (1971) 296.

71S1   Spender, M.R., Morrish, A.H.: Can. J. Phys. **49** (1971) 2659.

71S2   Sugier, A., Bloch, C.: Proc. 4th Int. Congr. Catal. **2** (1971) 238.

71T   Toda, M.: Proc. 3rd Conf. Solid State Devices, Tokyo **1971** 183.

71W   Wittekoek, S., Rinzema, G.: Phys. Status Solidi (**b**) **44** (1971) 849.

72B   Brüesch, P., D'Ambrogio, F.: Phys. Status Solidi (**b**) **50** (1972) 513.

72C   Carnall Jr., E., Pearlman, D., Coburn, T.J., Moser, F., Martin, T.W.: Mater. Res. Bull. **7** (1972) 1361.

72G   Goldstein, L., Gibart, P.: AIP Conf. Proc. **5** (1972) 883.

72L   Larsen, P.K., Wittekoek, S.: Phys. Rev. Lett. **29** (1972) 1597.

72N   Natsume, Yu., Kamimura, H.: Solid State Commun. **11** (1972) 875.

73A1   Ahrenkiel, R.K., Lee, T.H., Lyu, S.L., Moser, F.: Solid State Commun. **12** (1973) 1113.

73A2   Ahrenkiel, R.K., Coburn, T.J.: Appl. Phys. Lett. **22** (1973) 340.

73B   Belov, K.P., Tret'yakov, Yu.D., Gordeev, I.V., Koroleva, L.I., Ped'ko, A.V., Smirovskaya, E.I., Alferov, V.A., Saksonov, Yu.G.: Sov. Phys. Solid State **14** (1973) 1862.

73C   Coburn, T.J., Pearlman, D., Carnall Jr., E., Moser, F., Lee, T.H., Lyu, S.L., Martin, T.W.: AIP Conf. Proc. **10** (1973) 740.

73G1	Goldstein, L., Lyons, D.H., Gibart, P.: Solid State Commun. **13** (1973) 1503.
73G2	Gibart, P., Robbins, M., Lambrecht Jr., V.G.: J. Phys. Chem. Solids **34** (1973) 1363.
73G3	Gyorgy, E.M., Robbins, M., Gibart, P., Reed, W.A., Schnettler, F.J.: AIP Conf. Proc. **10** (1973) 1148.
73K	Konopka, D., Kozlowska, I., Chelkowski, A.: Phys. Lett. **44A** (1973) 289.
73L	Larsen, P.K., Voermans, A.B.: J. Phys. Chem. Solids **34** (1973) 645.
73N	Nimtz, G., Dietz, G.: Int. J. Magn. **5** (1973) 51.
73O	Osaka, S., Oka, T., Fujita, H.: J. Phys. Jpn. **34** (1973) 836.
73P1	Pidgeon, C.R., Dennis, R.B., Webb, J.S.: Surf. Sci. **37** (1973) 340.
73P2	Pearlman, D., Carnall Jr., E., Martin, T.W.: J. Solid State Chem. **7** (1973) 138.
73R1	Riedel, E., Horvath, E.: Mater. Res. Bull. **8** (1973) 973.
73R2	Riedel, E., Horvath, E.: Z. Anorg. Allg. Chem. **399** (1973) 219.
73W1	Watanabe, T.: J. Phys. Soc. Jpn. **34** (1973) 1695.
73W2	Watanabe, T.: Solid State Commun. **12** (1973) 355.
74A	Ahrenkiel, R.K., Coburn, T.J., Carnall Jr., E.: IEEE Trans. Magn. **10** (1974) 2.
74B	Barraclough, K.G., Lugscheider, W., Meyer, A., Schäfer, H., Treitinger, L.: Phys. Status Solidi **(a) 22** (1974) 401.
74G	Gibart, P., Goldstein, L., Dormann, J.L., Guittard, M.: J. Cryst Growth **24/25** (1974) 147.
74J	Jacobs, S.D., Teegarden, K.J., Ahrenkiel, R.K.: Appl. Opt. **13** (1974) 2313.
74K	Konopka, D., Slebarski, A., Chelkowski, A.: Acta Phys. Pol. **A46** (1974) 47.
74L1	Larsen, P.K.: Proc. Int. Conf. Magn. Moscow 1973 Nauka; Moscow (Publ.) **1974** V484.
74L2	Lyons, D., Goldstein, L., Gibart, P.: Proc. Int. Conf. Magn. Moscow 1973 Nauka; Moscow. (Publ.) **1974**, VI208.
74N	Nimtz, G.: Proc. Int. Conf. Magn. Moscow 1973 Nauka; Moscow (Publ.) **1974** V 489.
74P	Pearlman, D., Carnall Jr., E., Martin, W.: J. Solid State Chem. **9** (1974) 165.
74R	Robbins, M., Gibart, P., Johnson, D.W., Sherwood, R.C., Lambrecht Jr., V.G.: J. Solid State Chem. **9** (1974) 170.
74S	Slebarski, A., Konopka, D., Chelkowski, A.: Phys. Lett. **50A** (1974) 333.
74T	Tsukahara, S., Satoh, T., Tsushima, T.: J. Cryst. Growth **24/25** (1974) 158.
75A	Ahrenkiel, R.K., Lyu, S.L., Coburn, T.J.: J. Appl. Phys. **46** (1975) 894.
75B	Borukhovich, A.S., Marunya, M.S., Lobachevskaya, N.I., Bamburov, V.G., Gel'd, P.V.: Sov. Phys. Solid State **16** (1975) 1355.
75J	Jacobs, S.D.: J. Electr. Mater. **4** (1975) 223.
75L1	Lutz, H.D., Haeusler, H.: Ber. Bunsenges. Phys. Chem. **79** (1975) 604.
75L2	Lutz, H.D., Haeusler, H.: J. Solid State Chem. **13** (1975) 215.
75L3	Lotgering, F.K., van Diepen, A.M., Olijhoek, J.F.: Solid State Commun. **17** (1975) 1149.
75M	Masumoto, K., Nakatani, I., Umemura, H.: J. Jpn. Inst. Met. **39** (1975) 595.
75O1	Okónska-Kozlowska, I., Konopka, D., Jelonek, M., Heimann, J., Pietkiewicz, J., Chelkowski, A.: J. Solid State Chem. **14** (1975) 349.
75O2	Okuda, T., Ando, K., Hayashi, K., Sakurai, K., Kawanishi, K., Tsushima, T.: Applied Magnetism, 7th Symposium 4a-B13 **1975** 88.
75T	Tret'yakov, Yu.D., Vinnik, M.A., Saksonov, Yu.G., Kamyshova, V.K., Gordeev, I.V.: Sov. Phys. Solid State **17** (1975) 1184.
75V	Valiev, L.M., Kerimov, I.G., Babaev, S.Kh., Namazov, Z.M.: Inorg. Mater. **11** (1975) 176.
76G1	Göbel, H.: J. Mag. Magn. Mater. **3** (1976) 143.
76G2	Gibart, P., Goldstein, L., Brossard, L.: J. Mag. Magn. Mater. **3** (1976) 109.
76G3	Golik, L.L., Grigorovich, S.M., Kun'kova, Z.E., Ukrainskii, Yu.M., Shtykov, N.M.: Sov. Phys. Solid State **17** (1976) 1420.
76G4	Glidewell, C.: Inorg. Chim. Acta **19** (1976) L45.
76J	Jacobs, S.D., Teegarden, K.J.: Proc. 8th Ann. Symp. Opt. Mat. High Power Laser, NBS Boulder, Col. **1976**.
76K1	Kovaliev, V.I., Lisnyak, S.S.: Inorg. Mater. **12** (1976) 172.
76K2	Kun'koya, Z.E., Aminov, T.G., Golik, L.L., Elinson, M.I., Kalinnikov, V.T.: Sov. Phys. Solid State **18** (1976) 1212.
76K3	Koshizuka, N., Yokoyama, Y., Tsushima, T.: Solid State Commun. **18** (1976) 1333.
76P	Pink, H., Schäfer, H.: Siemens Forsch.- u. Entwickl.-Ber. **5** (1976) 283.
76T	Treitinger, L., Göbel, H., Pink, H.: Mater. Res. Bull. **11** (1976) 1375.
77E	Eivazov, E.A., Rustamov, A.G., Ibragimov, B.B.: Inorg. Mater. **13** (1977) 354.

77G1	Golik, L.L., Elinson, M.I., Kun'kova, Z.E., Novikov, L.N., Grigorovich, S.M., Ukrainskii, Yu.M.: Sov. Microelectron. **6** (1977) 152.
77G2	Grimes, N.W., Isaac, E.D.: Philos. Mag. **35** (1977) 503.
77K1	Kesler, Ya.A., Shchelkotunov, V.A., Tret'yakov, Yu.D., Kamyshova, V.K., Gordeev, I.V., Alferov, V.A.: Inorg. Mater. **13** (1977) 964.
77K2	Koshizuka, N., Yokoyama, Y., Tsushima, T.: Physica **B89** (1977) 214.
77N	Nikiforov, K.G., Gurevich, A.G., Radautsan, S.I., Tezlevan, V.E., Emiryan, L.M.: Sov. Phys. Solid State **20** (1977) 1096.
77S	Shchelkotunov, V.A., Danilov, V.N., Kesler, Ya.A., Kamyshova, V.K., Gordeev, I.V., Tret'yakov, Yu.D.: Inorg. Mater. **13** (1977) 1389.
77T	Tret'yakov, Yu.D., Gordeev, I.V., Kesler, Ya.A.: J. Solid State Chem. **20** (1977) 345.
78K	Koshizuka, N., Yokoyama, Y., Okuda, T., Tsushima, T.: J. Appl. Phys. **49** (1978) 2183.
78R	Radautsan, S.I., Nikiforov, K.G., Tezlevan, V.E.: Inorg. Mater. **14** (1978) 128.
78T	Treitinger, L., Barraclough, K.G., Feldtkeller, E.: Mater. Res. Bull. **13** (1978) 667.
78W	Watanabe, T., Nakada, I.: Jpn. J. Appl. Phys. **17** (1978) 1745.
79A	Ando, K., Nishihara, Y., Okuda, T., Tsushima, T.: J. Appl. Phys. **50** (1979) 1917.
79K	Koroleva, L.I., Shalimova, M.A.: Sov. Phys. Solid State **21** (1979) 266.
79T	Tazuke, Y.: Phys. Lett. **69A** (1979) 341.
80A	Ando, K.: Solid State Commun. **36** (1980) 165.
80K	Koshizuka, N., Uskioda, S., Tsushima, T.: Phys. Rev. **B21** (1980) 1316.
80T	Takeshi, K., Oguchi, T., Gondaira, K.I.: J. Phys. **C13** (1980) 1493.
81T	Tazuke, Y.: J. Phys. Soc. Jpn. **50** (1981) 413.

Physical property	Numerical value	Experimental conditions	Experimental method remarks	Ref.

10.2.3 Chalcogenides MCr$_2$X$_4$(M = Cd, Cu, Hg, Zn, Fe, Ni, V, Ba, Co; X = Se, Te)

The chromium spinels CdCr$_2$Se$_4$, CuCr$_2$Se$_4$, HgCr$_2$Se$_4$, ZnCr$_2$Se$_4$ and CuCr$_2$Te$_4$ (and their solid solutions) have been extensively treated in section 4.2.2 and 4.2.3 of Vol. III/12b. We include here only the most important data of these sections and refer to them for more information and references.

CdCr$_2$Se$_4$

Crystal structure

structure: spinel (Fig. 1 of section 10.2.2), space group O_h^7—Fd3m.

Ferromagnetic semiconductor, grown by flux [66H, 67B], liquid transport [67P1, 67P2, 69P] and vapor-liquid-transport [69N]. See also [70L, 71P, 74G].

phase diagram: Fig. 1

coordinates of atomic sites: see CdCr$_2$S$_4$ in section 10.2.2

lattice parameters (in Å):

a	10.745		The unit cell includes 14 atoms.	66H, 67B
	10.740		A structural change to a monoclinic phase is obtained	74P
	10.755		under pressure; it is accompanied by a change from	75M1, 77G3
	10.721	($u = 0.383$)	semiconducting to metallic behavior.	72B2, 76G2
d(Cd−Se)	2.44		For temperature dependence of a,	72B2
(Cr−Se)	2.71		see Fig. 2 of section 10.2.2;	68L
	2.54		for interatomic distances,	
(Cr−Cr)	3.78		see also [77G3] and [76G2].	72B2
			For lattice parameters of Cd$_x$Zn$_{1-x}$Cr$_2$Se$_4$, see [76W]; for CdCr$_2$(S$_{1-x}$Se$_x$)$_4$, see Fig. 90.	

Physical property	Numerical value	Experimental conditions	Experimental method remarks	Ref.

Electronic properties

energy gap (in eV):

E_g	1.7		temperature dependence: Fig. 2	71A2
	1.2		(see also below "Optical	70S2
	1.32		properties" Figs. 53···56, 61)	70A

energy diagrams and calculated band structures: Figs. 3···6

Lattice properties

For the phonon modes possible in the spinel lattice, see $CdCr_2S_4$ in section 10.2.2.

wavenumbers of Raman active modes (in cm^{-1}):

		line	T [K]	relative intensity at 50 K	half width cm^{-1}	assignment	
$\bar{\nu}$	84 (2)	A	300	3			70S3
	85 (2)		80		11		
	85 (2)		10				
	144 (2)	B	300	3			
	146 (2)		80		8		
	147 (2)		10				
	154 (1)	C	300	33	5.7 (5) ⎫		
	157 (1)		80		5.1 (5) ⎬ Γ_{12}^+		
	158 (1)		10		5.0 (5) ⎭		
	169.0 (15)	D	300	23			
	171.0 (15)		80		6.0 (10) ⎫ Γ_{25}^+		
	172.0 (15)		10		6.0 (5) ⎭		
	226 (2)	E	80	2.5			
	226 (2)		10				
	237.0 (15)	F	300	6.5			
	240.0 (15)		80		≈ 9 ⎫ Γ_1^+		
	241.0 (15)		10		⎭		
	291 (2)	G	80	6	≈ 12		
	291 (2)		10				
	300 (2)	H	80	7.4	≈ 12		
	300 (2)		10				

For Raman spectra, see Figs. 7, 8 and [78I2, 77K1]; for temperature dependence of Raman lines, see Figs. 8, 9; for influence of an electric field on Raman spectra, see [82S].

wavenumbers of infrared active modes (in cm^{-1}):

		T [K]		
$\bar{\nu}_{TO}$	291.6 (5)	79	reflectance	71L1
	270.9 (3)			
	288.1 (6)	300		
	266.2 (2)			
$\bar{\nu}_{LO}$	295.0 (5)	79	For temperature dependence of	
	286.2 (3)		two infrared active phonon	
	291.4 (6)	300	wavenumbers, see Fig. 10; for IR	
	281.3 (4)		spectra, see also [73P, 73W2].	

(continued)

Physical property	Numerical value	Experimental conditions	Experimental method remarks	Ref.
wavenumbers of infrared active modes (continued)				
$\bar{v}_{TO}$	290.0	80,	oscillator frequencies	71W1
	268.0	ferro-		
	189.0	magnetic		
	78.7			
	70.0			
	286.6	295,		
	264.5	para-		
	188.0	magnetic		
	74.5			
	293.8	90	shift near T_C: 0.4 cm^{-1}	76W
	289.3	300		
	276.6	90	shift near T_C: 0.6 cm^{-1}	
	271.2	300		
	191.8	90	shift near T_C: 0.6 cm^{-1}	
	189.2	300		
	78.3	90	shift near T_C: 0.4 cm^{-1}	
	75.5			

For Grüneisen parameter, see [76W].

Transport properties

CdCr$_2$Se$_4$ can be p-type or n-type depending on the method of preparation [67L2]. Substitution of monovalent elements such as Ag for Cd render the material p-type, while trivalent elements such as In replacing Cd render the material n-type [68W1, 68W2].

resistivities and Hall mobilities of several Cd$_{1-x}$In$_x$Cr$_2$Se$_4$ and Cd$_{1-x}$Ag$_x$Cr$_2$Se$_4$ solid solutions (from [82K]): (see also Fig. 16)

Composition	Type of conduction	Preparation (doping) method	ϱ Ω cm 300 K	ϱ Ω cm 77 K	μ_H cm^2 V^{-1} s^{-1} 300 K
Cd$_{0.96}$In$_{0.04}$Cr$_2$Se$_4$	n	during growth	$7.1 \cdot 10^3$	$3.2 \cdot 10^5$	
Cd$_{0.994}$In$_{0.006}$Cr$_2$Se$_4$	n	diffusion	$2.5 \cdot 10^2$	4.0	
Cd$_{0.988}$In$_{0.012}$Cr$_2$Se$_4$	n		47		
	n		2.3	$2 \cdot 10^{-1}$	1.4
Cd$_{0.976}$In$_{0.024}$Cr$_2$Se$_4$	n		$1.8 \cdot 10^{-1}$	$2.4 \cdot 10^{-1}$	56
Cd$_{0.91}$In$_{0.09}$Cr$_2$Se$_4$	n		$6.7 \cdot 10^{-3}$	$6.7 \cdot 10^{-3}$	65
Cd$_{0.952}$In$_{0.048}$Cr$_2$Se$_4$	n	in charge	32	8.8	0.47
CdCr$_2$Se$_4$	p		$4 \cdot 10^4$	$4 \cdot 10^6$	
Cd$_{0.999}$Ag$_{0.001}$Cr$_2$Se$_4$	p	during growth	72	$3.8 \cdot 10^4$	
Cd$_{0.995}$Ag$_{0.005}$Cr$_2$Se$_4$	p		$8 \cdot 10^{-1}$	$2.5 \cdot 10^3$	18
Cd$_{0.99}$Ag$_{0.01}$Cr$_2$Se$_4$	p		$3.2 \cdot 10^{-1}$	$4.4 \cdot 10$	
Cd$_{0.97}$Ag$_{0.03}$Cr$_2$Se$_4$	p		$3 \cdot 10^{-2}$	13	39

Annealing treatment: a sample with $\varrho = 7 \cdot 10^3\ \Omega$ cm and $E_A = 0.22$ eV annealed for 60 h at 600 °C with $p_{Se} = 13.3 \cdot 10^{-8}$ bar and quenched to RT showed a resistivity of 110 Ω cm with $E_A = 0.14$ eV [73L]. For the influence of different thermal treatment on the resistivity and activation energy of pure and doped CdCr$_2$Se$_4$, see also [77T1, 78T] and Fig. 11.

For temperature dependence of resistivity for pure and doped material, see Figs. 11···16 and [71T1, 75V3, 78T]; of conductivity: Figs. 17···21, 23, Fig. 75 of 10.2.2 and [67L1, 72K, 76I1, 81M]; of carrier concentration: Figs. 22···26; of Hall effect: Figs. 25, 26 and [70F, 70A, 76I1].

n-type conducting Cd chromium spinels show very large magnetoresistance, which is nearly independent of the orientation of the magnetic field. For magnetoconductivity and -resistance, see Figs. 27···30 and [70H, 70V, 72B3, 75M2, 77T1, 78T, 76I1,2, 78K2].

Böhm, Scharmann

Physical property	Numerical value	Experimental conditions	Experimental method remarks	Ref.

Optical properties

energies (in eV) for main optical transitions: (theoretical values, from [80K5])

For the assignment of the transitions, see Figs. 5, 6.

			assignment	
E(VB−CB)	1.91	paramagnetic phase	$\Gamma_{15} \to \Gamma_1$	
	2.52		$\Gamma_{15} \to \Gamma_{25'}$	
	2.98		$L_4 \to L_1$	
	3.50		$X_4 \to X_1$	
	3.36		$K_4 \to K_1$	
	1.66	ferromagnetic phase,	$\Gamma_{15} \to \Gamma_1$	
	2.31	spin ↑	$\Gamma_{15} \to \Gamma_{25'}$	
	2.74		$L_4 \to L_1$	
	3.26		$X_4 \to X_1$	
	3.12		$K_4 \to K_1$	
	2.11	ferromagnetic phase	$\Gamma_{15} \to \Gamma_1$	
	2.52	spin ↓	$\Gamma_{15} \to \Gamma_{25'}$	
	2.72		$L_4 \to L_1$	
	3.33		$X_4 \to X_1$	
	3.04		$K_4 \to K_1$	
E(VB→d$_\gamma$)	1.52···1.57	paramagnetic phase	$\Gamma_{15} \to d_\gamma$	
	1.53···2.04		$L_4 \to d_\gamma$	
	1.99···2.40		$X_4 \to d_\gamma$	
	1.91···2.35		$K_4 \to d_\gamma$	
	1.54···1.60	ferromagnetic phase,	$\Gamma_{15} \to d_\gamma$	
	1.54···2.06	spin ↑	$L_4 \to d_\gamma$	
	2.00···2.45		$X_4 \to d_\gamma$	
	1.89···2.35		$K_4 \to d_\gamma$	
E(d$_\varepsilon$→CB)	1.32···1.39	paramagnetic phase	$d_\varepsilon \to \Gamma_1$	
	1.88···1.95		$d_\varepsilon \to L_1$	
	1.91···2.01		$d_\varepsilon \to X_1$	
	1.93···2.04		$d_\varepsilon \to K_1$	
	1.05···1.14	ferromagnetic phase,	$d_\varepsilon \to \Gamma_1$	
	1.62···1.71	spin ↑	$d_\varepsilon \to L_1$	
	1.66···1.77		$d_\varepsilon \to X_1$	
	1.69···1.80		$d_\varepsilon \to K_1$	

Fundamental absorption edge between s conduction band and p valence band ≈ 1.8 eV [82Y]. Anomalous absorption edge is probably caused by transitions between the p valence band and the unfilled d-band (e$_g$) [82Y]. See also [80G].

dielectric constants:

		T [K]			
$\varepsilon(0)$	14.2	295			71W1
	10.28 (22)	300			71L1
	14.9	80			71W1
	10.27 (23)	79			71L1
$\varepsilon(\infty)$	11.0	295			71W1
	9.0 (2)	300			71L1
	11.3	80			71W1
	9.00 (23)	79			71L1

Physical property	Numerical value	Experimental conditions	Experimental method remarks	Ref.

figures and further references:

real and imaginary parts of the dielectric constant: Figs. 31, 38; absorption coefficient vs. energy: Figs. 32···36 and Fig. 20 of 10.2.2 (see also [73I]; absorption with polarized light, see also [76H]); reflectivity: Figs. 37···44 (see also Fig. 27 of 10.2.2); dichroism: Figs. 45, 46; Faraday rotation: Figs. 47, 48 (see also [76G3, 77G1]); thermoreflectance: Figs. 49, 50 (see also [76S2], for In doped samples, see [75T]); piezoreflectance: Figs. 51, 52; infrared transmission: [76W] (see also under "Lattice properties"); photoinduced optical density: [77H]; temperature dependence of absorption edge: Figs. 53···55 (see also [76S1]); refractive index: [73P]; photoconductivity: Figs. 57···61 (see also [71A1, 76H, 76S2]; for magnetic field dependence of photoconductivity, see [77G2]; for life time of charge carriers, see [75V3]); luminescence: Figs. 62···64 (for luminescence life time, see [81Y]; for magnetic field dependence, see [75V2]).

$CdCr_2Se_{4-x}S_x$: temperature dependence of absorption edge: Fig. 56 (see also Fig. 20 of 10.2.2)
$Cd_{1-x}Hg_xCr_2Se_4$: see under $HgCr_2Se_4$

Further properties

density:

d	5.68 g cm^{-3}			75G

Curie temperature:

T_C	129.5 K			77G3
	142 K			71L7

Debye temperature:

Θ_D	360 K	$T = 90$ K		75M1

heat capacity: temperature dependence, see Fig. 65

$CuCr_2Se_4$

This material exhibits metallic conduction and is ferromagnetic.
For physical properties, see Vol. III/12 b, section 4.2.2.2.

$CuCr_2X_3Y$ (X = S, Se, Te; Y = I, Br, Cl) proved to be semiconductors and ferromagnetics at RT [68M].
For data on $CuCr_2Se_{4-x}Br_x$ ($E_g \approx 0.9$ eV for x = 1), see also [71M2, 73V, 71L4].

$CuCr_2S_{4-x}Se_x$ (acc. to [79K])
lattice parameters:

a	9.88 Å	x = 0.5	for lattice parameters, interatomic	
	9.94 Å	1.0	distances, see also [73B], [73R2]	
	10.021 Å	1.5		

resistivities:

ϱ	0.04 Ω cm	x = 0.5	$T =$	semiconducting behavior for x = 1
	0.04 Ω cm	1.0	300 K	and x = 1.5 (see also Figs. 66, 67)
	0.01 Ω cm	1.5		

carrier concentrations:

p	$6 \cdot 10^{20}$ cm^{-3}	x = 1.0	$T =$	
	$6 \cdot 10^{20}$ cm^{-3}	1.5	100 K	

mobilities:

μ_p	0.15 cm^2/Vs	x = 1.0	$T =$	
	0.70 cm^2/Vs	1.5	100 K	

Physical property	Numerical value	Experimental conditions	Experimental method remarks	Ref.
activation energies:				
		T[K]		
E_A	1.5 meV	x = 1.0 77		
	50 meV	200		
	7.1 meV	1.5 77		
	66 meV	200		
Curie temperature:				
T_C	355 K	x = 0.5		
	310 K	1		
	350 K	1.5		

For semiconducting $Cu_{0.5}Fe_{0.5}Cr_2Se_{1.5}S_{2.5}$, see [73G].

$HgCr_2Se_4$

Crystal structure

structure: spinel (Fig. 1 of section 10.2.2), space group O_h^7—Fd3m.

Ferromagnetic semiconductor, grown by vapor transport [69L, 70T3] (see also [71P, 78G2]).

lattice parameters (in Å):				
a	10.743 (2)		temperature dependence: Fig. 68;	70T3
	10.725 (2)		see also Fig. 2 of section 10.2.2;	72B1
	10.7399	(u = 0.389)	for $Hg_xZn_{1-x}Cr_2Se_4$,	74K1
	10.753		see Fig. 87	77G3
$d(Cr-Cr)$	3.80			70M2
Electronic properties				
energy gap:				
E_g	0.44 eV	$T = 78$ K	optical measurements (see also	71L2
	0.84 eV	300 K	under "Optical properties"	
			(Figs. 84, 85)); for	
			$Hg_xZn_{1-x}Cr_2Se_4$, see Figs. 88, 89	
Transport properties				
resistivity:				
ϱ	1.2 Ω cm	$T = 300$ K	temperature dependence:	71L2
	310 Ω cm	77 K	Figs. 69···72; for In doped	
			samples, see also Fig. 86	
Hall mobility:				
μ_H	300···1200 cm²/Vs	$T = 4.2$ K	temperature dependence:	78G1
			Figs. 71, 72 (μ_H behavior differs	
			from μ_{opt} (Fig. 72)); for Hall effect	
			on $Hg_{1-x}Ag_xCr_2Se_4$, see [71M1]	
effective masses:				
m^*	$0.3\,m_0$	p-type	from Hall effect at RT	81S
	$0.35\,m_0$	n-type		
Seebeck coefficients (in μV/K):				
S	33	undoped	see also [71M1]	70T3
	229	1.0% Ag		
	181	1.55% Ag		
	24	1.0% In		
	23	2.0% In		

Physical property	Numerical value	Experimental conditions	Experimental method remarks	Ref.

magnetoresistance: Figs. 73, 74 (pure and doped samples) and [75M2]; for transport at ferromagnetic resonance conditions, see [70T1, 70T2]. n-type conducting $HgCr_2X_4$ show very large magnetoresistance, being nearly independent of the orientation of the *B*-field.

activation energies for conductivity in $HgCr_2S_{4-x}Se_x$ (in eV):

		x =	T [K]	
E_A	0.36	4	25···500	75O
	0.45	3.75	30···550	
	0.53	3.5	50···550	
	0.68	3	50···600	
	1.3	0.5	125···600	
	1.65	0.25	100···550	
	0.36	0	20···100	
	1.2	0	100···400	
	2.0	0	400···600	

For lattice parameters and X-ray diffraction of $HgCr_2S_{4-x}Se_x$, see [76L] and [73K], respectively.

Optical properties

absorption coefficient vs. energy and temperature: Figs. 75···77; reflectivity: Fig. 78; transmission: Fig. 79; Faraday effect: Fig. 80; Raman effect: Figs. 81, 82 (see also [82S]); cathodoluminescence: [81B]; photoconductivity: Figs. 83, 84; optical absorption edge, dependence on magnetic field: Fig. 85; $Hg_xZn_{1-x}Cr_2Se_4$: optical spectra: Fig. 97 (see also Figs. 88, 89 and [75W] for absorption and [73W2] for far infrared absorption); $Cd_{1-x}Hg_xCr_2Se_4$: cathodoluminescence [82A]

Further properties

Curie temperature:

T_C	106 K			77G3, 65B, 73A1
	120 K			71L7

Debye temperature:

Θ_D	540 K		"high temperature range"	76W

$ZnCr_2Se_4$

Crystal structure

structure: spinel (Fig. 1 of section 10.2.2), space group O_h^7—Fd3m; growth by vapor-liquid transport [69N].

lattice parameters (in Å):

a	10.494		temperature dependence: Fig. 2	69N
	10.4966		of section 10.2.2;	74K1
	10.440	($u = 0.385$)	for $ZnCr_2(S_{1-x}Se_x)_4$, see Fig. 90; for $Hg_xZn_{1-x}Cr_2Se_4$, see Fig. 87	72G
d(Cr−Cr)	3.712			77G3

Electronic properties

energy gap

E_g	1.285 eV		optical measurement (see also Fig. 95); for $Hg_xZn_{1-x}Cr_2Se_4$, see Figs. 88, 89	66B

Physical property	Numerical value	Experimental conditions	Experimental method remarks	Ref.

Transport properties

resistivity: Figs. 91, 92, Hall effect: Figs. 93, 94

The following table gives the resistivities in Ω cm at 300 K (in brackets at 77K) and the room-temperature values of the Hall coefficients (in cm^3/C), the carrier concentrations (in cm^{-3}), the Hall mobilities (in cm^2/Vs) and the Seebeck coefficients (in $\mu V/K$).

Physical property	Numerical value	Experimental conditions	Experimental method remarks	Ref.
ϱ	$40 \, (4.6 \cdot 10^7)$	sample 1	no doping, no annealing treatment	74W
R_H	130			
n	$4.8 \cdot 10^{16}$			
μ_H	3			
ϱ	$28 \, (1.2 \cdot 10^6)$		same crystal as above,	
R_H	200		annealing treatment:	
n	$3.1 \cdot 10^{16}$		Se 60 mg, 600 °C for 22 h	
μ_H	7			
ϱ	$43 \, (>10^9)$	sample 2	no doping, no annealing treatment	
R_H	160			
n	$3.9 \cdot 10^{16}$			
μ_H	4			
S	620			
ϱ	$50 \, (2.1 \cdot 10^4)$		as sample 2,	
R_H	300		annealing treatment:	
n	$2.0 \cdot 10^{16}$		Se 20 mg, 600 °C for 27 h	
μ_H	6			
ϱ	$20 \, (3.0 \cdot 10^7)$	sample 3	no doping, no annealing treatment	
R_H	120			
n	$5.2 \cdot 10^{16}$			
μ_H	6			
S	740			
ϱ	$21 \, (1.2 \cdot 10^6)$		as sample 3,	
R_H	120		annealing treatment:	
n	$5.2 \cdot 10^{16}$		Se 200 mg, 690 °C for 47 h	
μ_H	6			
ϱ	$330 \, (2.0 \cdot 10^8)$		as sample 3,	
R_H	160		annealing treatment:	
n	$3.9 \cdot 10^{15}$		in vacuum 555 °C for 68 h	
μ_H	5			
ϱ	$1.6 \cdot 10^3 \, (>10^9)$	sample 4	doped with 0.4 at% In,	
R_H	$1.9 \cdot 10^3$		no annealing treatment	
n	$3.3 \cdot 10^{15}$			
μ_H	1			
S	600			
ϱ	$200 \, (>10^9)$		as sample 4,	
			annealing treatment:	
			Se 200 mg, 535 °C for 53 h	
ϱ	$0.31 \, (8.4 \cdot 10^9)$	sample 5	doped with 0.24 at% Cu,	
R_H	0.52		no annealing treatment	
n	$1.2 \cdot 10^{19}$			
μ_H	2			
ϱ	$1.6 \, (4.7 \cdot 10^4)$		as sample 5,	
R_H	6.5		annealing treatment:	
n	$1 \cdot 10^{18}$		in vacuum 570 °C for 53 h	
μ_H	4			
ϱ	$9.0 \cdot 10^3 \, (>10^9)$	sample 6	doped with 0.6 at% Cd,	
			no annealing treatment	

Physical property	Numerical value	Experimental conditions	Experimental method remarks	Ref.

Optical properties

absorption coefficient: Fig. 95; magnetic field dependence of absorption: [71L6]; reflectivity: Figs. 100, 101; dielectric constant: Figs. 102, 103; refractive and absorption index: Fig. 102;
ZnCr$_2$(S$_{1-x}$Se$_x$)$_4$: IR spectra: Figs. 96, 99;
Zn$_{1-x}$Hg$_x$Cr$_2$(S, Se)$_4$: IR spectra: Figs. 97, 98 (see also [73W2]; for absorption, see also Figs. 88, 89 and [75W])

Further properties
density:

d	5.55 g cm^{-3}		X-ray	73M
	5.49 g cm^{-3}		pycnometric	

Néel temperature:

T_N	45.5 K		antiferrromagnetic with a helical	75P
	20 K		spin structure	77G3

FeCr$_2$Se$_4$
Crystal structure

structure: NiAs structure, hexagonal close packed array of anion with cations occupying only octahedral interstices, monoclinic symmetry.

lattice parameters:

a	6.242 Å	(6.27 Å)	(values in brackets acc. to [75V1])	73M
b	3.601 Å	(3.62 Å)		
c	10.872 Å	(11.85 Å)		
β	90° 48′	(90° 92′)		

density:

d	5.71 g cm^{-3}		X-ray	73M

Transport properties

temperature dependence of resistivity, Seebeck coefficient, Hall mobility: Figs. 37, 47 of section 10.2.2; Fig. 104 shows data on FeCr$_2$S$_{4-x}$Se$_x$

carrier concentration, Seebeck coefficient:

n	$0.4 \cdot 10^{20}$ cm^{-3}			75V1
S	25 μV/K		conflicting data, cf. Fig. 37	73M
			of section 10.2.2	

thermal conductivity (in 10^{-4} cal cm^{-1} s^{-1} K^{-1}):

		T [K]		
κ	0.023	85	polycrystalline material	73A2
	0.057	170		
	0.11	270		
	0.13	300		
	0.15	330		
	0.16	368		
	0.34	400		
	0.59	440		

For semiconducting Cu$_{0.5}$Fe$_{0.5}$Cr$_2$Se$_{1.5}$S$_{2.5}$, see [73G].

Physical property	Numerical value	Experimental conditions	Experimental method remarks	Ref.

$NiCr_2Se_4$

structure: monoclinic, space group C_{2h}^3—I2/m [63C]. See also [73A2].

lattice parameters (in Å):

a	6.22			63C
b	3.59			
c	11.53			
β	91°			

interatomic distance of transition metal: 2.88 Å [73A2].

electrical conductivity:

σ	36.4 Ω^{-1} cm^{-1}		polycrystalline material	73A2

thermal conductivity (in 10^{-4} cal cm^{-1} s^{-1} K^{-1}):

		T [K]		
κ	0.72	95	polycrystalline material	73A2
	2.56	210		
	4.28	298		
	4.56	312		
	4.41	330		
	5.89	350		
	7.37	390		
	10.48	440		

resistivity: (ac-measurement at 50 Hz) of pressed powders: $\varrho < 1$ Ω cm [71S].

VCr_2Se_4

structure: monoclinic [63C, 73A2].

lattice parameters (in Å):

a	6.28			63C
b	3.58			
c	11.73			
β	91°18′			

interatomic distance of the transition metal: 2.93 Å [73A2].

electrical conductivity:

σ	22.2 Ω^{-1} cm^{-1}		polycrystalline material	73A2

thermal conductivity (in 10^{-4} cal cm^{-1} s^{-1} K^{-1}):

		T [K]		
κ	0.24	95	polycrystalline material	73A2
	2.06	220		
	3.21	290		
	3.79	312		
	4.13	330		
	4.42	340		
	4.25	400		
	6.61	440		

$BaCr_2Se_4$

structure: hexagonal (nine formula units in the unit cell), space group C_6^1—P6.

lattice parameters:

a	22.939 Å			71O
c	3.629 Å			

Physical property	Numerical value	Experimental conditions	Experimental method remarks	Ref.

Curie temperature:

T_C 125 K   71O

resistivity: Fig. 105

CuCr$_2$Te$_4$

This compound exhibits metallic conduction and is ferromagnetic with a moment of about 5 μ_B per molecule. For physical properties, see Vol. III/12b, section 4.2.2.2.

CoCr$_2$Te$_4$

structure: monoclinic; growth by Bridgman method from polycrystalline specimens [77V].

lattice parameters:

a	6.70 Å			77V
b	3.76 Å			
c	12.53 Å			
β	90°20′			

transport data: conductivity and Seebeck coefficient: Fig. 106, Hall potential: Fig. 107

FeCr$_2$Te$_4$

structure: inverse spinel (NiAs structure).

lattice parameters:

a	6.70 Å			75V1
b	3.82 Å			
c	12.10 Å			
β	91.56°			

activation energy of conductivity, carrier concentration:

E_A	0.01 eV			77V
n	$\approx 10^{21}$ cm^{-3}			77V

Temperature dependence of conductivity, Seebeck coefficient and Hall mobility: Figs. 37 and 47 of section 10.2.2.

References for 10.2.3

63C   Chevreton, M.: Acta Crystallogr. **16** (1963) Suppl. A22 (Abstracts).
65B   Baltzer, P.K., Lehmann, H.W., Robbins, M.: Phys. Rev. Lett. **15** (1965) 493.
66B   Busch, G., Magyar, B., Wachter, P.: Phys. Lett. **23** (1966) 438.
66H   Harbeke, G., Pinch, H.: Phys. Rev. Lett. **17** (1966) 1090.
66P1   Plumier, R.J.: J. Appl. Phys. **37** (1966) 964.
66P2   Plumier, R.J.: J. Phys. (Paris) **27** (1966) 213.
67B   Berger, S.B., Pinch, H.: J. Appl. Phys. **38** (1967) 949.
67L1   Lehmann, H.W.: Phys. Rev. **163** (1967) 488.
67L2   LeCraw, R.C., von Philipsborn, H., Sturge, M.D.: J. Appl. Phys. **38** (1967) 965.
67P1   von Philipsborn, H.: J. Appl. Phys. **38** (1967) 955.
67P2   von Philipsborn, H.: Helv. Phys. Acta **40** (1967) 810.
68L   Lotgering, F.K., van Stapele, R.P.: Mater. Res. Bull. **3** (1968) 507.
68M   Miyatani, K., Wada, Y., Okamoto, F.: J. Phys. Soc. Jpn. **25** (1968) 369.
68W1   Wen, C.P., Hershenov, B., von Philipsborn, H., Pinch, H.: IEEE Trans. Magn. **4** (1968) 702.
68W2   Wen, C.P., Hershenov, B., von Philipsborn, H., Pinch, H.: Appl. Phys. Lett. **13** (1968) 188.
69L   Lehmann, H.W., Emmenegger, F.P.: Solid State Commun. **7** (1969) 965.
69N   von Neida, A.R., Shick, L.K.: J. Appl. Phys. **40** (1969) 1013.
69P   von Philipsborn, H.: J. Cryst. Growth **5** (1969) 135.
69R   Riedel, E., Horvath, E.: Z. Anorg. Allg. Chem. **371** (1969) 248.
70A   Amith, A., Friedman, L.: Phys. Rev. **B2** (1970) 434.
70F   Friedman, L.R., Amith, A.: IBM J. Res. Dev. **14** (1970) 289.

70H	Haas, C.: IBM J. Res. Dev. **14** (1970) 282.
70K1	Kanomata, T., Ido, H., Kaneko, T.: J. Phys. Soc. Jpn. **29** (1970) 332.
70K2	Kuse, D.: IBM J. Res. Dev. **14** (1970) 315.
70L	Lutz, H.D., Lovasz, Cs., Bertram, K.H., Sreckovic, M., Brinker, U.: Monatsh. Chem. **101** (1970) 519.
70M1	Miyatani, K., Takahashi, T., Minematsu, K., Osaka, S., Yoshida, K.: Proc. Int. Conf. Ferrites, Kyoto **1970**, University of Tokyo Press, Tokyo (1971) 607.
70M2	Motida, K., Miyahara, S.: J. Phys. Soc. Jpn. **29** (1970) 516.
70O	Okamoto, F., Takahashi, T., Wada, Y.: Oyo Butsuri (Appl. Phys. Jpn.) **39** (1970) 471.
70S1	Sato, K., Teranishi, T., Imamura, S.: Proceedings of the 10th International Conference on the Physics of Semiconductors, Cambridge, Mass. (USA) **1970** 247.
70S2	Shepherd, I.W.: Solid State Commun. **8** (1970) 1835.
70S3	Steigmeier, E.F., Harbeke, G.: Phys. Kondens. Mater. **12** (1970) 1.
70T1	Toda, M.: Trans. Inst. Electron. Commun. Eng. (Japan) **53** (1970) 397.
70T2	Toda, M.: Appl. Phys. Lett. **17** (1970) 1.
70T3	Takahashi, T.: J. Cryst. Growth **6** (1970) 319.
70V	Vural, B.: IBM J. Res. Dev. **14** (1970) 272.
71A1	Amith, A., Berger, S.B.: J. Appl. Phys. **42** (1971) 1472.
71A2	Ahrenkiel, R.K., Moser, F., Lyu, S.: J. Appl. Phys. **42** (1971) 1452.
71B1	Berger, S.B., Amith, A.: J. Phys. **32** (1971) 934.
71B2	Brüesch, P., Kalbfleisch, H., Lehmann, F.: Phys. Status Solidi (**b**) **46** (1971) K99.
71L1	Lee, T.H., Dietz, R.E.: J. Appl. Phys. **42** (1971) 1441.
71L2	Lee, T.H., Coburn, T., Gluck, R.: Solid State Commun. **9** (1971) 1821.
71L3	Lugscheider, W., Pink, H., Weber, K., Zinn, W.: Z. Angew. Physik. **32** (1971) 80.
71L4	Lee, T.H., Gluck, R.M., Ahrenkiel, R.K., Coburn, T.J.: AIP Conf. Proc. **5** (1971) 274.
71L5	Lee, T.H.: J. Appl. Phys. **42** (1971) 1441.
71L6	Lehmann, H.W., Harbeke, G., Pinch, H.: J. Phys. **32** (1971) 932.
71L7	Lotgering, F.K.: J. Phys. (Paris) **32** (1971) C1, 34.
71M1	Minematsu, K., Miyatani, K., Takahashi, T.: J. Phys. Soc. Jpn. **31** (1971) 123.
71M2	Miyatani, K., Minematsu, K., Wada, Y., Okamoto, F., Kato, K., Baltzer, P.K.: J. Phys. Chem. Solids **32** (1971) 142.
71O	Omloo, W.P.F.A.M., Bommerson, J.C., Heikens, H.H., Risselada, H., Vellinga, M.B., Van Bruggen, C.F., Haas, C. Jellinek, F.: Phys. Status Solidi (**a**) **5** (1971) 349.
71P	von Philipsborn, H.: J. Cryst. Growth **9** (1971) 296.
71S	Sugier, A., Bloch, C.: Proc. 4th Int. Congr. Catal. **2** (1971) 238.
71T1	Toda, M.: Proc. 3rd Conf. Solid State Devices, Tokyo **1971**, 183.
71T2	Takahashi, T., Minematsu, K., Miyatani, K.: J. Phys. Chem. Solids **32** (1971) 1007.
71W1	Wagner, V., Mitlehner, H., Geick, R.: Optics Commun. **2** (1971) 429.
71W2	Wakamura, K., Onari, S., Arai, T., Kudo, K.: J. Phys. Soc. Jpn. **31** (1971) 1845.
72B1	Broda, H., Konopka, D., Kozlowska, I.: Acta Phys. Pol. **A42** (1972) 255.
72B2	Brüesch, P., D'Ambrogio, F.: Phys. Status Solidi (**b**) **50** (1972) 513.
72B3	Balberg, I., Pinch, H.L.: Phys. Rev. Lett. **28** (1972) 909.
72G	Grimes, N.W.: Philos. Mag. **26** (1972) 1217.
72K1	Kamata, N., Yamazaki, S., Kabashima, S., Hattanda, T., Kawakubo, T.: Solid State Commun. **10** (1972) 905.
72K2	Kleinpenning, Th.G.M.: Physica **59** (1972) 370.
73A1	Arai, T., Wakaki, M., Onari, S., Kudo, K., Satoh, T., Tshushima, T.: J. Phys. Soc. Jpn. **34** (1973) 68.
73A2	Abdullaev, G.B., Aliev, G.M., Ivanova, V.A., Abdinov, D.Sh.: Heat Transfer-Sov. Res. **5** (1973) 30.
73B	Barraclough, K.G., Meyer, A.: J. Cryst. Growth **20** (1973) 212.
73F	Feldtkeller, E., Treitinger, L.: Int. J. Magnetism **5** (1973) 237.
73G	Gyorgy, E.M., Robbins, M., Gibart, P., Reed, W.A., Schmettler, F.J.: AIP Conf. Proc. **10** (1973) 1148.
73I	Itoh, T., Miyata, N., Narita, S.: Jpn. J. Appl. Phys. **12** (1973) 1265.
73K	Konopka, D., Kozlowska, I., Chelkowski, A.: Phys. Lett. **A 44** (1973) 289.
73L	Larsen, P.K., Voermans, A.B.: J. Phys. Chem. Solids **34** (1973) 645.
73M	Murashko, N.I., Obolonchik, V.A., Goryachev, Yu.M.: Izv. Akad. Nauk SSR Neorg. Mater. **8** (1972) 1669.

73P	Pearlman, D., Carnall Jr., E., Martin, T.W.: J. Solid State Chem. **7** (1973) 138.
73V	Valiev, L.M., Kerimov, I.G., Namazov, Z.M.: Phys. Status Solidi (**a**) **18** (1973) K117.
73W1	Wakamura, K., Arai, T., Onari, S., Kudo, K., Takahashi, T.: J. Phys. Soc. Jpn. **35** (1973) 1430.
73W2	Wagner, V.: Phys. Status Solidi (**b**) **55** (1973) K29.
74G	Gibart, P., Goldstein, L., Dormann, J.L., Guittard, M.: J. Cryst. Growth **24/25** (1974) 147.
74K1	Konopka, D., Slebarski, A., Chelkowski, A.: Acta Phys. Pol. **A46** (1974) 47.
74K2	Kanomata, T., Ido, H.: J. Phys. Soc. Jpn. **36** (1974) 1322.
74P	Prosser, V., Hlidek, P., Hoschl, P., Polivka, P., Zvara, M.: Czech. J. Phys. **B24** (1974) 1168.
74W	Watanabe, T.: J. Phys. Soc. Jpn. **37** (1974) 140.
75G	Gurevich, A.G., Ykovlev, Yu.M., Karpovich, V.I., Vinnik, M.A., Rubal'skaya, E.V.: Sov. Phys. Semicond. **9** (1975) 1.
75L	Lutz, H.D., Haeuseler, H.: J. Solid State Chem. **13** (1975) 215.
75M1	Marunya, M.S., Borukhovich, A.S., Bamburov, V.G., Lobachevskaya, N.I., Rokeakh, O.P., Gel'd, P.V.: Sov. Phys. Solid State **17** (1975) 776.
75M2	Manasse, F.K., Kyllonen, E.W.: J. Appl. Phys. **46** (1975) 977.
75O	Okonska-Kozlowska, I., Konopka, D., Jelonek, M., Heimann, J., Pietkiewicz, J., Chelkowski, A.: J. Solid State Chem. **14** (1975) 349.
75P	Plumier, R., Lecomte, M., Miédan-Gros, A., Sougi, M.: Phys. Lett. **A 55** (1975) 239.
75S	Stoyanova, S.G., Iliev, M.N., Stoyanova, S.P.: Phys. Status Solidi (**a**) **30** (1975) 133.
75T	Taniguchi, M., Kato, Y., Narita, S.: Solid State Commun. **16** (1975) 261.
75V1	Valiev, L.M., Kerimov, I.G., Babaev, S.Kh., Namazov, Z.M.: Inorg. Mater. **11** (1975) 213.
75V2	Veselago, V.G., Damaskin, I.A., Pyshkin, S.L., Radautsan, S., Tezlevan, V.E.: JETP Lett. **20** (1975) 149.
75V3	Veselago, V.G.: Colloq. CNRS **242** (1975) 295.
75W	Wakaki, M., Arai, T., Kudo, K.: Solid State Commun. **16** (1975) 679.
76A	Asaka, M., Matsumoto, S., Niimi, T.: Jpn. J. Appl. Phys. **15** (1976) 741.
76G1	Göbel, H.: J. Mag. Magn. Mater. **3** (1976) 143.
76G2	Glidewell, C.: Inorg. Chim. Acta **19** (1976) L45.
76G3	Golik, L.L., Grigorovitch, S.M., Elinson, M.I., Kunkova, Z.E., Ukrainskyi, Y.M.: Thin Solid Films **34** (1976) 279.
76H	Hlidek, P., Barvik, I., Prosser, V., Vaněček, M., Zvára, M.: Phys. Status Solidi (**b**) **75** (1976) K45.
76I1	Iizuka, T., Sugano, T.: Jpn. J. Appl. Phys. **15** (1976) 57.
76I2	Iizuka, T., Sugano, T.: J. Fac. Eng. (Univ. Tokyo) B **33** (1976) 373.
76K	Kun'kova, Z.E., Aminov, T.G., Golik, L.L., Elinson, M.I., Kalinnikov, V.T.: Sov. Phys. Solid State **18** (1976) 1212.
76L	Lutz, H.D., Okonsko-Kozlowska, I.: Z. Anorg. Allg. Chem. **427** (1976) 65.
76S1	Sakai, S., Sugano, T., Okabe, Y.: Jpn. J. Appl. Phys. **15** (1976) 2023.
76S2	Stoyanov, S.G., Iliev, M.N., Stoyanova, S.P.: Solid State Commun. **18** (1976) 1389.
76W	Wakamura, K., Arai, T., Kudo, K.: J. Phys. Soc. Jpn. **41** (1976) 130.
77B	Balberg, I., Maman, A.: Phys. Rev. **B 16** (1977) 4535.
77E	Eivazov, E.A., Rustamov, A.G., Gasumov, N.M., Gambarov, D.A.: Inorg. Mater. **13** (1977) 309.
77G1	Golik, L.L., Grigorovich, S.M., Elinson, M.I., Kun'kova, Z., Ukrainskii, Yu.M.: Sov. Phys. Solid State **19** (1977) 365.
77G2	Golik, L.L., Novikov, L.N., Elinson, M.I., Aminov, T.G., Kalinnikov, V.T., Shapsheva, N.P.: Sov. Phys. Solid State **18** (1977) 2156.
77G3	Grimes, N.W., Isaac, E.D.: Philos. Mag. **35** (1977) 503.
77H	Hlidek, P., Zvara, M., Prosser, V.: Phys. Status Solidi (**b**) **84** (1977) K119.
77K1	Koshizuka, N., Yokoyama, Y., Tsushima, T.: Solid State Commun. **23** (1977) 967.
77K2	Kiyosawa, T., Masumoto, K.: J. Phys. Chem. Solids **38** (1977) 609.
77S	Sato, K.: J. Phys. Soc. Jpn **43** (1977) 719.
77T1	Treitinger, L.: Thesis, Technical University, Vienna 1977.
77T2	Treitinger, L., Brendecke, H.: Mater. Res. Bull. **12** (1977) 1021.
77T3	Tretyakov, Yu.D., Gordeev, I.V., Kesler, Ya.A.: J. Solid State Chem. **20** (1977) 345.
77V	Valiev, L.M., Kerimov, I.G., Abdurragimov, A.A., Nabieva, N.B.: Inorg. Mater. **13** (1977) 613.
78E	Edel'man, I.S., Kononov, V.P., Vasil'ev, G.G., Dustmuradov, G., Suvorov, V.M.: Sov. Phys. Solid State **20** (1978) 2132.
78G1	Goldstein, L., Gibart, P., Selmi, A.: J. Appl. Phys. **49** (1978) 1474.

78G2	Gibart, P.: J. Cryst. Growth **43** (1978) 21.

78G2   Gibart, P.: J. Cryst. Growth **43** (1978) 21.
78I1   Iliev, M.N., Anastassakis, E., Arai, T.: Phys. Status Solidi (**b**) **86** (1978) 717.
78I2   Iliev, M., Güntherodt, G., Pink, H.: Solid State Commun. **27** (1978) 863.
78K1   Koshizuka, N., Yokoyama, Y., Okuda, T., Tsushima, T.: J. Appl. Phys. **49** (1978) 2183.
78K2   Kodama, K., Miimi, T.: J. Appl. Phys. **49** (1978) 679.
78S   Shalimova, M.A.: Sov. Phys. Solid State **20** (1978) 523.
78W   Wakaki, M., Arai, T.: Solid State Commun. **26** (1978) 757.
79B   Belov, K.P., Koroleva, L.I., Shalimova, M.A.: Sov. Phys. JETP **47** (1979) 1167.
79K1   Kalinnikov, V.T., Aminov, T.G., Golik, L.L., Novikov, L.N., Zhegalina, V.A., Shumilkin, N.S.: Inorg. Mater. **14** (1979) 1098.
79K2   Koroleva, L.I., Shalimova, M.A.: Sov. Phys. Solid State **21** (1979) 266.
79Z1   Zvara, M., Schlegel, A., Wachter, P.: J. Appl. Phys. **50** (1979) 7463.
79Z2   Zvara, M., Prosser, V., Schlegel, A., Wachter, P.: J. Magn. Magn. Mater. **12** (1979) 219.
80G   Golik, L.L., Kunkova, Z.E., Aminov, T.G., Kalinnikov, V.T.: Sov. Phys. Solid State **22** (1980) 512.
80K1   Koroleva, L.I., Tovmasyan, L.N., Shestakova, G.M., Kalinnikov, V.T., Babitsyna, A.A.: Phys. Met. Metallogr. (USSR) **48** (1980) 167.
80K2   Kodoma, K., Doi, S., Matsumura, T., Niimi, T.: Jpn. J. Appl. Phys. **19** (1980) 317.
80K3   Kodoma, K., Niimi, T.: Jpn. J. Appl. Phys. **19** (1980) 307.
80K4   Kremer, A.A., Golik, L.L.: Sov. Phys. Solid State **22** (1980) 130.
80K5   Kambara, T., Oguchi, T., Gondaira, K.I.: J. Phys. **C13** (1980) 1493.
80M   Merkulov, A.I., Radantsen, S.I., Tezlevan, V.E.: Sov. Phys. Semicond. **14** (1980) 600.
80Z   Zvara, M., Rode, O., Bok, J., Gardavsky, J., Heidek, P.: Phys. Status Solidi (**a**) **61** (1980) K141.
81B   Bogdankevich, O.V., Borisov, N.A., Veselago, V.G., Petsov, V.F., Faifer, V.H.: Sov. Phys. Solid State **23** (1981) 1066.
81M   Merkulov, A.I., Radantsan, S.F., Tezlevan, V.E.: Sov. Phys. Solid State **23** (1981) 1044.
81S   Samokhalov, A.A., Gizhevskii, B.A., Loshkareva, N.N., Arbuzova, T.I., Simonova, M.I., Chebotaev, N.M.: Sov. Phys. Solid State **23** (1981) 2016.
81Y   Yao, S.S., Pellegrino, F., Alfano, R.R., Miniscallo, W.J., Lempicki, A.: Phys. Rev. Lett. **46** (1981) 558.
82A   Auzin, V.B., Bogdankevich, O.V., Borisov, N.A., Veselago, V.G., Petsov, V.F., Faifer, V.N.: Sov. Phys. Solid State **24** (1982) 379.
82B   Berzhanskii, V.N., Chernov, V.K.: Sov. Phys. Solid State **24** (1982) 1357.
82K   Kotelnikova, A.M., Alenina, L.F., Yakovlev, Yu.M., Tudzhanova, I.N., Linev, Yu.A.: Sov. Phys. Solid State **24** (1982) 962.
82S   Samokhvalov, A.A., Sukhornkov, Yu.P.: JETP Lett. **35** (1982) 264.
82P   Pirogova, A.M., Shishkov, A.G., Ilicheva, E.N., Krasnozhen, L.A.: Sov. Phys. Solid State **24** (1982) 1263.
82W   Wakaki, M., Yamamoto, K., Onari, S., Arai, T.: Solid State Commun. **43**, (1982) 957.
82Y   Yao, S.S., Alfano, R.R.: Phys. Rev. Lett. **49** (1982) 69.
83Y   Yao, S.S., Alfano, R.R.: Phys. Rev. **B 27** (1983) 1180.

10.2.4 Transition-metal oxides

Many of the semiconducting transition-metal oxides are ferroelectrics. Their properties have been described in detail in Vol. III/16a of this series including the literature up to 1978. We restrict the presentation for these substances here to reproducing those data which are of most interest to the semiconductor physicist and adding some newer results. For further information and references, see Vol. III/16a.

For magnetic and related properties, the reader is referred to Vol. III/12a (perovskite type oxides), Vol. III/12b (oxides of the spinel type) and Vol. III/12c (hexagonal ferrites).

BaTiO₃

Crystal structure

structure:

high temperature phase:	hexagonal, space group D_{6h}^4—$P6_3/mmc$	55R1
phase I:	cubic, space group O_h^1—$Pm3m$	45M

(continued)

Physical property	Numerical value	Experimental conditions	Experimental method remarks	Ref.
structure (continued)				
phase II: tetragonal, space group C_{4v}^1—P4mm				46V, 46W
phase III: orthorhombic, space group C_{2v}^{14}—Amm2				49K,
phase IV: rhombohedral, space group C_{3v}^5—R3m				49R
Phases IV, III, II are ferroelectric, phase I is paraelectric.				
For the phase diagram of the system $BaO-TiO_2$, see Fig. 1.				
transition temperatures (in °C):				
T_{tr}	1460	I →HT phase		55R1
	130	II →I	ferroelectric Curie temperature	65J1,
			(in earlier papers, $\Theta_C \approx 120$)	69Y1
	5	III→II		49K,
	−90	IV→III		49R
lattice parameters (lengths in Å):				
a	3.996	phase I $T=120$°C	for temperature and pressure dependence of cell parameters, see Figs. 2···4	47M, 51R, 57S, 57J
a	3.9920	phase II,		
c	4.0361	$T=20$°C		
a	3.990	phase III,		
b	5.669	$T=-10$°C		
c	5.682			
a	4.001	phase IV,		
α	89° 51′	$T=-168$°C		

Electronic properties

Only little information is available about the band structure of $BaTiO_3$. Fig. 5 shows a calculated band structure (see also [80M] and [78B2]). Band gap energies and energies of higher optical transitions are:

band gap energies (in eV):

(see also Fig. 6 for the temperature dependence)

E_g	3.25	$T=403$ K	Faraday rotation	67B1
	3.11		Faraday rotation (different dispersion function used in analysis of results)	
	3.20		reflectivity peak (shoulder)	
	3.20		electroreflectance singularity	
	3.26		from absorption coefficient (value at $K=10^4$ cm^{-1})	

For effect of bias field on absorption edge, see [67G].

higher transition energies (in eV):

E	3.2	E_0	energies of fundamental	65C
	3.91	A_1	absorption edges, see Fig. 7	
	4.85	A_2		
	6.10	B_1		
	7.25	B_2		
	10.3	C_1		
	11.8	C_2		
	12.8	D		
	15	E		

Physical property	Numerical value	Experimental conditions	Experimental method remarks	Ref.
core level binding energies (in eV):				
		level:		
E	465.6 (1)	Ti $2p_{1/2}$	energies relative to the O 1s	78P2
	459.8 (1)	Ti $2p_{3/2}$	level taken at 531.0 eV,	
	92.5 (1)	Ba $4d_{3/2}$	X-ray photoelectron	
	89.8 (1)	Ba $4d_{5/2}$	spectroscopy	
	38.6 (1)	Ti $3p_{1/2}$		
	38.6 (1)	Ti $3p_{3/2}$		
	30.5 (1)	Ba 5s		
	23.0 (1)	O 2s		
	16.7 (1)	Ba $5p_{1/2}$		
	14.9 (1)	Ba $5p_{3/2}$		

Lattice properties

Dispersion curves for phonons obtained by inelastic neutron scattering are given in Fig. 8 for phase I and in Figs. 9···11 for phase II.

wavenumbers of vibrational modes (in cm^{-1}):

		assignment:		
$\bar{\nu}_{F_{1u}}$	36	E(TO, xy)	Raman scattering, data at 25°C	68D
	180		(accuracy $\pm 3\,cm^{-1}$);	
	486		in the third column the Raman	
	36	E(TO, z)	assignment of the modes is given;	
	180		for Raman spectra and temperature	
	518		dependence of mode wavenumbers,	
	715	E(LO, xy)	see Figs. 12···15	
	180		(see also [74H2, 72F3])	
	463			
	≈ 180	A_1(TO, xy)		
	180			
	470			
	727	A_1(LO, z)		
	178			
	470			
$\bar{\nu}_{F_{2u}}$	305	E(TO+LO, xy),		
		E(TO+LO, z),		
		B_1		

For elastic moduli, see [59H, 57M, 66B3, 73K, 69K].

Transport properties

The electrical conductivity depends on various factors as method of synthesis, crystal state, electrode material, aging etc. (for electric field dependence, see [62B]; for effect of oxygen pressure, see [76D1, 76D2, 76H3]; for effect of doping or reduction, see [69G, 77T, 78I]; for dependence on electrode material, see [60B, 62B]).

Typical data for transport properties are given in Figs. 16···18 (conductivity), Fig. 19 (carrier mobility), Figs. 20 and 21 (Seebeck coefficient).

Piezoelectrical data may be found in [63M1].

Optical properties

Optical spectra are given in Fig. 22 (far infrared), Figs. 23 and 24 (infrared), Figs. 25···28 (visible range), Figs. 29···32 (ultraviolet; see also [67F2] for Ta doped material).

The refractive indices are shown in Figs. 33···35 (see also [65B1, 70W1]; for birefringence, see [58M1, 68W]).

(continued)

Physical property	Numerical value	Experimental conditions	Experimental method remarks	Ref.

Optical properties (continued)

Luminescence spectra are shown in Figs. 36···39.

Electrooptical constants are given in [65J2, 69S3, 71J].

For Faraday rotation, see [67B1].

For dependence of the dielectric constant on frequency, temperature and pressure, see [64P2, 65J1, 68W, 76K2, 77C] and Figs. 18, 29.

Photoconductivity has been measured in [63C, 66H3, 81K1].

The effective mass for n-type $BaTiO_3$ is 6.5 (2) m_0 according to [67B2].

Further properties

melting point:

T_m	1618 °C			51W2

density:

d	6.02 g cm^{-3}	phase II	calculated	51R
	6.1 (1) g cm^{-3}	HT phase		48B

heat capacity: Fig. 40

thermal conductivity: Fig. 41, see also [67M]

$PbTiO_3$

Crystal structure

structure:

phase I (paraelectric): cubic, space group O_h^1—Pm3m				46M,
phase II (ferroelectric): tetragonal, space group C_{4v}^1—P4mm				51S3

transition temperatures:

T_{tr}	490 °C			46M, 51S3
	470 °C		ceramics	71I

Another phase transition was reported to occur at about -100 °C or -160 °C [55K, 69I, 73D2].

Phase diagram of the $PbO-TiO_2$ system: see Fig. 42.

lattice parameters (in Å):

a	3.9045	RT		46M
c	4.1524			
a	3.9072	RT	ceramics (1.25 mol% La_2O_3,	71I
c	4.1187		1.0 mol% MnO_2)	

For temperature and pressure dependence of lattice parameters, see Figs. 43···46.

Lattice properties

Dispersion curves for phonons obtained by inelastic neutron scattering are given in Figs. 47···50. Fig. 51 shows the temperature dependence of the lowest TO mode. Figs. 52···56 represent Raman spectra and the temperature and pressure dependence of vibrational modes obtained from these spectra (see also [78H3]).

For elastic constants, see [71G2, 71F].

Transport properties

The conductivity of ceramics is $10^{-5}···10^{-6}\,\Omega^{-1}\,m^{-1}$ at RT but depends very much on the sintering conditions of the ceramics (see e.g. [71I, 69G]). Figs. 57 and 58 show examples of the temperature dependence of transport parameters.

For piezoelectric coefficients, see [71G2, 72V, 71I].

Physical property	Numerical value	Experimental conditions	Experimental method remarks	Ref.

Optical properties

Optical spectra are shown in Figs. 59···62. Birefringence is shown in Fig. 63 (for refractive indices, see also [71P, 72S3]). The frequency dependence of the real and imaginary parts of the dielectric constants is given in Fig. 64 (for dielectric constants, see also [70R2, 71G2, 71I]).

Electrooptic constants are given in [71P].

A photoconductivity maximum at $\lambda \approx 410$ nm is reported [76M2] (see also [79M]).

Further properties

melting point:

T_m	1281 (3) °C			52R

densities:

d	7.96 (1) g cm^{-3}			52R
	7.9		single crystal	71G2
	7.87		ceramics with 1.25 mol% La_2O_3 and 1.0 mol% MnO_2	71I

heat capacity: Fig. 65
thermal conductivity: Fig. 66

NaNbO$_3$

Crystal structure

structure:

phase I:	cubic, space group O_h^1—Pm3m (paraelectric phase)			51W1
phase II:	tetragonal, space group D_{4h}^5—P4/mbm, $Z=2$			72G2
phase III:	orthorhombic, space group D_{2h}^{17}—Ccmm, $Z=8$			72G2
phase III':	orthorhombic, space group D_{2h}^{13}—Pnmm, $Z=8$			72A
phase III'':	orthorhombic, space group D_{2h}^{13}—Pmnm, $Z=24$			72A
phase IV:	orthorhombic, space group D_{2h}^{11}—Pbma, $Z=8$ (antiferroelectric phase)			61W
phase V:	trigonal, space group C_{3v}^6—R3c, $Z=2$ (ferroelectric phase)			71D

transition temperatures (in °C):

T_{tr}	643	II →I		73I
	572	III →II		
	520	III' →III		
	480	III''→III'		
	365	IV →III''		
	−200	V →IV		

lattice parameters (lengths in Å):

a_0	≈ 3.93	phase I		72G2
a	5.5639 (6)	phase II,		72G2
b	5.5639 (6)	$T=600$ °C		
c	3.9428			
a	7,8642	phase III,		72A
b	7.8550	$T=530$ °C		
c	7.8696			
a	7.8608	phase III'		72A
b	7.8556	$T=500$ °C		
c	7.8606			
a	≈ $2a_0$	phase III'',		71S2
b	≈ $6a_0$	$T>365$ °C		
c	≈ $2a_0$			

(continued)

Physical property	Numerical value	Experimental conditions	Experimental method remarks	Ref.
lattice parameters (continued)				
a	$2 \cdot (3.9150\,(2))$	phase IV,		73D1
b	$4 \cdot (3.8798\,(2))$	RT		
c	$2 \cdot (3.9150\,(2))$			
α	$90°$			
β	$90°40\,(3)'$			
γ	$90°$			
a, b, c	$2 \cdot (3.9083\,(5))$	phase V	measured at $-150\,°C$,	73D1
α	$89°13\,(1)'$		where phase V is metastable	

Physical properties

Energy gap was estimated to be 3.4 eV from photoconductivity data [77R1, 2].

conductivity: Figs. 67···69, polaron and drift mobility: Figs. 70, 71, Seebeck coefficient: Fig. 72. For pressure dependence of conductivity and Seebeck coefficient: [80P]. Dielectric constants, see [55C, 80P].

Refractive indices (phase IV): $n_\alpha = 2.13$, $n_\beta = 2.21$, $n_\gamma = 2.25$ for $\lambda = 589$ nm [78S2]; birefringence: Figs. 73, 74.

KNbO$_3$

Crystal structure

structure:

phase I:	cubic, space group O_h^1—Pm3m (paraelectric phase)	51W1
phase II:	tetragonal, space group C_{4v}^1—P4mm (ferroelectrical phase)	73H
phase III:	orthorhombic, space group C_{2v}^{14}—Bmm2 (ferroelectric phase)	63K
phase IV:	rhombohedral, space group C_{3v}^5—R3m (ferroelectric phase)	73H

transition temperatures (in °C):

T_{tr}	418 (5)	II →I		54S
	225	III→II		
	-10	IV→III		

lattice parameters (lengths in Å):

a	3.997	phase II,	for temperature dependence of	73H
c	4.063	$T = 270\,°C$	lattice parameters, see [54S]	
a	3.973	phase III,		73H
b	5.695	$T = 22\,°C$		
c	5.721			
a, b, c	4.016	phase IV,		73H
α, β, γ	89.817 (9)°	$T = -43\,°C$		

Electronic properties

The band gap is about 3.3 eV [76M3]. A calculated band structure and density of states is shown in Fig. 75. Fig. 76 shows transitions up to 40 eV in a XPS-spectrum.

Lattice properties

Phonon dispersion spectra are presented in Figs. 77 and 78, Raman spectra are shown in Figs. 79 and 80.
　　For elastic constants, see [74W2].

Transport and optical properties

conductivity: Figs. 81, 82 (for photoconductivity, see [78G3]); Seebeck coefficient: Fig. 83
reflectance: Fig. 84; refractive index, birefringence: Figs. 85, 86 (see also [74U1]); polariton dispersion curves: Fig. 87
　　For dielectric constants, see [56T, 74W2].
　　Piezoelectric and electrooptic properties are given in [74W2] and [74G3], respectively.

Physical property	Numerical value	Experimental conditions	Experimental method remarks	Ref.
Further properties				
melting point:				
T_m	1050 °C			55R2, 58M2
density:				
d	4.590 g cm^{-3}	$T = 25$ °C		52L
	4.62 g cm^{-3}			74W2
KTaO$_3$				
Crystal structure				
structure: cubic, space group O_h^1—Pm3m (perovskite) (ferroelectric phase)				65W, 64W1, 51V2
lattice parameter:				
a	3.9885 Å	RT	K at 1(a) position, Ta at 1(b) position, 3 O at 3(c) position	51V1, 51V2
For temperature dependence of a, see Fig. 88.				
linear thermal expansion coefficient:				
α	$5.3 \cdot 10^{-6}$ K^{-1}	$T = 283$ K		81W
	$3.9 \cdot 10^{-8}$ K^{-1}	4 K		
For temperature dependence of α, see [81W].				
Electronic properties				
A calculated band structure is shown in Fig. 89.				
energy gap (in eV):				
E_g	3.79	$T = 77$ K	Faraday effect, two data obtained	67B1,
	3.65		by different analyses	67F1
	3.77	$T = 286$ K	(see also Fig. 90 for electric	
	3.62		field dependence of E_g)	
	3.57		electroreflectance peaks	
	3.80			
	3.75		absorption edge	
	3.75		energy at $K \approx 10^4$ cm^{-1}	
	3.58		photoconductivity peak	65W
singularities in electroreflectance (in eV) (see Figs. 91, 92):				
E_{peak}	3.57	(100)-surface	energy gap (E$_1$ in Fig. 91)	67F2
	3.57	(110)-surface	(Ca doped sample)	
	3.80	(100)	E$_2$	
	3.77	(111)		
	3.80	(110)		
	4.40	(100)	A$_1$	
	4.45	(111)		
	4.47	(110)		
	4.88	(100)	A$_2'$	
	4.90	(111)		
	4.85	(110)		
	5.50	(100)	A$_2$	
	5.47	(111)		
	5.50	(110)		
For dependence of electroreflectance on polarization of incident light, see Fig. 93; for electroabsorption spectrum, see Fig. 94.				

Physical property	Numerical value	Experimental conditions	Experimental method remarks	Ref.

Lattice properties

Phonon dispersion curves and the temperature dependence of phonon modes are shown in Figs. 95···98 (see also [67P2]. For typical Raman spectra, see Figs. 99, 100 (see also [76L]). Fig. 99 contains values of various phonon wavenumbers.

For elastic properties, see [69B2, 68B].

Transport properties

Most data on transport properties have been obtained using oxygen-deficient semiconducting samples. For temperature dependence of resistivity, see Fig. 101. For dependence of conductivity on hydrostatic pressure, see [66W].

Hall mobility (in cm^2/Vs):

μ_H	30	RT		65W
	$8 \cdot 10^8\, T^{-3}$	$T > 100\,K$	see Fig. 102; for concentration dependence of μ_H, see Fig. 103	

Measurements on reduced single crystals with carrier concentrations between $3.5 \cdot 10^{17}$ and $1.3 \cdot 10^{19}\,cm^{-3}$ gave Hall mobilities between 27 and $31\,cm^2/Vs$ at 295 K and between 23000 and $3400\,cm^2/Vs$ at 4.2 K [65W]).

Seebeck coefficient:

S	$550\,\mu V/K$	RT	oxygen-deficient sample with $n = 3.5 \cdot 10^{17}\,cm^{-3}$	65W, 77K

effective mass:

m^*	$0.80\,(28)\,m_0$		from Seebeck coefficient	65W,
	$0.69\,(1)\,m_0$			77K, 70S4

Optical properties

absorption spectra: Figs. 104···106 (for optical absorption in impurity doped crystals, see [64W1]); reflectance: Figs. 107, 108.

Dielectric constants are given in [64W1, 65K, 65W, 67P2, 71A1, 73U, 74W2]; see also Fig. 109 for imaginary part of ε.

refractive index: Fig. 110

Faraday rotation: Fig. 111

For photoconductivity, see [65W, 73P2, 76O].

For luminescence measurements, see [65S2].

Electrooptic constants are given in [64G, 63G].

Further properties

thermal conductivity: Fig. 112

melting point:

T_m	1357 (3) °C			55R3, 56R

density:

d	$6.97\,g\,cm^{-3}$			58R, 64W2

Physical property	Numerical value	Experimental conditions	Experimental method remarks	Ref.

SrTiO$_3$

Crystal structure

structure:

phase I: cubic (pseudo-cubic?), space group O_h^1—Pm3m (paraelectric)   62R
phase II: tetragonal, space group D_{4h}^{18}—I4/mcm   62R,
        (phase II possibly consists of several individual phase [70L1])   67U

transition temperature:

T_{tr} 105 K   change with pressure, see [72F4]   62R

This phase transition is associated with zone-boundary phonon condensation at the R point of the Brillouin zone and was found by ESR [67U], Raman scattering [68F], Brillouin scattering [70L2], neutron scattering [69S2] and ultrasonic measurements [75O]. The existence of new phases induced by mechanical stress at low temperature was reported by [70M3, 71B4, 76U]).

lattice parameters:

a_0	3.905 Å	phase I, RT		64L2
a	$\approx \sqrt{2}\,a_0$	phase II	temperature dependence,	67U,
c	$\approx 2\,a_0$		see Fig. 113; thermal expansion,	73O,
$c/\sqrt{2}\,a-1$	$\approx 10^{-3}$		see also [69A2, 74O1]	79F2

Electronic properties

A calculated band structure is shown in Fig. 114.

energy gap (in eV):

		T [K]		
E_g	3.40	296	Faraday rotation	67B1
	3.43	77		
	3.21	296	Faraday rotation (different	
	3.26	77	dispersion function used in	
			analysis of results)	
	3.20	296	reflectivity peak (shoulder)	
	3.40	296	absorption	
	3.37	296	from absorption coefficient	
			(value at $K \approx 10^4$ cm^{-1})	

For temperature dependence of absorption edge, see Fig. 130.

higher transition energies (in eV):

(E_g	3.4)		all data from reflectivity,	65C,
E	4.0	A_1	see Fig. 115	78B2
	4.86	A_2		
	5.5	A_3		
	6.52	A_4		
	7.4	A_5		
	9.2	A_6		
	10.2	A_7		
	12.0	A_8		
	13.0	A_9		
	13.8	A_{10}		
	16.4	A_{11}		
	19.6	A_{12}		
	20.8	A_{13}		(continued)

Physical property	Numerical value	Experimental conditions	Experimental method remarks	Ref.

higher transition energies (continued)

E	21.6	A_{14}		65C,
	22.8	A_{15}		78B2
	23.4	A_{16}		
	24.3	A_{17}		

Lattice properties

Phonon dispersion curves and temperature dependence of phonon wavenumbers are given in Figs. 116$\cdots$121, Raman spectra are shown in Figs. 122$\cdots$125 (see also [76U, 67O, 76F2]. For elastic properties, see [70L3, 74S5, 75O, 82H].

Transport properties

resistivity: Fig. 126; Hall coefficient: Fig. 127; Hall mobility: Figs. 128, 129; Additional data concerning electrical conduction, see e.g. [64F, 67T, 70R3, 71A2, 71L, 73W, 75L3, 76P2, 78P1]. A density-of-states effective mass of 1.3 (2) m_0 was reported in [72M5, 70S4].

Data on piezoelectricity are given in [63S, 66T, 67R2]; for magnetoresistance, see [75K]; superconductivity in semiconducting $SrTiO_3$ is studied in [65S3, 66A, 67P1].

Optical properties

absorption, optical density, transmission, reflectance: Figs. 130$\cdots$137 (see also [66B2, 74M2, 79S1]); refractive index and dielectric constant: Figs. 137$\cdots$139 (for wavelength dependence of refractive index, see [65B3]; for dielectric constants, see also [66S2, 67H, 74F2]); luminescence spectra: Figs. 140$\cdots$142.

Data on piezooptic and electrooptic constants are given in [57G, 76A] and [70F3], respectively. For Faraday rotation, see [67B1].

Further properties

heat capacity: Fig. 143 (see also [74S2, 77H2])
thermal conductivity: Figs. 144, 145 (see also [74S3])

melting point:

T_m	2080 °C			71M

density:

d	5.13 g cm^{-3}			71M

$PbZrO_3$
Crystal structure
structure:

phase I: cubic, space group O_h^1—Pm3m (paraelectric phase)				50R, 51S2,
phase II: orthorhombic, space group C_{2v}^8—Pba2 (antiferroelectric phase)				51S5

transition temperature:

T_{tr}	230 °C			50R

lattice parameters (lengths in Å):

a_{mon}, c_{mon}	4.1604	phase II,	monoclinic cell parameters of	74S4,
b_{mon}	4.1111 (2)	RT	phase II	79F3
β	90°05′28(10)″			
a	5.8883 (3)		orthorhombic cell parameters of	
b	11.7581 (6)		phase II	
c	8.2222 (4)		for temperature dependence of cell parameters, see Figs. 146, 147	

Physical property	Numerical value	Experimental conditions	Experimental method remarks	Ref.

Physical properties

elastic constants: [55M]

conductivity: Fig. 148; Seebeck coefficient: Fig. 149; p-type conduction reported at high temperatures (340···600 °C) [72U]. Photoconductivity maximum at $\lambda = 360$ nm reported in [78P3]. Activation energies (energy gap?) quoted as 1.31 eV below 455 °C and 1.03 eV above 465 °C [74U2].

reflectance: Fig. 150; Raman spectrum: Fig. 151; real and imaginary parts of the dielectric constant: Fig. 152 (see also [75U, 78U, 70S5]); birefringence: Fig. 153 (see also [76F1] for electric field dependence); heat capacity: Fig. 154; thermal conductivity: Fig. 155

melting point:

T_m	1570 °C			67F3

CaTiO$_3$

Crystal structure

structure:

phase I: cubic, space group O_h^1—Pm3m (paraelectric phase)				54G
phase II: orthorhombic, space group D_{2h}^{16}—Pcmn				57K

transition temperature:

T_{tr}	1260 °C		Transition heat at phase transition: $2.30 \cdot 10^3$ J mol^{-1}.	46N

lattice parameters (in Å):

a	5.3670	phase II,	temperature dependence,	57K
b	7.6438	RT	Fig. 156	
c	5.4439			

Physical properties

vibrational modes (wavenumbers in cm^{-1}):

$\bar{\nu}$	117, 134, 162, 176, 189, 224, 260, 307, 320, 370, 440, 560			74K2

optical spectra: reflectivity, Fig. 157; refractive index, Fig. 158; dielectric constants: see [74K3, 77D]
heat capacity: Fig. 159

melting point:

T_m	1960 °C			62M2

density:

d	4.10 g cm^{-3}	RT		62M2
	3.94 g cm^{-3}			77D

LiVO$_3$

structure: monoclinic [72F1, 77H1, 73S1], space group C_{2h}^6—C2/c [73S1].
Growth by Czochralski method [72F1].

lattice parameters (lengths in Å):

a	9.542	(10.158)	(values in brackets from [73S1])	72F1
b	8.406	(8.4175)		
c	5.874	(5.8853)		
β	93.33°	(110.48°)		

density:

d	3.0005 g cm^{-3}		experimental	72F1
	3.00 g cm^{-3}		calculated	

transmission: Fig. 160

Physical property	Numerical value	Experimental conditions	Experimental method remarks	Ref.
Raman wavenumbers (in cm^{-1}) (linewidths in cm^{-1} in brackets):				
$\bar{\nu}$	527	(12)		72F1
	652	(12)		
	903	(6)		
	955	(6)		

LaVO$_3$
(cf. also section 10.3)

structure: tetragonal (perovskite, at 300 K, [74B1])

lattice parameters:				
a	5.547 Å		see also [74D]	74B1
c	7.851 Å			
conductivity:				
σ	$10^{-7}\ \Omega^{-1}\ \mathrm{cm}^{-1}$			74B2
activation energy for conductivity:				
E_A	0.171$\cdots$0.176 eV	$100 < T < 210$ K	semiconducting, p-type	74D,
	0.209 eV	$210 < T < 300$ K		75P
Seebeck coefficient:				
S	280 μV K^{-1}			74D, 75P
Néel temperature:				
T_N	137 K			74B2

heat capacity: temperature dependence: Fig. 161

Debye temperature:				
Θ_D	235 K			73B5, 74B2

Cs$_x$V$_3$O$_7$ (x = 0.35 and 0.3)
structure: space group C_{6h}^2—P6$_3$/m [77W1]
lattice parameters (in Å):

a	9.880	x = 0.35		77W1
c	3.605			
a	14.360	x = 0.3		77W2
c	3.611			

Measurements between 78 K and 400 K showed semiconducting behavior.

activation energies for conductivity:				
E_A	0.04 eV	$T > 200$ K, x = 0.35		77W1
	0.06 eV	$T < 200$ K		
	0.07 eV	x = 0.3		77W2

For x = 0.35 antiferromagnetism was observed between 78 K and 700 K with a Néel temperature of 200 K [77W1].

Physical property	Numerical value	Experimental conditions	Experimental method remarks	Ref.

$Co_3V_2O_8$

structure: orthorhombic

Ferromagnetic behavior with transition point at $T_C = 10$ K [70F1].

lattice parameters (in Å):

a	6.030			73S2
b	11.486			
c	8.312			

density:

d	4.693 g cm^{-3}		calculated	73S2
	4.70 g cm^{-3}		experimental	

MV_2O_4 are semiconductors for M = Co, Fe, Mg, Mn, Zn [74A]. FeV_2O_4 is semiconducting at 5.5 K [66H4], ZnV_2O_4 is a semiconductor acc. to [63R, 70G].

$MnVO_3$

structure: phase I: triclinic (ilmenite)

phase II: orthorhombic (perovskite), space group D_{2h}^{16}—Pbnm or C_{2v}^9—Pbn2$_1$ [71S1]

lattice parameters (lengths in Å):

a	5.072	phase I		71S1
b	5.550	($p < 45$ kbar)		
c	5.023			
α	90°0′			
β	118°38′			
γ	63°4′			
a	5.106	phase II		
b	5.265	($p > 45$ kbar)		
c	7.387			

magnetic Curie temperature:

T_C	70 K	phase I		71S1
	65 K	phase II		

resistivity:

ϱ		phase I,	pressed specimen	71S1
	$1.0 \cdot 10^4$ Ω cm	$T = 300$ K		
	$2.4 \cdot 10^7$ Ω cm	77 K		
		phase II,	semiconducting	
	17 Ω cm	$T = 300$ K		
	52 Ω cm	77 K		

$CuNbO_3$

structure: monoclinic (perovskite), space group C_2^1—P$_2$, C_s^1—Pm or C_{2h}^1—P2/m [70S1]

lattice parameters (lengths in Å):

a	3.836			70S1
b	10.432			
c	5.546			
β	94.66°			

resistivity:

ϱ	$3 \cdot 10^5$ Ω cm	$T = 25$ °C	orientation not specified	70S1

activation energy for conductivity:

E_A	0.60 eV	$T = 4.2 \cdots 298$ K	semiconducting	70S1

Physical property	Numerical value	Experimental conditions	Experimental method remarks	Ref.

SbNbO$_4$

structure: orthorhombic, space group C_{2v}^9—Pna2$_1$ [76R3].
  Growth by hydrothermal method.

lattice parameters (in Å):

a	5.53		phase transition at 605 °C	76R3
b	4.90			
c	11.89			

dielectric constant:

ε_{33}	240		see also [74I]	74P

spontaneous polarization, coercive field:

P_s	15 µC cm^{-2}	$T = 20$ °C		74I
E_c	10 kV cm^{-1}			76P1

ferroelectric Curie temperature:

Θ_C	403 °C			72L
	430 °C			79P

resistivity:

ϱ	$2 \cdot 10^{11}$ Ω cm		i along ferroelectric axis	72L

activation energy for conductivity:

E_A	0.56 eV	$T = 20 \cdots 260$ °C		72L

photo emf:

U_{ph}	$10^2 \cdots 10^4$ V			78F

For pyroelectric coefficients, see [79L, 79P]; for piezoelectric strain coefficient, see [72L].

Sr$_2$MNbO$_6$ (M = Ti, Zr, Hf)

structure: distorted perovskite (at 300 K).

resistivities (in Ω cm):

		T [K]		
ϱ(Sr$_2$TiNbO$_6$)	32	300	orientation not specified	76K1
	89	77		
ϱ(Sr$_2$ZrNbO$_6$)	500	300		
	1580	77		
ϱ(Sr$_2$HfNbO$_6$)	400	300		
	1260	77		

CuTaO$_3$

structure: isotopic with LiNbO$_3$ (perovskite), space group C_{3v}^6—R3c [70S1].

lattice parameters:

a	5.2356 Å			70S1
c	13.8631 Å			

resistivity:

ϱ	$1 \cdot 10^5$ Ω cm	$T = 25$ °C	orientation not specified	70S1

activation energy for conductivity:

E_A	0.44 eV		semiconducting	70S1

Physical property	Numerical value	Experimental conditions	Experimental method remarks	Ref.

CuTa$_2$O$_6$

structure: orthorhombic, space group D$_{2h}^1$—Pmmm [78V].

lattice parameters (in Å):

a	7.5228			78V
b	7.5248			
c	7.5199			

linear thermal expansion coefficient:

α	$8 \cdot 10^{-6}$ K^{-1}	$T = 500 \cdots$ 1000 °C	orientation not specified	75L2

resistivities:

ϱ	$2 \cdot 10^3$ Ω cm	RT	orientation not specified	75L2
	$7 \cdot 10^6$ Ω cm	$T = 140$ K		

K$_2$CrO$_4$

structure: orthorhombic, space group D$_{2h}^{16}$—Pnma [72M1, 78T1]

lattice parameters (in Å):

a	7.662		density $d = 2.74$ g cm^{-3}	78T1
b	5.919			
c	10.391			
d(Cr−O)	1.646			
d(Mn−O)	1.605			

Raman wavenumbers (in cm^{-1}):

		symmetry:		
$\bar{\nu}$	918	B$_{2g}$	internal modes, RT values	71C2
	903	A$_g$		
	881	B$_{2g}$		
	878	B$_{3g}$		
	876	B$_{1g}$		
	867	A$_g$		
	851	A$_g$, B$_{2g}$		
	396	A$_g$, B$_{2g}$		
	392	B$_{1g}$		
	387	B$_{3g}$		
	386	A$_g$, B$_{2g}$		
	350	B$_{1g}$, B$_{2g}$		
	346	B$_{3g}$		
	345	A$_g$		
	157	A$_g$		
	138	B$_{3g}$		
	119	B$_{1g}$, B$_{3g}$		
	116	A$_g$		
	114	B$_{2g}$		
	109	A$_g$		
	99	B$_{2g}$		
	93	A$_g$		
	91	B$_{3g}$		
	85	B$_{3g}$		
	83	A$_g$		
	67	B$_{1g}$		
	54	B$_{1g}$, B$_{2g}$		
	37	A$_g$		

Physical property	Numerical value	Experimental conditions	Experimental method remarks	Ref.
activation energy for conductivity (in eV):				
		T [°C]		
E_A	0.95	> 200	measured in air	73C
	0.77	> 200	air, doped with Na_3PO_4	
	1.05	> 200	air, doped with $BaCrO_4$	
	1.15	< 200	p-type, measured in vacuum	
	0.93	> 200	n-type	
	1.05	< 200	p-type, measured in oxygen	
	0.59	> 200	n-type	

temperature dependence of conductivity: Fig. 162

optical spectra: absorption, Figs. 163, 164; reflection: Fig. 165; an absorption band at $26\,800$ cm^{-1} is caused by a $^1A_1 - {}^1T_2$ ($t_1 - 2e$) transition [76I]

$PbCrO_3$
structure: perovskite, space group $O_h^1 - Pm3m$ [72C1]

lattice parameter:

a	4.0050 Å			72C1

Néel temperature:

T_N	160 K			72C1

resistivity:

ϱ	$2.6 \cdot 10^3$ Ω cm	RT		72C1

activation energy for conductivity:

E_A	0.27 eV	$T = 100 \cdots 300$ K	semiconducting	72C1

$PbMoO_4$
Crystal structure
structure: tetragonal, space group $C_{4h}^6 - I4_1/a$ [65L]

Growth by Czochralski and hydrothermal methods [69P, 61U, 70B1, 63D].

lattice parameters (in Å):

a	5.432 (5.4355)		values in brackets from [72S1]	65L
c	12.107 (12.108)			

Lattice properties
infrared wavenumbers (in cm^{-1}):

$\bar{v}$	786		assignment: $\bar{v}_3(A_u) + \bar{v}_2(E_u)$	72T
	374		$\bar{v}_2(A_u)$	
	307		$\bar{v}_4(E_u)$	
	272		$\bar{v}_4(A_u)$	
	125		transl. E_u, transl. A_u	
	110		rot. E_u	

linear expansion coefficients (in 10^{-6} K^{-1}):

α	12.61			75S1
α_{11}	10			71C4
α_{33}	25			

Physical property	Numerical value	Experimental conditions	Experimental method remarks	Ref.
sound velocity:				
$\dfrac{1}{v}\dfrac{\mathrm{d}v}{\mathrm{d}T}$	$-161\cdot10^{-6}$ K^{-1}		longitudinal wave along c axis	71C4
ultrasound velocities:				
v	$4.30\cdot10^5$ cm s^{-1}		quasi-longitudinal mode polarized in (001)-plane, measured by pulse-echo technique at 15 MHz	75F1
	$1.30\cdot10^5$ cm s^{-1}		quasi-shear mode, same polarization, see also [75G]	
elastic moduli (in 10^{10} N m^{-2}):				
c_{11}	10.9	RT	see also [75G]	75F1
c_{12}	6.8			
c_{13}	5.3			
c_{16}	-1.4			
c_{33}	9.2			
c_{44}	2.67			
c_{66}	3.37			
elastic compliances (in 10^{-12} m^2 N^{-1}):				
s_{11}	21.0	RT	see also [75G]	75F1
s_{33}	16.6			
s_{44}	37.5			
s_{66}	40.6			
s_{12}	-12.4			
s_{13}	-4.9			
s_{16}	13.5			
bulk modulus:				
B	$7.1\cdot10^{10}$ N m^{-2}			75F1
compressibilities (in 10^{-12} m^2 N^{-1}):				
κ_v	14.0		volume compressibility	75F1
$\kappa_{1,z}$	6.7		linear compressibility $\parallel$ c axis	
$\kappa_{1,xy}$	3.6		$\perp$ c axis	
elastooptic constants:				
p_{11}	0.24	$\lambda=632.8$ nm	see also [75G]	71C4, 78C
p_{12}	0.24			
p_{13}	0.255			
p_{16}	0.017			
p_{31}	0.175			
p_{33}	0.300			
p_{44}	0.067			
p_{45}	-0.01			
p_{61}	0.013			
p_{66}	0.05			
p_{13}	0.254	$\lambda=514.5$ nm		
p_{23}	0.308			

Transport properties

p-type conduction below 900 K [75L1].
conductivity vs. O_2 partial pressure: Fig. 166, vs. temperature: Fig. 167; ionic conductivity vs. temperature: Fig. 168

Physical property	Numerical value	Experimental conditions	Experimental method remarks	Ref.

Optical properties

optical spectra: transmission: Figs. 169, 170; Raman effect: Fig. 171; photoconductivity: Figs. 172, 173; luminescence [81B2]

dielectric constants:

$\varepsilon_{\parallel a}$	34.0			69B3
$\varepsilon_{\parallel c}$	40.6			

refractive indices:

		λ [μm]		
n_0 (n_e)	2.7191 (2.4429)	0.4046		74L
	2.5136 (2.3409)	0.4800		
	2.4754 (2.3185)	0.5086		
	2.4388 (2.2959)	0.5460		
	2.4160 (2.2814)	0.5770		
	2.3808 (2.2581)	0.6438		
	2.3072 (2.2066)	1.014		
	2.2808 (2.1871)	1.500		
	2.2679 (2.1769)	2.000		
	2.2577 (2.1686)	2.500		
	2.2474 (2.1603)	3.000		
	2.2418 (2.1556)	3.250		
	2.2360 (2.1507)	3.500		
	2.2297 (2.1455)	3.7500		
$\dfrac{1}{n}\dfrac{dn}{dT}$	$-30 \cdot 10^{-6}$ K^{-1} $(18 \cdot 10^{-6}$ K$^{-1})$			71C4

Further properties

density:

d	6.95 g cm^{-3}			69P

Debye temperature:

Θ_D	190 K		calculated from elastic constants	75S1

heat capacity:

C_p	≈ 0.5 J cm^{-3} K^{-1}			71C4

thermal conductivity:

$\kappa_{11} = \kappa_{33}$	15(2) mW cm^{-1} K^{-1}			78B1

PbWO$_4$

structure:

PbWO$_4$–I: tetragonal (scheelite type), space group C$_{4h}^6$–I4$_1$/a.

a	5.4622 Å		lattice parameter vs. temperature:	72B2,
c	12.048 Å		Fig. 174	72S1

PbWO$_4$–II: monoclinic (not wolframite type)
PbWO$_4$–III: monoclinic, space group C$_{2h}^5$–P2$_1$/n, high pressure phase

a	12.709 Å		exists above	76R2
b	7.048 Å		p[kbar]$=0.7+0.039\,T$ [°C]	
c	7.348 Å		$(T=350\cdots650$°C)	
β	90.57°			

Physical property	Numerical value	Experimental conditions	Experimental method remarks	Ref.

density:

d	$9.19 \, \text{g cm}^{-3}$			76R2

linear thermal expansion coefficients (in $10^{-6} \, \text{K}^{-1}$):

(first column $\alpha_\parallel$ (a, b-plane), second column $\alpha_\perp$ ($\parallel c$), third column: volume expansion coefficient β)

phase I:

				T [K]	78B1
$\alpha_{\parallel, \perp}, \beta$	9.2 (11)	22.1 (08)	40.5	100···200	
	11.7 (11)	25.8 (12)	49.2	200···300	
	12.7 (10)	26.8 (17)	52.2	300···450	
	13.8 (18)	28.1 (22)	55.7	450···600	
	15.0 (18)	28.4 (16)	58.4	600···800	

wavenumbers of Raman active modes (in cm^{-1}):

phase I:

$\bar{v}$	905	A_g	$\bar{v}_1$ (W−O stretch)	76S1
	766	B_g	$\bar{v}_3$ (W−O bend)	
	753	E_g		
	358	B_g, E_g	$\bar{v}_4$ (WO_4^{2-} bend)	
	328	A_g, B_g	$\bar{v}_2$ (WO_4^{2-} bend)	
	192	E_g	rotation (xy)	
	178	A_g	rotation (z)	
	63	E_g	Pb^{2+}−Pb^{2+} translation	
	54	B_g		
	90	E_g	WO_4^{2-}−WO_4^{2-} translation	
	78	B_g		

wavenumbers of infrared active modes (in cm^{-1}):

phase I:

$\bar{v}$	792	$\bar{v}_3(A_u) + \bar{v}_3(E_u)$	76S1
	377	$\bar{v}_2(A_u)$	
	301	$\bar{v}_4(E_u)$	
	262	$\bar{v}_4(A_u)$	
	116	translation (E_u)	
	94	rotation (E_u)	

optical spectra: transmission: Fig. 175; reflectivity: Fig. 176; photoconductivity: Fig. 177;

transport data: electrical conductivity: Fig. 178; ionic conductivity: Fig. 179;

$KMnO_4$

structure: orthorhombic, space group D_{2h}^{16}−Pnma [50C, 57R]

lattice parameters (in Å):

a	9.09		50C,
b	7.41		57R
c	5.72		

electrical conductivity: Fig. 180

Physical property	Numerical value	Experimental conditions	Experimental method remarks	Ref.

wavenumbers of Raman active phonons (in cm^{-1}):

$\bar{\nu}$	349		assignment: $\bar{\nu}_2$	71K
	396		$\bar{\nu}_4$	
	790		$2\bar{\nu}_4$	
	842.5		$\bar{\nu}_1$	
	904.5 ⎱		$\bar{\nu}_3$	
	914.5 ⎰			
	1682		$2\bar{\nu}_1$	
	1741 ⎱		$\bar{\nu}_1 + \bar{\nu}_3$	
	1758 ⎰			
	2520		$3\bar{\nu}_1$	
	2580		$2\bar{\nu}_1 + \bar{\nu}_3$	
	3358		$4\bar{\nu}_1$	
	4190		$5\bar{\nu}_1$	
	5015		$6\bar{\nu}_1$	

LaMnO$_3$

structure: orthorhombic (perovskite)

lattice parameters (in Å):

a	5.531		structure transformation at	76S2
b	5.602		523 K to cubic lattice	
c	7.742			

absorption spectrum: Fig. 181; heat capacity: Fig. 182. For IR absorption, see [70M1].

Néel temperature:

T_N	141 K			76S2
	130 K			70P

influence of oxygen partial pressure on annealing:

The decrease of oxygen partial pressure in the atmosphere of final annealing (preliminary annealing was carried out in air at 1253 K during 8 h) under 0.5 atm of the air leads to the (20 h) formation of LaMnO$_3$ with perovskite (cubic) structure. The following data have been measured under three different oxygen partial pressures [78K] (p_{O_2} in atm):

	LaMnO$_{3.079}$	LaMnO$_{3.000}$	LaMnO$_{2.947}$
$-\log p_{O_2}$	0	8.79	11.60
$a=c$	3.937 Å	3.993 Å	
b	3.877 Å	3.854 Å	
β	90°59′	92°16′	
σ at RT	(1.6 Ω^{-1} cm^{-1}	$6 \cdot 10^{-3}$ Ω^{-1} cm^{-1}	$8 \cdot 10^{-5}$ Ω^{-1} cm^{-1}

solid solutions La$_{1-x}$M$_x$MnO$_3$ with M = Ca, Sr, Ba, Pb:

resistivity (first column, in Ω cm) and **magnetoresistance** (second column, in %):

			T [K]		
$\varrho, \Delta\varrho/\varrho$	$1.2 \cdot 10^3$	-6.4	77	M = Ca, x = 0.1	73B3
	0.23		300		
	$8.4 \cdot 10^2$	-8.2	77	M = Ca, x = 0.2	
	0.24		300		
	71	-12.2	77	M = Sr, x = 0.1	
	0.32		300		

(continued)

Physical property	Numerical value		Experimental conditions	Experimental method remarks	Ref.
resistivity and **magnetoresistance** (continued)					
$\varrho,\ \Delta\varrho/\varrho$	0.1	-10.3	77	$M = Sr,\ x = 0.2$	73B3
	0.12	-0.18	300		
	0.066	-12.5	77	$M = Sr,\ x = 0.3$	
	0.10	-1.2	300		
	1.2	-14.6	77	$La_{0.6}Sr_{0.3}Ba_{0.1}MnO_3$	
	0.66	-1.1	300		
	$6.8 \cdot 10^3$	-8.8	77	$M = Ba,\ x = 0.1$	
	0.32		300		
	6.3	-13.1	77	$M = Ba,\ x = 0.2$	
	0.32		300		
	0.11	-22.0	77	$M = Pb,\ x = 0.3$	
	0.36				

For IR absorption, see [70M1]; for magnetic properties, see Landolt-Börnstein Vol. III/12a, p. 447ff.

$LiFe_5O_8$

structure: inverse spinel, space group O^6—$P4_332$ [77M2]

lattice parameters (in Å):

			preparation conditions (at 850°C, 15h):	
a	8.346		heated in O_2, 1 atm; α-$LiFe_5O_8$ + trace α-Fe_2O_3	77M1
	8.338		heated in air, 0.21 atm; α-$LiFe_5O_8$	
	8.330		heated in argon, less than 10^{-6} atm; α-$LiFe_5O_8$	

phase transition from α-phase to β-phase at 750°C [64S, 65A, 76M1]

density:

d	4.752 g cm^{-3}			73B1

transport data: conductivity, bulk and surface resistivity, magnetoresistance: Figs. 183···185
dielectric constant: Fig. 186
heat capacity: Fig. 187

$BaFe_{12}O_{19}$*)

structure: space group D_{6h}^4—$P6_3/mmc$

lattice parameters:

a	5.8 Å			69A1
c	23.18 Å			

density:

d	5 g cm^{-3}			69A1

thermal conductivity:

κ	4 W m^{-1} K^{-1}		orientation not specified	76C

magnetic Curie temperature:

T_C	440···460°C			76C

conductivity: Fig. 188; Peltier coefficient: Fig. 189

---

*) For a discussion of the properties of hexagonal ferrites, especially of their magnetic properties, see Landolt-Börnstein, Vol. III/4b and III/12c.

Physical property	Numerical value	Experimental conditions	Experimental method remarks	Ref.

$SrFe_{12}O_{19}$ *)

structure: space group $D_{6h}^4 - P6_3/mmc$

lattice parameters:

a	5.78 Å			69A1
c	22.98 Å			

conductivity: Fig. 190; Peltier coefficient: Fig. 191

$PbFe_{12}O_{19}$ *)

structure: space group: $D_{6h}^4 - P6_3/mmc$

lattice parameters:

a	5.78 Å			69A1
c	23.09 Å			

conductivity: Fig. 192; Peltier coefficient: Fig. 193

$LaCoO_3$ (see also Landolt-Börnstein, Vol. III/12a and section 10.3 of this volume)

structure: rhombohedral (perovskite), space group $D_{3d}^6 - R\bar{3}c$ (300···400 K), $C_3^4 - R\bar{3}$ (above 650 K) [67R1]

lattice parameters:

a	5.433 Å			79R
c	13.074 Å			

$LaCoO_3$ is semiconducting according to [69J], [69T].

Seebeck coefficient: Figs. 194, 196; resistivity: Figs. 195, 196; conductivity in solid solutions $LaCo_{1-x}M_xO_3$ with M = Mg, Te, Ni, Fe: Figs. 197···200; absorption spectrum: [75M]

$Ba_{0.5}OsO_3$

structure: body centered cubic [74S1]

lattice parameter:

a	9.33 Å			74S1

transport parameters:

ϱ	2 Ω cm	$T = 300$ K		74S3
E_A	0.21 eV			

$NiFe_2O_4$ **)

structure: inverse spinel, cubic, space group $O_h^7 - Fd3m$ [74H1]

lattice parameter:

a	8.3378 Å			71S3, 79S2

infrared absorption bands (wavenumbers, in cm^{-1}):

$\bar{\nu}$	593		transmission and reflection vs.	71G1
	404		wavenumber: Fig. 201	
	330			
	196			

conductivity:

σ	$0.22 \cdot 10^{-2}$ Ω^{-1} cm^{-1}	$T = 300$ K	see also [74C]	71C5

---

*) For a discussion of the properties of hexagonal ferrites, especially of their magnetic properties, see Landolt-Börnstein, Vol. III/4b and III/12c.

**) Data on ferrites of the spinel type are extensively presented in Landolt-Börnstein, Vol. III/12b.

Physical property	Numerical value	Experimental conditions	Experimental method remarks	Ref.

mobilities and carrier concentrations in the $Ni_xFe_{3-x}O_4$ system [70N]:

Composition (valence formula)	Mobility μ cm^2/Vs	Carrier concentration n $10^{20}/cm^3$	Transition (jump) frequency v_0 THz
$Fe^{3+}(Ni^{2+}Ni^{3+}_{0.002}Fe^{3+}_{0.998})O_4$ [a]	$4.08 \cdot 10^{-9}$	1.87	21
$Fe^{3+}(Ni^{2+}_{1.04}Fe^{2+}_{0.007}Fe^{3+}_{0.953})O_{3.977}$ [a]	$1.37 \cdot 10^{-8}$	0.96	48
$Fe^{3+}(Ni^{2+}_{1.03}Fe^{2+}_{0.035}Fe^{3+}_{0.935})O_{3.968}$ [a]	$2.97 \cdot 10^{-5}$	4.8	2.04
$Fe^{3+}(Ni^{2+}Fe^{2+}_{0.04}Fe^{3+}_{0.96})O_{3.98}$	$1.22 \cdot 10^{-5}$	5.5	2.68
$Fe^{3+}(Ni^{2+}_{0.918}Fe^{2+}_{0.049}\square_{0.011}Fe^{3+}_{1.022})O_4$	$2.44 \cdot 10^{-5}$	6.74	1.16
$Fe^{3+}(Ni^{2+}_{0.98}Fe^{2+}_{0.053}Fe^{3+}_{0.967})O_{3.974}$	$2.92 \cdot 10^{-5}$	7.28	1.66
$Fe^{3+}(Ni^{2+}_{0.88}Fe^{2+}_{0.06}\square_{0.02}Fe^{3+}_{1.04})O_4$	$1.825 \cdot 10^{-4}$	7.98	1.67
$Fe^{3+}(Ni^{2+}_{0.84}Fe^{2+}_{0.115}\square_{0.015}Fe^{3+}_{1.03})O_4$	$3.1 \cdot 10^{-4}$	15.8	2
$Fe^{3+}(Ni^{2+}_{0.99}Fe^{2+}_{0.12}Fe^{3+}_{0.89})O_{3.945}$	$5.04 \cdot 10^{-4}$	16.5	1.57
$Fe^{3+}(Ni^{2+}_{0.89}Fe^{2+}_{0.142}Fe^{3+}_{0.968})O_{3.986}$	$5.12 \cdot 10^{-4}$	19.85	1.8
$Fe^{+}(Ni^{2+}_{0.856}Fe^{2+}_{0.144}Fe^{3+})O_4$	$5.6 \cdot 10^{-4}$	19.8	1.37
$Fe^{3+}(Ni^{2+}_{0.92}Fe^{2+}_{0.205}Fe^{3+}_{0.875})O_{3.94}$	$1.23 \cdot 10^{-3}$	28.1	1.78
$Fe^{3+}(Ni^{2+}_{0.637}Fe^{2+}_{0.38}Fe^{3+}_{1.017})O_{3.992}$	$7.45 \cdot 10^{-3}$	60	1.72

[a]) Calculated from the Seebeck coefficient.

resistivity and conductivity of solid solutions based on $NiFe_2O_4$: Figs. 202···204; for Seebeck coefficient, see [74C].

heat capacity [74G1]:

T K	C_p $J\,mol^{-1}\,K^{-1}$	C_v $J\,mol^{-1}\,K^{-1}$
50	11.5	11.5
60	17.5	17.5
80	31.3	31.3
100	45.6	45.6
120	59.7	59.6
140	73.5	73.4
160	86.3	86.1
180	97.9	97.6
200	108.1	107.7
220	117.5	117.0
240	125.8	125.2
260	133.4	312.6
280	140.0	139.1
300	146.3	145.2

$CuFe_2O_4$ [**]

structure: cubic (spinel), space group O_h^7—Fd3m (quenched); tetragonal, space group D_{4h}^{19}—I4$_1$/amd (annealed) [50W]

lattice parameters (in Å):

a	8.389		quenched	71Z
	8.37		quenched	50B
	8.397	$(u = 0.381)$	$(= a^2 c)^{1/3}$ (tetragonal phase)	74O1
a	8.22		annealed	56P
c	8.70			

For effect of heat treatment and quenching temperature on lattice parameter, see [71Z]; see also [70M1].

resistivity, conductivity: Figs. 205, 206

[**]) Data on ferrites of the spinel type are extensively presented in Landolt-Börnstein, Vol. III/12b.

Physical property	Numerical value	Experimental conditions	Experimental method remarks	Ref.
density:				
d	5.42 g/cm^3		X-ray density (cubic phase)	47V
activation energy for conductivity:				
E_A	0.24 eV	$T < 350$ K	tetragonal structure	75B
	0.31 eV	$T > 350$ K		

CoAl$_2$O$_4$ **)
structure: cubic (spinel), space group O_h^7—Fd3m

lattice parameters (in Å):				
a	8.095			78T2
	8.1068			76G
	8.104			72M2
density:				
d	4.43 g cm^{-3}			72M2
refractive index:				
n	1.815			76Z

transmission: Fig. 207

conductivity, mobility, Seebeck coefficient: Fig. 208

activation energies:				
E_A	1.8 eV		for conductivity	72M4
	0.8 eV		for mobility	

Cd$_2$Os$_2$O$_7$
structure: cubic (pyrochlore), space group O_h^7—Fd3m

lattice parameter:				
a	10.16 Å		metal-semiconductor transition at 225 K	73S4, 75S3

resistivity, activation energy: Figs. 209, 210

Seebeck coefficient:				
S	$-300\ \mu$V K^{-1}	$T \ll 225$ K		74S2
	$+5\ \mu$V K^{-1}	$T \gg 225$ K		

magnetic susceptibility: [74S2]

Ca$_2$Os$_2$O$_7$
structure: face centered cubic

lattice parameter:				
a	10.22 Å			74S1
resistivity:				
ϱ	1 Ω cm	$T = 80$ K $\cdots$ 470 K	measured on pellets	74S1

**) Data on ferrites of the spinel type are extensively presented in Landolt-Börnstein, Vol. III/12 b.

Physical property	Numerical value	Experimental conditions	Experimental method remarks	Ref.
$Bi_2Os_2O_7$				
structure: cubic				
lattice parameter:				
a	9.36 Å			
resistivity (in 10^{-3} Ω cm):				
ϱ	1.7	$T = 300$ K		74S1
	0.6	4.2 K		
	1.9	140 K		
$Bi_2Pt_2O_7$				
structure: face centered cubic				
lattice parameter:				
a	10.36 Å			74S1
resistivity and activation energy:				
ϱ	8 Ω cm	$T = 300$ K		74S1
E_A	0.15 eV			
Ca_4PtO_6				
structure: I: orthorhombic, II: hexagonal				
lattice parameters (in Å):				
a	9.164	orthorhombic		75S2
b	9.235	modification		
c	6.502			
a	9.26	hexagonal		
c	11.26	modification		
resistivity:				
ϱ	$1.04 \cdot 10^2$ Ω cm	orthorhombic	orientation not specified; temperature dependence: Fig. 211	75S2
	$(4.80 \cdots 22.5) \cdot 10^3$ Ω cm	hexagonal		

References for 10.2.4

45M   Megaw, H.D.: Nature **155** (1945) 484.
46M   Megaw, H.D.: Proc. Phys. Soc. London **58** (1946) 133.
46N   Naylor, B.F., Cook, O.A.: J. Am. Chem. Soc. **68** (1946) 1003.
46S   Shomate, C.H.: J. Am. Chem. Soc. **68** (1946) 964.
46V   von Hippel, A., Breckenridge, R.G., Chesley, F.G., Tisza, L.: Ind. Eng. Chem. **38** (1946) 1097.
46W   Wul, B., Goldman, L.M.: C. R. Acad. Sci. USSR **51** (1946) 21.
47M   Miyake, S., Ueda, R.: J. Phys. Soc. Jpn. **2** (1947) 93.
47V   Verwey, E.J.W., Heilman, E.L.: J. Chem. Phys. **15** (1947) 174.
48B   Burbank, R.D., Evans Jr., H.T.: Acta Crystallogr. **1** (1948) 330.
49K   Kay, H.E., Vousden, P.: Philos. Mag. (7) **40** (1949) 1019.

49R	Rhodes, R.G.: Acta Crystallogr. **2** (1949) 417.
50B	Bertaut, F.: C. R. **230** (1950) 213.
50C	Crone Mc, W.C.: Anal. Chem. **22** (1950) 1459.
50R	Roberts, S.: J. Am. Ceram. Soc. **33** (1950) 63.
50W	Weil, L., Bertaut, F., Bochirol, L.: J. Phys. Radium **11** (1950) 208.
51R	Rhodes, R.G.: Acta Crystallogr. **4** (1951) 105.
51S1	Sawaguchi, E., Shirane, G., Takagi, Y.: J Phys. Soc. Jpn. **6** (1951) 333.
51S2	Sawaguchi, E., Maniwa, H., Hoshino, S.: Phys. Rev. **83** (1951) 1078.
51S3	Shirane, G., Hoshino, S.: J. Phys. Soc. Jpn. **6** (1951) 265.
51S4	Shirane, G., Sawaguchi, E.: Phys. Rev. **81** (1951) 458.
51S5	Shirane, G., Sawaguchi, E., Takagi, Y.: Phys. Rev. **84** (1951) 476.
51V1	Vousden, P.: Acta Crystallogr. **4** (1951) 68.
51V2	Vousden, P.: Acta Crystallogr. **4** (1951) 373.
51W1	Wood, E.A.: Acta Crystallogr. **4** (1951) 353.
51W2	Wood, E.A.: J. Chem. Phys. **19** (1951) 976.
52L	Lapitskii, A.V.: Zh. Obshch. Khim. **22** (1952) 379.
52R	Rogers, H.H.: Mass. Inst. Technol. Lab., Tech. Rep. Insulation Res., (1952) No. 56.
52S	Sawaguchi, E.: J. Phys. Soc. Jpn. **7** (1952) 110.
53S	Sawaguchi, E.: J. Phys. Soc. Jpn. **8** (1953) 615.
54G	Granicher, H., Jakits, O.: Nouvo Cimento (9) **11** (1954) Suppl. 480.
54H	Horie, T., Kawabe, K., Sawada, S.: J. Phys. Soc. Jpn. **9** (1954) 823.
54S	Shirane, G., Newnham, R.E., Pepinsky, R.: Phys. Rev. **96** (1954) 581.
55C	Cross, L.E., Nicholson, B.J.: Philos. Mag. (7) **46** (1955) 453.
55J	Jona, F., Shirane, G., Pepinsky, R.: Phys. Rev. **97** (1955) 1584.
55K	Kobayashi, J., Ueda, R.: Phys. Rev. **99** (1955) 1900.
55M	Marutake, M., Ikeda, T.: J. Phys. Soc. Jpn. **10** (1955) 424.
55R1	Rase, D.E., Roy, R.: J. Am. Ceram. Soc. **38** (1955) 102.
55R2	Reisman, A., Holtzberg, F.: J. Am. Chem. Soc. **77** (1955) 2115.
55R3	Reisman, A., Triebwasser, S., Holtzberg, F.: J. Am. Chem. Soc. **77** (1955) 4228.
56P	Prince, E., Trueting, R.G.: Acta Crystallogr. **9** (1956) 1025.
56R	Reisman, A., Holtzberg, F., Berkenblit, M., Berry, M.: J. Am. Chem. Soc. **78** (1956) 4514.
56S	Shirane, G., Pepinsky, R., Frazer, B.C.: Acta Crystallogr. **9** (1956) 131.
56T	Triebwasser, S.: Phys Rev. **101** (1956) 993.
57G	Giardini, A.A.: J. Opt. Soc. Am. **47** (1957) 726.
57J	Jona, F., Pepinsky, R.: Phys. Rev. **105** (1957) 861.
57K	Kay, H.F., Bailey, P.C.: Acta Crystallogr. **10** (1957) 219.
57M	Marutake, M., Ikeda, T.: J. Phys. Soc. Jpn. **12** (1957) 233.
57R	Ramaseshan, S., Venkatesan, K., Mani, N.V.: Proc. Indian Acad. Sci. Sect. **A46** (1957) 95.
57S	Shirane, G., Danner, H., Pepinsky, R.: Phys. Rev. **105** (1957) 856.
58I	Inuishi, Y., Uematsu, S.: J. Phys. Soc. Jpn. **13** (1958) 761.
58L	Linz, A., Herrington, K.: J. Chem. Phys. **28** (1958) 824.
58M1	Meyerhofer, D.: Phys. Rev. **112** (1958) 413.
58M2	Miller, C.E.: J. Appl. Phys. **29** (1958) 233.
58R	Reisman, A., Banks, E.: J. Am. Chem. Soc. **80** (1958) 1877.
59C	Casella, R.C., Keller, S.P.: Phys. Rev. **116** (1959) 1464.
59H	Huibregtse, E.J., Bessey, W.H., Drougard, M.E.: J. Appl. Phys. **30** (1959) 899.
60B	Branwood, A., Tredgold, R.H.: Proc. Phys. Soc. London **76** (1960) 93.
60Y	Yoshida, I.: J. Phys. Soc. Jpn. **15** (1960) 2211.
61H	Heywang, W.: Solid State Electron. **3** (1961) 51.
61U	Uitert van, L.G., Swanekamp, F.W., Preziosi, S.: J. Appl. Phys. **32** (1961) 1176.
61W	Wells, M., Megaw, H.D.: Proc. Phys. Soc. London **78** (1961) 1258.
62B	Branwood, A., Hughes, O.H., Hurd, J.D., Tredgold, R.H.: Proc. Phys Soc. London **79** (1962) 1161.
62G	Gurevich, V.M.: Thesis, of Candidate of Phys.-Math. Sciences, Moskva (1962) cited in [69G].
62K	Kabalkina, S.A., Vereshchagin, L.F., Shulenin, B.M.: Dokl. Akad. Nauk SSSR **144** (1962) 1019; Sov. Phys. Dokl. (English Transl.) **7** (1962) 527.
62M1	Makishima, S., Hasegawa, K., Shionoya, S.: J. Phys. Chem. Solids **23** (1962) 749.
62M2	Merker, L.: J. Am. Ceram. Soc. **45** (1962) 366.

62R	Rimai, L., deMars, G.A.: Phys. Rev. **127** (1962) 702.
62S	Spitzer, W.G., Miller, R.C., Kleinman, D.A., Howarth, L.E.: Phys. Rev. **126** (1962) 1710.
63C	Cox, G.A., Tredgold, R.H.: Phys. Lett. **4** (1963) 199.
63D	Demyanets, L.N., Garashina, L.S., Litvin, B.N.: Kristallografiya **8** (1963) 800.
63G	Geusic, J.E., Kurtz, S.K., Nelson, T.J., Wemple, S.H.: Appl. Phys. Lett. **2** (1963) 185.
63K	Katz, L., Ueda, R., Megaw, H.D.: Acta Crystallogr. **16** (1963) Suppl. A189.
63M1	Mattes, B.L.: J. Appl. Phys. **34** (1963) 682.
63M2	Miller, R.C., Kleinman, A.D., Savage, A.: Phys. Rev. Lett. **11** (1963) 146.
63M3	Myers, R.A.: Office of Naval Research, Contract No. 1866 (16); Tech. Rep. **433** (1963), Cruft Laboratory, Harvard University, Cambridge, Massachusetts.
63R	Rogers, D.B., Arnolt, R.J., Wold, A., Goodenough, J.B.: J. Phys. Chem. Solids **24** (1963) 347.
63S	Schmidt, G., Hegenbarth, E.: Phys. Status Solidi **3** (1963) 329.
64B1	Ballantyne, J.M.: Phys. Rev. **136** (1964) A429.
64B2	Brown, F., Taylor, C.E.: J. Appl. Phys. **35** (1964) 2554.
64C	Cowley, R.A.: Phys. Rev. **134** (1964) A981.
64F	Frederikse, H.P.R., Thurber, W.R., Hosler, W.R.: Phys. Rev. **134** (1964) A442.
64G	Geusic, J.E., Kurtz, S.K., van Uitert, L.G., Wemple, S.H.: Appl. Phys. Lett. **4** (1964) 141.
64H	Heikes, R.R., Miller, R.C., Mazelsky, R.: Physica **30** (1964) 1600.
64L1	Lawless, W.N., DeVries, R.C.: J. Appl. Phys. **35** (1964) 2638.
64L2	Lytle, F.W.: J. Appl. Phys. **35** (1964) 2212.
64P1	Perry, C.H., Khanna, B.N., Rupprecht, G.: Phys. Rev. **135** (1964) A408.
64P2	Poplavko, Yu.M.: Fiz. Tverd. Tela **6** (1964) 58; Sov. Phys. Solid State (English Transl.) **6** (1964) 45.
64R	Rimai, L., Deutsch, T., Silverman, B.D.: Phys. Rev. **133** (1964) A1123.
64S	Schieber, M.: J. Inorg. Nucl. Chem. **26** (1964) 1363.
64W1	Wemple, S.H.: Mass. Inst. Technol., Res. Lab. Electron., Tech. Rep. **425** (1964).
64W2	Winchell, A.N., Winchell H.: The Microscopic Character of Artificial Inorganic Solid Substances, 3-rd Ed., New York, Academic Press, **1964**.
65A	Anderson, J.C., Schieber, M.: Sci. Ceram. **2** (1965) 345.
65B1	Ballantyne, J.M.: Phys. Rev. **138** (1965) A646.
65B2	Bhide, V.G., Multani, M.S.: Phys. Rev. **139** (1965) A1983.
65B3	Bond, W.L.: J. Appl. Phys. **36** (1965) 1674.
65C	Cardona, M.: Phys. Rev. **140** (1965) A651.
65H	Hannon, D.M.: Bull. Am. Phys. Soc. (2) **10** (1965) 330.
65J1	Johnson, C.J.: Appl. Phys. Lett **7** (1965) 221.
65J2	Johnston, A.R., Weingar, J.M.: J. Opt. Soc. Am. **55** (1965) 828.
65K	Kahng, D., Wemple, S.H.: J. Appl. Phys. **36** (1965) 2925.
65L	Leciejewicz, J.: Z. Kristallogr. **121** (1965) 158.
65P	Perry, C.H., McCarthy, D.H., Rupprecht, G.: Phys. Rev. **138** (1965) A1537.
65S1	Sasaki, H.: Jpn. J. Appl. Phys. **4** (1965) 24.
65S2	Schaufele, R.F., Weber, M.J., Waugh, J.S.: Phys. Rev. **140** (1965) A872.
65S3	Schooley, J.F., Hosler, W.R., Ambler, E., Becker, J.H., Cohen, M.L., Koonce, C.S.: Phys. Rev. Lett. **14** (1965) 305.
65S4	Senhouse, L.S., Smith, G.E., DePaolis, M.V.: Phys. Rev. Lett. **15** (1965) 776.
65S5	Suemune, Y.: J. Phys. Soc. Jpn. **20** (1965) 174.
65W	Wemple, S.H.: Phys. Rev. **137** (1965) A1575.
66A	Ambler, E., Colwell, J.H., Hosler, W.R., Schooley, J.F.: Phys. Rev. **148** (1966) 280.
66B1	Baer, W.S.: Phys. Rev. Lett. **16** (1966) 729.
66B2	Barker Jr., A.S.: Phys. Rev. **145** (1966) 391.
66B3	Bechmann, R., Hearmon, R.F.S.: Landolt-Börnstein, Numerical Data and Functional Relationships in Science and Technology, New Series, Group III: Crystal and Solid State Physics, Vol. 1, Elastic, Piezoelectric, Piezooptic and Electrooptic Constants of Crystals, Springer: Berlin, Heidelberg, New York **1966**.
66C	Carlsson, L.: Acta Crystallogr. **20** (1966) 459.
66D	Dvorak, V., Fouskova, A., Glogar, P. (eds): Proceedings of the International Meeting on Ferroelectricity, held at Prague (1966) (in two volumes), Prague: Institute of Physics of the Czechoslovak Academy of Sciences **1966**.
66H1	Hauser, O., Schenk, M.: Phys. Status Solidi **18** (1966) 547.
66H2	Hegenbarth, E.: Article on p. 104 in Vol. I of [66D].

66H3	Hughe, O.H., Tredgold, R.H., Williams, R.H.: J. Phys. Chem. Solids **27** (1966) 79.
66H4	Hartmann-Boutron, F.: C.R. Acad. Sci. (Paris) Ser. **B263** (1966) 1131.
66S1	Sakudo, T., Unoki, H., Fujii, Y.: J. Phys. Soc. Jpn. **21** (1966) 2739.
66S2	Samara, G.A.: Phys. Rev. **151** (1966) 378.
66S3	Schooley, J.F., Thurber, W.R.: J. Phys. Soc. Jpn. **21** (1966) Suppl. 639.
66S4	Senhouse Jr., L.S., DePaolis Jr., M.V., Loomis, T.C.: Appl. Phys. Lett. **8** (1966) 173.
66T	Tufte, O.N., Stelzer, E.L.: Phys. Rev. **141** (1966) 675.
66W	Wemple, S.H., Jayaraman, A., DiDomenico Jr., M.: Phys. Rev. Lett. **17** (1966) 142.
67B1	Baer, W.S.: J. Phys. Chem. Solids **28** (1967) 677.
67B2	Bergulund, C.N., Baer, W.S.: Phys. Rev. **157** (1967) 358.
67F1	Fleury, P.A., Worlock, J.M.: Phys. Rev. Lett. **18** (1967) 665.
67F2	Frova, A., Boddy, P.J.: Phys. Rev. **153** (1967) 606.
67F3	Fushimi, S., Ikeda, T.: J. Am. Ceram. Soc. **50** (1967) 131.
67G	Gahwiller, Ch.: Phys. Kondens. Mater. **6** (1967) 269.
67H	Hegenbarth, E., Frenzel, C.: Cryogenics **7** (1967) 331.
67M	Mante, A.J.H., Volger, J.: Phys. Lett. A **24** (1967) 139.
67N	Nilsen, W.G., Skinner, J.G.: J. Chem. Phys. **47** (1967) 1413.
67O	O'Shea, D.C., Kolluri, R.V., Cummins, H.Z.: Solid State Commun. **5** (1967) 241.
67P1	Pfeiffer, E.R., Schooley, J.F.: Phys. Rev. Lett. **19** (1967) 783.
67P2	Perry, C.H., McNelly, T.F.: Phys. Rev. **154** (1967) 456.
67R1	Raccah, P.M., Goodenough, J.B.: Phys. Rev. **155** (1967) 932.
67R2	Rupprecht, G., Winter, W.H.: Phys. Rev. **155** (1967) 1019.
67S	Shirane, G., Nathans, R., Minkiewicz: Phys. Rev. **157** (1967) 396.
67T	Tufte, O.N., Chapman, P.W.: Phys. Rev. **155** (1967) 796.
67U	Unoki, H., Sakudo, T.: J. Phys. Soc. Jpn. **23** (1967) 546.
67Y	Yamamoto, H., Kakishima, S., Shionoya, S.: J. Phys. Soc. Jpn. **23** (1967) 1321.
68B	Barrett, H.H.: Phys. Lett. A **26** (1968) 217.
68D	DiDomenico Jr., M., Wemple, S.H., Porto, S.P.S., Bauman, R.P.: Phys. Rev. **174** (1968) 522.
68F	Fleury, P.A., Scott, J.F., Worlock, J.M.: Phys. Rev. Lett. **21** (1968) 16.
68S	Steigmeier, E.F.: Phys. Rev. **168** (1968) 523.
68W	Wemple, S.H., DiDomenico Jr., M., Camlibel, I.: J. Phys. Chem. Solids **29** (1968) 1797.
69A1	Aleshko-Ozhevskii, O.P., Faek, M.K., Yamzin, I.I.: Sov. Phys. Crystallogr. **14** (1969) 367.
69A2	Alefeld, B.: Z. Phys. **222** (1969) 155.
69B1	Bechmann, R., Hearmon, R.F.S., Kurtz, S.K.: Landolt-Börnstein, Numerical Data and Functional Relationships in Science and Technology, New Series, Group III: Crystal and Solid State Physics, Vol. 2, Elastic, Piezoelectric, Piezooptic, Electrooptic Constants, and Nonlinear Dielectric Susceptibilities of Crystals (Supplement and Extension to Vol. 1), Springer: Berlin, Heidelberg, New York **1969.**
69B2	Barrett, H.H.: Phys. Rev. **178** (1969) 743.
69B3	Brower, W.S.: J. Appl. Phys. **40** (1969) 4989.
69C	Cowley, R.A., Buyers, W.J.L., Dolling, G.: Solid St. Commun. **7** (1969) 181.
69G	Gurevich, V.M.: Elektroprovodnost' Segnetoelektrikov, Moskva: Izdatel'stvo Komiteta Standartov, Mer i Izmeritel'nykh Priboroy pri Sovete Ministrov SSSR 1969; Electric Conductivity of Ferroelectrics, Jerusalem: Israel Program for Scientific Translations 1971 (English Transl.).
69I	Ikegami, S., Ueda, I., Miyazawa, T.: J. Phys. Soc. Jpn. **26** (1969) 1324.
69J	Jonker, G.H.: Philips Res. Rept. **24** (1969) 1.
69K	Kudzin, A.Yu., Bunina, L.K., Grzhegorzhevskii, O.A.: Fiz. Tverd. Tela **11** (1969) 2397; Sov. Phys. Solid State (English Transl.) **11** (1970) 1939.
69P	Pinnow, D.A., van Uitert, L.G., Warner, A.W., Bonner, W.A.: Appl. Phys. Lett. **15** (1969) 83.
69S1	Sakowski-Cowley, A.C., Lukaszewicz, K., Megaw, H.D.: Acta Crystallogr. B **25** (1969) 851.
69S2	Shirane, G., Yamada, Y.: Phys. Rev. **177** (1969) 858.
69S3	Sonin, A.S., Perfilova, V.E.: Kristallografiya 14 (1969) 510; Sov. Phys. Crystallogr. (English Transl.) **14** (1969) 419.
69T	Tedmon, C.S., Spacil, H.S., Mitoff, S.P.: J. Electrochem. Soc. **116** (1969) 1170.
69Y1	Yagnik, C.M., Canner, J.P., Gerson, R., James, W.J.: J. Appl. Phys. **40** (1969) 4713.
69Y2	Yamada, Y., Shirane, G.: J. Phys. Soc. Jpn. **26** (1969) 396.
70A	Axe, J.D., Harada, J., Shirane, G.: Phys. Rev. B **1** (1970) 1227.
70B1	Bonner, W.A., Zydzik, G.J.: J. Cryst. Growth **7** (1970) 65.

70B2	Burns, G., Scott, B.A.: Phys. Rev. Lett. **25** (1970) 167.
70F1	Fuess, H., Bertaut, E.F., Pauthenet, R., Durif, A.: Acta Crystallogr. Sect. B **26** (1970) 2036.
70F2	Frederikse, H.P.R.: IBM J. Res. Dev. **14** (1970) 295.
70F3	Fujii, Y., Sakudo, T.: J. Appl. Phys. **41** (1970) 4118.
70G	Goodenough, J.B.: Mater. Res. Bull. **5** (1970) 621.
70L1	Land, C.E., Haertling, G.H.: Article on p. 96 in [70T1].
70L2	Laubereau, A., Zurek, R.: Z. Naturforsch. **25a** (1970) 391.
70L3	Luthi, B., Moran, T.J.: Phys. Rev. B **2** (1970) 1211.
70M1	Matsumoto, G.: J. Phys. Soc. Jpn. **29** (1970) 606.
70M2	Miller, R.C., Nordland, W.A.: Phys. Rev. B **2** (1970) 4896.
70M3	Muller, K.A., Berlinger, W., Slonczewski, J.C.: Phys. Rev. Lett. **25** (1970) 734.
70N	Nicolau, P., Bunget, I., Rosenberg, M., Belciu, I.: IBM J. Res. Dev. **14** (1970) 248.
70P	Pauthenet, R., Veyret, C.: J. Phys. (Paris) **31** (1970) 65.
70R1	Rehwald, W.: Solid State Commun. **8** (1970) 607.
70R2	Remeika, J.P., Glass, A.M.: Mater. Res. Bull. **5** (1970) 37.
70R3	Rozhdestvenskaya, M.V., Sheftel', I.T., Stogova, V.A., Kozyreva, M.S., Krayukhina, E.K.: Fiz. Tverd. Tela **12** (1970) 873; Sov. Phys. Solid State (English Transl.) **12** (1970) 674.
70S1	Sleight, A.W., Prewitt, C.T.: Mater. Res. Bull. **5** (1970) 207.
70S2	Shirane, G., Axe, J.D., Harada, J., Linz, A.: Phys. Rev. B **2** (1970) 3651.
70S3	Shirane, G., Axe, J.D., Harada, J., Remeika, J.P.: Phys. Rev. B **2** (1970) 155.
70S4	Sroubek, Z.: Phys. Rev. B **2** (1970) 3170.
70S5	Samara, G.A.: Phys. Rev. B **1** (1970) 3777.
70T1	Takagi, Y., (ed): Proceedings of the Second International Meeting on Ferroelectricity, held in Kyoto (1969), Tokyo: Physical Society of Japan 1970; J. Phys. Soc. Jpn. **28** (1970) Suppl.
70T2	Tornberg, N.E., Perry, C.H.: J. Chem. Phys. **53** (1970) 2946.
70W1	Wemple, S.H.: Phys. Rev. B **2** (1970) 2679.
70W2	Wiesendanger, E.: Ferroelectrics **1** (1970) 141.
71A1	Abel, W.R.: Phys. Rev. B **4** (1971) 2696.
71A2	Amodei, J.J., Roach, W.R.: J. Phys. Chem. Solids **32** (1971) Suppl. 93.
71B1	Belruss, V., Kalnajs, J., Linz, A., Folweiler, R.C.: Mater. Res. Bull. **6** (1971) 899.
71B2	Berre, B., Fossheim, K.: Article on p. 255 in [71S4].
71B3	Blazey, K.W.: Phys. Rev. Lett. **27** (1971) 146.
71B4	Burk, W.J., Pressley, R.J.: Solid State Commun. **9** (1971) 191.
71B5	Bursian, E.V., Girshberg, Ya.G., Starov, E.N.: Phys. Status Solidi (b) **46** (1971) 529.
71C1	Canner, J.P., Yagnik, C.M., Gerson, R., James, W.J.: J. Appl. Phys. **42** (1971) 4708.
71C2	Carter, R.L., Bricker, C.L.: Spectrochim. Acta **27a** (1971) 569.
71C3	Comes, R., Denoyer, F., Lambert, M.: J. Phys. (Paris) **32** (1971) Suppl. C-5a, 195.
71C4	Coquin, G.A., Pinnow, D.A., Warner, A.W.: J. Appl. Phys. **42** (1971) 2162.
71C5	Cuciureanu, E., Istrate, S., Rezlescu, N.: Phys. Status Solidi (a) **6** (1971) K37.
71D	Darlington, C.N.W.: Thesis, University of Cambridge (1971).
71F	Fox, A.J., Whipps, P.W.: Electronics Lett. **7** (1971) 139.
71G1	Grimes, N.W., Collett, A.J.: Nature (London) Phys. Sci. **230** (1971) 158.
71G2	Gavrilyachenko, V.G., Fesenko, E.G.: Kristallografiya **16** (1971) 640; Sov. Phys. Crystallogr. (English Transl.) **16** (1971) 549.
71H1	Harada, J., Axe, J.D., Shirane, G.: Phys. Rev. B **4** (1971) 155.
71H2	Hurst, J.J., Linz, A.: Mater. Res. Bull. **6** (1971) 163.
71I	Ikegami, S., Ueda, I., Nagata, T.: J. Acoust. Soc. Am. **50** (1971) 1060.
71J	Johnston, A.R.: J. Appl. Phys. **42** (1971) 3501.
71K	Kiefer, W., Bernstein, H.J.: Appl. Spectrosc. **25** (1971) 609.
71L	Lee, C., Yahia, J., Brebner, J.L.: Phys. Rev. B **3** (1971) 2525.
71M	Moses, A.J.: Optical Materials Properties in Handbook of Electronic Materials, Vol. 1 New York: IFI/Plenum **1971**.
71P	Pressley, R.J. (ed): Handbook of Lasers with Selected Data on Optical Technology, Cleveland: The Chemical Rubber Co. **1971**.
71S1	Syono, Y., Akimoto, S., Endoh, Y.: J. Phys. Chem. Solids **32** (1971) 243.
71S2	Sakowski-Cowely, A.C.: Thesis, University of Cambridge (1971).
71S3	Subramanyam, K.N.: J. Phys. C: Solid State Phys. **4** (1971) 2266.
71S4	Shirasaki, S.: Solid State Commun. **9** (1971) 1217.

71W	Westwood, W.D., Sadler, A.G.: Can. J. Phys. **49** (1971) 1103.
71Z	Zinovik, M.A., Shchepetkin, A.A., Chutarov, G.I.: Zh. Fiz. Khim. **45** (1971) 1824.
72A	Ahtee, M., Glazer, A.M., Megaw, H.D.: Philos. Mag. (8) **26** (1972) 995.
72B1	Bhide, V.G., Rajoria, D.S., Rao, G.R., Rao, C.N.R.: Phys. Rev. B **6** (1972) 1021.
72B2	Bayer, G.: J. Less-Common Met. **26** (1972) 255.
72C1	Chamberland, B.L., Moeller, C.W.: J. Solid State Chem. **5** (1972) 39.
72C2	Courtens, E.: Phys. Rev. Lett. **29** (1972) 1380.
72C3	Comes, R., Shirane, G.: Phys. Rev. B **5** (1972) 1886.
72F1	Feigelson, R.S., Martin, G.W., Johnson, B.C.: J. Cryst. Growth **13/14** (1972) 686.
72F2	Faughnan, B.W.: Phys. Rev. B **5** (1972) 4925.
72F3	Fontana, M.P., Lambert, M.: Solid State Commun. **10** (1972) 1.
72F4	Fossheim, K., Berre, B.: Phys. Rev. B **5** (1972) 3292.
72F5	Fukuda, T., Uematsu, Y: Jpn. J. Appl. Phys. **11** (1972) 163.
72G1	Godefroy, L., Godefroy, G. (Réducteurs): 2ème Congrès Européen de Ferroélectricité, Exposés et Communications présentés à Dijon (1971), J. Phys. (Paris) **33** (1972), Suppl. Colloque Ma C-2.
72G2	Glazer, A.M., Megaw, H.D.: Philos. Mag. (8) **25** (1972) 1119.
72L	Lobachev, A.N., Peskin, V.F., Popolitov, V.I., Syrkin, L., Feoktistova, N.M.: Sov. Phys. Solid State **14** (1972) 509.
72M1	McCinnety, J.A.: Acta Crystallogr. B **23** (1972) 2845.
72M2	Matveeva, N.G., Shelykh, A.I.: Phys. Status Solidi **50** (1972) 83.
72M3	Maguire, H.G., Ress, L.V.C.: Article on p. 173 in [72G1].
72M4	Matveeva, N.G., Shelykh, A.L.: Phys. Status Solidi (b) **50** (1972) 83.
72M5	Mattheiss, L.F.: Phys. Rev. B **6** (1972) 4740.
72M6	Mattheiss, L.F.: Phys. Rev. B **6** (1972) 4718.
72R	Redfield, D., Burke, W.J.: Phys. Rev. B **6** (1972) 3104.
72S1	Sleight, A.W.: Acta Crystallogr. B **28** (1972) 2809.
72S2	Svirina, E.P., Polivanova, E.N.: Sov. Phys. Semicond. **5** (1972) 1967.
72S3	Singh, S., Remeika, J.P., Potopowicz, J.R.: Appl. Phys. Lett. **20** (1972) 135.
72S4	Stirling, W.G.: J. Phys. C **5** (1972) 2711.
72T	Tarte, P., Liegeois-Duyckaerts, M.: Spectrochim. Acta **28A** (1972) 2029.
72U	Ujma, Z., Olech, J., Wrobel, Z.: Acta Phys. Pol. A **41** (1972) 179.
72V	Vieland, L.J.: J. Phys. Chem. Solids **33** (1972) 581.
73B1	Bandyopadhyay, G., Fulrath, R.M.: J. Am. Ceram. Soc. **57** (1973) 182.
73B2	Baszynski, J.: Acta Phys. Pol. A **43** (1973) 499.
73B3	Bayer, E., Schmelz, H.: Int. J. Magn. **5** (1973) 71.
73B4	Burns, G., Scott, B.A.: Phys. Rev. B **7** (1973) 3088.
73B5	Borukhovich, A.S., Bazuev, G.V., Shveikin, G.P.: Fiz. Tverd. Tela **15** (1973) 2203; Sov. Phys. Solid State (Engl. Transl.) **15** (1974) 1467.
73C	Cojocaru, I.N., Costea, T., Negoescu, I., Podeanu, G.: Z. Phys. Chem. **85** (1973) 225.
73D1	Darlington, C.N.W., Megaw, H.D.: Acta Crystallogr. B **29** (1973) 2171.
73D2	Doshi, P., Glass, J., Novotny, M.: Phys. Rev. B **7** (1973) 4260.
73E	Edmondson, D.R.: Solid State Commun. **12** (1973) 981.
73H	Hewat, A.W.: J. Phys. C **6** (1973) 2559.
73I	Ishida, K., Honjo, G.: J. Phys. Soc. Jpn. **34** (1973) 1279.
73K	Kashida, S., Hatta, I., Ikushima, A., Yamada, Y.: J. Phys. Soc. Jpn. **34** (1973) 997.
73M	Michel-Calendini, F.M., Mesnard, G.: J. Phys. C **6** (1973) 1709.
73O	Okazaki, A., Kawaminami, M.: Mater. Res. Bull. **8** (1973) 545.
73P1	Pasto, A.E., Condrate Sr, R.A.: J. Am. Ceram. Soc. **56** (1973) 436.
73P2	Petersen, P.E.: J. Appl. Phys. **44** (1973) 1240.
73S1	Shannon, R.D., Calvo, C.: Can. J. Chem. **51** (1973) 265.
73S2	Sauerbrei, E.E., Faggiani, R., Calvo, C.: Acta Crystallogr. B **29** (1973) 2304.
73S3	Samara, G.A., Morosin, B.: Phys. Rev. B **8** (1973) 1256.
73S4	Sarkozy, R.F., Chamberland, B.L.: Mater. Res. Bull. **8** (1973) 1351.
73U	Uwe, H., Unoki, H., Fujii, Y., Sakudo, T.: Solid State Commun. **13** (1973) 737.
73W	Wild, R.L., Rockar, E.M., Smith, J.C.: Phys. Rev. B **8** (1973) 3828.
74A	Arndt, D., Müller, K., Reuter, B., Riedel, E.: J. Solid State Chem. **10** (1974) 270.
74B1	Borukhovich, A.S., Bazuev, G.V., Shveikin, G.P.: Sov. Phys. Solid State **15** (1974) 1467.

74B2	Borukhovich, A.S., Bazuev, G.V.: Sov. Phys. Solid State **16** (1974) 181.
74B3	Binnig, G., Hoenig, H.E.: Solid State Commun. **14** (1974) 597.
74C	Constantin, C.: Rev. Roum. Phys. **19** (1974) 27.
74D	Dougier, P., Hagenmullen, P.: J. Solid State Chem. **11** (1974) 177.
74F1	Franke, V., Hegenbarth, E.: Phys. Status Solidi (a) **25** (1974) K 17.
74F2	Frenzel, C., Hegenbarth, E.: Phys. Status Solidi (a) **23** (1974) 517.
74G1	Grimes, N.W.: Proc. R. Soc. London **338** (1974) 209.
74G2	Grushevskii, Yu.A., Girshberg, Ya.G., Bursian, E.V.: Fiz. Tverd. Tela **16** (1974) 248; Sov. Phys. Solid State (English Transl.) **16** (1974) 161.
74G3	Gunter, P.: Opt. Commun. **11** (1974) 285.
74H1	Halasa, N.A., DePasquali, G., Drickamer, H.G.: Phys. Rev. B **10** (1974) 154.
74H2	Heiman, D., Ushida, D.: Phys. Rev. B **9** (1974) 2122.
74I	Ivanova, L.A., Popolitov, V.I., Stefanovich, S.Yu., Lobachev, A.N., Venevtsev, Yu.N.: Sov. Phys Crystallogr. **19** (1974) 356.
74K1	Kabanov, A.A., Zingel, E.M.: Izv. Akad. Nauk. SSSR Neorgan Mater. **10** (1974) 481.
74K2	Knyazev, A.S., Zakharov, V.P., Poplavko, Yu.M.: Opt. Spectrosc. **36** (1974) 556.
74K3	Knyazev, A.S., Poplavko, Yu.M., Zakharov, V.P., Alekseev, V.V.: Sov. Phys. Solid State **15** (1974) 2003
74L	Loiacono, G.M., Balascio, J.F., Bonner, R., Savage, A.: J. Cryst. Growth **21** (1974) 1.
74M1	Malakhovskii, A.V., Edelman, I.S., Gavrilin, V.P., Barinov, G.I.: Sov. Phys. Solid State **16** (1974) 266.
74M2	Mack, S.A., Handler, P.: Phys. Rev. B **9** (1974) 3415.
74O1	Okazaki, A., Kawaminami, M.: Ferroelectrics **7** (1974) 91.
74O2	Obi, Y.: Phys. Status Solidi (a) **25** (1974) 293.
74P	Popolitov, V.I., Ivanova, L.A., Stephanovitch, Yu.S., Chetchkin, V.V., Lobachev, A.N., Venevtsev, Yu.N.: Ferroelectrics **8** (1974) 519.
74R	Rubinchik, Ya.S., Prokudina, S.A., Pavlyuchenko, M.M.: Inorg. Mater. **9** (1974) 1732.
74S1	Sleight, A.W.: Mater. Res. Bull. **9** (1974) 1177.
74S2	Sleight, A.W., Gillson, J.L.,Weiher, J.F., Bindloss, W.: Solid State Commun. **14** (1974) 357.
74S3	Salamon, M.B., Garnier, P.R.: J. Phys. Chem. Solids **35** (1974) 851.
74S4	Shatalova, G.E., Filip'ev, V.S., Katsnel'son, L.M., Fesenko, E.G.: Kristallografiya **19** (1974) 412; Sov. Phys. Crystallogr. (English Transl.) **19** (1974) 257.
74S5	Sorse, G., Beige, H., Schmidt, G.: Phys. Status Solidi (a) **26** (1974) K 153.
74U1	Uematsu, Y.: Jpn. J. Appl. Phys. **13** (1974) 1362.
74U2	Ujma, Z.: Acta Phys. Pol. A **45** (1974) 773.
74W1	Wolters, M., Burggraaf, A.J.: Phys. Status Solidi (a) **24** (1974) 341.
74W2	Wiesendanger, E.: Ferroelectrics **6** (1974) 263.
74Y	Yacoby, Y., Cerdeira, F., Schmidt, M., Holzapfel, W.B.: Solid State Commun. **14** (1974) 1325.
75B	Belov, K.P., Goryaga, A.N., Antoshina, L.L.: Sov. Phys. Solid State **16** (1975) 1596.
75C	Cerdeira, F., Holzapfel, W.B., Bauerle, D.: Phys. Rev. B **11** (1975) 1188.
75F1	Farley, J.M., Saunders, G.A., Chung, D.Y.: J. Phys. C: Solid State Phys. **8** (1975) 780.
75F2	Fukumoto, T., Okamoto, A., Hattori, T., Mitsuishi, A.: Solid State Commun. **17** (1975) 427.
75G	Gabrielyan, V.T., Kludzin, V.V., Kulakov, S.V., Razzhivin, B.P.: Sov. Phys. Solid State **17** (1975) 388.
75I	Ikeda, T.: Solid Stae Commun. **16** (1975) 103.
75K	Kuchar, F., Frankus, P.: Solid State Commun. **16** (1975) 181.
75L1	Loo, W. van: J. Solid State Chem. **14** (1975) 359.
75L2	Longo, J.M., Sleight A.W.: Mater. Res. Bull. **10** (1975) 127.
75L3	Lee, C., Destry, J., Brebner, J.L.: Phys. Rev. B **11** (1975) 2299.
75M	Marx, R., Happ, H.: Phys. Status Solidi (b) **67** (1975) 181.
75O	Okai, B., Yoshimoto, J.: J. Phys. Soc. Jpn. **39** (1975) 162.
75P	Palanisamy, T., Gopalakrishnan, J., Sastri, M.V.C.: Z. Anorg. Allg. Chem. **415** (1975) 275.
75R	Rao, C.N.R., Parkash, O., Ganguly, P.: J. Solid State Chem. **15** (1975) 186.
75S1	Suryanarayana, S.V.: Ind. J. Pure Appl. Phys. **3** (1975) 55.
75S2	Shaplygin, I.S., Lazarev, V.B.: Mater. Res. Bull. **10** (1975) 903.
75S3	Sleight, A.W., Gillson, J.L., Weiher, J.F., Bindloss, W.: J. Solid State Chem. **12** (1975) 406.
75U	Ujma, Z., Handerek, T.: Phys. Status Solidi (a) **28** (1975) 489.
75V	Venero, A.F., Westrum Jr., F.: J. Chem. Thermodyn. **7** (1975) 693.

76A   Aso, K.: Jpn. J. Appl. Phys. **15** (1976) 1277.
76B1   Balkanski, M., Leite, R.C.C., Porto, S.P.S. (eds): Proceedings of the Third International Conference on Light Scattering in Solids, held in Campinas (1975), Paris: Flammarion Sciences 1976.
76B2   Bozinis, D.G., Hurrell, J.P.: Phys. Rev. B **13** (1976) 3109.
76C   Clark, A.F., Haynes, V.A., Deason, V.A., Trapani, R.J.: Cryogenics **16** (1976) 267.
76D1   Daniels, J.: Philips Res. Rep. **31** (1976) 505.
76D2   Daniels, J., Hardtl, K.H.: Philips Res. Rep. **31** (1976) 489.
76D3   Denoyer, F., Lambert, M., Comes, R., Currat, R.: Solid State Commun. **18** (1976) 441.
76F1   Fesenko, O.E., Smotrakov, V.G.: Ferroelectrics **12** (1976) 211.
76F2   Firnstein, L.A., Barbosa, G.A., Porto, S.P.S.: Article on p. 866 in [76B1].
76F3   Fujii, Y., Sakudo, T.: J. Phys. Soc. Jpn. **41** (1976) 888.
76G   Glidewell, C.: Inorg. Chim. Acta **19** (1976) L45.
76H1   Hatta, I., Ikushima, A.: J. Phys. Soc. Jpn. **41** (1976) 558.
76H2   Holman, R.L.: Ferroelectrics **14** (1976) 675.
76H3   Hennings, D.: Philips Res. Report. **31** (1976) 516.
76I   Itoh, Y.: J. Spectrosc. Soc. Jpn. **25** (1976) 83.
76K1   Kasimov, G.G., Vovkotrub, E.G., Miller, I.I.: Inorg. Mater. **12** (1976) 1227.
76K2   Kinoshita, K., Yamaji, A.: J. Appl. Phys. **47** (1976) 371.
76L   Lyons, K.B., Fleury, P.A.: Phys. Rev. Lett. **37** (1976) 161.
76M1   Mozhaev, A.P., Oleinikov, N.N., Tretyakov, Yu.D., Helmold, P.: Inorg. Mater. **12** (1976) 1662.
76M2   Malitskaya, M.A., Martynenko, M.A., Popov, Yu.M., Prokopalo, O.I., Fesenko, E.G.: Fiz. Tverd. Tela **18** (1976) 1386; Sov. Phys. Solid State (English Transl.) **18** (1976) 798.
76M3   Michel-Calendini, F.M., Castet, L.: Ferroelectrics **13** (1976) 367.
76O   Ohi, K., Iesaka, S.: J. Phys. Soc. Jpn. **40** (1976) 1371.
76P1   Pentegova, M.V., Sych, A.M., Tomashpolskii, Yu.Ya.: Inorg. Mater. **12** (1976) 801.
76P2   Perluzzo, G., Destry, J.: Can. J. Phys. **54** (1976) 1482.
76P3   Perry, C.H., Hayes, R.R., Tornberg, N.E.: Article on p. 812 in [76B1].
76P4   Prokopalo, O.I.: Ferroelectrics **14** (1976) 683.
76Q   Quittet, A.M., Bell, M.I., Krauzman, M., Raccah, P.M.: Phys. Rev. B **14** (1976) 5068.
76R1   Roy, B.N., Appalasami, S.: J. Cryst. Growth **36** (1976) 311.
76R2   Richter, P.W., Kruger, G.J., Pistorius, C.W.F.T.: Acta Crystallogr. B **32** (1976) 928.
76R3   Rannev, M.V., Shchedrin, B.M., Venevtsev, Yu.N.: Ferroelectrics **13** (1976) 523.
76S1   Stencel, J.M., Silberman, E., Springer, J.: Phys. Rev. B **14** (1976) 5435.
76S2   Sirota, N.N., Karavay, A.P., Pavlov, V.I.: Krist. Tech. **11** (1976) 861.
76U   Uwe, H., Sakudo, T.: Phys. Rev. B **13** (1976) 271.
76Z   Zhmurova, Z.I.: Sov. Phys. Crystallogr. **21** (1976) 127.
77B   Bernhardt, Hj.: Phys. Status Solidi (a) **40** (1977) 257.
77C   Clarke, R., Benguigui, L.: J. Phys. C **10** (1977) 1963.
77D   Degtyareva, E.V., Verba, L.I., Gulko, N.V., Kamenetskii, Yu.L., Gladkaya, N.V.: Izv. Akad. Nauk SSSR Neorgan. Mater. **13** (1977) 1048.
77F   Feltz, A., Langbein, H.: Ferroelectrics **15** (1977) 7.
77G   Ganguly, B.N., Nicol, M.: Phys. Status Solidi (b) **79** (1977) 617.
77H1   Hawthorne, F.C., Calvo, C.: J. Solid State Chem. **22** (1977) 157.
77H2   Hatta, I., Shiroishi, Y.: Phys. Rev. B **16** (1977) 1138.
77K   Kawashima, M., Abe, M., Tsukioka, M., Nomura, S.: Jpn. J. Appl. Phys. **16** (1977) 2049.
77M1   Matsui, T., Wagner Jr., J.B.: J. Electrochem. Soc. **124** (1977) 1141.
77M2   Mercier, M., Velleaud, G., Puvinel, J.: Physica **86···88B** (1977) 1089.
77P   Publishing Committee of the Electronic Ceramics (Editor): Semiconducting Barium Titanates, Tokyo: Giken Co., Ltd. 1977 (in Japanese).
77R1   Raevskii, I.P., Malitskaya, M.A., Prokopalo, O.I., Smotrakov, V.G., Fesenko, E.G.: Fiz. Tverd. Tela **19** (1977) 492; Sov. Phys. Solid State (English Transl.) **19** (1977) 283.
77R2   Raevskii, I.P., Malitkaya, M.A., Prokopalo, O.I., Smotrakov, V.G., Fesenko, E.G., Tsikhotskii, E.S.: Fiz. Tverd. Tela **19** (1977) 2033; Sov. Phys. Solid State (English Transl.) **19** (1977) 1189.
77S1   Sirota, N.N., Pavlov, V.I., Lashkov, E.S., Karavai, A.P.: Sov. Phys. Solid State **19** (1977) 1020.
77S2   Sirota, N.N., Karavay, A.P., Kofman, N.A., Pavlov, V.I.: Krist. Tech. **12** (1977) 945.
77S3   Scalabrin, A., Chaves, A.S., Shim, D.S., Porto, S.P.S.: Phys. Status Solidi (b) **79** (1977) 731.
77T   Toyoda, K.: Bibliography on Barium Titanate Semiconductors, p. 135 in [77P].
77W1   Waltersson, K., Forslund, B.: Acta Crystallogr. B **33** (1977) 775.

77W2   Waltersson, K., Forslund, B.: Acta Crystallogr. B **33** (1977) 780.
78B1   Beguemsi, T., Garnier, P., Weigel, D.: J. Solid State Chem. **25** (1978) 315.
78B2   Bauerle, D., Braun, W., Saile, V., Sprussel, G., Koch, E.E.: Z. Phys. B **29** (1978) 179.
78C   Chisty, I.L., Csillag, L., Kitaeva, V.F., Kroo, N., Soboloev, N.N.: Phys. Status Solidi (a) **47** (1978) 609.
78F   Fridkin, V.M., Popov, B.N.: Phys. Status Solidi (a) **46** (1978) 729.
78G1   Groenink, J.A., Van Wezep, D.A.: Phys. Status Solidi (a) **49** (1978) 651.
78G2   Geifman, I.N.: Phys. Status Solidi (b) **85** (1978) K 5.
78G3   Gunter, P., Micheron, F.: Ferroelectrics **18** (1978) 27.
78H1   Handerek, J., Wrobel, Z., Wojcik, K., Ujma, Z.: Ferroelectrics **18** (1978) 127.
78H2   Hastings, J.B., Shapiro, S.M., Frazer, B.C.: Phys. Rev. Lett. **40** (1978) 237.
78H3   Heiman, D., Ushida, S.: Phys. Rev. B **17** (1978) 3616.
78I   Ihrig, H., Hennings, D.: Phys. Rev. B **17** (1978) 4593.
78K   Kamata, K., Nakajima, T., Hayashi, T., Nakamura, T.: Mater. Res. Bull. **13** (1978) 49.
78O   Oeder, R., Scharmann, A., Schwabe, D., Vitt, B.: J. Cryst. Growth **43** (1978) 537.
78P1   Perluzzo, G., Destry, J.: Can. J. Phys. **56** (1978) 453.
78P2   Pertosa, P., Michel-Calendini, F.M.: Phys. Rev. B **17** (1978) 2011.
78P3   Prokopalo, O.I., Fesenko, E.G., Malitskaya, M.A., Popov, Yu.M., Smotrakov, V.G.: Ferroelectrics **18** (1978) 99.
78S1   Szabo, Z.G., Kamaras, K., Szebeni, Sz., Ruff, I.: Spectrochim. Acta **34A** (1978) 607.
78S2   Seidel, P., Bomas, H., Hoffmann, W.: Ferroelectrics **18** (1978) 243.
78T1   Toriumi, K., Saito, Y.: Acta Crystallogr. B **34** (1978) 3149.
78T2   Toriumi, K., Ozima, M., Akaogi, M., Saito, Y.: Act. Crystallogr. B **34** (1978) 1093.
78U   Ujma, Z., Handerek, J.: Acta Phys. Pol. A **53** (1978) 665.
78V   Vincent, H., Bochu, B., Aubert, J.J., Joubert, J.C., Marezio, M.: J. Solid State Chem. **24** (1978) 245.
79B   Botrous el Badramany, N.M., Mina, E.F., Merchant, H.D., Arafa, S., Poplawsky, R.P.: J. Am. Ceram. Soc. **62** (1979) 113.
79F1   Fontana, M.D., Dolling, G., Kugel, G.E., Carabatos, C.: Phys. Rev. B **20** (1979) 3850.
79F2   Fujishita, H., Shiozaki, Y., Sawaguchi, E.: J. Phys. Soc. Jpn. **46** (1979) 581.
79F3   Fujishita, H., Shiozaki, Y., Sawaguchi, E.: J. Phys. Soc. Jpn. **46** (1979) 1391.
79G   Groenink, J.A., Binsma, H.: J. Solid State Chem. **29** (1979) 227.
79H   Handerek, J., Aleksandrowicz, A., Badurski, M.: Acta Phys. Pol. A **56** (1979) 769.
79L   Leichenko, A.I., Popolitov, V.I., Peskin, V.F., Lobachev, A.N., Venevtsev, Yu.N.: Sov. Tech. Phys. Lett. **4** (1979) 463.
79M   Malitskaya, M.A., Popov, Yu.M., Prokopalo, O.J., Raevskii, J.P., Rogach, E.D.: Sov. Phys. Solid State **21** (1979) 1282.
79P   Pentegova, M.V., Tomashpoliskii, Y.Y.: Kristallografiya **24** (1979) 195.
79R   Ramadass, N., Gopalakrishnan, J., Sastri, M.V.C.: J. Less-Common Met. **65** (1979) 129.
79S1   Shablaev, S.I., Davieshevskii, A.M., Subashiev, V.K., Babashkin, A.A.: Sov. Phys. Solid State **21** (1979) 662.
79S2   Subramanyam, K.N., Khare, L.R.: Acta Crystallogr. B **35** (1979) 269.
79T1   Tsukioka, M., Tanaka, J., Miyazawa, Y., Mori, Y., Kojima, H., Ekara, S.: Solid State Commun. **32** (1979) 223.
79T2   Taylor, W., Murray, A.F.: Solid State Commun. **31** (1979) 937.
80L   Luspin, Y., Servoin, J.L., Gervais, F.: J. Phys. C **13** (1980) 3761.
80M   Michel-Calendini, F.M., Chermette, H., Weber, J.: J. Phys. C **13** (1980) 1427.
80P   Pisarski, M.: Phys. Status Solidi (b) **101** (1980) 635.
80S   Servoin, J.L., Luspin, Y., Gervais, F.: Phys. Rev. B **22** (1980) 5501.
80Y   Yalenbrovskii, M.A., Zamentin, V.J., Rabkin, L.M., Fesenko, E.G.: Sov. Phys. Solid State **22** (1980) 1440.
81B1   Brebner, J.L., Jandl, S., Lepine, Y.: Phys. Rev. B **23** (1981) 3816.
81B2   Bernhardt, Hj., Schnell, R.: Phys. Status Solidi (a) **64** (1981) 207.
81F   Freire, J.D., Katiyar, R.S.: Solid State Commun. **40** (1981) 903.
81I   Ihrig, H., Hengst, J.H.T., Keerk, M.: Z. Physik B **40** (1981) 301.
81K1   Käppler, W., Arlt, G.: Phys. Status Solidi (a) **63** (1981) 475.
81K2   Kulagin, N.A., Landar, S.V., Litvinov, L.A., Tolde, J.V.: Opt. Spectrosc. **50** (1981) 487.
81P   Pisarski, M.: Acta Phys. Pol. A **59** (1981) 303.

81T	Tsyashchenko, Yu.P., Krasnyanskii, G.E.: Phys. Status Solidi (b) **103** (1981) 391.
81W	White, G.K.: J. Phys. C **14** (1981) L 297.
82A1	Aguilar, M.: J. Phys. C **15** (1982) 3829.
82A2	Aguilar, M., Agullo-Lopez, F.: J. Appl. Phys. **53** (1982) 9009.
82B	Burns, G., Dacol, F.H.: Solid State Commun. **42** (1982) 9.
82G	Galzerani, J.C., Katiyar, R.S.: Solid State Commun. **41** (1982) 515.
82H	Höchli, U.T., Rohrer, H.: Phys. Rev. Lett. **48** (1982) 188.
82S	Schneider, E., Cressman, P.J., Holman, R.L.: J. Appl. Phys. **53** (1982) 4054.

Physical property	Numerical value	Experimental conditions	Experimental method remarks	Ref.

10.2.5 Further transition-metal sulfides, selenides and chlorides

Me_xZrS_2 (Me = Fe, Co, Ni)

lattice parameters for $Ni_{0.5}ZrS_2$ (in Å):

$d(Zr-Zr)$	3.642		space group $D_{3d}^3 - P\bar{3}m_1$	75T
$d(Zr-S)$	2.570			
$d(Ni-Ni)$	2.73			

For variation of parameters a and c, see [75T].
absorption coefficient for Ni_xZrS_2: Fig. 1
conductivity for all three compounds: Figs. 2···4

activation energies for conductivity (in eV):

$E_A(Fe_xZrS_2)$	0.32	x = 0.04		75T
	0.32	0.08		
	0.23	0.20		
	0.19	0.30		
	0.18	0.30		
	0.18	0.35		
$E_A(Co_xZrS_2)$	0.24	0.20		
	0.20	0.35		
	0.13	0.45		
$E_A(Ni_xZrS_2)$	0.34	0.25		
	0.28	0.35		
	0.13	0.50		
	0.10	0.60		

$CuCrS_2$ (semiconducting)

structure: rhombohedral, space group $D_{3d}^5 - R\bar{3}m$ [68B]

lattice parameters:

a	3.482 Å	see also [80K]; for positions of the	79N
c	18.686 Å	atoms and interatomic distances, see [68B]	

resistivity: temperature dependence, see Fig. 7
For energy band models, see [80K].
reflectivity: Fig. 5

Physical property	Numerical value	Experimental conditions	Experimental method remarks	Ref.
Néel temperatures (in K):				
T_N	43		phase I (non-stoichiometric, 10% of Cu atoms in interstitial positions)	79N
	36		phase II (Cu atoms ordered occupying half of the tetrahedral sites)	68B
	39		Temperature dependence of magnetic susceptibility: Fig. 6.	

For data on semiconducting $CuCrSe_2$, see [68B, 80K].

$CuFeS_2$ (semiconducting)

structure: tetragonal, space group $D_{2d}^{12}—I\bar{4}2d$ [73H].

lattice parameters:

a	5.292 Å			72A2
c	10.407 Å			

energy gap:

E_g	0.6 eV		absorption edge; for absorption spectrum, see Fig. 8	72A1

Three absorption bands have been observed at 1.0 eV, 2.1 eV and 3.2 eV due to transitions from the valence band to the Fe3d levels and the conduction band (4 s), respectively [80O].

   For optical constants n, k and dielectric constants, see [80O].
   For reflectivity, see Fig. 9.

resistivity: temperature dependence: Fig. 10

Hall mobility:

μ_H	7 cm^2 V^{-1} s^{-1}	$i \perp c$		74P

density:

d	4.19 g m^{-3}			74P

$CuFeS_2$ is antiferromagnetic ($T_N = 550$ °C [58D]).
   For semiconducting undoped and Fe doped $CuAlS_2$ and $CuGaS_2$, see [74K] and [80O].

Me_xNbS_2 (Me = Mn, Fe, Co, Ni)

Crystal structure

structure: sandwiches of hexagonally packed planes of metal between planes of chalcogen. The NbS_2 sandwiches stack so that Nb atoms lie directly on the top of one another and adjacent sandwiches are rotated by 60° with respect to each other. Thus two sandwiches are required to make up the unit cell, and there are empty sites between the layers which are either octahedrally or tetrahedrally coordinated by surrounding S atoms. The intercalate atoms (Mn, Fe, Co, Ni) sit in these octahedral sites between the layers.

lattice parameters (in Å):

a	3.25	$Mn_{0.3}NbS_2$		77F
c	12.57			
a	3.28	$Fe_{0.3}NbS_2$		
c	12.08			
a	3.26	$Co_{0.3}NbS_2$		
c	11.93			
a	3.28	$Ni_{0.3}NbS_2$		
c	11.80			

Physical property	Numerical value	Experimental conditions	Experimental method remarks	Ref.
Electronic properties				
energy gaps (in eV):				
E_g	0.85	$Mn_{0.3}NbS_2$	from photoemission	77F
	1.05	$Fe_{0.3}NbS_2$	($E_g \cong$ d-bandwidth)	
	0.90	$Co_{0.3}NbS_2$		
	0.85	$Ni_{0.3}NbS_2$		
Transport properties				
Hall coefficients (in $cm^3\,C^{-1}$), **carrier concentrations** (in cm^{-3}), **resistivities** (in Ω cm), **mobilities** (in cm^2/Vs):				
R_H	$9.7 \cdot 10^{-4}$	$Mn_{0.3}NbS_2$	orientation not specified	77F
n	$6.5 \cdot 10^{21}$			
ϱ	$1.5 \cdot 10^{-4}$			
μ	6.3			
R_H	$1.4 \cdot 10^{-3}$	$Fe_{0.3}NbS_2$		
n	$4.5 \cdot 10^{21}$			
ϱ	$3.2 \cdot 10^{-4}$			
μ	4.4			
R_H	$1.7 \cdot 10^{-3}$	$Co_{0.3}NbS_2$		
n	$3.8 \cdot 10^{21}$			
ϱ	$3.5 \cdot 10^{-4}$			
μ	5.1			
R_H	$9.4 \cdot 10^{-4}$	$Ni_{0.3}NbS_2$		
n	$6.6 \cdot 10^{21}$			
ϱ	$2.2 \cdot 10^{-4}$			
μ	4.3			
Magnetic properties				
effective magnetic moments, paramagnetic Curie temperatures:				
$Mn_{0.3}NbS_2$:				
p_{eff}	$5.0\ \mu_B$	$T > 100$ K		77F
	$6.2\ \mu_B$	$T < 100$ K		
Θ_p	58 K	$T > 100$ K		
	40 K	$T < 100$ K		
$Fe_{0.3}NbS_2$:				
p_{eff}	$6.3\ \mu_B$		from susceptibility $\chi_{\|c}$ and χ_{av}	
			($\chi_{av} = \frac{1}{3}\chi_{\|c} + \frac{2}{3}\chi_{\perp c}$)	
Θ_p	-115 K		from $\chi_{\|c}$	
	-300 K		from χ_{av}	

Tl_3VS_4

structure: cubic, space group T_d^3—$I\bar{4}3m$ [75I]

lattice parameter:

a	7.497 Å			75I

density:

d	$6.24\ g\,cm^{-3}$			75I

energy gap:

E_g	1.76 eV		photoconductivity	74S

Physical property	Numerical value	Experimental conditions	Experimental method remarks	Ref.
acoustic velocities (in cm s^{-1}):				
v_1	$2.46 \cdot 10^5$	longitudinal	propagation direction: [110]	75I
	$2.81 \cdot 10^5$		c axis	
v_t	$1.6 \cdot 10^5$	shear	propagation direction: [110]	
			polarization direction: [110]	
	$9.6 \cdot 10^4$		[110]	
			c axis	
	$8.73 \cdot 10^4$		c axis	
			[110]	

conductivity and photoconductivity: Fig. 12

optical transmission: Fig. 11

Cu_3VS_4

structure: cubic [79P], space group $T_d^1 - P\bar{4}3m$ [78A]

Cu_3VS_4 is a semiconducting material with a small number of mobile ions ($n_{ion} \approx 10^{17} \cdots 10^{20}$ cm^{-3})

lattice parameter:				
a	5.398 Å			78A

wavenumbers of Raman active modes (in cm^{-1}):				
$\bar{\nu}(F_2)$	149.00 (154.5)	$T = 300$ K		79P
$\bar{\nu}(F_2)$	203.00 (207.7)	(80 K)		
$\bar{\nu}(F_2)$	287.00 (291.5)			
$\bar{\nu}(E)$	300.50 (305.6)			
$\bar{\nu}(A_1)$	375.50 (376.6)			
$\bar{\nu}(F_2, TO)$	440.50 (443.9)			
$\bar{\nu}(F_2, LO)$	449.50 (452.2)			

conductivity:				
σ	$10^{-3} \cdots 10\ \Omega^{-1}$ cm^{-1}			78A

Hall mobility:				
μ_H	4 cm^2 V^{-1} s^{-1}			78A

resistivity of monoclinic Cu_xVS_2 with x = 0.65 and 0.75: Figs. 16, 17.

absorption, reflection: Figs. 13$\cdots$15

Na_xVS_2, Na_xVSe_2

According to [79B] two types exist:

Type I: octahedral coordination of V, trigonal prismatic coordination of Na; metallic conduction, Pauli-paramagnetic behavior, CDW-distortion.

Type II: octahedral coordination of V, octahedral coordination of Na; semiconductor, magnetic property: AF, Jahn-Teller distortion.

lattice parameters (in Å):				
$Na_{0.6}VS_2$:	$a = 3.346$,	$c = 21.02$	type I for x = 0.3$\cdots$1.0 (metallic)	79B
$NaVS_2$:	$a = 3.566$,	$c = 19.68$	type II for x = 1.0	
$Na_{0.6}VSe_2$:	$a = 3.482$,	$c = 22.21$	type I for x = 0.5$\cdots$0.6 (metallic)	
$NaVSe_2$:	$a = 3.735$,	$c = 20.639$	type II for x = 1.0	

According to [74W] the lattice is rhombohedral for x = 0.3$\cdots$1.0 with $a = 3.311$ Å and $c = 21.21$ Å.

There is an orthorhombic distortion at $T = 50$ K [78B, 80W]. The space group is $C_{2h}^3 - C2/m$ for $T < 50$ K.

Semiconducting behavior for $NaVSe_2$ and $NaVS_2$ (type II).

Physical property	Numerical value	Experimental conditions	Experimental method remarks	Ref.

figures: resistivity: Fig. 18, magnetic susceptibility, resistivity, Seebeck and Hall coefficient: Fig. 19, magnetic susceptibility: Fig. 20, magnetization and susceptibility: Figs. 21, 22

Carrier mobilities of 0.5 cm^2/Vs have been found in $Na_{0.8}VS_2$ at 4.2 K and in $Na_{0.6}VS_2$ at 275 K [79B].

CsZrCl$_6$

structure: space group O_h^5—Fm3m [76M]

lattice parameter:

a	10.420 Å			71T, 76M

transport data:

Conversion to semiconductor by heating in a saturated Cs metal vapor at 400 °C [71T].
Gd doping gives n-type conduction according to Hall and Seebeck measurements [74A].

ϱ	$6 \cdot 10^{-3} \,\Omega$ cm			71T
n	10^{19} cm^{-3}			
μ_H	100 cm^2 V^{-1} s^{-1}			

VFe$_2$Se$_4$

structure: space group C_{2h}^3—I2/m [72L]

lattice parameters (lengths in Å):

a	6.18	$T = 350$ K		72L
b	3.49			
c	11.56			
β	91°40′			

Néel temperature:

T_N	155 K			72L

electrical and thermal transport data (pressed samples):

σ	$1.08 \cdot 10^3 \,\Omega^{-1}$ cm^{-1}		temperature dependence: Fig. 23	74A
E_A	0.04 eV			
S	15 μV K^{-1}			
κ	$5.09 \cdot 10^{-3}$ cal cm^{-1} s^{-1} K^{-1}		temperature dependence: Fig. 24	

References for 10.2.5

58D   Donnay, G., Corliss, L.M., Donnay, J.D., Elliot, H., Hastings, J.M.: Phys. Rev. **112** (1958) 1917.

68B   Bongers, P.F., Van Bruggen, C.F., Koopstra, J., Omloo, W.P.F.A.M., Wiegers, G.A., Jellinek, F.: J. Phys. Chem. Solids **29** (1968) 977.

71T   Tzalmona, A., Kaplan, Z., Gabay, A.: Appl. Phys. Lett. **19** (1971) 295.

72A1   Adams, R., Beaulieu, R., Vassiliadis, M., Wold, A.: Mater. Res. Bull. **7** (1972) 87.

72A2   Adams, R., Russo, P., Arnott, R., Wold, A.: Mater. Res. Bull. **7** (1972) 93.

72L   Lambert-Andron, B., Berodias, G., Babot, D.: J. Phys. Chem. Solids **33** (1972) 87.

73H   Hall, S.R., Stewart, J.M.: Acta Crystallogr. **B29** (1973) 579.

74A   Akhmedov, N.R., Dzhalilov, N.Z., Aliev, G.M., Abdinov, D.Sh.: Inorg. Mater **10** (1974) 711.

74K   Kondo, K., Teranishi, T.: J. Phys. Soc. Jpn. **36** (1974) 311.

74P   Pitt, G.D., Vyas, M.K.R.: Solid State Commun. **15** (1974) 899.

74S   Shilova, M.V., Karpovich, I.A.: Inorg. Mater. **10** (1974) 475.

74T   Teranishi, T., Sato, K., Kondo, K.: J. Phys. Soc. Jpn. **36** (1974) 1618.

74W   Wiegers, G.A., van der Meer, R., van Heiningen, H., Kloosterbeer, H.J., Alberink, A.J.A.: Mater. Res. Bull. **9** (1974) 1261.

75I	Isaacs, T.J., Gottlieb, M., Daniel, M.R., Feichtner, J.D.: J. Electron. Mater. **4** (1975) 67.
75T	Trichet, L., Rouxel, J.: J. Solid State Chem. **14** (1975) 283.
76M	Maniv, S.: Appl. Cryst. **9** (1976) 245.
77F	Friend, R.H., Deal, A.R., Yoffe, A.D.: Philos. Mag. **35** (1977) 1269.
77L	Le Nagard N., Collin, G., Gorochov, O.: Mater. Res. Bull. **1** (1977) 975.
78A	Arribart, H., Sapoval, B., Gorochov, O., Le Nagard, N.: Solid State Commun. **26** (1978) 435.
78B	Bruggen van, C.F., Bloembergen, J.R., Bos-Alberink, A.J.A., Wiegers, G.A.: J. Less-Common Met. **60** (1978) 259.
79B	Bruggen van, C.F., Haas, C., Wiegers, G.A.: J. Solid State Chem. **27** (1979) 9.
79N1	Le Nagard, N., Collin, G., Gorochov, O.: Mater. Res. Bull. **14** (1979) 155.
79N2	Nagard, N. Le, Collin, G., Gorochov, O.: Mater. Res. Bull. **14** (1979) 1411.
79P	Petritis, D., Martinez, G.: Inst. Phys. Conf. Ser. **43** (1979) 677.
79Y	Yacobi, B.G., Boswell, F.W., Corbett, J.M.: Mater. Res. Bull. **14** (1979) 1033.
80K	Khumalo, F.S., Hughes, H.P.: Phys. Rev. **B 22**, (1980) 4066.
80O	Oguchi, T., Sator, K., Teranishi, T.: J. Phys. Soc. Jpn. **48** (1980) 123.
80W	Wiegers, G.A.: Physica **99 B** (1980) 151.
81P	Petritis, D., Martinez, G, Levy-Clement, C., Gorochov, O.: Phys. Rev. B **23** (1981) 6773.

10.3 Ternary rare earth compounds

This chapter lists the semiconducting properties of ternary rare earth compounds. For the magnetic and ferroelectric properties the reader is referred to the volumes III/4a, 12a and 16a of this series. The semiconducting properties of binary rare earth compounds are given in volume III/17g.

Within the oxides we concentrate on compounds, from which semiconducting type of conductivity has been reported. In some cases we have pointed out compounds with metallic behavior and also added compounds with unknown type of conductivity, in order to present the crystal systems in a systematic manner.

The sulfides, selenides, and tellurides generally have much smaller energy gaps when compared to the oxides. Typically, the fundamental absorption edge is located in the visible to infrared spectral region. Practically all compounds, besides a few metallic exceptions, show semiconducting behavior. Therefore, section 10.3.4 also includes compounds from which only lattice data are known. Y- and Sc-compounds complete this class of materials.

The tables and figures give room-temperature data if not otherwise specified. E_A is always the activation energy of the electrical conductivity.

Physical property	Numerical value	Experimental conditions	Experimental method remarks	Ref.

The observed magnitude and type of conductivity depends on the preparation method. For instance, the rare earth sesqui-sulfides have n-type conductivity when grown with the I$_2$-vapor-transport method [82S1] because of unavoidable incorporation of I$^-$-centers. Although n- and p-type conductivity as well as pn-junctions can be obtained by Cl$^-$ − and P^{3-}-diffusion, respectively at the S^{2-} site [82S1], one type of doping generaly will represent a problem for the preparation of high band gap materials. Due to the contribution of ionic bonding the crystals tend to self compensate which then results in relatively high electrical resistivity and low carrier mobility. Sulfur vacancies is another problem, which can be partially removed by annealing in S$_2$-atmosphere [82S1]. For sesqui-sulfides, therefore, the preparation of p-type material represents the problem. From the application point of view efficient phosphors can be made [81S]. Optical excitation and transfer to luminescent centers like Ce^{3+} can be very efficient, too [82S2]. However, due to the ionic contribution and the low mobilities for charge carriers, luminescent diodes will be very difficult to obtain.

One of the authors (E.-G. Scharmer) wishes to thank the "Deutsche Forschungsgemeinschaft" for support.

10.3.1 Oxides LnMO$_3$ with perovskite structure (M = Ti, V, Cr, Mn, Fe, Co, Ni)

10.3.1.1 LnTiO$_3$ compounds

LaTiO$_3$

Structure: cubic (O$_h^1$—Pm3m)			crystal structure: Fig. 1	54K
a	3.92 (1) Å			54K
σ	$\approx 10^2$ Ω^{-1} cm^{-1}	p-type	metallic	76G
	$1.25 \cdot 10^3$ Ω^{-1} cm^{-1}	n-type	metallic, temperature dependence of resistivity: Figs. 2, 3	78B
S	-18 μV K^{-1}		temperature dependence of Seebeck coefficient: Figs. 4, 5	78B
	$+15$ μV K^{-1}			76G

CeTiO$_3$

Structure: tetragonal				78B
a	5.590 Å			78B
c	7.868 Å			
σ	$\approx 10^2$ Ω^{-1} cm^{-1}	n-type	polycrystalline sample; resistivity does not depend on temperature: Fig. 2	78B

PrTiO$_3$

Structure: orthorhombic (D$_{2h}^{16}$—Pbnm)			crystal structure: Fig. 6	78B
tetragonal				65W
a	5.532 Å			78B
b	5.620 Å			
c	7.830 Å			
a	5.508 Å			65W
c	7.742 Å			
σ	50 Ω^{-1} cm^{-1}		polycrystalline sample (orthorhombic); resistivity does not depend on temperature: Fig. 2 Temperature dependence of Seebeck coefficient: Fig. 4.	78B

Physical property	Numerical value	Experimental conditions	Experimental method remarks	Ref.
NdTiO$_3$				
Structure: cubic (O_h^1—Pm3m)			crystal structure: Fig. 1	76G
orthorhombic (D_{2h}^{16}—Pbnm)			orthorhombic distortion: Fig. 6	78B
a	3.87 Å		cubic	76G
a	5.482 Å		orthorhombic	65W
b	5.521 Å			
c	7.728 Å			
E_A	0.06 eV		temperature dependence of resistivity: Figs. 2, 3	78B
σ	6.7 Ω^{-1} cm^{-1}	p-type		78B
S	$\approx 30\ \mu V\ K^{-1}$		temperature dependence of Seebeck coefficient: Figs. 4, 5 (N.B. Data for E_A, σ, S for cubic modification)	78B
	> 0, small			76G
SmTiO$_3$				
Structure: orthorhombic (D_{2h}^{16}—Pbnm)			crystal structure: Fig. 6	69M1
a	5.468 Å			69M1
b	5.665 Å			
c	7.737 Å			
E_A	0.15 eV		temperature dependence of resistivity: Figs. 2, 3	78B
σ	0.24 Ω^{-1} cm^{-1}	p-type		78B
S	$\approx 110\ \mu V\ K^{-1}$		temperature dependence of Seebeck coefficient: Figs. 4,5 (N.B. E_A, σ, S for polycrystalline samples)	78B
	> 0, small			76G
EuTiO$_3$				
Structure: cubic (O_h^1—Pm3m)			crystal structure: Fig. 1	75M1
a	7.810 Å			66H
GdTiO$_3$				
Structure: orthorhombic (D_{2h}^{16}—Pbnm)			crystal structure: Fig. 6	69M1
a	5.407 Å			69M1
b	5.667 Å			
c	7.692 Å			
E_A	0.19 eV			78B
σ	$3.6 \cdot 10^{-2}\ \Omega^{-1}$ cm^{-1}	p-type	temperature dependence of resistivity: Fig. 2	78B
S	$\approx 170\ \mu V\ K^{-1}$		temperature dependence of Seebeck coefficient: Fig. 4 (N.B. E_A, σ, S for polycrystalline samples)	78B
TbTiO$_3$				
Structure: orthorhombic (D_{2h}^{16}—Pbnm)			crystal structure: Fig. 6	69M1
a	5.388 Å			69M1
b	5.648 Å			
c	7.676 Å			
E_A	0.2 eV			78B
σ	$9.1 \cdot 10^{-3}\ \Omega^{-1}$ cm^{-1}	p-type	polycrystalline sample	78B

Physical property	Numerical value	Experimental conditions	Experimental method remarks	Ref.
DyTiO$_3$				
Structure: orthorhombic (D$_{2h}^{16}$—Pbnm)			crystal structure: Fig. 6	69M1
a	5.361 Å			69M1
b	5.695 Å			
c	7.647 Å			
σ	$\approx 10^{-1}\,\Omega^{-1}\,\mathrm{cm}^{-1}$		temperature dependence of resistivity: Fig. 3	76G
S	$\approx -140\,\mu\mathrm{V\,K}^{-1}$		temperature dependence of Seebeck coefficient: Fig. 5 (N.B. σ, S for polycrystalline samples)	76G
HoTiO$_3$				
Structure: orthorhombic (D$_{2h}^{16}$—Pbnm)			crystal structure: Fig. 6	69M1
a	5.339 Å			69M1
b	5.665 Å			
c	7.626 Å			
E_A	0.2 eV		temperature dependence of resistivity: Fig. 2	78B
σ	$1.1 \cdot 10^{-2}\,\Omega^{-1}\,\mathrm{cm}^{-1}$	p-type		78B
S	$\approx 260\,\mu\mathrm{V\,K}^{-1}$		temperature dependence of Seebeck coefficient: Fig. 4 (N.B. E_A, σ, S for polycrystalline samples)	78B
ErTiO$_3$				
Structure: orthorhombic (D$_{2h}^{16}$—Pbnm)			crystal structure: Fig. 6	69M1
a	5.318 Å			69M1
b	5.657 Å			
c	7.613 Å			
E_A	0.24 eV		temperature dependence of resistivity: Fig. 2	78B
σ	$7.1 \cdot 10^{-3}\,\Omega^{-1}\,\mathrm{cm}^{-1}$	p-type		78B
S	$\approx 300\,\mu\mathrm{V\,K}^{-1}$		temperature dependence of Seebeck coefficient: Fig. 4 (N.B. E_A, σ, S for polycrystalline samples)	78B
TmTiO$_3$				
Structure: orthorhombic (D$_{2h}^{16}$—Pbnm)			crystal structure: Fig. 6	69M1
a	5.306 Å			69M1
b	5.647 Å			
c	7.607 Å			
YbTiO$_3$				
Structure: orthorhombic (D$_{2h}^{16}$—Pbnm)			crystal structure: Fig. 6	69M1
a	5.293 Å			69M1
b	5.633 Å			
c	7.598 Å			
E_A	0.24 eV			78B
σ	$8.3 \cdot 10^{-3}\,\Omega^{-1}\,\mathrm{cm}^{-1}$	p-type	polycrystalline sample;	78B
	$\approx 10^{-1}\,\Omega^{-1}\,\mathrm{cm}^{-1}$		temperature dependence of resistivity: Figs. 2, 3	76G

Physical property	Numerical value	Experimental conditions	Experimental method remarks	Ref.
LuTiO$_3$				
Structure: orthorhombic (D$_{2h}^{16}$—Pbnm)			crystal structure: Fig. 6	69M1
a	5.274 Å			69M1
b	5.633 Å			
c	7.580 Å			

10.3.1.2 LnVO$_3$ compounds

LaVO$_3$ (see also section 10.2.4)

Physical property	Numerical value	Experimental conditions	Experimental method remarks	Ref.
Structure: tetragonal				74M
orthorhombic (D$_{2h}^{16}$—Pbnm) at RT, but tetragonal for $T < 139$ K				74D2
$a = b$	5.535 Å		tetragonal description	74M
c	7.830 Å			
a	5.542 Å		orthorhombic description	74D2
b	5.542 Å			
c	7.838 Å			
		T [K]		
a	7.868 Å	77	tetragonal	74D2
c	7.738 Å			
T_m	2353 K			66R
E_A	0.095 eV	100···130	temperature dependence of	76S1
	0.049 eV	130···300	resistivity: Figs. 7, 8, 9;	
	0.16 eV	300···670	resistivity at room	66R
	0.12 eV	4···300	temperature: Fig. 10	75S
	0.13 eV			76G
	0.171···0.176 eV	$100 < T < 210$		74D2,
	0.209 eV	$210 < T < 300$		75P1
σ	2.3 Ω^{-1} cm^{-1}		p-type semiconductor	76S1
p	$3 \cdot 10^{19}$ cm^{-3}	800		76W
μ_p	0.15 cm^2 V^{-1} s^{-1}	800		76W
m_p	17 m_o			76W
S	≈ 150 µV K^{-1}		temperature dependence of	75S
	280 µV K^{-1}		Seebeck coefficient: Figs. 11, 12	74D2, 75P1
	330 µV K^{-1}			76W, 76G
E_g	≈ 1 eV		indirect absorption edge (N.B. Transport data for polycrystalline samples)	75D, 76W

La$_{1-x}$Sr$_x$VO$_3$ ($0 \leqq x \leqq 0.4$)

Physical property	Numerical value	Experimental conditions	Experimental method remarks	Ref.
E_A	0.02 eV	$x = 0.05$	p-type conduction; temperature dependence of conductivity for various x: Fig. 13	75S
S	36 µV K^{-1}	$x = 0.15$	pressed polycrystalline pellet; temperature dependence of Seebeck coefficient for various x: Fig. 14	76W
	150 µV K^{-1}	0.1	For X-ray diffraction, see [57K]; Anderson transition at x = 0.25.	

Physical property	Numerical value	Experimental conditions	Experimental method remarks	Ref.
CeVO$_3$				
Structure: tetragonal				74M
	for orthorhombic description (space group D_{2h}^{16}—Pbnm), see e.g. [56 B]			
a	5.519 Å			74M
c	7.809 Å			
E_A	0.041 eV	$T = 120 \cdots 170$ K	electrical resistivity at	76S1
	0.054 eV	$170 \cdots 300$ K	room temperature: Fig. 10; temperature dependence of resistivity: Fig. 15; p-type semiconductor	
PrVO$_3$				
Structure: orthorhombic (D_{2h}^{16}—Pbnm)			crystal structure: Fig. 6	74M
a	5.472 Å			74M
b	5.529 Å			
c	7.774 Å			
E_A	0.065 eV	$T = 100 \cdots 180$ K	electrical resistivity at	76S1
	0.103 eV	$180 \cdots 300$ K	room temperature: Fig. 10; temperature dependence of resistivity: Fig. 15; p-type semiconductor	
NdVO$_3$				
Structure: orthorhombic (D_{2h}^{16}—Pbnm)			crystal structure: Fig. 6	74M
a	5.451 Å			74M
b	5.575 Å			
c	7.740 Å			
E_A	0.16 eV			76G
	0.060 eV	$T = 100 \cdots 130$ K	electrical resistivity at	76S1
	≈ 0	$130 \cdots 300$ K	room temperature: Fig. 10; temperature dependence of resistivity: Figs. 9, 15; p-type semiconductor	
SmVO$_3$				
Structure: orthorhombic (D_{2h}^{16}—Pbnm)			crystal structure: Fig. 6	74M
a	5.394 Å			74M
b	5.581 Å			
c	7.684 Å			
E_A	0.106 eV	$T = 100 \cdots 175$ K	electrical resistivity at	76S1
	0.130 eV	$175 \cdots 300$ K	room temperature: Fig. 10;	
	0.18 eV		temperature dependence of resistivity: Fig. 15	76G
S	535 µV K^{-1}		p-type semiconductor, polycrystalline sample	76G
EuVO$_3$				
Structure: orthorhombic (D_{2h}^{16}—Pbnm)			crystal structure: Fig. 6	74M
a	5.362 Å			74M
b	5.599 Å			
c	7.651 Å			
E_A	0.099 eV	$T = 110 \cdots 175$ K	electrical resistivity at	76S1
	0.132 eV	$175 \cdots 300$ K	room temperature: Fig. 10;	
	0.18 eV		p-type semiconductor; temperature dependence of resistivity: Fig. 9	76G

(continued)

Physical property	Numerical value	Experimental conditions	Experimental method remarks	Ref.
EuVO$_3$ (continued)				
S	$610\ \mu V\ K^{-1}$		polycrystalline sample; temperature dependence of Seebeck coefficient: Fig. 12	76G
GdVO$_3$				
Structure: orthorhombic (D$_{2h}^{16}$—Pbnm)			crystal structure: Fig. 6	74M
a	$5.342\ \text{Å}$			74M
b	$5.604\ \text{Å}$			
c	$7.637\ \text{Å}$			
E_A	$0.100\ \text{eV}$	$T = 100 \cdots 165\ \text{K}$	electrical resistivity at	76S1
	$0.141\ \text{eV}$	$165 \cdots 300\ \text{K}$	room temperature: Fig. 10;	
	$0.20\ \text{eV}$		temperature dependence of	76G
	$0.96\ \text{eV}$	$T > 520\ \text{K}$	resistivity: Figs. 9, 15;	75P1
	$0.23\ \text{eV}$	$T < 520\ \text{K}$	p-type semiconductor	
σ	$10^{-6}\ \Omega^{-1}\ cm^{-1}$	p-type	(N.B. Transport data for	75P1
S	$560\ \mu V\ K^{-1}$		polycrystalline samples)	76G
	$400\ \mu V\ K^{-1}$			75P1
TbVO$_3$				
Structure: orthorhombic (D$_{2h}^{16}$—Pbnm)			crystal structure: Fig. 6	74M
a	$5.325\ \text{Å}$			74M
b	$5.606\ \text{Å}$			
c	$7.614\ \text{Å}$			
E_A	$0.101\ \text{eV}$	$T = 105 \cdots 185\ \text{K}$	electrical resistivity at	76S1
	0.116eV	$185 \cdots 300\ \text{K}$	room temperature: Fig. 10; p-type semiconductor	
DyVO$_3$				
Structure: orthorhombic (D$_{2h}^{16}$—Pbnm)			crystal structure: Fig. 6	74M
a	$5.299\ \text{Å}$			74M
b	$5.594\ \text{Å}$			
c	$7.593\ \text{Å}$			
E_A	$0.095\ \text{eV}$	$T = 100 \cdots 190\ \text{K}$	electrical resistivity at	76S1
	$0.125\ \text{eV}$	$190 \cdots 300\ \text{K}$	room temperature: Fig. 10	
S	$575\ \mu V\ K^{-1}$		p-type semiconductor, polycrystalline sample Temperature dependent absorption spectroscopy: [70W].	76G
HoVO$_3$				
Structure: orthorhombic (D$_{2h}^{16}$—Pbnm)			crystal structure: Fig. 6	74M
a	$5.276\ \text{Å}$			74M
b	$5.592\ \text{Å}$			
c	$7.576\ \text{Å}$			
E_A	$0.102\ \text{eV}$	$T = 120 \cdots 180\ \text{K}$	electrical resistivity at	76S1
	$0.128\ \text{eV}$	$180 \cdots 300\ \text{K}$	room temperature: Fig. 10; p-type semiconductor	
ErVO$_3$				
Structure: orthorhombic (D$_{2h}^{16}$—Pbnm)			crystal structure: Fig. 6	74M
a	$5.256\ \text{Å}$			74M
b	$5.581\ \text{Å}$			
c	$7.559\ \text{Å}$			
E_A	$0.121\ \text{eV}$	$T = 125 \cdots 190\ \text{K}$	electrical resistivity at room	76S1
	$0.173\ \text{eV}$	$190 \cdots 300\ \text{K}$	temperature: Fig. 10; temperature dependence of resistivity: Fig. 15; p-type semiconductor	

Physical property	Numerical value	Experimental conditions	Experimental method remarks	Ref.
TmVO₃				
Structure: orthorhombic (D_{2h}^{16}—Pbnm)			crystal structure: Fig. 6	74M
a	5.237 Å			74M
b	5.573 Å			
c	7.545 Å			
E_A	0.122 eV	$T = 130\cdots185$ K	electrical resistivity at	76S1
	0.166 eV	$185\cdots300$ K	room temperature: Fig. 10; p-type semiconductor	
YbVO₃				
Structure: orthorhombic (D_{2h}^{16}—Pbnm)			crystal structure: Fig. 6	74M
a	5.223 Å			74M
b	5.564 Å			
c	7.537 Å			
E_A	0.125 eV	$T = 100\cdots190$ K	electrical resistivity at	76S1
	0.169 eV	$190\cdots300$ K	room temperature: Fig. 10; p-type semiconductor;	
	0.17 eV		temperature dependence of resistivity: Fig. 9	76G
S	525 μV K⁻¹		polycrystalline sample; temperature dependence of Seebeck coefficient: Fig. 12	76G
LuVO₃				
Structure: orthorhombic (D_{2h}^{16}—Pbnm)			crystal structure: Fig. 6	74M
a	5.214 Å			74M
b	5.561 Å			
c	7.530 Å			
E_A	0.127 eV	$T = 120\cdots185$ K	electrical resistivity at	76S1
	0.171 eV	$185\cdots300$ K	room temperature: Fig. 10; p-type semiconductor	

10.3.1.3 LnCrO₃ compounds

Physical property	Numerical value	Experimental conditions	Experimental method remarks	Ref.
LaCrO₃				
Structure: orthorhombic (D_{2h}^{16}—Pbnm)			crystal structure: Fig. 6	57G
rhombohedral for $T > 540$ K				73T
a	5.477 Å			57G
b	5.514 Å			
c	7.755 Å			
a	5.479 Å			63Q
b	5.515 Å			
c	7.753 Å			
a	5.477 Å			73T
b	5.514 Å			
c	7.768 Å			
		T [K]		
a	5.502 Å	610		73T
α	66° 10′			
a	5.51 Å	$550 < T < 1300$	rhombohedral high-temperature phase: Fig. 16	67R1
c	13.34 Å			
a	3.92 Å	1500	cubic high-temperature phase: Fig. 1	

(continued)

Physical property	Numerical value	Experimental conditions	Experimental method remarks	Ref.
LaCrO$_3$ (continued)		T[K]		
T_m	2790 K			65F1
E_A	0.6 eV	$300\cdots800$		67R1
	0.55 eV	$300\cdots800$		69M2
	0.22 eV	$300\cdots800$		71S
	0.18 eV	$77\cdots300$		77W
	0.25 eV	$T<350$		80T1
	0.28 eV	$350\cdots685$		
	0.22 eV	$685\cdots1000$		
σ	$\approx 10^{-2}\,\Omega^{-1}\,cm^{-1}$	p-type	temperature dependence of conductivity: Figs. 17, 18, 19; Hall coefficient: Fig. 20	77W
p	$5\cdot10^{19}\,cm^{-3}$			77W
μ_p	$5.2\cdot10^{-4}\,cm^2\,V^{-1}\,s^{-1}$			77W
	$5\cdot10^{-3}\,cm^2\,V^{-1}\,s^{-1}$			69M2
S	$520\,\mu V\,K^{-1}$		temperature dependence of Seebeck coefficient: Figs. 21, 22	77W
E_g	4.7 eV			70S, 71S
			Infrared absorption: Fig. 23; XPS spectra: Figs. 24, 25, 26. (N.B. All data for transport properties are from polycrystalline samples)	

La$_{0.8}$Mg$_{0.2}$CrO$_3$
XPS spectra: Figs. 24, 25

La$_{1-x}$Sr$_x$CrO$_3$

σ	$\approx 0.1\,\Omega^{-1}\,cm^{-1}$	p-type, $x=0.05$	temperature dependence of conductivity: Fig. 17	77W
	$\approx 1\,\Omega^{-1}\,cm^{-1}$	p-type, 0.1		
	$\approx 2\,\Omega^{-1}\,cm^{-1}$	p-type, 0.2		
μ_p	$5.2\cdot10^{-4}\,cm^2\,V^{-1}\,s^{-1}$	$0\leqq x\leqq0.2$		77W
	$5\cdot10^{-3}\,cm^2\,V^{-1}\,s^{-1}$	$0\leqq x\leqq0.2$		69M2
S	$\approx 280\,\mu V\,K^{-1}$	$x=0.05$, RT	temperature dependence of Seebeck coefficient: Fig. 21	77W
	$\approx 220\,\mu V\,K^{-1}$	0.1, RT		
	$\approx 170\,\mu V\,K^{-1}$	0.2, RT		
κ	$5.1\cdot10^{-2}\,W\,m^{-1}\,K^{-1}$	$T=1100\cdots 2000\,K$		69M2
α	$9\cdot10^{-6}\,K^{-1}$	$T=300\cdots 1000\,K$		69M2
			XPS spectra: Figs. 24, 25. (N.B. All data for polycrystalline samples)	

La$_{0.85}$Ba$_{0.15}$Cr$_{1-x}$Fe$_x$O$_3$
Cr and Fe conductivity [54J]

CeCrO$_3$
Structure: orthorhombic (D$_{2h}^{16}$—Pbnm)   crystal structure: Fig. 6   63Q

a	5.475 Å			63Q
b	5.475 Å			
c	7.740 Å			

Physical property	Numerical value	Experimental conditions	Experimental method remarks	Ref.
PrCrO$_3$				
Structure: orthorhombic (D$_{2h}^{16}$—Pbnm)			crystal structure: Fig. 6	63Q
a	5.448 Å			63Q
b	5.479 Å			
c	7.718 Å			
T_m	≈ 2690 K			65F1
NdCrO$_3$				
Structure: orthorhombic (D$_{2h}^{16}$—Pbnm)			crystal structure: Fig. 6	63Q
a	5.425 Å			63Q
b	5.478 Å			
c	7.694 Å			
T_m	≈ 2675 K			65F1
E_A	0.29 eV	$T = 300 \cdots 625$ K	temperature dependence of	80T1
	0.22 eV	$625 \cdots 1000$ K	ac conductivity: Fig. 18;	
			p-type semiconductor	
S	≈ 1.15 mV K^{-1}*)		polycrystalline sample;	80T1
			temperature dependence of	
			Seebeck coefficient: Fig. 22	
			For absorption spectra,	
			see [72H, 75H1].	
SmCrO$_3$				
Structure: orthorhombic (D$_{2h}^{16}$—Pbnm)			crystal structure: Fig. 6	63Q
a	5.367 Å			63Q
b	5.508 Å			
c	7.643 Å			
a	5.364 Å			74T
b	5.507 Å			
c	7.645 Å			
T_m	≈ 2655 K			65F1
E_A	0.36 eV	$T = 300 \cdots 625$ K	temperature dependence of	80T1
	0.20 eV	$625 \cdots 1000$ K	ac conductivity: Fig. 18;	
			p-type semiconductor	
S	≈ 1 mV K^{-1}*)		polycrystalline sample;	80T1
			temperature dependence of	
			Seebeck coefficient: Fig. 22	
Θ_D	216 K			68D
			For magneto-optical properties,	
			see [70T1].	
EuCrO$_3$				
Structure: orthorhombic (D$_{2h}^{16}$—Pbnm)			crystal structure: Fig. 6	63Q
a	5.340 Å			63Q
b	5.515 Å			
c	7.622 Å			
a	5.3395 Å			74T
b	5.5130 Å			
c	7.6252 Å			
GdCrO$_3$				
Structure: orthorhombic (D$_{2h}^{16}$—Pbnm)			crystal structure: Fig. 6	63Q
a	5.312 Å			63Q
b	5.252 Å			
c	7.606 Å			

*) The authors of [80T1] used opposite definition for sign of S.

(continued)

Physical property	Numerical value	Experimental conditions	Experimental method remarks	Ref.
GdCrO$_3$ (continued)				
a	5.315 Å		For ESR measurements, see [75M3], for Raman spectra (temperature dependence), see [75U].	74T
b	5.5271 Å			
c	7.6082 Å			
T_m	≈ 2640 K			65F1
TbCrO$_3$				
Structure: orthorhombic (D$_{2h}^{16}$—Pbnm)			crystal structure: Fig. 6	63Q
a	5.291 Å			63Q
b	5.618 Å			
c	7.576 Å			
a	5.2894 Å			74T
b	5.5141 Å			
c	7.5744 Å			
DyCrO$_3$				
Structure: orthorhombic (D$_{2h}^{16}$—Pbnm)			crystal structure: Fig. 6	63Q
a	5.265 Å			63Q
b	5.520 Å			
c	7.552 Å			
a	5.2675 Å			74T
b	5.5224 Å			
c	7.5557 Å			
T_m	2610 K			65F1
E_A	0.27 eV			71S
		T [°C]		
σ	$3.1 \cdot 10^{-4}\,\Omega^{-1}\,cm^{-1}$	200, p-type	pressed polycrystalline pellets	71S
	$3.6 \cdot 10^{-3}\,\Omega^{-1}\,cm^{-1}$	500, p-type		
	$6.6 \cdot 10^{-3}\,\Omega^{-1}\,cm^{-1}$	700, p-type		
			For magneto-optical properties, see [70T1]; for absorption spectra, see [70T1] and [71U]; thermodynamic properties are given in [68D].	
HoCrO$_3$				
Structure: orthorhombic (D$_{2h}^{16}$—Pbnm)			crystal structure: Fig. 6	63Q
a	5.243 Å			63Q
b	5.519 Å			
c	7.538 Å			
a	5.2434 Å			74T
b	5.5213 Å			
c	7.5375 Å			
T_m	≈ 2595 K			65F1
E_A	0.33 eV			71S
		T [°C]		
σ	$9 \cdot 10^{-5}\,\Omega^{-1}\,cm^{-1}$	200, p-type	pressed polycrystalline pellets; temperature dependence of ac conductivity: Fig. 27	71S
	$1.41 \cdot 10^{-3}\,\Omega^{-1}\,cm^{-1}$	500, p-type		
	$3.1 \cdot 10^{-3}\,\Omega^{-1}\,cm^{-1}$	700, p-type	For optical absorption of Cr^{3+} (^{2}E state), see [70M1]; for dielectric properties, see [68C].	

Physical property	Numerical value	Experimental conditions	Experimental method remarks	Ref.
ErCrO₃				
Structure: orthorhombic (D_{2h}^{16}—Pbnm)			crystal structure: Fig. 6	63Q
a	5.223 Å			63Q
b	5.516 Å		Er^{3+} ground state splitting	
c	7.519 Å		[70C1]; optical absorption of	
a	5.2208 Å		Cr^{3+} (^2E state) [70M1], see	74T
b	5.5141 Å		also [70M3], [75C1], [75H2].	
c	7.5154 Å		For dielectric properties,	
T_m	≈ 2600 K		see [68D].	65F1
TmCrO₃				
Structure: orthorhombic (D_{2h}^{16}—Pbnm)			crystal structure: Fig. 6	63Q
a	5.209 Å			63Q
b	5.508 Å			
c	7.500 Å			
a	5.2063 Å			74T
b	5.5039 Å			
c	7.4972 Å			
YbCrO₃				
Structure: orthorhombic (D_{2h}^{16}—Pbnm)			crystal structure: Fig. 6	63Q
a	5.195 Å			63Q
b	5.510 Å			
c	7.490 Å			
T_m	≈ 2590 K			65F1
E_A	0.37 eV			71S
		T [°C]		
σ	$3 \cdot 10^{-5}\,\Omega^{-1}\,cm^{-1}$	200, p-type	pressed polycrystalline pellets	71S
	$9.2 \cdot 10^{-4}\,\Omega^{-1}\,cm^{-1}$	500, p-type		
	$1.62 \cdot 10^{-3}\,\Omega^{-1}\,cm^{-1}$	700, p-type		
			For absorption spectra (magnon sidebands at 4.2 K), see [70T1]. Dielectric properties are given in [68R].	
LuCrO₃				
Structure: orthorhombic (D_{2h}^{16}—Pbnm)			crystal structure: Fig. 6	63Q
a	5.176 Å			63Q
b	5.497 Å			
c	7.475 Å			
a	5.1739 Å		For dielectric properties, see	74T
b	5.4987 Å		[68R]; for absorption spectra,	
c	7.4743 Å		see [75K].	

10.3.1.4 LnMnO₃ compounds

Physical property	Numerical value	Experimental conditions	Experimental method remarks	Ref.
LaMnO₃				
Structure: orthorhombic (D_{2h}^{16}—Pbnm)			crystal structure: Fig. 6	75V
a	5.536 Å		2% Mn^{4+}	75V
b	5.726 Å			
c	7.697 Å			
			For dependence of structural phase transition temperature on Mn^{3+} content, see [59W].	

(continued)

Physical property	Numerical value	Experimental conditions	Experimental method remarks	Ref.
LaMnO$_3$ (continued)				
		T [K]		
E_A	0.13 eV	300$\cdots$550		71S
	0.33 eV	$T < 740$		66G
σ	$10^{-4}\,\Omega^{-1}\,cm^{-1}$		polycrystalline sample	66J
	$4.7 \cdot 10^{-2}\,\Omega^{-1}\,cm^{-1}$	473, p-type	17% Mn^{4+}; pressed poly-	71S
	$9.3 \cdot 10^{-2}\,\Omega^{-1}\,cm^{-1}$	773, p-type	crystalline pellet; temperature	
	$0.106\,\Omega^{-1}\,cm^{-1}$	973, p-type	dependence of ac conductivity: Fig. 19	
μ_p	$2 \cdot 10^{-4}\,cm^2\,V^{-1}\,s^{-1}$		pressed polycrystalline pellet For IR absorption (400$\cdots$ 4000 cm^{-1}), see [70M4].	71S

LaMn$_{1-x}$Co$_x$O$_3$ (see also LaCo$_{1-x}$Mn$_x$O$_3$, subsection 10.3.1.6)
For phase diagram; see Fig. 28.

σ	$2.5 \cdot 10^{-2}\,\Omega^{-1}\,cm^{-1}$	x = 0.5	room temperature resistivity as a function of x: Fig. 29	66J

LaMn$_{0.75}$Mo$_{0.25}$O$_3$ [80S]
Structure: orthorhombic

a	5.607 Å			
b	5.675 Å			
c	7.934 Å			
E_A	0.3 eV			
σ	$5 \cdot 10^{-5}\,\Omega^{-1}\,cm^{-1}$	$T = 50\,°C$	pressed polycrystalline pellet	

La$_{1-x}$Sr$_x$MnO$_3$
Resistivity of mixed crystals as a function of x: Fig. 30.

La$_{0.85}$Ba$_{0.15}$Mn$_{1-x}$Cr$_x$O$_3$ (see also (La, Y) (Co, Mn)O$_3$, subsection 10.3.1.6)
Resistivity of mixed crystals as a function of x: Fig. 31.

La$_{1-x}$Ca$_x$Fe$_{1-x}$Mn$_x$O$_3$
Semiconducting properties according to [70B]; $\mu \approx 10^{-8} \cdots 10^{-5}\,cm^2\,V^{-1}\,s^{-1}$.

La$_{0.85}$Ba$_{0.15}$Mn$_{1-x}$Fe$_x$O$_3$
Resistivity of mixed crystals as a function of x: Fig. 32 ($E_A = 0.15 \cdots 0.45$ eV) [54J].

La$_{0.85}$Y$_{0.15}$Mn$_{1-x}$Co$_x$O$_3$

σ	$0.25 \cdots 10^{-4}\,\Omega^{-1}\,cm^{-1}$		polycrystalline samples; ϱ dependence on x and heat treatment of sample: Fig. 33. Seebeck coefficient as a function of x: Fig. 34.	66J

CeMnO$_3$ [68Q]
Structure: orthorhombic (D$_{2h}^{16}$—Pbnm)   crystal structure: Fig. 6

a	5.537 Å			
b	5.557 Å			
c	7.812 Å			

PrMnO$_3$ [68Q]
Structure: orthorhombic (D$_{2h}^{16}$—Pbnm)   crystal structure: Fig. 6

a	5.545 Å			
b	5.787 Å			
c	7.575 Å			

Physical property	Numerical value	Experimental conditions	Experimental method remarks	Ref.
NdMnO₃				
Structure: orthorhombic (D_{2h}^{16}—Pbnm)			crystal structure: Fig. 6	68Q
a	5.380 Å			68Q
b	5.854 Å			
c	7.557 Å			
a	5.414 Å			73M
b	5.829 Å			
c	7.551 Å			
SmMnO₃				
Structure: orthorhombic (D_{2h}^{16}—Pbnm)			crystal structure Fig. 6	68Q
a	5.359 Å			68Q
b	5.843 Å			
c	7.482 Å			
a	5.358 Å			73M
b	5.825 Å			
c	7.483 Å			
EuMnO₃				
Structure: orthorhombic (D_{2h}^{16}—Pbnm)			crystal structure: Fig. 6	68Q
a	5.338 Å			68Q
b	5.842 Å			
c	7.453 Å			
a	5.336 Å			73M
b	5.842 Å			
c	7.451 Å			
GdMnO₃				
Structure: orthorhombic (D_{2h}^{16}—Pbnm)			crystal structure: Fig. 6	68Q
a	5.313 Å			68Q
b	5.853 Å			
c	7.432 Å			
a	5.310 Å			73M
b	5.840 Å			
c	7.430 Å			
TbMnO₃ [68Q]				
Structure: orthorhombic (D_{2h}^{16}—Pbnm)			crystal structure: Fig. 6	
a	5.297 Å			
b	5.831 Å			
c	7.403 Å			
DyMnO₃				
Structure: orthorhombic (D_{2h}^{16}—Pbnm)			crystal structure: Fig. 6	68Q
a	5.275 Å			68Q
b	5.828 Å			
c	7.375 Å			
a	5.272 Å			73M
b	5.795 Å			
c	7.379 Å			
HoMnO₃				
Structure: hexagonal (C_{6v}^{3}—P6₃cm)				70P
orthorhombic (D_{2h}^{16}—Pbnm)			crystal structure: Fig. 6	66W
(high pressure phase)				
a	6.136 Å			70P
c	11.42 Å			(continued)

Physical property	Numerical value	Experimental conditions	Experimental method remarks	Ref.
HoMnO$_3$ (continued)				
a	5.27 Å		high-temperature and high-	66W
b	5.84 Å		pressure preparation	
c	7.36 Å		(1000 °C, 35···40 · 10^8 Pa)	
a	5.26 Å		high-pressure phase	73W
b	5.84 Å			
c	7.35 Å			
E_A	0.57 eV			71S
		T [°C]		
σ	$1 \cdot 10^{-5}\,\Omega^{-1}\,cm^{-1}$	200, p-type	pressed polycrystalline pellet;	71S
	$5.6 \cdot 10^{-4}\,\Omega^{-1}\,cm^{-1}$	500, p-type	temperature dependence of	
	$1.5 \cdot 10^{-3}\,\Omega^{-1}\,cm^{-1}$	700, p-type	conductivity: Fig. 27	
			For dielectric constant,	
			see Fig. 34α.	
ErMnO$_3$				
Structure: hexagonal (C$_{6v}^3$—P6$_3$cm)				63Y
orthorhombic (D$_{2h}^{16}$—Pbnm)			crystal structure: Fig. 6	67W
(high-pressure phase)				
a	6.115 Å			63Y
c	11.41 Å			
a	5.24 Å		high-pressure preparation	67W
b	5.82 Å			
c	7.335 Å			
d	7.282 g cm^{-3}		hexagonal phase	63Y
			For dielectric constant,	
			see Fig. 34α.	
TmMnO$_3$				
Structure: hexagonal (C$_{6v}^3$—P6$_3$cm)				63Y
orthorhombic (high-pressure phase)				67W
a	6.062 Å			63Y
c	11.40 Å			
a	5.23 Å		high-pressure preparation	67W
b	5.81 Å			
c	7.32 Å			
YbMnO$_3$				
Structure: hexagonal (C$_{6v}^3$—P6$_3$cm)				63Y
orthorhombic (high-pressure phase)				67W
a	6.070 Å			63Y
c	11.395 Å			
a	5.22 Å		high-pressure preparation	67W
b	5.81 Å		(1000 °C, 35···40 · 10^8 Pa)	73W
c	7.30 Å			
E_A	0.73 eV			71S
σ	$3.1 \cdot 10^{-4}\,\Omega^{-1}\,cm^{-1}$	$T = 500$ °C, p-type	pressed polycrystalline pellet	71S
	$2.9 \cdot 10^{-3}\,\Omega^{-1}\,cm^{-1}$	$T = 700$ °C, p-type		
			For dielectric constant,	
			see Fig. 34α.	

Physical property	Numerical value	Experimental conditions	Experimental method remarks	Ref.
LuMnO$_3$				
Structure: hexagonal (C$_{6v}^3$—P6$_3$cm)				63Y
orthorhombic (high-pressure phase)				67W
a	6.042 Å			63Y
c	11.37 Å			
a	5.205 Å		high-pressure preparation	67W
b	5.79 Å			
c	7.31 Å			
			For dielectric properties, see [67B].	

10.3.1.5 LnFeO$_3$ compounds

Physical property	Numerical value	Experimental conditions	Experimental method remarks	Ref.
LaFeO$_3$				
Structure: orthorhombic (D$_{2h}^{16}$—Pbnm)			crystal structure: Fig. 6;	56G
rhombohedral above 980 °C				59D
a	5.556 Å			56G
b	5.565 Å			
c	7.862 Å			
a	5.553 Å			71M3
b	5.563 Å			
c	7.862 Å			
d	6.63 g cm^{-3}			71R1, 71R2
		T [°C]		
E_A	0.39 eV	$T < 470$		71S
	0.22 eV	$T > 470$		
σ	2.5 · 10^{-3} Ω^{-1} cm^{-1}	500, p-type	pressed polycrystalline pellet;	71S
	4.4 · 10^{-3} Ω^{-1} cm^{-1}	700, p-type	temperature dependence of ac conductivity: Fig. 19	
			For absorption spectra, birefringence and Faraday rotation, see [70W2] and [71C1].	
LaFe$_{0.75}$Mo$_{0.25}$O$_3$ [80S]				
Structure: orthorhombic				
a	5.594 Å			
b	5.636 Å			
c	7.940 Å			
E_A	0.14 eV			
σ	3.2 · 10^{-3} Ω^{-1} cm^{-1}	$T = 50$ °C, n-type	pressed polycrystalline pellets; temperature dependence of conductivity: Fig. 35	
S	≈ −45 µV K^{-1}		temperature dependence of Seebeck coefficient: Fig. 36	
La$_{1-x}$Ca$_x$Fe$_{1-x}$Mn$_x$O$_3$ [70B]				
Structure: orthorhombic for x < 0.3, cubic for 0.3 < x < 0.7				
orthorhombic modification:				
a	5.502 Å	x = 0.2		
b	5.512 Å	0.2		
c	7.790 Å	0.2		
cubic modification:				
a	3.877 Å	x = 0.3		
	3.858 Å	0.4	98% Mn^{4+}	
	3.820 Å	0.6	98.1% Mn^{4+}	
	3.796 Å	0.7	98.2% Mn^{4+}	(continued)

Physical property	Numerical value	Experimental conditions	Experimental method remarks	Ref.
La$_{1-x}$Ca$_x$Fe$_{1-x}$Mn$_x$O$_3$ [70B] (continued)				
E_A	0.51 eV	$x = 0.2$		
	0.44 eV	0.3		
	0.39 eV	0.4		
	0.26 eV	0.6		
	0.21 eV	0.7		
σ	$8.3 \cdot 10^{-7}\,\Omega^{-1}\,cm^{-1}$	$x = 0.2$	p-type	
	$8.3 \cdot 10^{-6}\,\Omega^{-1}\,cm^{-1}$	0.3	p-type	
	$6.7 \cdot 10^{-5}\,\Omega^{-1}\,cm^{-1}$	0.4	p-type above 330 K, n-type below 330 K	
	$5 \cdot 10^{-3}\,\Omega^{-1}\,cm^{-1}$	0.6		
	$3.3 \cdot 10^{-2}\,\Omega^{-1}\,cm^{-1}$	0.7		
μ_p	$1.2 \cdot 10^{-8}\,cm^2\,V^{-1}\,s^{-1}$	$x = 0.3$		
	$6 \cdot 10^{-8}\,cm^2\,V^{-1}\,s^{-1}$	0.4		
	$1 \cdot 10^{-5}\,cm^2\,V^{-1}\,s^{-1}$	0.6		
	$4.8 \cdot 10^{-5}\,cm^2\,V^{-1}\,s^{-1}$	0.7		

La$_{1-x}$Sr$_x$FeO$_3$

Phase diagram of La$_{1-x}$Sr$_x$Mn$_{1-y}$Fe$_y$O$_3$: Fig. 37; resistivities of mixed crystals: Fig. 38.
Conductivity and optical absorption go through a maximum at $x = 0.5$ [57W2, 60W].

La$_{0.85}$Ba$_{0.15}$Fe$_x$Cr$_{1-x}$O$_3$

Cr and Fe conductivity [54J]

La$_{0.85}$Ba$_{0.15}$Fe$_x$Mn$_{1-x}$O$_3$

E_A	0.15 eV	$x = 0$	resistivities of mixed crystals:	54J
	0.27 eV	1	Fig. 32	
	0.45 eV	0.85		

CeFeO$_3$

Structure: orthorhombic				59D
a	5.541 Å			59D
b	5.577 Å			
c	7.809 Å			

PrFeO$_3$

Structure: orthorhombic (D$_{2h}^{16}$—Pbnm)			crystal structure Fig. 6	70M2
a	5.482 Å		For absorption spectra,	70M2
b	5.578 Å		birefringence and Faraday	
c	7.786 Å		rotation, see [70W2], [71C2].	
d	6.79 g cm^{-3}		For birefringence on	71R1
			Pr$_{1-x}$Sm$_x$FeO$_3$ ($x = 0.4$),	
			see [71C3].	

PrFe$_{0.75}$Mo$_{0.25}$O$_3$ [80S]

Structure: orthorhombic				
a	5.488 Å			
b	5.613 Å			
c	7.902 Å			
E_A	0.15 eV			
σ	$3.1 \cdot 10^{-3}\,\Omega^{-1}\,cm^{-1}$	$T = 50\,°C$, n-type	pressed polycrystalline pellet	

Physical property	Numerical value	Experimental conditions	Experimental method remarks	Ref.
NdFeO$_3$				
Structure: orthorhombic (D$_{2h}^{16}$—Pbnm)			crystal structure: Fig. 6	70M2
a	5.453 Å		For absorption spectra,	70M2
b	5.584 Å		birefringence and Faraday	
c	7.768 Å		rotation, see [70W2], [71C2];	
d	7.01 g cm^{-3}		Zeeman splitting of ground	71R1
			doublet: [73H1]. For magneto-	
			elastic properties (anomaly	
			of Young's modulus at 130 K),	
			see [71B1].	
NdFe$_{0.75}$Mo$_{0.25}$O$_3$ [80S]				
Structure: orthorhombic				
a	5.471 Å			
b	5.619 Å			
c	7.861 Å			
E_A	0.19 eV			
σ	$2.6 \cdot 10^{-3} \Omega^{-1}$ cm^{-1}	$T = 50\,°C$, n-type	pressed polycrystalline pellet	
SmFeO$_3$				
Structure: orthorhombic (D$_{2h}^{16}$—Pbnm)			crystal structure: Fig. 6	70M2
a	5.400 Å		Complex polar Kerr effect:	70M2
b	5.597 Å		Fig. 40.	
c	7.711 Å		For absorption, birefringence	
d	7.26 g cm^{-3}		and Faraday rotation,	71R1
			see [70W2], [71C2], [75A]	
			and [70T2].	
SmFe$_{0.75}$Mo$_{0.25}$O$_3$ [80S]				
Structure: orthorhombic				
a	5.400 Å			
b	5.622 Å			
c	7.767 Å			
E_A	0.2 eV			
σ	$3 \cdot 10^{-3} \Omega^{-1}$ cm^{-1}	$T = 50\,°C$, n-type	pressed polycrystalline pellet; temperature dependence of conductivity: Fig. 35	
EuFeO$_3$				
Structure: orthorhombic (D$_{2h}^{16}$—Pbnm)			crystal structure: Fig. 6	70M2
a	5.372 Å		Complex polar Kerr effect:	70M2
b	5.606 Å		Figs. 40, 41; reflectivity spectrum:	
c	7.685 Å		Fig. 42; complex dielectric	
a	5.371 Å		function and complex refractive	72M
b	5.598 Å		index: Fig. 43.	
c	7.681 Å			
d	7.34 g cm^{-3}			71R1
EuFe$_{0.75}$Mo$_{0.25}$O$_3$ [80S]				
Structure: orthorhombic				
a	5.374 Å			
b	5.633 Å			
c	7.708 Å			
E_A	0.2 eV			
σ	$2 \cdot 10^{-3} \Omega^{-1}$ cm^{-1}	$T = 50\,°C$, n-type	pressed polycrystalline pellet	

Physical property	Numerical value	Experimental conditions	Experimental method remarks	Ref.
GdFeO$_3$				
Structure: orthorhombic (D$_{2h}^{16}$—Pbnm)			crystal structure: Fig. 6	70M2
a	5.349 Å			70M2
b	5.611 Å			
c	7.669 Å			
d	7.52 g cm^{-3}			71R1
E_A	0.89 eV	$T < 385\,°C$	p-type semiconductor;	71S
	1.10 eV	$T > 385\,°C$	temperature dependence of anisotropic conductivity: Fig. 39. Complex polar Kerr effect: Fig. 40. For absorption and Faraday rotation, see [70W2] and [70T2], respectively. For absorption as a function of heat treatment, see [68G].	
GdFe$_{0.75}$Mo$_{0.25}$O$_3$ [80S]				
Structure: orthorhombic				
a	5.373 Å			
b	5.647 Å			
c	7.709 Å			
E_A	0.23 eV			
σ	$1.8 \cdot 10^{-3}\,\Omega^{-1}\,cm^{-1}$	$T = 50\,°C$, n-type	pressed polycrystalline pellet; temperature dependence of conductivity: Fig. 35	
TbFeO$_3$				
Structure: orthorhombic (D$_{2h}^{16}$—Pbnm)			crystal structure: Fig. 6	65E
a	5.326 Å		Complex polar Kerr effect: Fig. 40.	65E
b	5.602 Å		For absorption and Faraday	
c	7.635 Å		rotation, see [70W2] and	
d	7.66 g cm^{-3}		[70T2], respectively.	71R2
TbFe$_{0.75}$Mo$_{0.25}$O$_3$ [80S]				
Structure: orthorhombic				
a	5.368 Å			
b	5.641 Å			
c	7.699 Å			
E_A	0.23 eV			
σ	$9.1 \cdot 10^{-4}\,\Omega^{-1}\,cm^{-1}$	$T = 50\,°C$, n-type	pressed polycrystalline pellet	
DyFeO$_3$				
Structure: orthorhombic (D$_{2h}^{16}$—Pbnm)			crystal structure: Fig. 6	65E
a	5.302 Å			65E
b	5.598 Å			
c	7.623 Å			
d	7.84 g cm^{-3}			71R2
Δn	0.034	$\lambda = 633$ nm	birefringence; for Δn vs. λ, see [70C2]. For absorption spectra, see [70W2]; for Faraday rotation, see [70T2] and [72C]. Complex polar Kerr effect: Fig. 40.	70C2

Physical property	Numerical value	Experimental conditions	Experimental method remarks	Ref.
DyFe$_{0.75}$Mo$_{0.25}$O$_3$ [80S]				
Structure: orthorhombic				
a	5.338 Å			
b	5.606 Å			
c	7.646 Å			
E_A	0.26 eV			
σ	$5.9 \cdot 10^{-4}\,\Omega^{-1}\,cm^{-1}$	$T = 50\,°C$, n-type	pressed polycrystalline pellet	
HoFeO$_3$				
Structure: orthorhombic (D$_{2h}^{16}$—Pbnm)			crystal structure: Fig. 6	65E
a	5.278 Å			65E
b	5.591 Å			
c	7.602 Å			
d	7.96 g cm^{-3}			71R2
		T [°C]		
E_A	1.10 eV	$T < 367$		71S
	0.31 eV	$T > 367$		
σ	$8.4 \cdot 10^{-4}\,\Omega^{-1}\,cm^{-1}$	500, p-type	pressed polycrystalline pellets; temperature dependence of conductivity: Fig. 27	71S
	$2.2 \cdot 10^{-3}\,\Omega^{-1}\,cm^{-1}$	700, p-type		
Δn	0.042	$\lambda = 633$ nm	birefringence; for Δn vs. λ, see [70C2]. For absorption spectra, see [70W2]; for Faraday rotation, see [70C2] and [70T2]. Complex polar Kerr effect: Fig. 40. Zeeman splitting: [74W].	70C2
HoFe$_{0.75}$Mo$_{0.25}$O$_3$ [80S]				
Structure: orthorhombic				
a	5.324 Å			
b	5.593 Å			
c	7.632 Å			
E_A	0.3 eV			
σ	$1.4 \cdot 10^{-4}\,\Omega^{-1}\,cm^{-1}$	$T = 50\,°C$, n-type	pressed polycrystalline pellet	
ErFeO$_3$				
Structure: orthorhombic (D$_{2h}^{16}$—Pbnm)			crystal structure: Fig. 6	65E
a	5.263 Å		Complex polar Kerr effect: Figs. 40, 41; absorption spectrum: Fig 44.	65E
b	5.582 Å			
c	7.591 Å			
d	8.08 g cm^{-3}		For absorption spectra, see also [70W2] and [70A1]; for Faraday rotation, see [70T2] and [73H2]. Zeeman splitting is discussed in [69W]. For magneto-elastic effect, see [74G2].	71R1

Physical property	Numerical value	Experimental conditions	Experimental method remarks	Ref.
ErFe$_{0.75}$Mo$_{0.25}$O$_3$ [80S]				
Structure: orthorhombic				
a	5.287 Å			
b	5.611 Å			
c	7.681 Å			
E_A	0.3 eV			
σ	$1.4 \cdot 10^{-4} \, \Omega^{-1} \, cm^{-1}$	$T = 50\,°C$, n-type	pressed polycrystalline pellet	
TmFeO$_3$				
Structure: orthorhombic (D$_{2h}^{16}$—Pbnm)			crystal structure: Fig. 6	65E
a	5.251 Å		Complex polar Kerr effect: Fig. 40; 4f absorption spectra of Tm^{3+}: Figs. 45, 46, 47; Zeeman splitting of Tm^{3+} ground state: Figs. 48, 49.	65E
b	5.576 Å			
c	7.584 Å			
d	8.15 g cm^{-3}		For absorption spectra, see also e.g. [70A1] and [70M5]; for Faraday rotation, see [70T2] and [75C2]. For anomalies of elastic wave velocity and Young's modulus, see [74G2] and [71B1].	71R2
TmFe$_{0.75}$Mo$_{0.25}$O$_3$ [80S]				
Structure: orthorhombic				
a	5.265 Å			
b	5.579 Å			
c	7.583 Å			
E_A	0.32 eV			
σ	$5.3 \cdot 10^{-5} \, \Omega^{-1} \, cm^{-1}$	$T = 50\,°C$, n-type	pressed polycrystalline pellet	
YbFeO$_3$				
Strcuture: orthorhombic (D$_{2h}^{16}$—Pbnm)			crystal structure: Fig. 6	65E
a	5.233 Å			65E
b	5.557 Å			
c	7.570 Å			
d	8.35 g cm^{-3}			71R2
		T [°C]		
E_A	1.05 eV	$T > 700$		71S
σ	$1 \cdot 10^{-5} \, \Omega^{-1} \, cm^{-1}$	500, p-type	pressed polycrystalline pellet	71S
	$9.3 \cdot 10^{-4} \, \Omega^{-1} \, cm^{-1}$	700, p-type		
Δn	0.042	$\lambda = 633$ nm	birefringence; for Δn vs. λ, see [70C2]. Complex polar Kerr effect: Figs. 40, 41. For absorption spectra, see [70W2] and [71B2]; for IR absorption, see [70A2]; Faraday rotation is discussed in [70T2] and [73H2].	70C2

Physical property	Numerical value	Experimental conditions	Experimental method remarks	Ref.
YbFe$_{0.75}$Mo$_{0.25}$O$_3$ [80S]				
Structure: orthorhombic				
a	5.252 Å			
b	5.543 Å			
c	7.554 Å			
E_A	0.37 eV			
σ	$4.8 \cdot 10^{-5}\,\Omega^{-1}\,cm^{-1}$	$T = 50\,°C$, n-type	pressed polycrystalline pellet; temperature dependence of conductivity: Fig. 35	
LuFeO$_3$				
Structure: orthorhombic (D$_{2h}^{16}$—Pbnm)			crystal structure: Fig. 6	65E
a	5.213 Å		Complex polar Kerr effect:	65E
b	5.547 Å		Figs. 40, 41.	
c	7.765 Å		For absorption and Faraday rotation, see [70W2], [70A1] and [70T2], respectively; for reflection, see [72B2].	
LuFe$_{0.75}$Mo$_{0.25}$O$_3$ [80S]				
Structure: orthorhombic				
a	5.218 Å			
b	5.630 Å			
c	7.469 Å			
E_A	0.36 eV			
σ	$4 \cdot 10^{-5}\,\Omega^{-1}\,cm^{-1}$	$T = 50\,°C$, n-type	pressed polycrystalline pellet	

10.3.1.6 LnCoO₃ compounds

LaCoO$_3$				
Structure: $T < 375\,°C$: rhombohedral (D$_{3d}^6$—R$\bar{3}$c); $T > 375\,°C$: rhombohedral (C$_{3i}^2$—R$\bar{3}$)				67R2, 67M
a	5.436 Å		temperature dependence of lattice parameters: Fig. 50	67R2, 67M
α	60°48′			
			Localized-electron → collective-electron first order phase transition at 937 °C [67R2]; electronic state of Co ion [72B3].	
E_A	0.1 eV	$T \leqq 200$ K	Band scheme: Fig. 51.	72B1
	0.2 eV	200 K $< T$ < 400 K		
σ	$\approx 10\,\Omega^{-1}\,cm^{-1}$	p-type semi-conductor	temperature dependence of resistivity: Fig. 52	72B1, 64H
S	350 μV K^{-1}		temperature dependence of Seebeck coefficient: Fig. 53.	72B1, 64H
	570 μV K^{-1}		For thermal conductivity, see [64G2]. Optical absorption spectrum: Fig. 54; $^5A_1$-splitting of Co^{3+}: Fig. 55; XPS spectra of Co-2p: Fig. 56.	

Physical property	Numerical value	Experimental conditions	Experimental method remarks	Ref.
La$_{1-x}$Sr$_x$CoO$_3$ [62G]				
E_A	0.13 eV	$x = 0.002$	activation energy for charge	
	0.10 eV	0.005	carrier density; see also	
	0.07 eV	0.01	Fig. 57	
	0.04 eV	0.02		
σ	4 Ω^{-1} cm^{-1}	0.005, p-type	temperature dependence of	
	16 Ω^{-1} cm^{-1}	0.02, p-type	conductivity at various	
			Sr concentrations: Fig. 58	
			(see also Fig. 52 and [75B])	
			Seebeck coefficient as a	
			function of Sr content: Fig. 59	
			(see also Fig. 53); temperature	
			dependence of Seebeck	
			coefficient: Fig. 60.	
μ_p	0.28 cm^2 V^{-1} s^{-1}	$\leqq 0.05$	temperature dependence of	
			mobility: Fig. 63	
R_H	0.017 cm^3 C^{-1}	0.005	Hall coefficient as a function	
			of Sr content: Fig. 61;	
			temperature dependence of	
			Hall coefficient: Fig. 62	
La$_{1-x}$Th$_x$CoO$_3$ [62G]				
σ	≈ 1 Ω^{-1} cm^{-1}	$x = 0.02$, n-type	temperature dependence of	
	≈ 3 Ω^{-1} cm^{-1}	0.05, n-type	conductivity: Fig. 58	
μ_n	$2.1 \cdot 10^{-2}$ cm^2 V^{-1} s^{-1}	$x \leqq 0.05$		
			Seebeck coefficient as a	
			function of Th content: Fig. 59;	
			temperature dependence of	
			Seebeck coefficient: Fig. 60.	
LaCo$_{1-x}$Mn$_x$O$_3$ (see also LaMn$_{1-x}$Co$_x$O$_3$, subsection 10.3.1.4) **[66J]**				
σ	0.22 Ω^{-1} cm^{-1}	$x = 0.01$, p-type	polycrystalline sample;	
	1 Ω^{-1} cm^{-1}	0.12, n-type	resistivity as a function of	
			Mn content: Fig. 64	
S	380 μV K^{-1}	$x = 0.001$	Seebeck coefficient as a	
	-275 μV K^{-1}	0.005	function of Mn content: Fig. 65	
(La,Y)(Co,Mn)O$_3$ (see La$_{0.85}$Y$_{0.15}$Mn$_{1-x}$Co$_x$O$_3$, subsection 10.3.1.4)				
LaCo$_{0.75}$Mo$_{0.25}$O$_3$ [80S]				
Structure: orthorhombic				
a	5.554 Å			
b	5.590 Å			
c	7.958 Å			
E_A	0.22 eV			
σ	$8.3 \cdot 10^{-4}$ Ω^{-1} cm^{-1}	$T = 50$°C, n-type	pressed polycrystalline pellet	
			Temperature dependence of	
			Seebeck coefficient: Fig. 36.	
LaCo$_{0.75}$W$_{0.25}$O$_3$ [80S]				
Structure: orthorhombic				
a	5.586 Å			
b	5.628 Å			
c	7.983 Å			
E_A	0.33 eV			
σ	$1.2 \cdot 10^{-4}$ Ω^{-1} cm^{-1}	$T = 50$°C, n-type	pressed polycrystalline pellet	

Physical property	Numerical value	Experimental conditions	Experimental method remarks	Ref.

10.3.1.7 LnNiO$_3$ compounds

LaNiO$_3$
metallic [75O]

LaNi$_{0.75}$Mo$_{0.25}$O$_3$ [80S]
Structure: cubic (O$_h^5$—Fm3m)

$2a$	7.868 Å		suggestion of superlattice	
E_A	0.11 eV			
σ	0.63 Ω^{-1} cm^{-1}	$T = 50\,°C$, p-type		
			Temperature dependence of Seebeck coefficient: Fig. 36.	

LaNi$_{0.75}$W$_{0.25}$O$_3$ [80S]
Structure: cubic (O$_h^5$—Fm3m)

$2a$	7.884 Å		superlattice	
E_A	0.23 eV			
σ	$2.5 \cdot 10^{-5}$ Ω^{-1} cm^{-1}	$T = 50\,°C$, p-type		
			Temperature dependence of Seebeck coefficient: Fig.36.	

10.3.2 Other oxides with transition-metal elements

La$_2$NiO$_4$
Structure: tetragonal (D$_{4h}^{17}$—I4/mmm)

				58R
a	3.855 (1) Å			58R
c	12.652 (3) Å			
S	$-5\,\mu$V K^{-1}	$T = 500$ K	polycrystalline sample; temperature dependence of Seebeck coefficient: Fig. 2 Metallic above 500 K [73G]. Temperature dependence of resistivity: Fig. 1.	73G1

La$_2$(WO$_4$)$_3$
Structure: monoclinic (C$_{2h}^6$—A2/a)

				65N
a	11.65 Å			65N
b	11.83 Å			
c	7.89 Å			
β	109°47′			
T_m	1338 (10) K			65N
		T [K]		
E_A	2.24 eV	600···950, n-type	temperature dependence of conductivity: Fig. 3	79D
	2.40 eV	1050···1200, p-type		
μ_n	1.15···0.81 cm^2 V^{-1} s^{-1}	600···950	polycrystalline pellet Temperature dependence of Seebeck coefficient: Fig. 4.	79D

Physical property	Numerical value	Experimental conditions	Experimental method remarks	Ref.
$Ce_2(WO_4)_3$				
Structure: monoclinic (C_{2h}^6—A2/a)				65N
a	11.59 Å			65N
b	11.69 Å			
c	7.79 Å			
β	109° 38′			
T_m	1338 (10) K			65N
E_A	1.9 eV	$T = 600\cdots800$ K, n-type	temperature dependence of conductivity: Fig. 3	79D
μ_n	$1.3\cdots1.05$ cm^2 V^{-1} s^{-1}	$600\cdots800$ K	polycrystalline pellet Temperature dependence of Seebeck coefficient: Fig. 5.	79D
Pr_2CuO_4				
Structure: tetragonal (K_2NiF_4-type)				74G1
a	3.96 Å			65F3
c	12.23 Å			
E_A	0.36 eV	$T = 100\cdots392$ °C	temperature dependence of	74G1
	0.11 eV	$392\cdots900$ °C	conductivity: Fig. 6 For further transport properties, see [73G2] and [73K].	
$Pr_2(WO_4)_3$				
Structure: monoclinic (C_{2h}^6—A2/a)				65N
a	11.56 Å			65N
b	11.64 Å			
c	7.77 Å			
β	109° 39′			65N
T_m	1383 (10) K			
		T [K]		
E_A	2.12 eV	$600\cdots950$, n-type	temperature dependence of conductivity: Fig. 3	79D
	2.48 eV	$1000\cdots1200$, p-type		
μ_n	$0.32\cdots0.23$ cm^2 V^{-1} s^{-1}	$600\cdots950$	polycrystalline sample Temperature dependence of Seebeck coefficient: Fig. 7.	79D
Nd_2NiO_4				
Structure: monoclinic				57W1
tetragonal				61F
a	3.92 Å		monoclinic setting	57W1
b	6.16 Å			
c	3.77 Å			
β	92.4°			
a	3.81 Å		tetragonal setting	61F
c	12.31 Å			
S	-15 μV K^{-1}	$T = 500$ K	polycrystalline sample; metallic above 500 K Temperature dependence of resistivity: Fig. 1; temperature dependence of Seebeck coefficient: Fig. 2. For further transport data, see [73G2].	73G1

Physical property	Numerical value	Experimental conditions	Experimental method remarks	Ref.
Nd$_2$CuO$_4$				
Structure: tetragonal (K$_2$NiF$_4$-type)				73G1, 61F
a	3.94 Å			61F
c	12.15 Å			
E_A	0.28 eV	$T = 100 \cdots 478$ K	temperature dependence of	74G1
	0.82 eV	$478 \cdots 900$ K	conductivity: Figs. 6, 8	
			Temperature dependence of Seebeck coefficient: Fig. 9. For further transport data, see [73G2] and [73K].	
Nd$_2$(WO$_4$)$_3$				
Structure: monoclinic (C$_{2h}^6$—A2/a)				65N
a	11.51 Å			65N
b	11.59 Å			
c	7.75 Å			
β	109° 40′			
T_m	1408 (10) K			65N
		T [°C]		
E_A	2.50 eV	$600 \cdots 1000$, p-type	temperature dependence of conductivity: Fig. 3	79D
	2.60 eV	$1050 \cdots 1200$, p-type		
μ_n	$0.89 \cdots 0.42$ cm^2 V^{-1} s^{-1}	$600 \cdots 1000$	polycrystalline pellet	79D
	$0.99 \cdots 0.82$ cm^2 V^{-1} s^{-1}	$1050 \cdots 1200$		
μ_p	$1.13 \cdots 0.53$ cm^2 V^{-1} s^{-1}	$600 \cdots 1000$		79D
	$1.29 \cdots 1.06$ cm^2 V^{-1} s^{-1}	$1050 \cdots 1200$		
			Temperature dependence of Seebeck coefficient: Fig. 10.	
Sm$_2$CuO$_4$				
Structure: tetragonal (K$_2$NiF$_4$-type)				73G1, 61F
a	3.91 Å			61F
c	11.93 Å			
E_A	0.18 eV	$T = 100 \cdots 484$ K	temperature dpendence of	74G1
	0.82 eV	$484 \cdots 900$ K	conductivity: Figs. 6, 8 Temperature dependence of Seebeck coefficient: Fig. 9. For transport data, see also [73G2], [73K].	
Sm$_2$(WO$_4$)$_3$				
Structure: monoclinic (C$_{2h}^6$—A2/a)				65N
a	11.42 Å			65N
b	11.47 Å			
c	7.68 Å			
β	109° 39′			
T_m	1428 (10) K			65N
E_A	2 eV	$T = 600 \cdots 1200$ K, p-type	temperature dependence of conductivity: Fig. 3	79D
μ_p	$1.59 \cdots 0.95$ cm^2 V^{-1} s^{-1}	$T = 600 \cdots 1200$ K	polycrystalline pellet	79D
			Temperature dependence of Seebeck coefficient: Fig. 11.	

Physical property	Numerical value	Experimental conditions	Experimental method remarks	Ref.
Eu_2CuO_4				
Structure: tetragonal (K_2NiF_4-type)				73G1, 65F3
a	3.91 Å			65F3
c	11.92 Å			
E_A	0.14 eV	$T = 100 \cdots 547\,°C$	temperature dependence of	74G1
	0.83 eV	$547 \cdots 900\,°C$	conductivity: Fig. 6	
$EuNb_2O_6$				
Structure: tetragonal				75F
a	12.345 Å			75F
c	3.860 Å			
E_A	0.15 eV	$T = 723$ K		76S2
σ	$\approx 10^{-2}\,\Omega^{-1}\,cm^{-1}$		polycrystalline sample;	76S2
	$0.54\,\Omega^{-1}\,cm^{-1}$	723 K	temperature dependence of conductivity: Fig. 12	
$Eu_{1.2}Nb_2O_6$				
Structure: tetragonal (W bronze type)				75F
E_A	0.05 eV			76S2
	0.15 eV	$T = 723$ K		
σ	$3\,\Omega^{-1}\,cm^{-1}$		polycrystalline sample;	76S2
	$80\,\Omega^{-1}\,cm^{-1}$	723 K	temperature dependence of conductivity: Fig. 12	
$Eu_xSr_{1-x}Nb_4O_{11}$ [77S1]				
Structure: tetragonal				
a	52.50 Å	$x = 0$	lattice parameters vs.	
c	7.88 Å	0	composition x: Fig. 13	
a	52.60 Å	0.5		
c	7.65 Å	0.5		
a	52.53 Å	1		
c	7.65 Å	1		
E_A	0.41 eV	$x = 0$		
	0.43 eV	0.5		
	0.28 eV	1		
σ	$2.9 \cdot 10^{-4}\,\Omega^{-1}\,cm^{-1}$	$x = 0$, n-type	polycrystalline pellet;	
	$3.9 \cdot 10^{-3}\,\Omega^{-1}\,cm^{-1}$	0.5, n-type	temperature dependence of	
	$1.3 \cdot 10^{-2}\,\Omega^{-1}\,cm^{-1}$	1, n-type	conductivity at various composition x: Fig. 14	
$EuTa_2O_6$				
Structure: orthorhombic				75F
a	12.351 Å			75F
b	12.418 Å			
c	3.854 Å			
E_A	0.10 eV			76S2
	0.55 eV	$T = 723$ K		
σ	$10^{-7}\,\Omega^{-1}\,cm^{-1}$		polycrystalline sample;	76S2
	$10^{-4}\,\Omega^{-1}\,cm^{-1}$	723 K	temperature dependence of conductivity: Fig. 12	
$Eu_xSr_{1-x}Ta_4O_{11}$ [77S1]				
Structure: tetragonal				
a	52.65 Å	$x = 0$	lattice parameters vs.	
c	7.74 Å	0	composition: Fig. 15	

(continued)

Physical property	Numerical value	Experimental conditions	Experimental method remarks	Ref.
$Eu_xSr_{1-x}Ta_4O_{11}$ [77S1] (continued)				
a	52.70 Å	$x = 0.5$		
c	7.77 Å	0.5		
a	52.60 Å	1		
c	7.75 Å	1		
σ	$\approx 10^{-8}\,\Omega^{-1}\,cm^{-1}$		x not specified, polycrystalline pellet	
$EuWO_4$				
Structure: scheelite-type (C_{4h}^6)				71M2
a	5.411 Å			71M2
c	11.936 Å			
E_g	≈ 3.1 eV		from conductivity	74L
		T [K]		
E_A	0.08 eV	680		74L
	0.3 eV	725		
σ	$10^{-8}\,\Omega^{-1}\,cm^{-1}$	$< 500, \perp c$	temperature dependence of conductivity: Fig. 16	74L
μ_{pol}	$10^{-2}\,cm^2\,V^{-1}\,s^{-1}$		polaron mobility	74L
m^{**}	$\approx 100\,m_o$	> 725	polaronic mass	74L
$\varepsilon(0)$	7.3	$f = 450$ kHz	temperature dependence of dielectric constant: Fig. 17	74L
$\varepsilon(\infty)$	3.24			
$Eu_2(WO_4)_3$				
Structure: monoclinic (C_{2h}^6—A2/a)				65N
a	11.40 Å			65N
b	11.43 Å			
c	7.65 Å			
β	109° 37′			
T_m	1438 (10) K			65N
		T [K]		
E_A	1.84 eV	600···790, p-type	temperature dependence of conductivity: Figs. 3, 18, 19	79D
	2.40 eV	900···1200, p-type		
μ_p	5.70··· 4.64 $cm^2\,V^{-1}\,s^{-1}$	600···790	polycrystalline sample	79D
m^{**}	418.4 m_o		polaronic mass	76L
$\varepsilon(0)$	17.5 (5)	$\parallel c$	temperature dependence of dielectric constant: Fig. 20	76L
$\varepsilon(\infty)$	5.0 (5)	$\parallel c$		
Gd_2CuO_4				
Structure: tetragonal (K_2NiF_4-type)				74G1, 61F
a	3.89 Å			61F
c	11.85 Å			
		T [°C]		
E_A	0.90 eV	100···287	temperature dependence of conductivity: Fig. 6	74G1
	0.40 eV	287···582		
	0.80 eV	582···900		
			For transport data, see also [73K].	

Physical property	Numerical value	Experimental conditions	Experimental method remarks	Ref.
$Gd_2(WO_4)_3$				
Structure: orthorhombic (F2/d)				65N
a	7.67 Å			65N
b	11.41 Å			
c	21.44 Å			
T_m	1503 (10) K			65N
E_A	2.36 eV	$T = 600 \cdots 1200$ K	polycrystalline sample; temperature dependence of conductivity: Fig. 3	79D
μ_n	$0.14 \cdots 0.05$ cm^2 V^{-1} s^{-1}			79D
μ_p	$0.46 \cdots 0.16$ cm^2 V^{-1} s^{-1}			79D
			Temperature dependence of Seebeck coefficient: Fig. 21.	
$Gd_2(W_{2/3}V_{4/3})O_7$ [79S]				
Structure: cubic (O_h^7—Fd3m)				
a	10.234 Å			
E_A	0.18 eV			
σ	10^{-3} Ω^{-1} cm^{-1}	p-type	polycrystalline sample; temperature dependence of conductivity: Fig. 22	
S	62 μV K^{-1}			
$Tb_2(W_{2/3}V_{4/3})O_7$ [79S]				
Structure: cubic (O_h^7—Fd3m)				
a	10.207 Å			
E_A	0.21 eV			
σ	$4.8 \cdot 10^{-4}$ Ω^{-1} cm^{-1}	p-type	polycrystalline sample	
S	81 μV K^{-1}			
$Dy_2(W_{2/3}V_{4/3})O_7$ [79S]				
Structure: cubic (O_h^7—Fd3m)				
a	10.191 Å			
E_A	0.20 eV			
σ	$5.3 \cdot 10^{-4}$ Ω^{-1} cm^{-1}	p-type	polycrystalline sample	
S	83 μV K^{-1}		temperature dependence of Seebeck coefficient: Fig. 22	
$Ho_2(W_{2/3}V_{4/3})O_7$ [79S]				
Structure: cubic (O_h^7—Fd3m)				
a	10.165 Å			
E_A	0.23 eV			
σ	$3.4 \cdot 10^{-4}$ Ω^{-1} cm^{-1}	p-type	polycrystalline sample	
S	98 μV K^{-1}			
$Er_2(W_{2/3}V_{4/3})O_7$ [79S]				
Structure: cubic (O_h^7—Fd3m)				
a	10.137 Å			
E_A	0.24 eV			
σ	$3.6 \cdot 10^{-4}$ Ω^{-1} cm^{-1}	p-type	polycrystalline sample	
S	115 μV K^{-1}			

Physical property	Numerical value	Experimental conditions	Experimental method remarks	Ref.
$(Er_xYb_{1-x})_2V_2O_7$ [77S2]				
Structure: cubic (O_h^7—Fd3m)			Lattice constant vs. composition x: Fig. 23.	
E_A	0.032 eV	$T = 90\cdots273$ K, x = 0.3		
	0.42 eV	$T = 273\cdots350$ K, x = 0.3		
σ	$5.62 \cdot 10^{-4}\,\Omega^{-1}\,cm^{-1}$	n-type, x = 0.3	polycrystalline sample	
$Tm_2V_2O_7$ [77S2]				
Structure: cubic (O_h^7—Fd3m)				
a	9.9575 Å			
E_A	0.042 eV	$T = 90\cdots273$ K		
	0.44 eV	$273\cdots350$ K		
σ	$1.09 \cdot 10^{-3}\,\Omega^{-1}\,cm^{-1}$	n-type	polycrystalline sample; temperature dependence of conductivity: Fig. 24	
$Tm_2V_{4/3}W_{2/3}O_7$ [79S]				
Structure: cubic (O_h^7—Fd3m)				
a	10.118 Å			
E_A	0.24 eV			
σ	$3.4 \cdot 10^{-4}\,\Omega^{-1}\,cm^{-1}$	p-type	polycrystalline sample	
S	103 µV K^{-1}			
$Yb_2V_2O_7$ [77S2]				
Structure: cubic (O_h^7—Fd3m)				
a	9.9346 Å			
E_A	0.047 eV	$T = 90\cdots273$ K	polycrystalline sample; temperature dependence of conductivity: Fig. 24; n-type conductivity	
	0.44 eV	$273\cdots350$ K		
$Yb_2V_{4/3}W_{2/3}O_7$ [79S]				
Structure: cubic (O_h^7—Fd3m)				
a	10.093 Å			
E_A	0.24 eV			
σ	$2.3 \cdot 10^{-4}\,\Omega^{-1}\,cm^{-1}$	p-type	polycrystalline sample; temperature dependence of conductivity and Seebeck coefficient: Fig. 22	
S	126 µV K^{-1}			
$Lu_2V_2O_7$ [77S2]				
Structure: cubic (O_h^7—Fd3m)				
a	9.9231 Å			
E_A	0.043 eV	$T = 90\cdots273$ K		
	0.42 eV	$273\cdots350$ K		
σ	$6.37 \cdot 10^{-4}\,\Omega^{-1}\,cm^{-1}$	n-type	polycrystalline sample; temperature dependence of conductivity: Fig. 24	

Physical property	Numerical value	Experimental conditions	Experimental method remarks	Ref.

10.3.3 RE oxynitrides

$EuO_{1-x}N_x$ [80E]
Structure: cubic (NaCl-type)

| | | | Lattice parameter vs. composition x: Fig. 1; temperature dependence of lattice parameter: Fig. 2; Mössbauer spectra: Fig. 3; temperature dependence of conductivity: Fig. 4; temperature dependence of Seebeck coefficient: Fig. 5; density of states: Fig. 6; variation of activation energy vs. composition x: Fig. 7. | |

$Eu_{1-x}Nd_xO_{1-x}N_x$
Structure: cubic (NaCl-type)

E_A	0.82 eV	x = 0.05	Lattice parameter vs. composition x: Figs. 1, 8; temperature dependence of lattice parameter: Fig. 9; temperature dependence of conductivity: Fig. 10; temperature dependence of Seebeck coefficient: Fig. 11; variation of activation energy vs. composition x: Fig. 7, density of states: Fig. 6.	80E
	0.18 eV	0.18		79C

$Eu_{1-x}Gd_xO_{1-x}N_x$
Structure: cubic (NaCl-type)

E_A	0.3 eV	x = 0.25	Lattice parameter vs. composition x: Fig. 1; temperature dependence of lattice parameter: Fig. 12; temperature dependence of conductivity: Fig. 13; variation of activation energy vs. composition x: Fig. 7; density of states: Fig. 6.	80E
				79C

Physical property	Numerical value	Experimental conditions	Experimental method remarks	Ref.

10.3.4 RE sulfides, selenides, tellurides
(containing elements of groups I a, b; II a, b; III a, IV a, b)

10.3.4.1 Compounds with I a elements

LiHoS$_2$

Structure: rhombohedral (NaFeO$_2$-type)				79F
a	3.898 Å			81R
c	18.68 Å			

LiErS$_2$

Structure: rhombohedral (NaFeO$_2$-type)				79F
a	3.875 Å			81R
c	18.63 Å			

LiYbS$_2$

Structure: rhombohedral				79F
a	3.842 Å			81R
c	18.54 Å			

NaLaS$_2$

Structure: cubic (NaCl-type)				79F
a	5.881 Å			81R

NaCeS$_2$

Structure: cubic (NaCl-type)				79F
a	5.832 Å			81R

NaPrS$_2$

Structure: cubic (NaCl-type)				79F
a	5.770 Å			81R

NaNdS$_2$

Structure: cubic (NaCl-type)				79F
a	5.803 Å			81R

NaSmS$_2$

Structure: rhombohedral (NaFeO$_2$-type)				79F
a	4.056 Å			81R
c	19.87 Å			

NaEuS$_2$

Structure: rhombohedral (NaFeO$_2$-type)				79F
a	4.042 Å			81R
c	19.92 Å			

NaGdS$_2$

Structure: rhombohedral (NaFeO$_2$-type)				79F
a	4.009 Å			81R
c	19.87 Å			

NaTbS$_2$

Structure: rhombohedral (NaFeO$_2$-type)				79F
a	3.989 Å			81R
c	19.87 Å			

Physical property	Numerical value	Experimental conditions	Experimental method remarks	Ref.
NaDyS$_2$				
Structure: rhombohedral (NaFeO$_2$-type)				79F
a	3.978 Å			81R
c	19.92 Å			
NaHoS$_2$				
Structure: rhombohedral (NaFeO$_2$-type)				79F
a	3.945 Å			81R
c	19.86 Å			
NaErS$_2$				
Structure: rhombohedral (NaFeO$_2$-type)				79F
a	3.939 Å			81R
c	19.98 Å			
NaYS$_2$				
Structure: rhombohedral (NaFeO$_2$-type)				79F
a	3.968 Å			81R
c	19.89 Å			
NaLaSe$_2$				
Structure: rhombohedral (NaFeO$_2$-type)				79F
a	4.348 Å			81R
c	20.79 Å			
NaCeSe$_2$				
Structure: rhombohedral (NaFeO$_2$-type)				79F
a	4.303 Å			81R
c	20.76			
NaPrSe$_2$				
Structure: rhombohedral (NaFeO$_2$-type)				79F
a	4.268 Å			81R
c	20.78 Å			
NaNdSe$_2$				
Structure: rhombohedral (NaFeO$_2$-type)				79F
a	4.250 Å			81R
c	20.82 Å			
NaSmSe$_2$				
Structure: rhombohedral (NaFeO$_2$-type)				79F
a	4.202 Å			81R
c	20.80 Å			
NaEuSe$_2$				
Structure: rhombohedral (NaFeO$_2$-type)				79F
a	4.211 Å			81R
c	20.88 Å			
NaGdSe$_2$				
Structure: rhombohedral (NaFeO$_2$-type)				79F
a	4.166 Å			81R
c	20.82 Å			

Physical property	Numerical value	Experimental conditions	Experimental method remarks	Ref.
NaTbSe$_2$				
Structure: rhombohedral (NaFeO$_2$-type)				79F
a	4.142 Å			81R
c	20.80 Å			
NaDySe$_2$				
Structure: rhombohedral (NaFeO$_2$-type)				79F
a	4.124 Å			81R
c	20.83 Å			
NaHoSe$_2$				
Structure: rhombohedral (NaFeO$_2$-type)				79F
a	4.107 Å			81R
c	20.83 Å			
NaErSe$_2$				
Structure: rhombohedral (NaFeO$_2$-type)				79F
a	4.091 Å			81R
c	20.80 Å			
NaYSe$_2$				
Structure: rhombohedral (NaFeO$_2$-type)				79F
a	4.118 Å			81R
c	20.82 Å			
KLaS$_2$				
Structure: rhombohedral (NaFeO$_2$-type)				79F
a	4.464 Å			81R
c	21.895 Å			
KCeS$_2$				
Structure: rhombohedral (NaFeO$_2$-type)				79F
a	4.223 Å			81R
c	21.80 Å			
KPrS$_2$				
Structure: rhombohedral (NaFeO$_2$-type)				79F
a	4.185 Å			81R
c	21.75 Å			
KNdS$_2$				
Structure: rhombohedral (NaFeO$_2$-type)				79F
a	4.160 Å			81R
c	21.83 Å			
KSmS$_2$				
Structure: rhombohedral (NaFeO$_2$-type)				79F
a	4.107 Å			81R
c	21.76 Å			
KEuS$_2$				
Structure: rhombohedral (NaFeO$_2$-type)				79F
a	4.093 Å			81R
c	21.85 Å			

Huber, Scharmer

Physical property	Numerical value	Experimental conditions	Experimental method remarks	Ref.
KGdS$_2$				
Structure: rhombohedral (NaFeO$_2$-type)				79F
a	4.075 Å			81R
c	21.89 Å			
KTbS$_2$				
Structure: rhombohedral (NaFeO$_2$-type)				79F
a	4.051 Å			81R
c	21.87 Å			
KDyS$_2$				
Structure: rhombohedral (NaFeO$_2$-type)				79F
a	4.030 Å			81R
c	21.83 Å			
KYS$_2$				
Structure: rhombohedral (NaFeO$_2$-type)				79F
a	4.0225 Å			81R
c	21.856 Å			
KHoS$_2$				
Structure: rhombohedral (NaFeO$_2$-type)				79F
a	4.0095 Å			81R
c	21.80 Å			
KErS$_2$				
Structure: rhombohedral (NaFeO$_2$-type)				79F
a	3.993 Å			81R
c	21.77 Å			
KYbS$_2$				
Structure: rhombohedral (NaFeO$_2$-type)				79F
a	3.964 Å			81R
c	21.82 Å			
RbLaS$_2$				
Structure: rhombohedral (NaFeO$_2$-type)				79F
a	4.292 Å			81R
c	22.89 Å			
RbCeS$_2$				
Structure: rhombohedral (NaFeO$_2$-type)				79F
a	4.246 Å			81R
c	22.80 Å			
RbPrS$_2$				
Structure: rhombohedral (NaFeO$_2$-type)				79F
a	4.222 Å			81R
c	22.87 Å			
RbNdS$_2$				
Structure: rhombohedral (NaFeO$_2$-type)				79F
a	4.189 Å			81R
c	22.89 Å			

Physical property	Numerical value	Experimental conditions	Experimental method remarks	Ref.
RbSmS$_2$				
Structure: rhombohedral (NaFeO$_2$-type)				79F
a	4.141 Å			81R
c	22.86 Å			
RbEuS$_2$				
Structure: rhombohedral (NaFeO$_2$-type)				79F
a	4.119 Å			81R
c	22.84 Å			
RbGdS$_2$				
Structure: rhombohedral (NaFeO$_2$-type)				79F
a	4.098 Å			81R
c	22.88 Å			
RbTbS$_2$				
Structure: rhombohedral (NaFeO$_2$-type)				79F
a	4.070 Å			81R
c	22.80 Å			
CsLaS$_2$				
Structure: rhombohedral (NaFeO$_2$-type)				79F
a	4.306 Å			81R
c	24.08 Å			
CsCeS$_2$				
Structure: rhombohedral (NaFeO$_2$-type)				79F
a	4.262 Å			81R
c	23.99 Å			

10.3.4.2 Compounds with IIa elements

Physical property	Numerical value	Experimental conditions	Experimental method remarks	Ref.
CaLa$_2$S$_4$				
Structure: cubic (Th$_3$P$_4$-type)			Raman spectrum: Fig. 1.	73Y
a	8.687 Å			81R
CaCe$_2$S$_4$				
Structure: cubic (Th$_3$P$_4$-type)			Green-yellow cathodoluminescence: Fig. 2.	73Y
a	8.615 Å			81R
CaPr$_2$S$_4$				
Structure: cubic (Th$_3$P$_4$-type)			Raman spectrum: Fig. 1.	73Y
a	8.578 Å			81R
CaNd$_2$S$_4$				
Structure: cubic (Th$_3$P$_4$-type)			Raman spectrum: Fig. 1.	73Y
a	8.533 Å			81R
CaSm$_2$S$_4$				
Structure: cubic (Th$_3$P$_4$-type)			Raman spectrum: Fig. 1.	73Y
a	8.578 Å			81R
CaGd$_2$S$_4$				
Structure: cubic (Th$_3$P$_4$-type)			Raman spectrum: Fig. 1.	73Y
a	8.423 Å			81R

Physical property	Numerical value	Experimental conditions	Experimental method remarks	Ref.
$CaTb_2S_4$				
Structure: cubic (Th_3P_4-type)				73Y
a	8.400 Å			81R
$CaDy_2S_4$				
Structure: cubic (Th_3P_4-type)				73Y
a	8.376 Å			81R
$CaHo_2S_4$				
Structure: orthorhombic				65F2
a	12.90 Å			81R
b	13.04 Å			
c	3.86 Å			
d	5.10 g cm^{-3}			81R
$CaEr_2S_4$				
Structure: orthorhombic				65F2
a	12.87 Å			81R
b	13.01 Å			
c	3.85 Å			
d	5.18 g cm^{-3}			81R
$CaTm_2S_4$				
Structure: orthorhombic				65F2
a	12.85 Å			81R
b	12.98 Å			
c	3.84 Å			
d	5.25 g cm^{-3}			81R
$CaYb_2S_4$				
Structure: orthorhombic				65F2
a	12.82 Å			81R
b	12.95 Å			
c	3.83 Å			
d	5.37 g cm^{-3}			81R
$CaLu_2S_4$				
Structure: orthorhombic			Dark yellow electroluminescence	65F2
a	12.82 Å		was reported.	73Y
b	12.95 Å			
c	3.83 Å			
d	5.41 g cm^{-3}			81R
CaY_2Se_4				
Structure: rhombohedral (Yb_3Se_4-type)				64G1
a	8.29 Å			81R
α	59° 16′			
d	4.43 g cm^{-3}		pycnometric density	81R
	4.47 g cm^{-3}		X-ray	
$CaHo_2Se_4$				
Structure: rhombohedral				64G1
a	8.30 Å			81R
α	59° 02′			
d	5.75 g cm^{-3}		from X-ray measurements	81R

Physical property	Numerical value	Experimental conditions	Experimental method remarks	Ref.
$CaEr_2Se_4$				
Structure: rhombohedral				64G1
a	8.28 Å			81R
α	58°54′			
d	5.75 g cm^{-3}		pycnometric density	81R
	5.86 g cm^{-3}		X-ray	
$CaTm_2Se_4$				
Structure: rhombohedral				64G1
a	8.27 Å			81R
α	58°42′			
d	5.99 g cm^{-3}		from X-ray measurements	81R
$CaYb_2Se_4$				
Structure: rhombohedral				64G1
a	8.264 Å			81R
α	58°31′			
d	5.82 g cm^{-3}		pycnometric density	81R
	6.04 g cm^{-3}		X-ray	
$CaLu_2Se_4$				
Structure: rhombohedral				64G1
a	8.261 Å			81R
α	58°24′			
d	6.10 g cm^{-3}		from X-ray measurements	81R
$CaDy_2Te_4$				
Structure: rhombohedral				69P
a	7.673 Å			81R
α	33°02′			
d	6.08 g cm^{-3}			81R
$CaHo_2Te_4$				
Structure: rhombohedral				69P
a	7.680 Å			81R
α	32°54′			
d	6.15 g cm^{-3}			81R
$CaEr_2Te_4$				
Structure: rhombohedral				69P
a	7.673 Å			81R
α	32°50′			
d	6.22 g cm^{-3}			81R
$CaTm_2Te_4$				
Structure: rhombohedral				69P
a	7.677 Å			81R
α	32°44′			
d	6.27 g cm^{-3}			81R
$CaLu_2Te_4$				
Structure: rhombohedral				69P
a	7.677 Å			81R
α	32°38′			
d	6.42 g cm^{-3}			81R

Physical property	Numerical value	Experimental conditions	Experimental method remarks	Ref.
CaY_2Te_4				
Structure: rhombohedral				69P
a	7.676 Å			81R
α	32° 56′			
d	4.87 g cm^{-3}			81R
$SrLa_2S_4$				
Structure: cubic (Th_3P_4-type)			Raman spectrum: Fig. 3.	79F
a	8.790 Å			81R
$SrCe_2S_4$				
Structure: cubic (Th_3P_4-type)				79F
a	8.718 Å			81R
$SrPr_2S_4$				
Structure: cubic (Th_3P_4-type)			Raman spectrum: Fig. 3.	79F
a	8.682 Å			81R
$SrNd_2S_4$				
Structure: cubic (Th_3P_4-type)			Raman spectrum: Fig. 3;	79F
a	8.649 Å		complex dielectric constant: Fig. 4; reflectance curve: Fig. 5.	81R
$SrSm_2S_4$				
Structure: cubic (Th_3P_4-type)			Raman spectrum: Fig. 3.	79F
a	8.595 Å			81R
$SrGd_2S_4$				
Structure: cubic (Th_3P_4-type)				79F
a	8.556 Å			81R
$SrTb_2S_4$				
Structure: orthorhombic ($CaFe_2O_4$-type)				79F
a	11.96 Å			81R
b	14.35 Å			
c	4.01 Å			
d	5.16 g cm^{-3}		X-ray density	81R
$SrDy_2S_4$				
Structure: orthorhombic ($CaFe_2O_4$-type)				79F
a	11.91 Å			81R
b	14.29 Å			
c	3.99 Å			
d	5.29 g cm^{-3}		X-ray density	81R
$SrHo_2S_4$				
Structure: orthorhombic ($CaFe_2O_4$-type)				79F
a	11.89 Å			81R
b	14.27 Å			
c	3.98 Å			
d	5.37 g cm^{-3}		X-ray density	81R

Physical property	Numerical value	Experimental conditions	Experimental method remarks	Ref.
SrEr$_2$S$_4$				
Structure: orthorhombic (CaFe$_2$O$_4$-type)				79F
a	11.84 Å			81R
b	14.22 Å			
c	3.97 Å			
d	5.47 g cm^{-3}		X-ray density	81R
SrTm$_2$S$_4$				
Structure: orthorhombic (CaFe$_2$O$_4$-type)				79F
a	11.80 Å			81R
b	14.17 Å			
c	3.94 Å			
d	5.58 g cm^{-3}		X-ray density	81R
SrYb$_2$S$_4$				
Structure: orthorhombic (CaFe$_2$O$_4$-type)				79F
a	11.76 Å			81R
b	14.13 Å			
c	3.90 Å			
d	5.76 g cm^{-3}		X-ray density	81R
SrLu$_2$S$_4$				
Structure: orthorhombic (CaFe$_2$O$_4$-type)				79F
a	11.71 Å			81R
b	14.09 Å			
c	3.90 Å			
d	5.84 g cm^{-3}		X-ray density	81R
SrY$_2$S$_4$				
Structure: orthorhombic				79F
a	11.97 Å			81R
b	14.34 Å			
c	3.99 Å			
d	3.77 g cm^{-3}		pycnometric density	81R
	3.82 g cm^{-3}		X-ray	
SrLa$_2$Se$_4$				
Structure: cubic (Th$_3$P$_4$-type)				64G1
a	9.124 Å			81R
SrCe$_2$Se$_4$				
Structure: cubic (Th$_3$P$_4$-type)				64G1
a	9.060 Å			81R
SrPr$_2$Se$_4$				
Structure: cubic (Th$_3$P$_4$-type)				64G
a	9.019 Å			81R
SrNd$_2$Se$_4$				
Structure: cubic (Th$_3$P$_4$-type)				64G1
a	8.989 Å			81R
SrSm$_2$Se$_4$				
Structure: cubic (Th$_3$P$_4$-type)				64G1
a	8.931 Å			81R

Physical property	Numerical value	Experimental conditions	Experimental method remarks	Ref.
SrGd$_2$Se$_4$				
Structure: cubic (Th$_3$P$_4$-type)				64G1
a	8.895 Å			81R
SrTb$_2$Se$_4$				
Structure: orthorhombic (CaFe$_2$O$_4$-type)				64G1
a	12.50 Å			81R
b	14.97 Å			
c	4.17 Å			
d	5.94 g cm^{-3}		pycnometric density	81R
	6.14 g cm^{-3}		X-ray	
SrDy$_2$Se$_4$				
Structure: orthorhombic (CaFe$_2$O$_4$-type)				64G1
a	12.49 Å			81R
b	14.92 Å			
c	4.15 Å			
d	6.05 g cm^{-3}		pycnometric density	81R
	6.25 g cm^{-3}		X-ray	
SrEr$_2$Se$_4$				
Structure: orthorhombic (CaFe$_2$O$_4$-type)				64G1
a	12.48 Å			81R
b	14.81 Å			
c	4.12 Å			
d	6.24 g cm^{-3}		pycnometric density	81R
	6.44 g cm^{-3}		X-ray	
SrYb$_2$Se$_4$				
Structure: orthorhombic (CaFe$_2$O$_4$-type)				64G1
a	12.47 Å			81R
b	14.71 Å			
c	4.08 Å			
d	6.53 g cm^{-3}		pycnometric density	81R
	6.65 g cm^{-3}		X-ray	
SrLu$_2$Se$_4$				
Structure: orthorhombic (CaFe$_2$O$_4$-type)				64G1
a	12.76 Å			81R
b	14.67 Å			
c	4.06 Å			
d	6.60 g cm^{-3}		pycnometric density	81R
	6.74 g cm^{-3}		X-ray	
SrY$_2$Se$_4$				
Structure: orthorhombic (CaFe$_2$O$_4$-type)				64G1
a	12.49 Å			81R
b	14.88 Å			
c	4.15 Å			
d	4.75 g cm^{-3}		pycnometric density	81R
	5.00 g cm^{-3}		X-ray	
BaLa$_2$S$_4$				
Structure: cubic (Th$_3$P$_4$-type)				79F
a	8.917 Å			81R

Physical property	Numerical value	Experimental conditions	Experimental method remarks	Ref.
BaCe$_2$S$_4$				
Structure: cubic (Th$_3$P$_4$-type)				79F
a	8.864 Å			81R
BaPr$_2$S$_4$				
Structure: cubic (Th$_3$P$_4$-type)				79F
a	8.817 Å			81R
BaNd$_2$S$_4$				
Structure: cubic (Th$_3$P$_4$-type)				79F
a	8.793 Å			81R
BaNd$_2$S$_4$				
Structure: orthorhombic (CaFe$_2$O$_4$-type)				79F
a	12.29 Å			81R
b	14.78 Å			
c	4.14 Å			
d	4.83 g cm^{-3}		from X-ray measurements	81R
BaSm$_2$S$_4$				
Structure: orthorhombic (CaFe$_2$O$_4$-type)				79F
a	12.24 Å			81R
b	14.73 Å			
c	4.12 Å			
d	4.92 g cm^{-3}		pycnometric density	81R
	4.98 g cm^{-3}		X-ray	
BaGd$_2$S$_4$				
Structure: orthorhombic (CaFe$_2$O$_4$-type)				79F
a	12.21 Å			81R
b	14.65 Å			
c	4.08 Å			
d	5.13 g cm^{-3}		pycnometric density	81R
	5.19 g cm^{-3}		X-ray	
BaTb$_2$S$_4$				
Structure: orthorhombic (CaFe$_2$O$_4$-type)				79F
a	12.16 Å			81R
b	14.55 Å			
c	4.06 Å			
d	5.32 g cm^{-3}		from X-ray measurements	81R
BaDy$_2$S$_4$				
Structure: orthorhombic (CaFe$_2$O$_4$-type)				79F
a	12.09 Å			81R
b	14.50 Å			
c	4.04 Å			
d	5.46 g cm^{-3}		pycnometric density	81R
	5.48 g cm^{-3}		X-ray	
BaHo$_2$S$_4$				
Structure: orthorhombic (CaFe$_2$O$_4$-type)				79F
a	12.07 Å			81R
b	14.47 Å			
c	4.02 Å			
d	5.50 g cm^{-3}		from X-ray measurements	81R

Huber, Scharmer

Physical property	Numerical value	Experimental conditions	Experimental method remarks	Ref.
$BaEr_2S_4$				
Structure: orthorhombic ($CaFe_2O_4$-type)				79F
a	11.99 Å			81R
b	14.42 Å			
c	3.99 Å			
d	5.82 g cm^{-3}		pycnometric density	81R
	5.58 g cm^{-3}		X-ray	
$BaTm_2S_4$				
Structure: orthorhombic ($CaFe_2O_4$-type)				79F
a	11.96 Å			81R
b	14.40 Å			
c	3.98 Å			
d	5.75 g cm^{-3}		from X-ray measurements	81R
$BaYb_2S_4$				
Structure: orthorhombic ($CaFe_2O_4$-type)				79F
a	11.91 Å			81R
b	14.33 Å			
c	3.96 Å			
d	5.90 g cm^{-3}		pycnometric density	81R
	5.94 g cm^{-3}		X-ray	
$BaLu_2S_4$				
Structure: orthorhombic ($CaFe_2O_4$-type)				79F
a	11.90 Å			81R
b	14.32 Å			
c	3.96 Å			
d	5.99 g cm^{-3}		from X-ray measurements	81R
BaY_2S_4				
Structure: orthorhombic ($CaFe_2O_4$-type)				79F
a	12.16 Å			81R
b	14.56 Å			
c	4.06 Å			
d	4.10 g cm^{-3}		pycnometric density	81R
	4.03 g cm^{-3}		X-ray	
$BaLa_2Se_4$				
Structure: cubic (Th_3P_4-type)				64G1
a	9.258 Å			81R
$BaCe_2Se_4$				
Structure: cubic (Th_3P_4-type)				64G1
a	9.186 Å			81R
$BaPr_2Se_4$				
Structure: cubic (Th_3P_4-type)				64G1
a	9.150 Å			81R
$BaNd_2Se_4$				
Structure: cubic (Th_3P_4-type)				64G1
a	9.120 Å			81R

Physical property	Numerical value	Experimental conditions	Experimental method remarks	Ref.
BaSm$_2$Se$_4$				
Structure: orthorhombic (CaFe$_2$O$_4$-type)				64G1
a	12.75 Å			81R
b	14.32 Å			
c	4.26 Å			
d	6.00 g cm^{-3}		pycnometric density	81R
	6.02 g cm^{-3}		X-ray	
BaGd$_2$Se$_4$				
Structure: orthorhombic (CaFe$_2$O$_4$-type)				64G1
a	12.74 Å			81R
b	15.20 Å			
c	4.24 Å			
d	6.11 g cm^{-3}		pycnometric density	81R
	6.21 g cm^{-3}		X-ray	
BaDy$_2$Se$_4$				
Structure: orthorhombic (CaFe$_2$O$_4$-type)				64G1
a	12.73 Å			81R
b	15.11 Å			
c	4.22 Å			
d	6.63 g cm^{-3}		pycnometric density	81R
	6.77 g cm^{-3}		X-ray	
BaEr$_2$Se$_4$				
Structure: orthorhombic (CaFe$_2$O$_4$-type)				64G1
a	12.69 Å			81R
b	14.99 Å			
c	4.18 Å			
d	6.48 g cm^{-3}		pycnometric density	81R
	6.58 g cm^{-3}		X-ray	
BaYb$_2$Se$_4$				
Structure: orthorhombic (CaFe$_2$O$_4$-type)				64G1
a	12.59 Å			81R
b	14.87 Å			
c	4.13 Å			
d	6.83 g cm^{-3}		pycnometric density	81R
	6.86 g cm^{-3}		X-ray	
BaLu$_2$Se$_4$				
Structure: orthorhombic (CaFe$_2$O$_4$-type)				64G1
a	12.59 Å			81R
b	14.82 Å			
c	4.10 Å			
d	6.97 g cm^{-3}		from X-ray measurements	81R
BaY$_2$Se$_4$				
Structure: orthorhombic (CaFe$_2$O$_4$-type)				64G1
a	12.72 Å			81R
b	15.10 Å			
c	4.22 Å			
d	5.23 g cm^{-3}		pycnometric density	81R
	5.17 g cm^{-3}		X-ray	

Physical property	Numerical value	Experimental conditions	Experimental method remarks	Ref.

10.3.4.3 Compounds with IVb elements

ZrLa$_2$S$_5$

Structure: orthorhombic (D$_{2h}^{16}$—Pnam)

				74D1, 81R
a	11.4864 (5) Å			74D1,
b	7.3894 (3) Å			81R
c	8.2167 (5) Å			
d	5.05 g cm^{-3}			74D1
E_A	1.4 eV	$T = 298 \cdots 667$ K		74D1
σ	$10^{-15}\,\Omega^{-1}\,\mathrm{cm}^{-1}$		polycrystalline sample	74D1

ZrSm$_2$S$_5$ [74D1]

Structure: orthorhombic (D$_{2h}^{16}$—Pnam)

a	11.491 (1) Å
b	7.2848 (5) Å
c	7.9092 (9) Å

ZrHo$_2$S$_5$ [74D1]

Structure: orthorhombic (D$_{2h}^{16}$—Pnam)

a	11.4781 (8) Å
b	7.211 (1) Å
c	7.7154 (5) Å

ZrEr$_2$S$_5$ [74D1]

Structure: orthorhombic (D$_{2h}^{16}$—Pnam)

a	11.4664 (7) Å
b	7.1958 (4) Å
c	7.6810 (6) Å

ZrLa$_2$Se$_5$ [74D1, 81R]

Structure: orthorhombic (D$_{2h}^{16}$—Pnam)

a	12.058 (3) Å
b	7.719 (2) Å
c	8.505 (2) Å

ZrSm$_2$Se$_5$ [74D1]

Structure: orthorhombic (D$_{2h}^{16}$—Pnam)

a	12.026 (2) Å
b	7.607 (1) Å
c	8.203 (1) Å

ZrGd$_2$Se$_5$ [74D1]

Structure: orthorhombic (D$_{2h}^{16}$—Pnam)

a	12.064 (8) Å
b	7.6028 (5) Å
c	8.1379 (7) Å

ZrTb$_2$Se$_5$ [74D1]

Structure: orthorhombic (D$_{2h}^{16}$—Pnam)

a	12.020 (4) Å
b	7.604 (3) Å
c	8.138 (3) Å

Physical property	Numerical value	Experimental conditions	Experimental method remarks	Ref.

HfCe$_2$S$_5$ [74D1]

Structure: orthorhombic (D$_{2h}^{16}$—Pnam)

a	11.4177 (6) Å			
b	7.3389 (4) Å			
c	8.1277 (5) Å			
E_A	0.5 eV	$T = 298 \cdots 400$ K		
σ	$< 10^{-10}\ \Omega^{-1}\ \mathrm{cm}^{-1}$			

HfSm$_2$S$_5$ [74D1]

Structure: orthorhombic (D$_{2h}^{16}$—Pnam)

a	11.4340 (6) Å			
b	7.2724 (4) Å			
c	7.8964 (4) Å			
d	6.45 g cm^{-3}			
E_A	0.05 eV	$T = 77 \cdots 400$ K		
σ	$1.6\ \Omega^{-1}\ \mathrm{cm}^{-1}$			

HfHo$_2$S$_5$ [74D1]

Structure: orthorhombic (D$_{2h}^{16}$—Pnam)

a	11.415 (3) Å
b	7.204 (2) Å
c	7.777 (2) Å

HfEr$_2$S$_5$ [74D1]

Structure: orthorhombic (D$_{2h}^{16}$—Pnam)

a	11.442 (2) Å
b	7.1932 (8) Å
c	7.681 (2) Å

HfLa$_2$Se$_5$ [74D1]

Structure: orthorhombic (D$_{2h}^{16}$—Pnam)

a	11.919 (1) Å
b	7.680 (1) Å
c	8.524 (1) Å

HfCe$_2$Se$_5$ [74D1]

Structure: orthorhombic (D$_{2h}^{16}$—Pnam)

a	11.886 (1) Å
b	7.6540 (7) Å
c	8.4292 (7) Å

10.3.4.4 Compounds with Ib elements (only Cu)

CuLaS$_2$

Structure: monoclinic (C$_{2h}^5$—P2$_1$/b)

a	6.50 Å			79F
b	6.91 Å			81R
c	7.30 Å			
β	98° 53′			

Huber, Scharmer

Physical property	Numerical value	Experimental conditions	Experimental method remarks	Ref.
CuCeS$_2$				
Structure: monoclinic (C$_{2h}^5$—P2$_1$/b)				79F
a	6.47 Å			81R
b	6.88 Å			
c	7.25 Å			
β	98°60′			
CuPrS$_2$				
Structure: monoclinic (C$_{2h}^5$—P2$_1$/b)				79F
a	6.42 Å			81R
b	6.86 Å			
c	7.21 Å			
β	98°43′			
CuNdS$_2$				
Structure: monoclinic (C$_{2h}^5$—P2$_1$/b)				79F
a	6.43 Å			81R
b	6.83 Å			
c	7.17 Å			
β	98°88′			
CuSmS$_2$				
Structure: monoclinic (C$_{2h}^5$—P2$_1$/b)				79F
a	6.39 Å			81R
b	6.79 Å			
c	7.10 Å			
β	98°17′			
CuGdS$_2$				
Structure: monoclinic (C$_{2h}^5$—P2$_1$/b)				79F
a	6.36 Å			81R
b	6.75 Å			
c	7.05 Å			
β	98°67′			
CuTbS$_2$				
Structure: monoclinic (C$_{2h}^5$—P2$_1$/b)				79F
a	6.74 Å			81R
b	6.72 Å			
c	7.02 Å			
β	98°67′			
Cu$_3$SmS$_3$ [81R]				
Structure: monoclinic				
a	6.55 Å			
b	7.22 Å			
c	7.00 Å			
β	98°			
d	4.64 g cm^{-3}		pycnometric density	
	4.64 g cm^{-3}		X-ray	
T_m	>1523 K			
E_g	1.81 eV		from conductivity	
S	365 µV K^{-1}		temperature dependence: Fig. 7	
			Temperature dependence of conductivity: Fig. 6.	

Physical property	Numerical value	Experimental conditions	Experimental method remarks	Ref.
Cu$_3$GdS$_3$ [81R]				
Structure: monoclinic				
a	6.39 Å			
b	6.78 Å			
c	7.07 Å			
β	98°			
d	5.86 g cm^{-3}		pycnometric density	
	5.80 g cm^{-3}		X-ray	
T_m	1513 K			
E_g	1.62 eV		from conductivity	
S	$\approx 425\ \mu$V K^{-1}		temperature dependence: Fig. 7 (deviating data in table and Fig. 7) Temperature dependence of conductivity: Fig. 6.	
Cu$_3$TbS$_3$ [81R]				
Structure: trigonal (P$\bar{3}$)				
a	3.89 Å		hexagonal setting	
c	6.384 Å			
d	5.90 g cm^{-3}		pycnometric density	
	5.94 g cm^{-3}		X-ray	
T_m	>1523 K			
E_g	1.62 eV		from conductivity	
S	412 μV K^{-1}			
Cu$_3$DyS$_3$ [81R]				
Structure: trigonal (P$\bar{3}$)				
a	3.88 Å		hexagonal setting	
c	6.366 Å			
d	5.94 g cm^{-3}		pycnometric density	
	5.99 g cm^{-3}		X-ray	
T_m	1523 K			
E_g	1.60 eV		from conductivity	
S	340 μV K^{-1}			
Cu$_3$YS$_3$ [81R]				
Structure: trigonal (P$\bar{3}$)				
a	3.90 Å		hexagonal setting	
c	6.35 Å			
d	4.70 g cm^{-3}		pycnometric density	
	4.90 g cm^{-3}		X-ray	
T_m	>1523 K			
E_g	1.78 eV		from conductivity	
S	260 μV K^{-1}			
Cu$_3$HoS$_3$ [81R]				
Structure: trigonal (P$\bar{3}$)				
a	3.876 Å		hexagonal setting	
c	6.36 Å			
d	5.96 g cm^{-3}		pycnometric density	
	6.00 g cm^{-3}		X-ray	
T_m	>1523 K			
E_g	1.65 eV		from conductivity	
S	240 μV K^{-1}			

Physical property	Numerical value	Experimental conditions	Experimental method remarks	Ref.
Cu_3YbS_3 [81R]				
Structure: trigonal (P$\bar{3}$)				
a	3.86 Å		hexagonal setting	
c	6.384 Å			
d	6.40 g cm^{-3}		pycnometric density	
	6.10 g cm^{-3}		X-ray	
T_m	1423 K			
E_g	1.56 eV		from conductivity	
S	420 µV K^{-1}		temperature dependence: Fig. 7	
			Temperature dependence of conductivity: Fig. 6.	
Cu_3LuS_3 [81R]				
Structure: trigonal (P$\bar{3}$)				
a	3.87 Å		hexagonal setting	
c	6.270 Å			
d	6.50 g cm^{-3}		pycnometric density	
	6.47 g cm^{-3}		X-ray	
T_m	>1523 K			
E_g	1.50 eV		from conductivity	
S	380 µV K^{-1}			
Cu_3ScS_3 [81R]				
Structure: trigonal (P$\bar{3}$)				
a	3.76 Å		hexagonal setting	
c	5.930 Å			
d	4.67 g cm^{-3}		pycnometric density	
	4.38 g cm^{-3}		X-ray	
T_m	>1523 K			
E_g	1.86 eV		from conductivity	
S	≈ 550 µV K^{-1}		temperature dependence: Fig. 7 (deviating data in table and Fig. 7)	
			Temperature dependence of conductivity: Fig. 6.	
Cu_3SmSe_3 [81R]				
Structure: monoclinic (C_{2h}^4—P2/b)				
a	6.996 Å		(monoclinic angle not given)	
b	7.29 Å			
c	7.29 Å			
d	6.51 g cm^{-3}		pycnometric density	
	6.53 g cm^{-3}		X-ray	
T_m	1383 K			
E_g	0.15 eV		from conductivity	
S	186 µV K^{-1}			
			Temperature dependence of conductivity: Fig. 8.	

Physical property	Numerical value	Experimental conditions	Experimental method remarks	Ref.
Cu_3GdSe_3 [81R]				
Structure: trigonal ($P\bar{3}$)				
a	4.060 Å		hexagonal setting	
c	6.83 Å			
d	6.69 g cm^{-3}		pycnometric density	
	6.69 g cm^{-3}		X-ray	
T_m	>1523 K			
E_g	0.14 eV		from conductivity	
S	176 μV K^{-1}			
			Temperature dependence of conductivity: Fig. 8.	
Cu_3TbSe_3 [81R]				
Structure: trigonal ($P\bar{3}$)				
a	4.050 Å		hexagonal setting	
c	6.68 Å			
d	6.76 g cm^{-3}		pycnometric density	
	6.80 g cm^{-3}		X-ray	
T_m	>1523 K			
E_g	0.16 eV		from conductivity	
S	170 μV K^{-1}			
Cu_3DySe_3 [81R]				
Structure: trigonal ($P\bar{3}$)				
a	4.053 Å		hexagonal setting	
c	6.582 Å			
d	6.97 g cm^{-3}		pycnometric density	
	6.97 g cm^{-3}		X-ray	
T_m	>1523 K			
E_g	0.20 eV		from conductivity	
S	225 μV K^{-1}			
			Temperature dependence of conductivity: Fig. 8.	
Cu_3YSe_3 [81R]				
Structure: trigonal ($P\bar{3}$)				
a	4.060 Å		hexagonal setting	
c	6.61 Å			
d	5.77 g cm^{-3}		pycnometric density	
	5.80 g cm^{-3}		X-ray	
T_m	>1523 K			
E_g	0.88 eV		from conductivity	
S	440 μV K^{-1}			
Cu_3HoSe_3 [81R]				
Structure: trigonal ($P\bar{3}$)				
a	4.030 Å		hexagonal setting	
c	6.57 Å			
d	7.10 g cm^{-3}		pycnometric density	
	7.14 g cm^{-3}		X-ray	
T_m	>1523 K			
E_g	0.16 eV		from conductivity	
S	250 μV K^{-1}			

Physical property	Numerical value	Experimental conditions	Experimental method remarks	Ref.
Cu_3YbSe_3 [81R]				
Structure: trigonal ($P\bar{3}$)				
a	3.961 Å		hexagonal setting	
c	6.536 Å			
d	7.55 g cm^{-3}		pycnometric density	
	7.60 g cm^{-3}		X-ray	
T_m	1573 K			
E_g	0.20 eV		from conductivity	
S	160 µV K^{-1}			
			Temperature dependence of conductivity: Fig. 8.	
Cu_3ScSe_3 [81R]				
Structure: trigonal ($P\bar{3}$)				
a	3.854 Å		hexagonal setting	
c	6.264 Å			
d	6.496 g cm^{-3}		pycnometric density	
	6.50 g cm^{-3}		X-ray	
T_m	>1523 K			
E_g	0.30 eV		from conductivity	
S	630 µV K^{-1}			
			Temperature dependence of conductivity: Fig. 8.	
Cu_3SmTe_3 [81R]				
Structure: trigonal ($P\bar{3}$)				
a	4.220 Å		hexagonal setting	
c	7.300 Å			
d	6.700 g cm^{-3}		pycnometric density	
	6.609 g cm^{-3}		X-ray	
E_g	0.23 eV		from conductivity	
S	145 µV K^{-1}			
n	1.73 · 10^{20} cm^{-3}			
R_H	0.036 cm^3 C^{-1}			
			Temperature dependence of conductivity: Fig. 9.	
Cu_3GdTe_3 [81R]				
Structure: trigonal ($P\bar{3}$)				
a	4.212 Å		hexagonal setting	
c	7.286 Å			
d	7.269 g cm^{-3}		pycnometric density	
	7.220 g cm^{-3}		X-ray	
E_g	0.52 eV		from conductivity	
S	160 µV K^{-1}			
n	1.9 · 10^{19} cm^{-3}			
R_H	0.059 cm^3 C^{-1}			
			Temperature dependence of conductivity: Fig. 9.	

Physical property	Numerical value	Experimental conditions	Experimental method remarks	Ref.

Cu_3TbTe_3 [81R]

Structure: trigonal ($P\bar{3}$)

a	4.210 Å		hexagonal setting	
c	7.280 Å			
d	7.083 g cm^{-3}		pycnometric density	
	7.252 g cm^{-3}		X-ray	
E_g	0.46 eV		from conductivity	
S	225 μV K^{-1}			
n	$2 \cdot 10^{20}$ cm^{-3}			
R_H	0.082 cm^3 C^{-1}			

Cu_3DyTe_3 [81R]

Structure: trigonal ($P\bar{3}$)

a	4.200 Å		hexagonal setting	
c	7.272 Å			
d	7.200 g cm^{-3}		pycnometric density	
	7.330 g cm^{-3}		X-ray	
E_g	0.34 eV		from conductivity	
S	450 μV K^{-1}			
n	$1.8 \cdot 10^{19}$ cm^{-3}			
R_H	0.098 cm^3 C^{-1}			
			Temperature dependence of conductivity: Fig. 9.	

Cu_3YTe_3 [81R]

Structure: trigonal ($P\bar{3}$)

a	4.300 Å		hexagonal setting	
c	7.240 Å			
d	6.40 g cm^{-3}		pycnometric density	
	6.30 g cm^{-3}		X-ray	
E_g	0.72 eV		from conductivity	
S	350 μV K^{-1}			
n	$2.3 \cdot 10^{20}$ cm^{-3}			
R_H	0.027 cm^3 C^{-1}			
			Temperature dependence of conductivity: Fig. 9.	

Cu_3HoTe_3 [81R]

Structure: trigonal ($P\bar{3}$)

a	4.190 Å		hexagonal setting	
c	7.210 Å			
d	7.37 g cm^{-3}		pycnometric density	
	7.42 g cm^{-3}		X-ray	
E_g	0.26 eV		from conductivity	
S	284 μV K^{-1}			
n	$1.6 \cdot 10^{18}$ cm^{-3}			
R_H	0.860 cm^3 C^{-1}			

Cu_3ErTe_3 [81R]

Structure: trigonal ($P\bar{3}$)

a	4.180 Å		hexagonal setting	
c	7.160 Å			
d	7.58 g cm^{-3}		pycnometric density	
	7.56 g cm^{-3}		X-ray	

(continued)

Physical property	Numerical value	Experimental conditions	Experimental method remarks	Ref.
Cu$_3$ErTe$_3$ [81R] (continued)				
E_g	0.24 eV		from conductivity	
S	250 μV K^{-1}			
n	2.4 $\cdot$ 10^{20} cm^{-3}			
R_H	0.049 cm^3 C^{-1}			
Cu$_3$TmTe$_3$ [81R]				
Structure: trigonal (P$\bar{3}$)				
a	4.230 Å		hexagonal setting	
c	7.110 Å			
d	7.43 g cm^{-3}		pycnometric density	
	7.46 g cm^{-3}		X-ray	
E_g	0.14 eV		from conductivity	
S	225 μV K^{-1}			
n	2.8 $\cdot$ 10^{20} cm^{-3}			
R_H	0.087 cm^3 C^{-1}			
Cu$_5$GdS$_4$ [76R]				
Structure: hexagonal				
a	11.7 Å			
c	6.50 Å			
d	5.20 g cm^{-3}		pycnometric density	
	5.25 g cm^{-3}		X-ray	
S	205 μV K^{-1}		temperature dependence: Fig. 10	
σ	0.3 (Ω cm)$^{-1}$		temperature dependence: Fig. 11 (single crystal, orientation not specified)	
Cu$_5$TbS$_4$ [76R]				
Structure: hexagonal				
a	11.68 Å			
c	6.47 Å			
d	5.39 g cm^{-3}		pycnometric density	
	5.30 g cm^{-3}		X-ray	
S	427 μV K^{-1}		temperature dependence: Fig. 10	
n	9.3 $\cdot$ 10^{17} cm^{-3}			
R_H	6.735 cm^3 C^{-1}			
σ	0.4 (Ω cm)$^{-1}$		temperature dependence: Fig. 11 (single crystal, orientation not specified)	
Cu$_5$DyS$_4$ [76R]				
Structure: hexagonal				
a	11.65 Å			
c	6.45 Å			
d	5.45 g cm^{-3}		pycnometric density	
	5.32 g cm^{-3}		X-ray	
S	609 μV K^{-1}		temperature dependence: Fig. 10	
σ	3 $\cdot$ 10^{-2} (Ω cm)$^{-1}$		temperature dependence: Fig. 11 (single crystal, orientation not specified)	

Physical property	Numerical value	Experimental conditions	Experimental method remarks	Ref.
Cu_5HoS_4 [76R]				
Structure: hexagonal				
a	11.63 Å			
c	6.46 Å			
d	5.50 g cm^{-3}		pycnometric density	
	5.39 g cm^{-3}		X-ray	
E_g	0.53 eV		from conductivity	
S	-840 μV K^{-1}		temperature dependence: Fig. 10	
n	$2.3 \cdot 10^{20}$ cm^{-3}			
R_H	0.027 cm^3 C^{-1}			
σ	$\approx 2.6 \cdot 10^{-2}$ (Ω cm)$^{-1}$		temperature dependence: Fig. 11 (deviating data in table and Fig. 11) (single crystal, orientation not specified)	
Cu_5LuS_4 [76R]				
Structure: hexagonal				
a	11.59 Å			
c	6.42 Å			
d	5.77 g cm^{-3}		pycnometric density	
	5.60 g cm^{-3}		X-ray	
E_g	0.50 eV		from conductivity	
S	-725 μV K^{-1}		temperature dependence: Fig. 10	
n	$1.3 \cdot 10^{16}$ cm^{-3}			
R_H	4.796 cm^3 C^{-1}			
σ	$\approx 1.24 \cdot 10^{-2}$ (Ω cm)$^{-1}$		temperature dependence: Fig. 11 (deviating data in table and Fig. 11) (single crystal, orientation not specified)	
Cu_5GdSe_4 [76R]				
Structure: hexagonal				
a	11.92 Å			
c	6.72 Å			
d	6.12 g cm^{-3}		pycnometric density	
	6.05 g cm^{-3}		X-ray	
E_g	0.64 eV		from conductivity	
S	325 μV K^{-1}		temperature dependence: Fig. 12	
n	$10.6 \cdot 10^{19}$ cm^{-3}			
R_H	0.059 cm^3 C^{-1}			
σ	$\approx 3 \cdot 10^{-2}$ (Ω cm)$^{-1}$		temperature dependence: Fig. 13 (deviating data in table and Fig. 13) (single crystal, orientation not specified)	

Physical property	Numerical value	Experimental conditions	Experimental method remarks	Ref.

Cu_5TbSe_4 [76R]

Structure: hexagonal

a	11.90 Å		
c	6.72 Å		
d	6.35 g cm^{-3}		pycnometric density
	6.38 g cm^{-3}		X-ray
E_g	1.04 eV		from conductivity
S	464 μV K^{-1}		temperature dependence: Fig. 12
n	$1.77 \cdot 10^{19}$ cm^{-3}		
R_H	0.353 cm^3 C^{-1}		
σ	$\approx 1 \cdot 10^{-3}$ (Ω cm)$^{-1}$		temperature dependence: Fig. 13 (deviating data in table and Fig. 13) (single crystal, orientation not specified)

Cu_5YbSe_4 [76R]

Structure: hexagonal

a	12.12 Å		
c	6.40 Å		
d	6.70 g cm^{-3}		pycnometric density
	6.68 g cm^{-3}		X-ray
E_g	0.80 eV		from conductivity
S	≈ 115 μV K^{-1}		temperature dependence: Fig. 12 (deviating data in table and Fig. 12)
n	$11.6 \cdot 10^{19}$ cm^{-3}		
R_H	0.054 cm^3 C^{-1}		
σ	$\approx 8 \cdot 10^{-2}$ (Ω cm)$^{-1}$		temperature dependence: Fig. 13 (deviating data in table and Fig. 13) (single crystal, orientation not specified)

Cu_5LuSe_4 [76R]

Structure: hexagonal

a	12.10 Å		
c	6.32 Å		
d	6.75 g cm^{-3}		pycnometric density
	6.70 g cm^{-3}		X-ray
E_g	1.00 eV		from conductivity
S	206 μV K^{-1}		
n	$8.7 \cdot 10^{19}$ cm^{-3}		
R_H	0.072 cm^3 C^{-1}		
σ	$\approx 1 \cdot 10^{-3}$ (Ω cm)$^{-1}$		temperature dependence: Fig. 13 (deviating data in table and Fig. 13) (single crystal, orientation not specified)

Physical property	Numerical value	Experimental conditions	Experimental method remarks	Ref.
Cu_5GdTe_4 [76R]				
Structure: hexagonal				
a	12.63 Å			
c	6.86 Å			
d	6.65 g cm^{-3}		pycnometric	
	6.60 g cm^{-3}		X-ray	
S	325 μV K^{-1}			
σ	5.00 (Ω cm)$^{-1}$		(single crystal, orientation not specified)	
Cu_5DyTe_4 [76R]				
Structure: hexagonal				
a	12.62 Å			
c	6.83 Å			
d	7.08 g cm^{-3}		pycnometric	
	6.99 g cm^{-3}		X-ray	
S	54.7 μV K^{-1}			
σ	510.6 (Ω cm)$^{-1}$			

10.3.4.5 Compounds with IIb elements

$ZnTm_2S_4$ [73Y]				
Structure: orthorhombic (Er_2MnS_4-type)				
a	7.734 Å			
b	13.227 Å			
c	6.263 Å			
E_g	3.6 eV		from optical absorption	
$ZnYb_2S_4$ [73Y]				
Structure: orthorhombic (Er_2MnS_4-type)				
E_g	≈2.5 eV		from optical absorption	
$ZnLu_2S_4$ [73Y]				
Structure: orthorhombic (Er_2MnS_4-type)				
a	7.658 Å		Blue-green cathodoluminescence detected: Fig. 2.	
b	13.183 Å			
c	6.246 Å			
E_g	3.7 eV		from optical absorption	
$ZnSc_2S_4$ [73Y]				
Structure: spinel				
a	10.483 Å			
E_g	2.1 eV		from optical absorption; optical absorption coefficient: Fig. 14	

Doping with Ag, Cr, Ga causes n-type conductivity. See also [68T] (see below).

$ZnSc_2S_4$:Ag [73Y]				
ϱ	0.71 Ω cm		n-type conductivity	
n	$1.2 \cdot 10^{18}$ cm^{-3}			
μ_n	7.8 cm^2 V^{-1} s^{-1}			

Physical property	Numerical value	Experimental conditions	Experimental method remarks	Ref.
$ZnSc_2S_4:Ga$ [73Y]				
ϱ	$2.53\ \Omega\ cm$		n-type conductivity	
n	$8.10^{17}\ cm^{-3}$			
μ_n	$3\ cm^2\ V^{-1}\ s^{-1}$			
$ZnSc_2S_4:Cr$ [73Y]				
ϱ	$0.31\ \Omega\ cm$		n-type conductivity	
n	$5\cdot10^{17}\ cm^{-3}$			
μ_n	$40\ cm^2\ V^{-1}\ s^{-1}$			
$CdLa_2S_4$ [73Y]				
Structure: cubic (Th_3P_4-type)				
a	$8.7033\ \text{Å}$			
E_g	$\approx 2.6\ eV$		from optical absorption Weak-green cathodoluminescence observed [73Y].	
$CdPr_2S_4$ [73Y]				
Structure: trigonal				
E_g	$\approx 2.1\ eV$		from optical absorption	
$CdNd_2S_4$ [73Y]				
Structure: trigonal				
a	$4.6\ \text{Å}$			
c	$8.0\ \text{Å}$			
$CdDy_2S_4$ [73Y]				
E_g	$\approx 2.5\ eV$		from optical absorption	
$CdHo_2S_4$				
Structure: spinel				73Y
a	$11.1674\ \text{Å}$		Dark red cathodoluminescence	73Y
	$11.240\ \text{Å}$		observed.	72F
u	0.377			
$CdEr_2S_4$				
Structure: spinel				73Y
a	$11.1347\ \text{Å}$			73Y
	$11.198\ \text{Å}$			72F
u	0.378			
E_g	$\approx 1.8\ eV$		from optical absorption	73Y
$CdTm_2S_4$				
Structure: spinel				73Y
a	$11.09\ \text{Å}$			
	$11.092\ \text{Å}$			64S
	$11.10\ \text{Å}$			75P2
E_g	$\approx 2.4\ eV$		from optical absorption Compound exhibits electro-luminescence [64S].	73Y

Physical property	Numerical value	Experimental conditions	Experimental method remarks	Ref.
$CdYb_2S_4$ [73Y]				
Structure: spinel				
a	11.0684 Å			
E_g	≈ 2.5 eV		from optical absorption	
$CdLu_2S_4$				
Structure: spinel				73Y
a	11.045 Å		Dark orange cathodoluminescence	73Y
	10.945 Å		observed [73Y].	72F
u	0.378			72F
CdY_2S_4				
Structure: spinel				73Y
a	11.172 Å		Brown cathodoluminescence	64S
	11.196 Å		observed [73Y]; compound shows electroluminescence [64S].	
$CdSc_2S_4$ [73Y]				
Structure: spinel				
a	10.733 Å			
E_g	2.3 eV		from optical absorption; optical absorption coefficient: Fig. 15 Brown cathodoluminescence observed.	
$CdSc_2S_4$:Ag [73Y]				
ϱ	$4.5 \cdot 10^6$ Ω cm			
$CdSc_2S_4$:Li [73Y]				
ϱ	$1.1 \cdot 10^3$ Ω cm		n-type conductivity	
n	$1.4 \cdot 10^{14}$ cm^{-3}			
μ_n	41 cm^2 V^{-1} s^{-1}			
$CdSc_2S_4$:Mg [73Y]				
ϱ	$2.5 \cdot 10^8$ Ω cm			
$CdSc_2S_4$:Al [73Y]				
ϱ	$3.2 \cdot 10^7$ Ω cm			
$CdSc_2S_4$:Ga [73Y]				
ϱ	$4.5 \cdot 10^2$ Ω cm		n-type conductivity	
n	$2.8 \cdot 10^{15}$ cm^{-3}			
μ_n	4.9 cm^2 V^{-1} s^{-1}			
$CdSc_2S_4$:Sn [73Y]				
ϱ	$1.8 \cdot 10^7$ Ω cm			
$CdSc_2S_4$:Pb [73Y]				
ϱ	$8.3 \cdot 10^7$ Ω cm			
$CdSc_2S_4$:P [73Y]				
ϱ	$4.8 \cdot 10^7$ Ω cm			
$CdSc_2S_4$:Cr [73Y]				
ϱ	$7 \cdot 10^4$ Ω cm		p-type conductivity	
p	$2 \cdot 10^{12}$ cm^{-3}			
μ_p	43 cm^2 V^{-1} s^{-1}			

Physical property	Numerical value	Experimental conditions	Experimental method remarks	Ref.
CdY_2Se_4				
Structure: spinel				81R, 64S
a	11.650 Å			81R
	11.654 Å		Compound exhibits electro-luminescence.	64S
$CdGd_2Se_4$ [81R]				
Structure: Th_3P_4-type				
a	8.860 Å			
$CdDy_2Se_4$ [64S, 81R]				
Structure: spinel				
a	11.659 Å		Compound exhibits luminescence.	
$CdHo_2Se_4$				
Structure: spinel				81R, 72F
a	11.638 Å			81R
u	0.376			72F
d	10.9 g cm^{-3}			72F
$CdEr_2Se_4$				
Structure: spinel				81R, 72F
a	11.603 Å			81R
u	0.378			72F
$CdYb_2Se_4$				
Structure: spinel				81R, 64S
a	11.530 Å		Compound exhibits electro-luminescence [64S], see also [75P2].	81R, 64S
	11.54 Å			74P
$YbYb_4S_7$ [81R]				
Structure: monoclinic (C_{2h}^3—C2/m)				
a	12.60 Å			
b	3.79 Å			
c	11.29 Å			
β	104°			
d	6.92 g cm^{-3}		pycnometric density	
	6.88 g cm^{-3}		X-ray	
$CdYb_4S_7$ [81R]				
Structure: monoclinic (C_{2h}^3—C2/m)				
a	12.55 Å			
b	3.79 Å			
c	11.28 Å			
β	104°			
d	6.49 g cm^{-3}		pycnometric density	
	6.56 g cm^{-3}		X-ray	

Physical property	Numerical value	Experimental conditions	Experimental method remarks	Ref.

CdYb$_4$Se$_7$ [81R]

Structure: monoclinic (C$_{2h}^3$—C2/m)

a	13.36 Å			
b	4.02 Å			
c	11.97 Å			
β	105°			
d	7.16 g cm^{-3}		pycnometric density	
	7.25 g cm^{-3}		X-ray	

ZnYb$_4$S$_7$ [81R]

Structure: monoclinic (C$_{2h}^3$—C2/m)

a	12.57 Å			
b	3.80 Å			
c	11.32 Å			
β	105°			
d	6.34 g cm^{-3}		pycnometric density	
	6.34 g cm^{-3}		X-ray	

CdEr$_4$S$_7$ [81R]

Structure: monoclinic (C$_{2h}^3$—C2/m)

a	12.58 Å			
b	3.80 Å			
c	11.34 Å			
β	106°			
d	6.45 g cm^{-3}		pycnometric density	
	6.57 g cm^{-3}		X-ray	

CdEr$_4$Se$_7$ [81R]

Structure: monoclinic (C$_{2h}^3$—C2/m)

a	13.30 Å			
b	4.01 Å			
c	11.95 Å			
β	105°			
d	7.08 g cm^{-3}		pycnometric density	
	7.20 g cm^{-3}		X-ray	

CdHo$_4$S$_7$ [81R]

Structure: monoclinic (C$_{2h}^3$—C2/m)

a	12.60 Å			
b	3.80 Å			
c	11.35 Å			
β	106°			
d	6.41 g cm^{-3}		pycnometric density	
	6.33 g cm^{-3}		X-ray	

CdHo$_4$Se$_7$ [81R]

Structure: monoclinic (C$_{2h}^3$—C2/m)

a	13.32 Å			
b	4.02 Å			
c	11.96 Å			
β	105°			
d	7.85 g cm^{-3}		pycnometric density	
	7.75 g cm^{-3}		X-ray	

Physical property	Numerical value	Experimental conditions	Experimental method remarks	Ref.
CdTm$_4$S$_7$ [81R]				
Structure: monoclinic (C_{2h}^3—C2/m)				
a	12.79 Å			
b	3.18 Å			
c	11.42 Å			
β	105°			
d	6.90 g cm^{-3}		pycnometric density	
	6.01 g cm^{-3}		X-ray	

10.3.4.6 Compounds with III a elements

Physical property	Numerical value	Experimental conditions	Experimental method remarks	Ref.
GaLaS$_3$				
Structure: unknown				79F
d	3.78 g cm^{-3}		Temperature dependence of conductivity: [81R]; emission and excitation spectra of Ce doped LaGaS$_3$: Fig. 16.	81R
Ga$_{10/3}$La$_6$S$_{14}$				
Structure: hexagonal (C_6^6—P6$_3$)				79F
a	10.15 Å			81R
c	6.08 Å			
GaCeS$_3$				
Structure: unknown				79F
d	3.97 g cm^{-3}		Temperature dependence of conductivity: [81R].	81R
Ga$_{10/3}$Ce$_6$S$_{14}$				
Structure: hexagonal (C_6^6—P6$_3$)				79F
a	10.03 Å			81R
c	6.08 Å			
Ga$_2$EuS$_4$				
Structure: orthorhombic (D_{2h}^{24}—Fddd)			Ga$_2$YbS$_4$ also reported: [76A].	79F, 79R
a	10.22 Å		Semiconductor according to [80R].	76A
b	6.11 Å			
c	10.36 Å			
a	20.727 Å			79R
b	12.197 Å			
c	20.454 Å			
d	4.3 g cm^{-3}			79R, 76A
T_m	1493 K			74D3
	1488 K			
σ	$\approx 3.2 \cdot 10^{-6}$ Ω^{-1} cm^{-1}	p-type	temperature dependence of conductivity: Fig. 17	76A
S	70 μV K^{-1}		(deviating data in table and Fig. 17), Seebeck coefficient: Fig. 18, Hall coefficient: Fig. 19, carrier mobility: Fig. 20	76A

Physical property	Numerical value	Experimental conditions	Experimental method remarks	Ref.
Ga$_2$EuSe$_4$				
Structure: orthorhombic (D$_{2h}^{24}$—Fddd)			Ga$_2$YbSe$_4$ also reported: [76A].	79F, 80R
a	10.67 Å			76A,
b	6.37 Å			74D3
c	10.80 Å			
T_m	1383 K			76A
	1366···1387 K			74D3
σ	$3.16 \cdot 10^{-5}\,\Omega^{-1}\,cm^{-1}$	p-type	temperature dependence of	76A
ϱ	$7 \cdot 10^6\,\Omega\,cm$		conductivity: Fig. 17,	74D3
S	$58\,\mu V\,K^{-1}$		Seebeck coefficient: Fig. 18,	76A
			Hall coefficient: Fig. 19,	
			and carrier mobility: Fig. 20	
Ga$_2$EuTe$_4$				
Structure: orthorhombic				79F
a	11.06 Å			76A
b	6.60 Å			
c	11.00 Å			
T_m	1283 K			76A
σ	$\approx 10^{-5}\,\Omega^{-1}\,cm^{-1}$	p-type	temperature dependence of	76A
S	$\approx 48\,\mu V\,K^{-1}$		conductivity: Fig. 17	76A
			(deviating data in table	
			and Fig. 17),	
			Seebeck coefficient: Fig. 18	
			(deviating data in table	
			and Fig. 18),	
			Hall coefficient: Fig. 19	
In$_2$EuS$_4$				
Structure: orthorhombic			In$_2$YbS$_4$ also reported: [76A].	79F, 74D3
a	10.49 Å			76A,
b	6.50 Å			74D3
c	10.36 Å			
σ	$\approx 3.1 \cdot 10^{-5}$ $\Omega^{-1}\,cm^{-1}$	p-type	temperature dependence of conductivity: Fig. 17	76A 74D3
ϱ	$\geqq 10^{11}\,\Omega\,cm$		(deviating data in table	76A
S	$\approx 55\,\mu V\,K^{-1}$		and Fig. 17),	
			Seebeck coefficient: Fig. 18	
			(deviating data in table	
			and Fig. 18),	
			Hall coefficient: Fig. 19,	
			and carrier mobility: Fig. 20	
In$_2$EuSe$_4$				
Structure: orthorhombic			In$_2$YbSe$_4$ also reported: [76A].	79F, 74D3
a	11.10 Å			76A,
b	6.70 Å			74D3
c	10.88 Å			
T_m	1283 K			76A
	1285 K			74D3
				(continued)

Physical property	Numerical value	Experimental conditions	Experimental method remarks	Ref.
In$_2$EuSe$_4$ (continued)				
σ	$\approx 3.05 \cdot 10^{-5}$ Ω^{-1} cm^{-1}	p-type	temperature dependence of conductivity: Fig. 17 (deviating data in table and Fig. 17),	76A 74D3
ϱ	$\geqq 10^{12}\,\Omega$ cm			
S	$\approx 55\,\mu$V K^{-1}		Seebeck coefficient: Fig. 18 (deviating data in table and Fig. 18), Hall coefficient: Fig. 19, and carrier mobility: Fig. 20	76A
In$_2$EuTe$_4$				
Structure: orthorhombic				79F
a	11.72 Å			76A
b	6.92 Å			
c	11.36 Å			
T_m	1263 K			76A
σ	$2 \cdot 10^{-5}\,\Omega^{-1}$ cm^{-1}	p-type	temperature dependence of conductivity: Fig. 17, Seebeck coefficient: Fig. 18, Hall coefficient: Fig. 19	76A
S	43 μV K^{-1}			76A

10.3.4.7 Compounds with IVa elements

(N.B. E_g: optical energy gap)

La$_2$GeSe$_5$				
E_g	1.7 eV		Temperature dependence of conductivity and Hall coefficient: Fig. 21; diffuse reflection spectra: Fig. 22.	78M1
La$_2$SnSe$_5$				
E_g	1.65 eV			78M1
Ce$_2$GeSe$_5$				
E_g	1.55 eV			78M1
Ce$_2$SnSe$_5$				
E_g	1.52 eV			78M1
Pr$_2$GeSe$_5$				
E_g	1.85 eV		Temperature dependence of conductivity and Hall coefficient: Fig. 21; diffuse reflection spectra: Fig. 22.	78M1
Pr$_2$SnSe$_5$				
E_g	1.65 eV		Diffuse reflection spectra: Fig. 23.	78M1
Nd$_2$GeSe$_5$				
E_g	1.90 eV		Temperature dependence of conductivity and Hall coefficient: Fig. 21.	78M1

Physical property	Numerical value	Experimental conditions	Experimental method remarks	Ref.
Nd_2SnSe_5				
E_g	1.70 eV		Diffuse reflection spectra: Fig. 23.	78M1
Sm_2GeSe_5				
E_g	2.00 eV		Temperature dependence of conductivity and Hall coefficient: Fig. 21.	78M1
Sm_2SnSe_5				
E_g	1.88 eV			78M1
Gd_2GeSe_5				
E_g	1.90 eV		Temperature dependence of conductivity and Hall coefficient: Fig. 21.	78M1
Gd_2SnSe_5				
E_g	1.70 eV		Diffuse reflection spectra: Fig. 23.	78M1

10.3.5 References for 10.3

54J   Jonker, G.H.: Physica **20** (1954) 1118.
54K   Kestigan, M., Ward, R.: J. Am. Chem. Soc. **76** (1954) 6027.
56B   Bertaut, F., Forrat, F.: J. Phys. Radium **17** (1956) 129.
56G   Geller, S., Wood, E.A.: Acta Crystallogr. **9** (1956) 563.
57G   Geller, S.: Acta Crystallogr. **10** (1957) 243.
57K   Kestigan, M., Dickinson, F.G., Ward, R.: J. Am. Chem. Soc. **79** (1957) 5598.
57W1   Wold, A., Post, B., Banks, E.: J. Am. Chem. Soc. **79** (1957) 4911.
57W2   Watanabe, H.: J. Phys. Soc. Jpn. **12** (1957) 515.
58R   Rabenau, A., Eckerlin, P.: Acta Crystallogr. **11** (1958) 304.
59D   Dalziel, J.A.W.: J. Chem. Soc. London **1959**, 1993.
59W   Wold, A., Arnott, R.: J. Phys. Chem. Solids **9** (1959) 176.
60W   Waugh, J.S.: Tech. Report 152, MIT Lab. for Insulation Research (1960).
61F   Foëx, M.: Bull. Soc. Chim. France **1961**, 109.
62G   Gerthsen, P., Hardtl, K.H.: Z. Natuforsch. **17a** (1962) 514.
63Q   Quezel-Ambrunaz, S., Mareschal, M.: Bull. Soc. Fr. Mineral. Crystallogr. **86** (1963) 204.
63Y   Yakel, H.L., Koehler, W.C., Bertaut, E.F., Forrat, E.F.: Acta Crystallogr. **16** (1963) 957.
64G1   Golabi, S.M., Flahaut, J., Domange, L.: C.R. Acad. Sc. Paris **259** (1964) 820.
64G2   Gerthsen, P., Kettel, G.: J. Phys. Chem. Solids **25** (1964) 1023.
64H   Heikes, R.R., Miller, R.C., Mazelsky, R.: Physica **30** (1964) 1600.
64S   Suchow, L., Stemle, N.R.: J. Electrochem. Soc. **111** (1964) 191.
65E   Eibschütz, M.: Acta Crystallogr. **19** (1965) 337.
65F1   Foëx, M.: C.R. Acad. Sc. Paris **260** (1965) 6389.
65F2   Flahaut, J., Guittard, M., Patrie, M., Pardo, M.P., Golabi, S.M., Domange, L.: Acta Crystallogr. **19** (1965) 14.
65F3   Frushour, R.H., Vorres, K.S.: Research Prog. Report (1965) U.S. Atomic Energy Comm.
65N   Nassau, K., Levinstein, H.J., Loiacono, G.M.: J. Phys. Chem. Solids **26** (1965) 1805.
65W   Werner, W.: Thesis, Technische Universität Berlin (1965).
66C   Coeuré, Ph., Guinet, P., Peuzin, J.C., Buisson, G., Bertaut, E.F.: Article on p. 332 in Vol. I of Dvorak, V., Fouskova, A., Glogar, P. (eds.): Proceedings of the International Meeting on Ferroelectricity, held at Prague (1966) (in two volumes), Prague: Institute of Physics of the Czechoslovak Academy of Sciences 1966.

66G	Goodenough, J.B.: J. Appl. Phys. **37** (1966) 1415.
66H	Holzapfel, H., Sieler, J.: Z. Anorg. Allgem. Chem. **343** (1966) 174.
66J	Jonker, G.H.: J. Appl. Phys. **37** (1966) 1424.
66R	Rogers, D.B., Ferretti, A., Ridgley, D.H., Amott, R.J., Goodenough, J.B.: J. Appl. Phys. **37** (1966) 1431.
66W	Waintal, A., Capponi, J.J., Bertaut, E.F.: Solid State Commun. **4** (1966) 125.
67B	Bertaut, E.F., Lissalde, F.: Solid State Commun. **5** (1967) 173.
67M	Menyuk, N., Dwight, K., Raccah, P.M.: J. Phys. Chem. Solids **28** (1967) 549.
67R1	Ruiz, J.S., Anthony, A.-M., Foëx, M.M.: C.R. Acad. Sc. Paris **264** (1967) 1271.
67R2	Raccah, P.M., Goodenough, J.B.: Phys. Rev. **155** (1967) 932.
67W	Waintal, A., Chenavas, J.: Mater. Res. Bull **2** (1967) 819.
68C	Coeuré, P.: Solid State Commun. **6** (1968) 129.
68D	de Combarieu, A., Mareschal, J., Michel, J.C., Sivardiere, J.: Solid State Commun. **6** (1968) 257.
68G	Gyorgy, E.M., Remeika, J.P., Wood, D.L.: J. Appl. Phys. **39** (1968) 3499.
68Q	Quezel-Ambrunaz, S.: Bull. Soc. Fr. Mineral. Crystallogr. **91** (1968) 339.
68R	Rao, G.V., Chandrashekhar, G.V., Rao, C.N.R.: Solid State Commun. **6** (1968) 177.
68T	Tressler, R.E., Hummel, F.A., Stubican, V.S.: J. Am. Ceram. Soc. **51** (1968) 648.
69K	Kahn, F.J., Pershan, P.S., Remeika, J.P.: Phys. Rev. **186** (1969) 891.
69M1	McCarthy, G.J., White, W.B., Roy, R.: Mater. Res. Bull. **4** (1969) 251.
69M2	Meadowcroft, D.B.: Br. J. Appl. Phys. **2** (1969) 1225.
69P	Pardo, M.P., Flahaut, J.: Bull. Soc. Chim. Fr. **1969**, 6.
69W	Wood, D.L., Holmes, L.M., Remeika, J.P.: Phys. Rev. **185** (1969) 689.
70A1	Antonov, A.V., Balbashov, A.M., Chervonenkis, A.Ya.: Fiz. Tverd. Tela **12** (1970) 1724; Sov. Phys. Solid State (English Transl.) **12** (1970) 1363.
70A2	Aring, K.B., Sievers, A.J.: J. Appl. Phys. **41** (1970) 1197.
70B	Banks, E., Tashima, N.: J. Appl. Phys. **41** (1970) 1186.
70C1	Courths, R., Hüfner, S., Pelzl, J., van Uitert, L.G.: Solid State Commun. **8** (1970) 1163.
70C2	Coeuré, P., Challeton, D.: Solid State Commun. **8** (1970) 1345.
70M1	Meltzer, R.S.: Phys. Rev. B **2** (1970) 2398.
70M2	Marezio, M., Remeika, J.P., Dernier, P.D.: Acta Crystallogr. **B 26** (1970) 2008.
70M3	Meltzer, R.S., Moos, H.W.: J. Appl. Phys. **41** (1970) 1240.
70M4	Matsumoto, G.: J. Phys. Soc. Jpn. **29** (1970) 606.
70M5	Malozemoft, A.P., White, R.L.: Solid State Commun. **8** (1970) 665.
70P	Pauthenet, R., Veyret, C.: J. Phys. (Paris) **31** (1970) 65.
70S	Subba Rao, G.V., Ferraro, J.R., Rao, C.N.R.: J. Appl. Spectrosc. **24** (1970) 436.
70T1	Tsushima, K., Aoyagi, K.: J. Appl. Phys. **41** (1970) 1238.
70T2	Tabor, W.J., Anderson, A.W., Van Uitert, L.G.: J. Appl. Phys. **41** (1970) 3018.
70W1	Wright, J.C., Moos, H.W.: J. Appl. Phys. **41** (1970) 1244.
70W2	Wood, D.L., Remeika, J.P., Kolb, E.D.: J. Appl. Phys. **41** (1970) 5315.
71B1	Belov, K.P., Kadomtseva, A.M.: Usp. Fiz. Nauk **103** (1971) 577; Sov. Phys. Usp. (English Transl.) **14** (1971) 154.
71B2	Balbashov, A.M., Chervonenkis, A.Y., Antonov, A.V., Bakhteuzov, V.E.: Izv. Akad. Nauk SSSR, Ser. Fiz. **35** (1971) 1243; Bull. Acad. Sci. USSR, Phys. Ser. (English Transl.) **35** (1971) 1135.
71C1	Clark, R.H., Moulton, W.G.: Phys. Rev. **B 5** (1971) 788.
71C2	Clover, R.B., Wentworth, C., Mroczkowski, S.S.: IEEE Trans. MAG **7** (1971) 480.
71C3	Clover, R.B., Rayl, M., Gutman, D.: AIP Conf. Proc. **5** (1971) 264.
71M1	Malozemoff, A.P.: J. Phys. Chem. Solids **32** (1971) 1669.
71M2	McCarthy, G.J.: Mater. Res. Bull. **6** (1971) 31.
71M3	Marezio, M., Dernier, P.D.: Mater. Res. Bull. **6** (1971) 23.
71R1	Robbins, M., Pierce, R.D., Wolfe, R.: J. Phys. Chem. Solids **32** (1971) 1789.
71R2	Robbins, M., Pierce, R.D., Wolfe, R.: J. Appl. Phys. **42** (1971) 1563.
71S	Subba Rao, G.V., Wanklyn, B.M., Rao, C.N.R.: J. Phys. Chem. Solids **32** (1971) 345.
71U	Uesaka, Y., Tsujikawa, J., Aoyaki, K., Tsushima, K., Sugano, S.: J. Phys. Soc. Jpn. **31** (1971) 1380.
72B1	Bhide, V.G., Rajoria, D.S., Rama Rao, G., Rao, C.N.R.: Phys. Rev. **B 6** (1972) 1021.
72B2	Blazey, K.W.: AIP Conf. Proc. **10** (1972) 735.
72B3	Bhide, V.G., Rajoria, D.S., Reddy, Y.S.: Phys. Rev. Lett. **28** (1972) 1133.

72C	Chetkin, MV., Didosyan, Y.S., Akhutkina, A.I.: Fiz. Tverd. Tela **13** (1971) 3414; Sov. Phys. Solid State (English Transl.) **13** (1972) 2871.
72F	Fujii, H., Okamoto, T., Kamigaichi, T.: J. Phys. Soc. Jpn. **32** (1972) 1432.
72H	Hornreich, R.M., Komet, Y.: Solid State Commun. **11** (1972) 969.
72M	McCarthy, G.J., Fischer, R.D.: J. Solid State Chem. **4** (1972) 340.
73G1	Ganguly, P., Rao, C.N.R.: Mater. Res. Bull. **8** (1973) 405.
73G2	Goodenough, J.B.: Mater. Res. Bull. **8** (1973) 423.
73H1	Hornreich, R.M., Yaeger, J.: Intern. J. Magnetism **4** (1973) 71.
73H2	Hasegawa, M., Daido, K., Saito, M.: Jpn. J. Appl. Phys. **12** (1973) 1904.
73K	Kenjo, T., Yajima, S.: Bull. Chem. Soc. Jpn. **46** (1973) 1329.
73M	McCarthy, G.J., Gallagher, P.V., Sipe, C.: Mater. Res. Bull. **8** (1973) 1277.
73T	Terao, N.: C.R. Acad. Sci. (Paris) Ser. C **276** (1973) 5.
73W	Wood, V.E., Austin, A.E., Collings, E.W., Brog, K.C.: J. Phys. Chem. Solids **34** (1973) 859.
73Y	Yim, W.M., Fan, A.K., Stofko, E.J.: J. Electrochem. Soc. **120** (1973) 441.
74D1	Donohue, P.C., Jeitschko, W.: Mater. Res. Bull. **9** (1974) 1333.
74D2	Dougier, P., Hagenmuller, P.: J. Solid State Chem. **11** (1974) 177.
74D3	Donohue, P.C., Hanlon, J.E.: J. Electrochem. Soc. **121** (1974) 137.
74G1	George, A.M., Gopalakrishnan, I.K., Karkhanavala, M.D.: Mater. Res. Bull. **9** (1974) 721.
74G2	Grishmanovskii, A.N., Lemanov, V.V., Smolenskii, G.A., Balbashov, A.M., Chervonenkis, A.Y.: Fiz. Tverd. Tela **16** (1974) 1426; Sov. Phys. Solid State (English Transl.) **16** (1974) 916.
74L	Lal, H.B., Dar, N., Kumar, A.: J. Phys. C **7** (1974) 4335.
74M	McCarthy, G.J., Sipe, C.A., McIlvried: Mater. Res. Bull. **9** (1974) 1279.
74P	Pokrzywnicki, S., Czopnik, A., Wróbel, B., Pawlak, L.: Phys. Status Solidi (b) **64** (1974) 685.
74T	Tamaki, T., Yamaura, R., Tsushima, K.: Single Crystal Growth and Weak Ferromagnetism of Rare Earth Orthochromites, NHK Tech. J. **26** (1974) 57 (in Japanese).
74W	Walling, J.C., White, R.L.: Phys. Rev. **B 10** (1974) 4748.
75A	Abe, M., Kimura, T., Nomura, S.: Jpn. J. Appl. Phys. **14** (1975) 1507.
75B	Bhide, V.G., Rajoria, D.S., Rao, C.N.R., Rama Rao, G., Jadhao, V.G.: Phys. Rev. **B 12** (1975) 2832.
75C1	Cook, W.R. Jr.: J. Am. Ceram. Soc. **58** (1975) 151.
75C2	Chetkin, M.V., Shcherbakov, Y.I., Volenko, A.P., Shevchuk, L.D.: Zh. Eksperin. i Teor. Fiz. **67** (1974) 1027; Sov. Phys. JETP (English Transl.) **40** (1975) 509.
75D	Dougier, P.: PhD Thesis, Univ. of Bordeaux **1975**.
75F	Fayolle, J.P., Studer, F., Desgardin, G., Reveau, B.: J. Solid State Chem. **13** (1975) 57.
75H1	Hornreich, R.M., Komet, Y., Nolan, R., Wanklyn, B.M., Yaeger, J.: Phys. Rev. **B 12** (1975) 5094.
75H2	Hasson, A., Hornreich, R.M., Komet, Y.: Phys. Rev. **B 12** (1975) 5051.
75K	Kajiura, M., Aoyagi, K., Tamaki, T.: J. Phys. Soc. Jpn. **39** (1975) 1572.
75M1	McCarthy, G.J., Greedan, J.E.: Inorg. Chem. **14** (1975) 772.
75M2	Marx, R., Happ, H.: Phys. Status Solidi (b) **67** (1975) 181.
75M3	Marchand, A.: C.R. Acad. Sci. (Paris) Ser. **B 280** (1975) 41.
75O	Obayashi, H., Kudo, T.: Jpn. J. Appl. Phys. **14** (1975) 330.
75P1	Palanisamy, T., Gopalakrishnan, J., Sastri, M.V.C.: Z. Anorg. Allgem. Chem. **415** (1975) 275.
75P2	Pokrzywnicki, S., Czopnik, A.: Phys. Status Solidi (b) **70** (1975) K85.
75S	Sayer, M., Chen, R., Fletcher, R., Mansingh, A.: J. Phys. C **8** (1975) 2059.
75U	Udagawa, M., Kohn, K., Koshizuka, N., Tsushima, T., Tsushima, K.: Solid State Commun. **16** (1975) 779.
75V	Voorhoeve, R.J.H., Remeika, J.P., Trimble, L.E., Cooper, A.S., SiSalvo, F.D., Gallagher, P.K.: J. Solid State Chem. **14** (1975) 395.
76A	Aliev, O.M., Kurbanov, T.K., Rustamov, P.G., Alidzhanov, M.A., Salmanov, S.M.: Izv. Akad. Nauk SSSR, Neorg. Mater. **12** (1976) 1944.
76G	Ganguly, P., Parkash, O., Rao, C.N.R.: Phys. Status Solidi (a) **36** (1976) 669.
76L	Lal, H.B., Dar, N., Lundgren, L.: J. Phys. Soc. Jpn. **41** (1976) 1216.
76R	Rustamov, P.G., Aliev, O.M., Guseinov, G.G., Alidzhanov, M.A., Agaev, A.B.: Izv. Akad. Nauk SSSR, Neorg. Mater. **12** (1976) 1192.
76S1	Sakai, T., Adachi, G., Shiokawa, J.: Mater. Res. Bull. **11** (1976) 1295.
76S2	Studer, F., Fayolle, J.P., Raveau, B.: Mater. Res. Bull. **11** (1976) 1125.
76W	Webb, J.B., Sayer, M.: J. Phys. C **9** (1976) 4151.

77P	Provenzano, P.L., Boldish, S.I., White, W.B.: Mater. Res. Bull. **12** (1977) 939.
77S1	Sato, K., Ishino, T., Adachi, G., Shiokawa, J.: Mater. Res. Bull. **12** (1977) 789.
77S2	Shin-ike, T., Adachi, G., Shiokawa, J.: Mater. Res. Bull. **12** (1977) 1149.
77W	Webb, J.B., Sayer, M., Mansingh, A.: Can. J. Phys. **55** (1977) 1725.
78B	Bazuev, G.V., Shveikin, G.P.: Izv. Akad. Nauk SSSR, Neorg. Mater. **14** (1978) 267.
78M1	Murguzow, M.I.: Fiz. Tekh. Poluprovodn. **12** (1978) 1823; Sov. Phys. Semicond. (English Transl.) **12** (1978) 1080.
78M2	Murguzow, M.I.: Fiz. Tekh. Poluprovodn. **12** (1978) 1825; Sov. Phys. Semicond. (English Transl.) **12** (1978) 1081.
79C	Chevalier, B., Demazeau, G., Etourneau, J., Hagenmuller, P.: Phys. Status Solidi (b) (1979) K63.
79D	Dar, N., Lal, H.B.: Mater. Res. Bull. **14** (1979) 1263.
79F	Flahaut, J.: Handbook on the Physics and Chemistry of Rare Earths, Gschneidner, K.A., Eyring, L.R. (eds.), North Holland, Amsterdam Vol. 4, **1979**.
79K	Khattak, C.P., Wang, F.F.Y.: "Perovskites and Garnets" in: Handbook on the Physics and Chemistry of Rare Earths, Gschneidner, K.A., Eyring, L.R. (eds.), North Holland, Amsterdam Vol. 3, **1979**, p. 525.
79R	Roques, R., Rimet, R., Declercq, J.P., Germain, G.: Acta Crystallogr. **B 35** (1979) 555.
79S	Subramanian, M.A., Aravamudan, G., Subba Rao, G.V.: Mater. Res. Bull. **14** (1979) 1457.
80E	Etourneau, J., Chevalier, B., Hagenmuller, P., Georges, G.: J. Phys. **41** (1980) C5-193.
80H	Howng, W.-Y., Thorn, R.J.: J. Phys. Chem. Solids **41** (1980) 75.
80M	Marx, R.: Phys. Status Solidi (b) **99** (1980) 555.
80R	Rimet, R., Buder, R., Schlenker, C., Zanchetta, J.V.: J. Mag. Magn. Mater. **15–18** (1980) 987.
80S	Subramanian, M.A., Subba Rao, G.V.: J. Solid State Chem. **31** (1980) 329.
80T1	Tripathi, A.K., Lal, H.B.: Mater. Res. Bull. **15** (1980) 233.
80T2	Takeda, T., Machida, Y.: Jpn. J. Appl. Phys. **19** (1980) 1575.
81M	Main, J.G., Robins, G.A., Demazeau, G.: J. Phys. C **14** (1981) 3633.
81R	Rustamov, P.G., Aliev, O.M., Kurbanov, T.Kh.: Ternary Chalcogenides of Rare Earth Elements, Mamedov, Kh.S. (ed.), Elm, Baku **1981**.
81S	Scharmer, E.-G., Leiß, M., Huber, G.: J. Lumin. **24/25** (1981) 751.
82S1	Scharmer, E.-G.: Thesis, Universität Hamburg **1982**.
82S2	Scharmer, E.-G., Leiß, M., Huber, G.: J. Phys. C **15** (1982) 1071.

10.4 Further ternary compounds

Besides the main groups of semiconducting ternary compounds presented in sections 10.1 to 10.3 a huge amount of other ternaries are known to show semiconducting behavior. Properties of such substances are compiled in this section. For many compounds data on semiconductor properties are scarce. Here we refrained from presenting all other known properties. Furthermore the data are often rather inaccurate or even conflicting, so that experimental details as orientation of the sample, polarization of fields etc. have only limited informative value. Thus, the material presented in this chapter can only give a first introduction to the wealth of further semiconducting ternaries and the reader is asked to consult the original papers for further information. – In the tables all values are at RT if not otherwise stated.

10.4.1 I_x–IV_y–VI_z compounds

I_2–IV–VI_3 compounds:　see section 10.1.4.
I_8–IV–VI_6 compounds:　Ag_8SiS_6, Ag_8GeS_6, Ag_8SnS_6, Ag_8SiSe_6, Ag_8GeSe_6, Ag_8SnSe_6,
　　　　　　　　　　　　　Ag_8SiTe_6, Ag_8GeTe_6, Cu_8SiS_6, Cu_8GeS_6, Cu_8SiSe_6, Cu_8GeSe_6.
I_4–IV_3–VI_5 compounds:　$Cu_4Ge_3S_5$, $Cu_4Ge_3Se_5$, $Cu_4Sn_3Se_5$.
Others: Cu_4SnS_4.

10.4.1.0 Structure

The compounds treated in this section can be thought of as ternary compounds occurring in the phase diagrams of the pseudobinary systems $(I_2$–$VI)$ – $(IV$–$VI_2)$ and $(I_2$–$VI)$ – $(IV$–$VI)$:

$$I_2–IV–VI_3 = (I_2–VI) + (IV–VI_2);\quad I_8–IV–VI_6 = 4(I_2–VI) + (IV–VI_2);$$
$$I_4–IV_3–VI_5 = 2(I_2–VI) + 3(IV–VI),\quad I_4–IV–VI_4 = 2(I_2–VI) + (IV–VI_2).$$

Such phase diagrams are shown in Figs. 1···8.

The **I_2–IV–VI_3 compounds** – as ternary analogues of the II–VI compounds – have already been treated in section 10.1.4.

The **I_8–IV–VI_6 compounds** are isoelectronic analogues to the II_3–V_2 compounds. They exhibit a pronounced low-temperature polymorphism with a fcc high-temperature phase (γ-phase, space group O^2—$P4_2 32$) and six low-temperature phases: α, α', α'', β, β', β''. From these α'' is cubic (space group T^3—$I23$ or T^5—$I2_1 3$). The structure of the other phases is not clear, β' and β'' may be cubic or monoclinic with $a \approx b \approx c$, $\beta \approx 90°$. For further information compare the references given and Landolt-Börnstein, NS, Vols. III/6 and III/14 (in preparation).

Ag_8GeS_6 and Ag_8SnS_6 are known as the minerals **argyrodite** and **canfieldite,** respectively. The following structure data are known (Ag-compounds: [68G, 81K], Cu compounds: [65H, 74K, 82A]):

Substance	Temperature of phase transformation [°C]	Lattice parameters (in Å):		
		at 25°C	of γ-phase at	T [°C]
Ag_8SiS_6	234 ($\alpha'' \to \gamma$)	21.00	10.63	250
Ag_8GeS_6	223 ($\alpha'' \to \gamma$)	21.19	10.70	240
Ag_8SnS_6	172 ($\alpha'' \to \gamma$)	21.43	10.85	200
Ag_8SiSe_6	10 ($\alpha' \to \beta''$)	10.87	10.97	150
	40 ($\beta'' \to \gamma$)			
Ag_8GeSe_6	−4 ($\alpha' \to \beta'$)	10.95	10.99	65
	48 ($\beta' \to \gamma$)			
Ag_8SnSe_6	83 ($\beta' \to \gamma$)	11.07	11.12	200
Ag_8SiTe_6	−78 ($\alpha \to \beta$)		11.515	20
	−10 ($\beta \to \gamma$)			
Ag_8GeTe_6*)	−52 ($\alpha \to \beta$)		11.570	20
	−29 ($\beta \to \gamma$)			
Cu_8SiS_6			9.76	20
Cu_8GeS_6	55 ($\beta' \to \gamma$)	9.90	9.909 (5)	60
Cu_8SiSe_6			10.17	20
Cu_8GeSe_6	56			

———————

*) Acc. to [81K] there are three phase transitions at $−99°C$ ($\alpha' \to \alpha$), $−49°C$ ($\alpha \to \beta$), $−28°C$ ($\beta \to \gamma$).

Physical property	Numerical value	Experimental conditions	Experimental method remarks	Ref.

For the pressure dependence of the transition temperatures, see [70P].

Among the **I_4–IV_3–VI_5 compounds** $Cu_4Ge_3S_5$ has a tetragonal lattice with $a = 5.30$ Å and $c = 10.48$ Å at RT, $Cu_4Ge_3Se_5$ has a fcc lattice with $a = 5.53$ Å at RT [77D]. The structure of $Cu_4Sn_3Se_5$ is not known. Phase diagrams of the systems Cu_2S–GeS, Cu_2Se–GeSe and Cu_2Se–SnSe are shown in Fig. 9.

Cu_4SnS_4 has an orthorhombic lattice with $a = 13.70(1)$ Å, $b = 7.750(5)$ Å, $c = 6.454(5)$ Å at room temperature. At $T = -41\,°C$ a phase transition occurs without a change of the lattice parameters [74K].

10.4.1.1 Physical properties

I_8–IV–VI_6 compounds

Ag_8SiS_6:

T_m	940 °C			68G
d_{th}	6.21 g/cm³			65H
d_{exp}	5.99 g/cm³	$T = 298$ K		68G

Ag_8GeS_6 (argyrodite):

		T [K]		
E_g	1.39 eV	293	"red edge" of photoconductivity spectrum (Fig. 11)	72O
	1.41 eV	295	fundamental absorption edge	76K, 77K
	1.48(5) eV	293	fundamental absorption edge (Fig. 12)	75O
dE_g/dT	$-8.5 \cdot 10^{-4}$ eV K⁻¹	α''-phase	temperature shift of absorption edge (Figs. 10, 11)	77K
	$-6.6 \cdot 10^{-4}$ eV K⁻¹	α''-phase	photoconductivity	72O
	$-5 \cdot 10^{-4}$ eV K⁻¹	γ-phase	absorption edge	77K
ΔE_g	-55 meV		step in E_g at $\alpha'' \rightarrow \gamma$ phase transition	77K
$\hbar\omega_i$	93 meV	330	infrared absorption bands (phonon transitions)	75B
	96 meV			
	98 meV			
σ	$10^{-3}\,\Omega^{-1}$ cm⁻¹	280		76P
κ	$3 \cdot 10^{-3}$ W cm⁻¹ K⁻¹			76P
T_m	955 °C			68G
d_{th}	6.30 g/cm³			65H
d_{exp}	6.21 g/cm³	298		68G

Ag_8SnS_6 (canfieldite):

		T [K]		
E_g	1.28 eV	295	fundamental absorption edge	76K, 77K
	1.434 eV	293	"red edge" of photoconductivity (Fig. 11)	72O
	1.39(2) eV	293	fundamental absorption edge (Fig. 12)	75O
dE_g/dT	$-4 \cdot 10^{-4}$ eV K⁻¹	α''-phase	photoconductivity	72O
	$-(5\cdots6) \cdot 10^{-4}$ eV K⁻¹		absorption edge (Figs. 10, 11)	75O
	$-5 \cdot 10^{-4}$ eV K⁻¹	γ-phase		76K, 77K
ΔE_g	-60 meV		step in E_g at $\alpha'' \rightarrow \gamma$ phase transition	77K
σ	$10^{-3}\,\Omega^{-1}$ cm⁻¹	280		76P
S	-1200 µV K⁻¹			76P
κ	$3.2 \cdot 10^{-3}$ W cm⁻¹ K⁻¹			76P

(continued)

Physical property	Numerical value	Experimental conditions	Experimental method remarks	Ref.
Ag_8SnS_6 (continued)				
T_m	839 °C			68G
d_{th}	6.31 g/cm³			65H
d_{exp}	6.28 g/cm³	$T = 298$ K		68G
Ag_8SiSe_6:				
		T [K]		
E_g	0.97 eV	295	absorption edge (Fig. 10)	76B1, 77K
dE_g/dp	$1.6 \cdot 10^{-6}$ eV cm² kg⁻¹	β''- and α'-phase	pressure shift of absorption edge (Fig. 14)	76B1, 77K
ΔE_g	60 meV		step in E_g at $\beta'' \to \alpha'$ phase transition	76B1, 77K
$\hbar\omega_i$	70 meV 80 meV 104 meV	330	infrared absorption bands (phonon transitions, Fig. 13)	75B
T_m	930 °C			68G
d_{th}	7.06 g/cm³			68G
d_{exp}	6.95 g/cm³	298		68G
Ag_8GeSe_6:				
		T [K]		
E_g	0.84⋯0.88 eV	300 β'-phase	conductivity, photoconductivity, absorption edge (Figs. 10, 12, 15)	76B1, 77O, 77K
dE_g/dp	$4 \cdot 10^{-6}$ eV cm² kg⁻¹ $3 \cdot 10^{-6}$ eV cm² kg⁻¹ $2 \cdot 10^{-6}$ eV cm² kg⁻¹	α'-phase β'-phase γ-phase	pressure shift of absorption edge (Figs. 14, 16)	77K
dE_g/dT	$-5 \cdot 10^{-4}$ eV K⁻¹ $-4.8 \cdot 10^{-4}$ eV K⁻¹ $-(5\cdots8) \cdot 10^{-4}$ eV K⁻¹	α'-phase β'-phase γ-phase	temperature shift of absorption edge (Fig. 10).	76K, 77K
ΔE_{g1}	-70 meV		step at $\alpha' \to \beta'$ phase transition	77K
ΔE_{g2}	-13 meV		step at $\beta' \to \gamma$ phase transition	77K
$\hbar\omega_i$	55 meV 62 meV 71 meV	330	infrared absorption bands (phonon transitions, Fig. 13)	75B
T_m	902 °C			68G
d_{th}	7.13 g/cm³			68G
d_{exp}	7.07 g/cm³	298		68G
Ag_8SnSe_6:				
		T [K]		
E_g	0.83 eV	295	absorption edge	76B1
dE_g/dp	$2.5 \cdot 10^{-6}$ eV cm² kg⁻¹	β'-phase	pressure shift of absorption edge (Fig. 14)	77K
	$1.8 \cdot 10^{-6}$ eV cm² kg⁻¹	γ-phase		76B2
dE_g/dT	$-5 \cdot 10^{-4}$ eV K⁻¹	γ-phase	temperature shift of absorption edge	76K, 77K
ΔE_g	-55 meV		step at $\beta' \to \gamma$ phase transition	77K
$\hbar\omega_i$	53 meV 57 meV 63 meV	330	infrared absorption band (phonon transition, Fig. 13)	75B
σ	$2 \cdot 10^{-2}\cdots70\ \Omega^{-1}$ cm⁻¹	280	data on several samples (temperature dependence of conductivity, Fig. 17)	76P (continued)

Physical property	Numerical value	Experimental conditions	Experimental method remarks	Ref.
Ag_8SnSe_6 (continued)				
S	$-160\cdots-730\ \mu V\ K^{-1}$			76P
κ	$(3.1\cdots3.5)\cdot10^{-3}$ $W\ cm^{-1}\ K^{-1}$			76P
T_{perit}	$735\,°C$		temperature of peritectic decomposition	68G
d_{th}	$7.12\ g/cm^3$			68G
d_{exp}	$7.01\ g/cm^3$	$T=298$ K		68G
Ag_8SiTe_6:				
T_m	$870\,°C$			68G
d_{th}	$7.21\ g/cm^3$			68G
d_{exp}	$7.23\ g/cm^3$	$T=298$ K		68G
Ag_8GeTe_6:				
		T [K]		
E_g	0.43 eV	293	absorption edge	75B
	0.47 eV	295	absorption edge	77K
dE_g/dT	$<1\cdot10^{-4}\ eV\ K^{-1}$	β-phase	temperature shift of absorption edge (Fig. 10)	77K
	$1\cdots2\cdot10^{-4}\ eV\ K^{-1}$	γ-phase		
ΔE_{g1}	-45 meV		step at $\alpha\rightarrow\beta$ phase transition	77K
ΔE_{g2}	$+18$ meV		step at $\beta\rightarrow\gamma$ phase transition	77K
T_{perit}	$645\,°C$		temperature of peritectic decomposition	68G
d_{th}	$7.31(7)\ g/cm^3$			81K
d_{exp}	$7.22\ g/cm^3$	298		81K
Cu_8SiS_6:				
d_{th}	$5.20\ g/cm^3$			65H
d_{exp}	$5.01\ g/cm^3$	$T=298$ K		65H
Cu_8GeS_6:				
E_g	0.10 eV	β'-phase	for conductivity measurements, see Fig. 18	74K
	0.04 eV	γ-phase		
T_m	$980(3)\,°C$			74K
d_{th}	$5.28\ g/cm^3$			74K
d_{exp}	$5.97\ g/cm^3$	$T=298$ K		74K
Cu_8SiSe_6:				
d_{th}	$6.27\ g/cm^3$			65H
d_{exp}	$5.97\ g/cm^3$			65H

For some data on Cu_8GeSe_6, see [82A] and Fig. 19.

Other I_x–IV_y–VI_z compounds

The compounds $Cu_4Ge_3S_5$, $Cu_4Ge_3Se_5$ and $Cu_4Sn_3Se_5$ (melting points $T_m=675(3)\,°C$, $615\,°C$ and $600\,°C$, respectively) are proven as semiconductors in [77D]. Hall mobilities are in the range of 10 to 300 cm^2/Vs.

Cu_4SnS_4:				
E_g	0.03 eV	high-temperature phase	conductivity (Fig. 20), Hall effect	74K
	0.11 eV	low-temperature phase		
μ_p	3 cm^2/Vs	$T=300$ K		74K

10.4.1.2 References for 10.4.1

65H   Hahn, H., Schulze, H., Sechser, L.: Naturwissenschaften **52** (1965) 451.

68G   Gorochov, O.: Bull. Soc. Chim. France **6** (1968) 2263.

70P   Pistorius, C.W.F.T., Gorochov, O.: High Temp. – High Pressure **2** (1970) 31.

72O   Osipishin, I.S., Butsko, N.I., Gasii, B.I., Zhezhnich, I.D.: Sov. Phys. Semicond. **6** (1972) 974 (transl. from Fiz. Tekh. Poluprovodn. **6** (1972) 1121).

74K   Khanafer, M., Gorochov, O., Rivet, J.: Mater. Res. Bull. **9** (1974) 1543.

75B   Bendorius, R., Irzikevicius, A., Kinduris, A., Tsvetkova, E.V.: Phys. Status Solidi (a) **28** (1975) K 125.

75O   Osipishin, I.S., Gasii, B.I., Butsko, N.I.: Sov. Phys. Semicond. **8** (1975) 1045 (transl. from. Fiz. Tekh. Poluprovodn. **8** (1974) 1609).

76B1   Bendorius, R.A., Kinduris, A.S., Tsvetkova, E.V., Shileika, A.Yu.: Inorg. Mater. (USSR) **12** (1976) 1437; (transl. from Izv. Akad. Nauk SSSR, Neorg. Mater. **12** (1976) 1745).

76B2   Bendorius, R., Kinduris, A., Shileika, A.: High Temp. – High Pressure **7** (1976) 695.

76K   Kinduris, A.S., Bendorius, R.A., Senulene, D.B.: Sov. Phys. Semicond. **10** (1976) 916 (transl. from. Fiz. Tekh. Poluprovodn. **10** (1976) 1544).

76P   Petrov, A.V., Orlov, V.M., Zaitsev, V.K., Feigel'man: Sov. Phys. Solid State **17** (1976) 2407 (transl. from Fiz. Tverd. Tela **17** (1975) 3703).

77D   Dovletov, K., Tashliev, K., Rozyeva, K.A., Ashirov, A., Anikin, A.V.: Inorg. Mater. (USSR) **13** (1977) 889 (transl. from Izv. Akad. Nauk SSSR, Neorg. Mater. **13** (1977) 1092).

77K   Kinduris, A., Shileika, A.: 3rd Int. Conf. on Ternary Compounds, Edinburgh **1977,** The Institute of Physics, London 1977, p. 67.

77O   Osipishin, I.S.: Sov. Phys. Semicond. **11** (1977) 102 (transl. from Fiz. Tekh. Poluprovodn. **11** (1977) 181).

81K   Katty, A., Gorochov, O., Letoffe, J.M.: J. Solid State Chem. **38** (1981) 259.

82A   Aliev, M.I., Arasly, D.G., Dzhabrailov, T.G.: Sov. Phys. Solid State **24** (1982) 150 (transl. from Fiz. Tverd. Tela **24** (1982) 268).

10.4.2 I_x–V_y–VI_z compounds

I–V–VI_2 compounds: $AgAsS_2$, $AgAsSe_2$, $AgAsTe_2$, $AgSbS_2$, $AgSbSe_2$, $AgSbTe_2$, $AgBiS_2$, $AgBiSe_2$, $AgBiTe_2$, $CuSbS_2$, $CuSbSe_2$, $CuSbTe_2$, $CuBiSe_2$, $CuBiTe_2$; $MSbS_2$; $MSbSe_2$ with M = Li, Na, K, Rb, Cs.

I_3–V–VI_3 compounds: Ag_3AsS_3, Ag_3SbS_3.

I_3–V–VI_4 compounds: see section 10.1.5.

10.4.2.0 Structure, chemical bond

I–V–VI_2 compounds

All $AgSbX_2$ and $AgBiX_2$ compounds with X = S, Se, Te crystallize (at least in their high-temperature modification) in the face centered cubic NaCl structure with I- and V-atoms distributed statistically between the close-packed layers of VI-atoms [59G].

lattice parameters:

Substance	a [Å]	T [°C]	Ref.
$AgSbS_2$**)	5.6514(5)	> 403	77B2
$AgSbSe_2$	5.786(3)	25	59G
$AgSbTe_2$	6.078(3)	25	59G
$AgBiS_2$**)	5.648(3)*)	25	59G
	5.682(3)	200	59G
	5.693(3)	243	59G
$AgBiSe_2$**)	5.832(3)*)	25	59G
	5.887(3)	300	59G
$AgBiTe_2$**)	6.155(3)*)	25	59G

*) extrapolation from values of solid solutions
**) high-temperature phase

Below 380 °C **AgSbS$_2$** crystallizes in a monoclinic α-phase (space group: C^3_{2h}—A2/m or Aa) with $a=$ 13.2269(13) Å, $b=4.4112(5)$ Å, $c=12.8798(11)$ Å, $β=98.48(1)$ ° [77B2].

In the **AgBiX$_2$** compounds further phases have been reported [58W2, 59G, 77M]:

intermediate phase: rhombohedral structure (space group D^5_{3d}—R$\bar{3}$m). This is an ordered structure, where planes containing alternately I-, VI-, V-, VI-atoms stacked normally to a [111]-direction of the fcc-lattice.

The existence of this phase is proved at least for **AgBiSe$_2$** in the range 120 °C$<T<$287 °C with the lattice parameters $a=7.022$ Å, $α=34°40'$ (or $a=4.184$ Å, $c=19.67$ Å in the hexagonal description).

For **AgBiTe$_2$** the transition temperatures are most probably at 428 °C [58W2] and 130 °C [58Z].

room-temperature phase: hexagonal structure (space group D^3_{3d}—P$\bar{3}$m1) or monoclinic structure belonging to a subgroup of P$\bar{3}$m1. Here the I- and V-atoms are slightly displaced relative to their positions in the intermediate phase.

lattice parameters (room-temperature phase):

	a [Å]	c [Å]	d_X [g/cm^3]	Ref.
AgBiS$_2$ *)	4.07(2)	19.06(5)	6.94	59G
AgBiSe$_2$	4.18(2)	19.67(5)	7.94	59G
AgBiTe$_2$ **)	4.37(2)	20.76(5)	8.30	59G

*) for different data, see [75K]
**) not thermodynamically stable

For phase diagrams of solid solutions between AgSbS$_2$, AgSbSe$_2$, AgSbTe$_2$, AgBiSe$_2$ and AgBiTe$_2$, see [58W2] and [75K].

Not much is known about other I–V–VI$_2$ compounds:

AgAsS$_2$ (the mineral smithite) has a monoclinic lattice (C^6_{2h}) (lattice parameters at RT: $a=14.02$ Å, $c=9.15$ Å) [75G1]. For phase diagrams for several cross-sections of the Ag–As–S system, see [71K].

CuAsSe$_2$ ($T_m=415$ °C) crystallizes below 300 K in the sphalerite structure ($a=5.75$ Å), above 300 K in the wurtzite structure [68I].

CuSbS$_2$ (chalcostibite) is isostructural with Sb$_2$S$_3$. It crystallizes below 366 K in an orthorhombic modification (space group probably C^9_{2v}—Pna2$_1$). The lattice consists of SbS$_2^+$-chains along the b axis, which are linked by Cu-atoms into double plane layers. At 366 K a phase transition $C^9_{2v} \rightarrow D^{16}_{2h}$ occurs [76G3].

CuSbSe$_2$ crystallizes in the orthorhombic lattice (D^{16}_{2h}—Pnma) (lattice parameters at RT: $a=6.40$ Å, $b=$ 3.95 Å, $c=15.33$ Å) [64I].

CuBiSe$_2$ is reported to crystallize in the fcc structure with $a=5.69$ Å [58Z].

CuSbTe$_2$ and **CuBiTe$_2$** possess a Bi$_2$Te$_3$-like hexagonal structure with $a=4.22$ Å, $c=29.9$ Å and $a=4.35$ Å, $c=30.1$ Å (at RT), respectively [58Z].

NaSbSe$_2$ is mentioned as a semiconducting I–V–VI$_2$ compound crystallizing in the fcc structure with $a=$ 5.96 Å [77B1]. Semiconducting films of **LiSbS$_2$(–Se$_2$)**, **NaSbS$_2$(–Se$_2$)**, **KSbS$_2$(–Se$_2$)** and **CsSbS$_2$(–Se$_2$)** have been investigated in [68L, 69Z].

I$_3$–V–VI$_3$ compounds

Ag$_3$AsS$_3$ and Ag$_3$SbS$_3$ crystallize in a non-centrosymmetric uniaxial structure (space group C^6_{3v}—R3c). Each As(Sb)-atom forms pyramidal bonds to three S-atoms. Each S-atom is bound to two Ag-atoms. For phase transitions at $T=24$ K and 56 K in Ag$_3$AsS$_3$, see e.g. [76N, 78A, 79S], at 9.7 K and ≈ 140 K in Ag$_3$SbS$_3$, see [83E].

lattice parameters (in hexagonal description, RT values):

Substance	a [Å]	c [Å]	Ref.
Ag$_3$AsS$_3$	10.80	8.69 *)	73D
Ag$_3$SbS$_3$	11.058	8.698	73G

*) For temperature dependence, see [78A].

Physical property	Numerical value	Experimental conditions	Experimental method remarks	Ref.

10.4.2.1 Physical properties of I–V–VI$_2$ compounds

AgAsS$_2$:

		T [K]		
$E_{g, ind}$	2.084 eV	295 polarization $\perp$ crystal axis	indirect absorption edge, (Fig. 1)	74G
	2.108 eV	polarization $\parallel$ crystal axis	(for identification of the phonons participating in the indirect transitions see [74G])	
$E_{g, dir}$	2.14 eV	293	maximum in the spectral distribution of photoconductivity (Fig. 2, see also Fig. 3)	75G1
	> 2.22 eV	295	direct absorption edge (Fig. 3)	74G
ϱ	$9 \cdot 10^9\ \Omega$ cm	293		75G1
T_m	419 °C			75G1

electroabsorption spectrum: Fig. 4

photoemission data, see [76G1], data on amorphous samples, see [74G, 75G1, 76G1]. Raman and infrared reflection spectra analyzed in [82S]; for wavenumbers of normal modes, see Fig. 5

AgAsSe$_2$:

E_g	$\approx 0.8 \cdots 1.0$ eV			58W1
T_m	390 °C			58W1

AgAsTe$_2$:

E_g	$\approx 0.8 \cdots 1.0$ eV			58W1
T_m	325 °C			58W1

AgSbS$_2$:

		T [K]		
E_g	1.73 eV	300 K	optical energy gap, Fig. 6	77B2
	2.06 eV	85 K	photoconductivity (for photocurrent vs. wavelength, see Fig. 7)	80V
dE_g/dT	$-1.96 \cdot 10^{-3}$ eV/K	$T < 653$ K		77B2
	$-1.58 \cdot 10^{-4}$ eV/K	$653 \cdots 676$ K		
	$-4.95 \cdot 10^{-4}$ eV/K	$T > 676$ K		
$E_A(\sigma)$	0.89 eV	$T < 476$ K	activation energy of conductivity; the electrical conductivity increases exponentially with rising temperature, see Fig. 8, also Fig. 9	79V
	1.64 eV	$T > 476$ K		
$\mu_{H, p}$	0.24 cm^2/Vs	300		79V

For a reflectivity spectrum, see Fig. 3; for temperature dependence of the Seebeck coefficient at high temperature, see Fig. 10.

T_m	512 (2) °C			77B2
d_{th}	5.42 g/cm^3		X-ray density; for temperature dependence of experimental density, see Fig. 11	59G

Physical property	Numerical value	Experimental conditions	Experimental method remarks	Ref.
AgSbSe$_2$:				
$E_{g,th}$	0.58···0.62 eV		from conductivity	68A
μ_p	1500 cm^2/Vs			68A

For temperature dependence of transport parameters, see Figs. 9, 10, 12; for doping dependence of S, see [60H].

Physical property	Numerical value	Experimental conditions	Experimental method remarks	Ref.
κ	$1.1 \cdot 10^{-3}$ cal/K cm s		practically independent of temperature	62P
Θ_D	175 K			62P
α	$23 \cdot 10^{-6}$ K^{-1}	295···675 K		62P
T_m	636 °C			62P
d_{th}	6.60 g/cm^3		for $d_{exp}(T)$, see Fig. 11	59G

AgSbTe$_2$:

All data below 410 K are uncertain (and sometimes controversial) since most material contains two phases (AgSbTe$_2$ + Ag$_2$Te?) [63N, 62P, 58Z, 60H, 60W, 60A]. A recent investigation [80B] presents several optical spectra (Figs. 13···15) and discusses a tentative band structure obtained by comparison with PbTe.

Physical property	Numerical value	T [K]	Experimental method remarks	Ref.
σ	160 Ω^{-1} cm^{-1}	300		58Z
μ_p	75 cm^2/Vs	300		58Z
	35 cm^2/Vs	300	temperature dependence proportional $T^{0.5}$	60W
κ	$1.7 \cdot 10^{-3}$ cal/K cm s	80···400		60H
S	230 mV/K		for doping dependence of S, see [60H]	58Z
α	$23 \cdot 10^{-6}$ K^{-1}	295···375		58Z
T_m	561 °C			68K
d_{th}	7.12 g/cm^3		for $d_{exp}(T)$, see Fig. 11	59G

For AgSbTe$_2$ – PbTe solid solutions, see [79B].

AgBiS$_2$:

Physical property	Numerical value	T [K]	Experimental method remarks	Ref.
E_g	0.9 eV		from reflectivity measurements (see Fig. 3)	75G1
σ	$4 \cdot 10^3$ Ω^{-1} cm^{-1}	293		75G1
T_m	810 °C			75G1
d_{th}	7.02 g/cm^3	300		59G
	6.90 g/cm^3	473		
	6.86 g/cm^3	516		

AgBiSe$_2$:

Physical property	Numerical value	T [K]	Experimental method remarks	Ref.
σ	180 Ω^{-1} cm^{-1}	293		58Z
S	−80 mV/K			58Z
T_m	762 °C			58Z
d_{th}	7.95 g/cm^3	300		59G
	7.72 g/cm^3	573		

AgBiTe$_2$:

Physical property	Numerical value	Experimental conditions	Experimental method remarks	Ref.
E_g	0.075 eV	high-temperature phase	estimate from Hall coefficient of quenched sample	62P
	0.16 eV	room-temperature phase	estimate from Hall coefficient of annealed sample	
	0.05···0.08 eV	room-temperature phase(?)	from heat conductivity (see Figs. 16 and 17)	(continued)

Physical property	Numerical value	Experimental conditions	Experimental method remarks	Ref.
AgBiTe₂ (continued)				
σ	$1300\ \Omega^{-1}\,\mathrm{cm}^{-1}$	$T = 293\ \mathrm{K}$	cf. Fig. 16 for σ, S, κ (conflicting	58Z
S	$-55\ \mathrm{mV/K}$		data)	58Z
κ	$1.4\cdots1.8\cdot10^{-3}$ $\mathrm{cal/K\,cm\,s}$			58Z
α	$20\cdot10^{-6}\ \mathrm{K}^{-1}$	$T > 120\,°\mathrm{C}$	indication of a phase transi-	58Z
	$25\cdot10^{-6}\ \mathrm{K}^{-1}$	$T < 120\,°\mathrm{C}$	tion between the intermediate and the room-temperature phase	
T_m	$520\,°\mathrm{C}$		NaCl (high-temperature) phase	58Z
d_th	$8.14\ \mathrm{g/cm^3}$			59G

CuSbS₂:

This is a ferro- and piezoelectric semiconductor with an unusually high dielectric constant. The conductivity shows a discontinuity at the phase transition temperature $T_\mathrm{tr} = 366\ \mathrm{K}$ connected with a change of the activation energy from 0.26 eV below T_tr to 0.79 eV above T_tr (0.83 eV according to [68A]) (Fig. 18). $T_\mathrm{m} = 535\,°\mathrm{C}$ [76G3]. Temperature dependence of density: Fig. 11.

CuSbSe₂:

$E_\mathrm{g,th}$	$0.83\ \mathrm{eV}$			68A
σ	$4\ \Omega^{-1}\,\mathrm{cm}^{-1}$			58Z
S	$400\ \mathrm{mV/K}$		(see also Fig. 10)	58Z
μ_p	$5\ \mathrm{cm^2/Vs}$			58Z
T_m	$480\,°\mathrm{C}$			68A

For temperature dependence of density, see Fig. 11.

CuSbTe₂:

σ	$3000\ \Omega^{-1}\,\mathrm{cm}^{-1}$			58Z
S	$30\ \mathrm{mV/K}$			58Z
κ	$3.2\cdot10^{-3}\ \mathrm{cal/K\,cm\,s}$			62P
α	$20.5\cdot10^{-6}\ \mathrm{K}^{-1}$	$T = 20\cdots300\,°\mathrm{C}$		62P
Θ_D	$175\ \mathrm{K}$			62P
T_m	$530\,°\mathrm{C}$			62P

CuBiSe₂:

σ	$1200\ \Omega^{-1}\,\mathrm{cm}^{-1}$			58Z
S	$30\ \mathrm{mV/K}$			58Z
α	$20.8\cdot10^{-6}\ \mathrm{K}^{-1}$	$T < 260\,°\mathrm{C}$	polymorphic transition	58Z
	$25\cdot10^{-6}\ \mathrm{K}^{-1}$	$T > 260\,°\mathrm{C}$		
T_m	$585\,°\mathrm{C}$			58Z

CuBiTe₂:

σ	$2000\ \Omega^{-1}\,\mathrm{cm}^{-1}$			58Z
S	$30\ \mathrm{mV/K}$			58Z
α	$23\cdot10^{-6}\ \mathrm{K}^{-1}$	$T = 20\cdots400\,°\mathrm{C}$		58Z
T_m	$520\,°\mathrm{C}$			58Z

CuAsS₂ and **CuSbS₂** are called semiconductors in [57W] without further information ($T_\mathrm{m} = 625\,°\mathrm{C}$ and 415 °C, respectively).

NaSbSe₂ is reported to be a photoelectric semiconducting material with $E_\mathrm{g} \approx 1.3\ \mathrm{eV}$ at RT and $T_\mathrm{m} = 740\,°\mathrm{C}$ [77B1]. For other compounds of the type (Li, Na, K, Rb, Cs)(Sb, Bi)(S, Se)₂, see e.g. [70G1, 70G2, 70S].

Physical property	Numerical value	Experimental conditions	Experimental method remarks	Ref.

10.4.2.2 Physical properties of I_3–V–VI_3 compounds

Ag_3AsS_3 (proustite) and **Ag_3SbS_3** (pyrargite) are important for non-linear optical applications (especially optical mixing [67H]). Both semiconductors are transparent over a wide spectral range. They are pyro- and piezoelectric. By its non-centrosymmetric uniaxial structure they have a large refractive index and a large birefringence [69B]. For elastic constants, electromechanical coupling factors and pyroelectric coefficients, see Landolt-Börnstein, NS, Vols. III/11 and III/18.

Cu_3AsS_3 (tennantite) and **Cu_3SbS_3** (tetrahedrite, skinnerite) ($T_m = 640\,°C$ and $555\,°C$, respectively) are called semiconductors in [57W] without presentation of data (see also [80W]). For elastic constants of Cu_3AsS_3, see [81B].

Ag_3AsS_3:

band structure: Fig. 19, Brillouin zone: Fig. 20

Physical property	Numerical value	T [K] / Experimental conditions	Experimental method remarks	Ref.
$E_{g,\,ind}$	2.012 eV	300, $E \parallel c$	for participating phonons and	71D
	2.004 eV	$E \perp c$	tentative identification of the	
$E_{g,\,dir}$	2.156 eV	$E \parallel c$	structure of the reflection	71D
	2.125 eV	$E \perp c$	spectrum, see [71D] (Figs. 21 and 22)	
$E_{g,\,th}$	1.86 eV		from conductivity, cf. Fig. 23	75G2
dE_g/dT	$-3.48 \cdot 10^{-4}$ eV/K	$77 \cdots 300$	absorption edge (Fig. 21)	71D
σ_{ion}	$\approx 10^{-5}\,\Omega^{-1}\,cm^{-1}$		ionic conductivity by Ag^+-ions	69B, 69D
σ_{el}	$0.53 \cdot 10^{-5}\,\Omega^{-1}\,cm^{-1}$	300, $\parallel c$ axis	electronic conductivity, measured with ac (10 kHz)	74B1
	$1.22 \cdot 10^{-5}\,\Omega^{-1}\,cm^{-1}$	$\perp c$ axis	temperature dependence of electronic conductivity, see Fig. 23; see also [81Z]	69B, 69D

For activation energy of conductivity in the impurity region, see Fig. 23 and [75G2].

Physical property	Numerical value	Experimental conditions	Experimental method remarks	Ref.
$\varepsilon(0)$	21.4	295, $E \parallel c$	very high apparent dielectric	75R
	44.5	$E \perp c$	constants are observed at low	
$\varepsilon(\infty)$	6.3	$E \parallel c$	frequencies (≈ 1000) due to	75R
	7.45	$E \perp c$	space charge effects involving ionic conduction and electrode processes [74B2]; see also [79P]	
$n_o - n_e$	> 0.2	$\lambda = 0.6 \cdots 4.6\,\mu m$	birefringence, for details, see [67H]	67H
χ	$-0.332 \cdot 10^{-6}\,cm^3/g$	300, $\parallel c$ axis		74B1
	$-0.481 \cdot 10^{-6}\,cm^3/g$	$\perp c$ axis		
T_m	480 °C			57W
	490 °C			75G1

Further lattice properties

Elastic, piezoelectric and dielectric coefficients are listed in [82O], photoelastic constants in [75E], non-linear acoustic properties are discussed in [81V1, 81V2, 82B]; coefficient of linear expansion: Fig. 25; heat capacity: [82Z].

Further optical properties

Range of transparency: $0.6 \cdots 13\,\mu m$ (see Fig. 24 and [70F]), reflection spectrum: Fig. 3, Raman and IR active phonon wavenumbers are listed in [75R, 79S, 80T, 83R], Brillouin scattering is studied in [81S].

Physical property	Numerical value	Experimental conditions	Experimental method remarks	Ref.
Ag_3SbS_3:		T [K]		
$E_{g,th}$	1.77 eV		from conductivity	75G2
E_g	1.93 eV	300, synthetic single crystal	absorption edge	73G
	2.06 eV	film	see Fig. 26	76G2
dE_g/dT	$-8 \cdot 10^{-4}$ eV/K	film		76G2
σ	$0.3 \cdot 10^{-4}\,\Omega^{-1}\,cm^{-1}$	300, $\parallel c$ axis	measured with ac (10 KHz); temperature dependence of conductivity and carrier activation energies, see Fig. 27 and [75G2]	74B1
	$0.6 \cdot 10^{-4}\,\Omega^{-1}\,cm^{-1}$	$\perp c$ axis		
$\varepsilon(0)$	27	300		73G
χ	$-0.316 \cdot 10^{-6}\,cm^3/g$	300, $\parallel c$ axis		74B1
	$-0.401 \cdot 10^{-6}\,cm^3/g$	$\perp c$ axis		
T_m	473(3) °C			73G

Range of transparency: 0.7⋯14 µm (Fig. 28, [70F]); for properties of amorphous films, see [71P].
For a discussion of low-temperature Raman spectra, see [83E].
The elastic, piezoelectric and dielectric coefficients are listed in [82O]; coefficient of linear expansion: Fig. 25.

10.4.2.3 References for 10.4.2

57W   Wernick, J.H., Benson, K.E.: J. Phys. Chem. Solids **3** (1957) 157.
58W1   Wernick, J.H., Geller, S., Benson, K.E.: J. Phys. Chem. Solids **4** (1958) 154.
58W2   Wernick, J.H., Geller, S., Benson, K.E.: J. Phys. Chem. Solids **7** (1958) 240.
58Z   Zhuse, V.P., Sergeeva, V.M., Shtrum, E.L.: Sov. Phys. Tech. Phys. **3** (1958) 1925 (transl. from Zh. Tekhn. Fiz. **28** (1958) 2093).
59G   Geller, S., Wernick, J.H.: Acta Crystallogr. **12** (1959) 46.
60A   Armstrong, R.W., Faust, J.W., Tiller, W.A.: J. Appl. Phys. **31** (1960) 1954.
60H   Haake, G., Poganski, S.: Proc. Int. Conf. Phys. Semicond., Prague **1960,** Publ. House of the Acad. Sci., Prague **1960,** p. 999.
60W   Wolfe, R.W., Wernick, J.H., Hazko, S.E.: J. Appl. Phys. **31** (1960) 1959.
62P   Petrov, A.V., Shtrum, E.L.: Sov. Phys. Solid State **4** (1962) 1061 (transl. from Fiz. Tverd. Tela **4** (1962) 1442).
63N   Nensberg, E.D., Shtrum, E.L.: Sov. Phys. Solid State **5** (1963) 2463 (transl. from Fiz. Tverd. Tela **5** (1963) 3357).
64I   Imamov, R.M., Pinsker, Z.G., Ivcenko, A.I.: Kristallografija **9** (1964) 853.
67H   Hulme, K.F., Jones, O., Davies, P.H., Hobden, M.V.: Appl. Phys. Lett. **10** (1967) 133.
68A   Abdullaev, G.B., Mal'sagov, A.U., Glazov, V.M.: Inorg. Mater. **4** (1968) 1082 (transl. from Izv. Akad. Nauk SSSR, Neorg. Mater. **4** (1968) 1233).
68I   Imamov, R.M., Petrov, I.I.: Kristallografija **13** (1968) 412.
68K   Krestovnikov, A.N., Mal'sagov, A.U., Glazov, V.M.: Inorg. Mater. **4** (1968) 119 (transl. from Izv. Akad. Nauk SSSR, Neorg. Mater. **4** (1968) 144).
68L   Lushnaya, N.P., Berul', S.I., Finkel'shtein, Ya.G.: Inorg. Mater. **4** (1968) 286 (transl. from Ivz. Akad. Nauk SSSR, Neorg. Mater. **4** (1968) 342).
69B   Bardsley, W., Davies, P.H., Hobden, M.V., Hulme, K.F., Jones, O., Pomeroy, W., Warner, J.: Opto-electronics **1** (1969) 29.
69D   Davis, P.H., Elliott, C.T., Hulme, K.F.: Brit. J. Appl. Phys. Ser. 2, **2** (1969) 165.
69Z   Zorina, E.L., Gnidash, N.I., Finkel'shtein, Ya.G., Verul', S.I., Lushnaya, N.P.: Inorg. Mater. **5** (1969) 1788 (transl. from Izv. Akad. Nauk SSSR, Neorg. Mater. **5** (1969) 2099).
70F   Feichtner, J.D., Johannes, R., Roland, G.W.: Appl. Optics **9** (1970) 1716.

70G1	Gnidash, N.I., Sukhorukova, L.N., Kuznetsov, M.S., Finkel'shtein, Ya.G., Berul', S.I., Luzhnaya, N.P., Bazakutsa, V.A.: Inorg. Mater. **6** (1970) 208 (transl. from Izv. Akd. Nauk SSSR, Neorg. Mater. **6** (1970) 237).
70G2	Golovei, M.I., Berul', S.I., Luzhnaya, N.P., Peresh, E.Yu.: Inorg. Mater. **6** (1970) 961 (transl. from Izv. Akad. Nauk SSSR, Neorg. Mater. **6** (1970) 1100).
70S	Sobolev, V.V., Berul', S.I., Vorob'ev, V.G., Finkel'shtein, Ya.G., Luzhnaya, N.P.: Inorg. Mater. **6** (1970) 1350 (transl. from Izv. Akad. Nauk SSSR, Neorg. Mater. **6** (1970) 1532).
71D	Dovgii, Ya.O., Butsko, N.I., Korolyshin, V.N., Moroz, E.T.: Sov. Phys. Solid State **13** (1971) 995 (transl. from Fiz. Tverd. Tela **13** (1971) 1202).
71K	Kovaleva, I.S., Popova, L.D., Luzhnaya, N.P., Sukhankina, V.V., Antonova, L.I.: Inorg. Mater **7** (1971) 1340 (transl. from Izv. Akad. Nauk SSSR, Neorg. Mater. **7** (1971) 1512).
71P	Popova, L.D., Voinova, L.G., Luzhnaya, N.P., Kovaleva, I.S., Bazakutsa, V.A.: Inorg. Mater. **7** (1971) 278 (transl. from Izv. Akad. Nauk SSSR, Neorg. Mater **7** (1971) 317).
73D	Dovgii, Ya.O., Korolyshin, V.N., Moroz, E.T.: Sov. Phys. Dokl. **17** (1973) 1070 (transl. from Dokl. Akad. Nauk SSSR **207** (1972) 71).
73E	Esayan, S.Kh., Lemanov, V.V., Rez, I.S., Shakin, O.V.: Sov. Phys. Solid State **15** (1973) 627 (transl. from Fiz. Tverd. Tela **15** (1973) 907).
73G	Golovei, M.I., Gurjan, M.I., Olekseyuk, I.D., Rez, I.S., Voroshilov, Yu.V., Roman, I.Y.: Krist. Tech. **8** (1973) 453.
74B1	Butsko, N.I., Pidorya, M.M., Krushel'nitskaya, T.D.: Sov. Phys. Crystallogr. **18** (1974) 540 (transl. from Kristallografiya **18** (1973) 855).
74B2	Byer, H.H., Bobb, L.C.: J. Appl. Phys. **45** (1974) 3738.
74G	Golovach, I.I., Slivka, V.Yu., Dovgoshei, N.I., Syrbu, N.N., Bogdanova, A.V., Golovei, M.I.: Sov. Phys. Semicond. **9** (1974) 834 (transl. from Fiz. Tekh. Poluprovodn. **9**(1974) 1260).
75G1	Golovach, I.I., Dovgoshei, N.I., Slivka, V.Yu., Suslikov, L.M., Golovei, M.I., Bogdanova, A.V.: Inorg. Mater. **11** (1975) 820 (transl. from Izv. Akad. Nauk SSSR, Neorg. Mater. **11** (1975) 956).
75G2	Gurzan, M.I., Golovei, M.I., Bodnar, M.P., Chepur, D.V.: Inorg. Mater. **11** (1975) 1149 (transl. from Izv. Akad. Nauk SSSR, Neorg. Mater. **11** (1975) 1349).
75K	Kovaleva, I.S., Tokbaeva, K.A., Antonova, L.I., Luzhnaya, N.P.: Inorg. Mater. **11** (1975) 136 (transl. from Izv. Akad. Nauk SSSR, Neorg. Mater. **11** (1975) 163).
75R	Riccius, H.D., Carey, P.R., Siimann, O.: Phys. Status Solidi (**b**) **72** (1975) K99.
76B	Belyaev, A.D., Baisa, D.F., Bondar, A.V., Machulin, V.F., Miselyuk, E.G.: Sov. Phys. Solid State **18** (1976) 1018 (transl. from Fiz. Tverd. Tela **18** (1976) 1749).
76G1	Golovach, I.I., Slivka, V.Yu., Matyashovskii, V.V., Dovgoshei, N.I., Bentsa, V.M., Golovei, M.I.: Sov. Phys. Solid State **18** (1976) 1930 (transl. from Fiz. Tverd. Tela **18** (1976) 3313).
76G2	Golovei, M.I., Kovach, E.T., Dovgoshei, I.I., Chepur, D.V., Stefanovich, V.A., Lada, A.V., Loya, V.Yu.: Inorg. Mater. **12** (1976) 847 (transl. from Izv. Akad. Nauk SSSR, Neorg. Mater. **12** (1976) 1011).
76G3	Grigas, I., Mozgova, N.N., Orlyukas, A., Samulenis, V.: Sov. Phys. Crystallogr. **20** (1976) 741 (transl. from Kristallografiya **20** (1976) 1226).
76N	Novik, V.K., Drozhdin, S.N., Popova, T.V., Koptsik, V.A., Gavrilova, N.D.: Sov. Phys. Solid State **17** (1976) 2286 (transl. from Fiz. Tverd. Tela **17** (1975) 3499).
77B1	Bazakutsa, V.A., Lazarev, V.B., Zozulya, L.P., Gnidarch, N.I., Kul'chitskaya, A.K., Salow, A.V.: Inorg. Mater. **13** (1977) 971 (transl. from Izv. Akad. Nauk SSSR, Neorg. Mater. **13** (1966) 1198).
77B2	Bohac, P., Orliukas, A., Gäumann, A., Girgis, K.: Helv. Phys. Acta **50** (1977) 853.
77M	Manolikas, C., Spyridelis, J.: Mater. Res. Bull. **12** (1977) 907.
78A	Abdikamalov, B.A., Ivanov, V.I., Shekhtman, V.Sh., Shmyt'ko, I.M.: Sov. Phys. Solid State **20** (1978) 1711 (transl. from Fiz. Tverd. Tela **20** (1978) 1968).
79B	Borisova, L., Dimitrova, S.: Phys. Status Solidi (**a**) **53** (1979) 403.
79P	Popova, T.V., Gavrilova, N.D., Novik, V.K., Koptsik, V.A., Gurzan, M.I., Voroshilov, Yu.V.: Sov. Phys. Solid State **21** (1979) 45 (transl. from Fiz. Tverd. Tela **21** (1979) 76).
79S	Smooenskii, G.A., Sinii, I.G., Kuz'minov, E.G., Godovikov, A.A.: Sov. Phys. Solid State **21** (1979) 1343 (transl. from Fiz. Tverd. Tela **21** (1979) 2332).
79V	Valyukenas, V.I., Orlyukas, A.S., Sakals, A.P., Mikolaitis, V.A.: Sov. Phys. Solid State **21** (1979) 1409 (transl. from Fiz. Tverd. Tela **21** (1979) 2449).
80B	Baleva, M.: Phys. Status Solidi (**b**) **101** (1980) 389.
80T	Taylor, W., Paul, G.L.: Solid State Commun. **35** (1980) 829.

80V	Valyukenas, V.I., Orlyukas, A.S., Roizentok, V.S.: Sov. Phys. Semicond. **14** (1980) 343 (transl. from Fiz. Tekh. Poluprovodn. **14** (1980) 581).
80W	Whitfield, H.J.: Solid State Commun. **33** (1980) 747.
81B	Babushkin, A.N., Zlokazov, V.B., Zadvorkin, S.M., Kobelev, L.Ya., Kuznetsov, Yu.S.: Fiz. Tverd. Tela **23** (1981) 3705.
81S	Smolenskii, G.A., Sinii, I.G., Prokhorova, S.D., Godovikov, A.A., Laikho, R., Levola, T., Karae-myaki, E.: Sov. Phys. Solid State **23** (1981) 1178 (transl. from Fiz. Tverd. Tela **23** (1981) 2017).
81V1	Vil'chinskas, Sh.P., Zarembo, L.K., Novik, V.K., Serdobol'skaya, O.Yu.: Sov. Phys. Solid State **23** (1981) 1062 (transl. from Fiz. Tverd. Tela **23** (1981) 1821).
81V2	Vil'chinskas, Sh.P., Zarembo, L.K., Serdobol'skaya, O.Yu., Novik, V.K.: Sov. Phys. Solid State **23** (1981) 814 (transl. from Fiz. Tverd. Tela **23** (1981) 1395).
81Z	Zlokazov, V.B., Kobelev, L.Ya., Karpachev, S.V.: Sov. Phys. Dokl. **26** (1981) 684 (transl. from Dokl. Akad. Nauk SSSR **259** (1981) 344).
82B	Belyaev, A.D., Gololobov, Yu.P., Machulin, V.F., Miselyuk, E.G.: Sov. Phys. Solid State **24** (1982) 1077 (transl. from Fiz. Tverd. Tela **24** (1982) 1886).
82O	O'Hara, C., Shorrocks, N.M., Whatmore, R.W., Jones, O.: J. Phys. D **15** (1982) 1289.
82S	Slivka, V.Yu., Vysochanskii, Yu.M., Stefanovich, V.A., Gerasimenko, V.S., Chepur, D.V.: Sov. Phys. Solid State **24** (1982) 392 (transl. from Fiz. Tverd. Tela **24** (1982) 696).
82Z	Zlokazov, V.B., Babushkin, A.N., Kobelev, L.Ya., Gorin, Yu.F.: Sov. Phys. Solid State **24** (1982) 335 (transl. from Fiz. Tverd. Tela **24** (1982) 597).
83E	Ewen, P.J.S., Taylor, W.: Solid State Commun. **45** (1983) 227.
83R	Rebane, L.A., Haller, K.E.: Sov. Phys. Solid State **24** (1983) 1335 (transl. from Fiz. Tverd. Tela **24** (1982) 2351).

10.4.3 II_x–III_y–VI_z compounds

II–III$_2$–VI$_4$ compounds: see section 10.1.6.

II$_3$–III$_2$–VI$_6$ compounds: see section 10.1.7.

II–III–VI$_2$ compounds: $CdInS_2$, $CdInSe_2$, $CdInTe_2$, $CdTlS_2$, $CdTlSe_2$, $CdTlTe_2$, $HgTlS_2$.

10.4.3.0 Structure

We restrict the discussion in this section to II–III–VI$_2$ compounds. II–III$_2$–VI$_4$ and II$_3$–III$_2$–VI$_6$ compounds have already been treated in sections 10.1.6 and 10.1.7. There exist some semiconducting ternary intermediate phases in the II–III–VI-system of the type II_m–III_2–VI_{m+3}. $Zn_2In_2S_5$ and $Zn_3In_2S_6$ have properties very similar to $ZnIn_2S_4$ [71D, 72S, 69R, 70R].

The II–III–VI$_2$ compounds crystallize in a trigonal (α) modification and a tetragonal (β) modification. The lattice of the α-phase consists of a hexagonal close-packed arrangement of VI-atoms with II-atoms and III-atoms located in layers in the octahedral sites. The space group is D_{3d}^3—P$\bar{3}$m1 or C_{3v}^1—P3m1 [67G]. The exact structure of the β-phase is not yet known. There are eight molecules in the unit cell. A tentative determination of the space group yielded D_{4h}^{17}—I4/mmm [66G].

lattice parameters:

	Substance	a [Å]	c [Å]	Ref.
Trigonal phase:	$CdInS_2$	3.603	6.825	69G1
	$CdTlS_2$	3.645	6.825	67G
	$CdTlSe_2$	3.723	7.073	69G2
	$CdTlTe_2$	3.890	7.220	69G2
Tetragonal phase:	$CdInS_2$	11.586	6.522	69G1
	$CdInSe_2$	12.152	7.14	69G1
	$CdInTe_2$	12.612	7.434	69G1
	$CdTlS_2$	11.784	6.668	69G1
	$CdTlSe_2$	12.174	7.212	69G1
	$CdTlTe_2$	12.669	7.528	69G1
	$ZnInTe_2$	12.18	6.09	69G1
	$HgTlS_2$	12.20(5)	6.60(2)	66G

Physical property	Numerical value	Experimental conditions	Experimental method remarks	Ref.

10.4.3.1 Physical properties

All nine Tl compounds (Hg, Cd, Zn)Tl(S_2, Se_2, Te_2) have been synthesized and proved to display semi-conducting properties [66G]. As to the In compounds only the CdIn(S_2, Se_2, Te_2) compounds have been investigated [69G1]. No detailed investigation of ZnTl(S_2, Se_2, Te_2) and of HgTlSe$_2$(Te$_2$) has been published.

β-CdInS$_2$ [69G1]:

$E_{g, th}$	1.70···1.74 eV		temperature dependence of con-	
dE_g/dT	$-1.43 \cdot 10^{-4}$ eV K^{-1}		ductivity and Hall coefficient	
			(Fig. 1 and 2a)	
m_n	0.172 m_0		analysis of transport measurements	
m_p	0.44 m_0			
d_{th}	4.421 g/cm^3			
d_{exp}	4.420 g/cm^3			

For Hall mobilities above room temperature, see Figs. 1, 2.

β-CdInSe$_2$ [69G1]:

$E_{g, th}$	1.40···1.42 eV		temperature dependence of con-	
dE_g/dT	$-1.03 \cdot 10^{-4}$ eV K^{-1}		ductivity, Hall effect, thermo-	
m_p	0.23 m_0		electric power (Fig. 2)	
d_{th}	4.97 g/cm^3			
d_{exp}	4.970 g/cm^3			

For Hall mobilities above room temperature, see Fig. 2.

β-CdInTe$_2$ [69G1]:

$E_{g, th}$	1.10···1.12 eV		temperature dependence of con-	
dE_g/dT	$-2.54 \cdot 10^{-4}$ eV K^{-1}		ductivity and Hall coefficient	
m_n	0.08 m_0		(Fig. 2)	
m_p	0.18 m_0			
d_{th}	5.41 g/cm^3			
d_{exp}	5.550 g/cm^3			

For Hall mobilities above room temperature, see Fig. 2.

α-CdTlS$_2$:

		T [K]		
E_g	1.46 eV	0	temperature dependence	67G,
			of conductivity (Fig. 4)	69G2
	1.52 eV		of Hall coefficient (Figs. 4, 6)	
	1.56 eV	300	spectral dependence of	
			photoconductivity (Fig. 3)	
dE_g/dT	$-1.3 \cdot 10^{-4}$ eV K^{-1}			67G,
				69G2
m_p	0.6 m_0		analysis of transport	69G2
m_n	0.19 m_0		measurements	69G2
μ_p	4200 cm^2/Vs	300	Hall mobility of a polycrystalline	69G2
			sample. For temperature depend-	
			ence of the Hall mobility of a single	
			crystal see Figs. 4c, 6b.	
T_m	≈ 600 °C			69G2
n_i	$1.08 \cdot 10^{19}$ cm^{-3}	730		6962
d_{th}	5.06 g cm^{-3}			69G1
d_{exp}	5.05 g cm^{-3}			69G1

Measurements of electrical and thermal conductivity, Hall effect and thermoelectric power: Figs. 4···6.

Physical property	Numerical value	Experimental conditions	Experimental method remarks	Ref.
α-CdTlSe$_2$:				
$E_{g,th}$	0.40 eV		conductivity, Hall effect (Fig. 6)	69G2
dE_g/dT	$-4.1 \cdot 10^{-4}$ eV K^{-1}			69G2
m_n	0.11 m_0		analysis of transport measurements	69G2
m_p	0.65 m_0			69G2
d_{th}	5.91 g cm^{-3}			69G1
d_{exp}	5.89 g cm^{-3}			69G1

Measurement of transport coefficients and of the Hall mobility, see Fig. 6.

α-CdTlTe$_2$:				
$E_{g,th}$	0.18 eV		conductivity, Hall effect (Fig. 6)	69G2
d_{th}	6.28 g cm^{-3}			69G1
d_{exp}	6.40 g cm^{-3}			69G1

Measurements of transport coefficients and of Hall mobility, see Fig. 6.

HgTlS$_2$:				
E_g	1.28 eV		conductivity (Fig. 7)	66G
	1.25 eV	$T = 300$ K	maximum in the spectral distribution of photoconductivity (Fig. 8)	68G
dE_g/dT	$-2.46 \cdot 10^{-4}$ eV K^{-1}		shift of absorption edge	68G
d_{th}	6.34 g cm^{-3}			68G

n-and p-type conduction has been found; thermal conductivity: Fig. 9.

10.4.3.2 References for 10.4.3

66G   Guseinov, G.D., Ismailov, M.Z., Talybov, A.G.: Phys. Status Solidi **18** (1966) 929.
67G   Guseinov, G.D., Ismailov, M.Z., Guseinov, G.G.: Mater. Res. Bull. **2** (1967) 765.
68G   Guseinov, G.D., Ismailov, M.Z., Talybov, A.G.: Inorg. Mater. **4** (1968) 440 (transl. from Izv. Akad. Nauk SSSR, Neorg. Mater. **4** (1968) 514).
69G1 Guseinov, G.D., Abdullaev, G.B., Kerimova, E.M., Gamidov, R.S., Guseinov, G.G.: Mater. Res. Bull. **4** (1969) 807.
69G2 Guseinov, G.D., Guseinov, G.G., Ismailov, M.Z., Godzhaev, E.M.: Inorg. Mater. **5** (1969) 27 (transl. from Izv. Akad. Nauk SSSR, Neorg. Mater. **5** (1969) 33).
69R   Radautsan, S.I., Donika, F.G., Kyosse, G.A., Mustya, I.G., Zhitar, V.F.: Phys. Status Solidi **34** (1969) K129.
70R   Radautsan, S.I., Donika, F.G., Kyosse, G.A., Mustya, I.G.: Phys. Status Solidi **37** (1970) K123.
71D   Damaskin, I.A., Donika, F.G., Mustya, I.G., Pyshkin, S.L., Radautsan, S.I.: Sov. Phys. Semicond. **4** (1971) 1723 (transl. from Fiz. Tekh. Poluprovodn. **4** (1971) 2009).
72S   Sobolev, V.V.: Inorg. Mater. **8** (1972) 21 (transl. from Izv. Akad. Nauk SSSR, Neorg. Mater. **8** (1972) 26).

10.4.4 III$_x$–V$_y$–VI$_z$ compounds

III–V–VI$_2$ compounds: TlPS$_2$, TlAsSe$_2$, TlSbS$_2$, TlSbSe$_2$, TlSbTe$_2$, TlBiS$_2$, TlBiSe$_2$, TlBiTe$_2$. Other III$_x$–V$_y$–VI$_z$ compounds are listed below.

The **III–V–VI$_2$ compounds** can be considered as analogues to the IV–VI compounds.

TlSbTe$_2$ and the three Bi compounds **TlBiS$_2$(Se$_2$, Te$_2$)** crystallize in a rhombohedral structure (space group D$_{3d}^5$–R$\bar{3}$m). The lattice parameters are given for TlSbTe$_2$ and TlBiTe$_2$ as $a = 8.177$ (10) Å, $\alpha = 31°25(15)'$ and $a = 8.137$ (10) Å, $\alpha = 32°18(15)'$, respectively, [61H]. For TlBiS$_2$ the lattice parameters are given in hexagonal description as $a = 4.15$ (3) Å, $c = 10.91$ (4) Å [74B2] and for TlBiSe$_2$ as $a = 4.24$ Å, $c = 22.33$ Å [63M].

Physical property	Numerical value	Experimental conditions	Experimental method remarks	Ref.

TlAsS$_2$ crystallizes in a monoclinic structure (C_{2h}^5—P2$_1$/c) with $a = 6.11$ Å, $b = 11.33$ Å, $c = 12.27$ Å, $\beta = 104.2°$ [59Z].

TlSbS$_2$ crystallizes in a disordered NaCl-structure (O_h^5—Fm3m, lattice parameter $a = 5.87 \cdots 5.94$ Å [59S]) like many of the I–V–VI$_2$ compounds (see section 10.4.2).

TlSbSe$_2$ has an orthorhombic structure with $a = 4.20$ Å, $b = 9.0$ Å, $c = 24.0$ Å according to [75B].

These compounds – as well as many other III$_x$–V$_y$–VI$_z$ compounds with semiconducting properties – occur as ternary phases in the phase diagrams of the pseudo-binary systems (III$_x$V$_y$)–(III$_u$VI$_v$) and (III$_x$V$_y$)–(V$_u$VI$_v$). Systems investigated are:

System	Ternary compound	Ref.
GaSb—GaTe (Fig. 1)	Ga$_6$Sb$_5$Te	66K
InSb—InTe (Fig. 2)	In$_6$Sb$_5$Te, In$_7$SbTe$_6$ (another compound In$_4$SbTe$_3$ [59G] has not been confirmed in later experiments [66K])	66K
InAs—InTe (Fig. 3)	In$_8$As$_5$Te$_3$	66K
In$_2$O$_3$—Sb$_2$O$_3$	In$_2$Sb$_4$O$_9$, InSbO$_4$	75V
In$_2$S$_3$—Sb$_2$S$_3$ (Fig. 4)	InSbS$_3$, InSb$_3$S$_6$	77G, 73K
In$_2$Se$_3$—Sb$_2$S$_3$	InSbSe$_3$	77G
In$_2$Te$_3$—Sb$_2$Te$_3$ (Fig. 5)	In$_7$Sb$_3$Te$_{15}$	66K
In$_2$Te$_3$—Bi$_2$Te$_3$ (Fig. 6)	In$_4$Bi$_6$Te$_{15}$	66K
Tl$_2$S—As$_2$S$_3$	TlAsS$_2$, Tl$_3$AsS$_3$, Tl$_4$As$_2$S$_5$, Tl$_6$As$_4$S$_9$	74B1
Tl$_2$S—Sb$_2$S$_3$ (Fig. 7)	TlSbS$_2$, Tl$_3$SbS$_3$, TlSb$_5$S$_8$	74B1
Tl$_2$S—Bi$_2$S$_3$ (Fig. 8)	TlBiS$_2$, Tl$_4$Bi$_2$S$_5$	69D1
Tl$_2$Se—As$_2$Se$_3$ (Fig. 9)	TlAsSe$_2$, Tl$_3$AsSe$_3$	74B1, 69D, 74R
Tl$_2$Se—Sb$_2$Se$_3$ (Fig. 10)	TlSbSe$_2$, Tl$_5$SbSe$_4$, Tl$_9$SbSe$_6$	73G, 75B
Tl$_2$Te—Sb$_2$Te$_3$ (Fig. 11)	TlSbTe$_2$, Tl$_9$SbTe$_6$	77B1
Tl$_2$Te—Bi$_2$Te$_3$	TlBiTe$_2$ (no other compound found in [66S]), TlBiTe$_3$ [63B] not confirmed, Tl$_9$BiTe$_6$ cited in [74B1]	

Solid solutions of the type TlSb$_{1-x}$Bi$_x$Te$_2$, (PbTe)$_{1-x}$(Tl(Sb, Bi)Te$_2$)$_{x/2}$, (SnTe)$_{1-x}$(Tl(Sb, Bi)Te$_2$)$_{x/2}$ are described in [62M].

The following table summarizes details on the physical properties of several III$_x$–V$_y$–VI$_z$ compounds:

Some details on **TlPS$_2$** and other Tl-P compounds are given in [68B].

TlAsS$_2$:

E_g	1.3 eV	$T = 300$ K		77B2
T_m	260 °C			77B2
d	5.53 g/cm^3			59Z

TlSbS$_2$:

		T [K]		
$E_{g,th}$	1.42 eV		temperature dependence of conductivity	69D1
	1.7 eV			74B1
$E_{g,ind}$	1.556 eV	300	reflectivity (allowed indirect transition, see Fig. 12)	70S
	1.54 eV	300	electroabsorption	70B
$E_{g,dir}$	1.69 eV	300	reflectivity (allowed direct transition, see Figs. 12, 13)	70B
$\hbar\omega_{LA}$	33.2 meV	300	analysis of reflectivity spectra of Fig. 16	75S
$\hbar\omega_{TA}$	9.3 meV			
$\hbar\omega_{LO}$	35.6 meV			
$\hbar\omega_{TO}$	38.8 meV			

(continued)

Physical property	Numerical value	Experimental conditions	Experimental method remarks	Ref.
TlSbS$_2$ (continued)				
$\varepsilon(\infty)$	10.732	300	from dispersion of $n_\infty = \sqrt{\varepsilon(\infty)}$, see Fig. 14	70S
$\varepsilon(0)$	11.891	300		75S
ϱ	$10^9\,\Omega\,cm$	293		74Z
T_m	484 (2) °C			74B1
d_{th}	6.40⋯6.18 g/cm^3			59S

For photoconductivity spectrum, see Fig. 15 and [74Z, 76C].

TlSbSe$_2$:

No semiconductor data are available. For phonon absorption bands and their interpretation, see Fig. 6 and [75S].

TlSbTe$_2$:

Some qualitative results about transport measurements in [74G].

TlBiS$_2$:

E_g	0.40 eV			69D1
T_m	740 °C			69D1

For further properties (phase transitions under pressure etc.), see [74B2, 68D].

TlBiSe$_2$:

E_g	0.28 eV			69D1
T_m	720 °C			69D1

TlBiTe$_2$:

Highly degenerate semiconductor [72J] with Hall mobilities of electrons of 64 cm^2/V s at 300 K and 128 cm^2/V s at 4.2 K, $n = 5 \cdot 10^{19}$ cm^{-3}, $T_m = 535$ °C, $d_{exp} = 8.06$ g/cm^3 [61H]. For conductivity and Hall coefficient, see Fig. 17, see also [81V]. Superconductivity in TlBiTe$_2$ has been reported in [70H].

Ga$_6$Sb$_5$Te [66K]:

E_g	0.65 eV	$T = 300$ K	transmission, see Fig. 18	
	0.80 eV	0 K	conductivity, see Fig. 19	
ϱ	3.8 Ω cm	$p = 2 \cdot 10^{17}$ cm^{-3}		
μ_p	9.5 cm^2/V s	$T = 300$ K		
κ_L	$7.0 \cdot 10^{-2}$ W/cm K			

In$_6$Sb$_5$Te [66K]:

ϱ	$0.7 \cdot 10^{-3}\,\Omega$ cm	$n = 8 \cdot 10^{18}$ cm^{-3}		
μ_n	1200 cm^2/V s	$T = 300$ K		
κ_L	$3.1 \cdot 10^{-2}$ W/cm K			

In$_7$SbTe$_6$ [66K]:

ϱ	$2.5 \cdot 10^{-3}\,\Omega$ cm	$p = 5 \cdot 10^{18}$ cm^{-3}		
μ_p	440 cm^2/V s	$T = 300$ K		
κ_L	$3.6 \cdot 10^{-2}$ W/cm K			

There is little information about the other substances listed at the beginning of this section. We refer to the literature given there.

References for 10.4.4

59G	Goryunova, N.A., Radautsan, S.I., Kiosse, G.A.: Sov. Phys. Solid State **1** (1959) 1702 (transl. from Fiz. Tverd. Tela **1** (1959) 1958).
59S	Semiletov, S.A., Man, L.I.: Sov. Phys. Crystallogr. **4** (1959) 383 (transl. from Kristallografiya **4** (1959) 414).
59Z	Zemann, A., Zemann, J.: Acta Crystallogr. **12** (1959) 1002.
61H	Hockins, E.F., White, J.G.: Acta Crystallogr. **14** (1961) 328.
62M	Mazelsky, R., Lubell, M.S.: J. Phys. Chem. **66** (1962) 1408.
63B	Borisova, L.A., Efremova, M.V., Vlasov, V.V.: Dokl. Akad. Nauk SSSR **149** (1963) 117.
63M	Man, L.I., Semiletov, S.A.: Sov. Phys. Crystallogr. **7** (1963) 686 (transl. from Kristallografiya **7** (1962) 844).
66K	Kurata, K., Hirai, T.: Solid State Electron. **9** (1966) 633.
66S	Spitzer, D.P., Sykes, J.A.: J. Appl. Phys. **37** (1966) 1563.
68B	Batsanov, S.S., Rigin, V.I., Derbeneva, S.S.: Inorg. Mater. **4** (1968) 117 (transl. from Izv. Akad. Nauk SSSR, Neorg. Mater. **4** (1968) 143).
68D	Dembovskii, S.A., Lisovskii, L.G., Bunin, V.M.: Inorg. Mater. **4** (1968) 115 (transl. from Izv. Akad. Nauk SSSR, Neorg. Mater. **4** (1968) 140).
69D1	Dembovskii, S.A., Lisovskii, L.G., Bunin, V.M., Kanishcheva, A.S.: Inorg. Mater. **5** (1969) 1724 (transl. from Izv. Akad. Nauk SSSR, Neorg. Mater **5** (1968) 2023).
69D2	Dembovskii, S.A., Kirilenko, V.V., Khvorostenko, A.V.: Russ. J. Inorg. Chem. **14** (1969) 1347.
70B	Botgros, I.V., Stepanov, G.I., Chinik, B.S., Stratan, G.I., Kantser, Ch.T.: Sov. Phys. Solid State **12** (1970) 495 (transl. from Fiz. Tverd. Tela **12** (1970) 643).
70H	Hein, R.A., Swiggard, E.M.: Phys. Rev. Lett. **24** (1970) 53.
70S	Stepanov, G.I., Botgros, I.V., Chinik, B.S., Dontsoi, P.I.: Sov. Phys. Solid State **12** (1970) 1423 (transl. from Fiz. Tverd. Tela **12** (1970) 1797).
72J	Jensen, J.D., Burke, J.R., Ernst, D.W., Allgaier, R.S.: Phys. Rev. **B 6** (1972) 319.
73G	Gäumann, A., Bohac, P.: J. Less-Common Met. **31** (1973) 314.
73K	Kompanichenko, N.M., Tchans, I.S., Suchenko, B.D., Sheka, I.A., Lugin, W.N.: Zh. Neorg. Khim. **18** (1973) 1080.
74B1	Bohac, P., Brönnimann, E., Gäumann, A.: Mater. Res. Bull. **9** (1974) 1033.
74B2	Brandt, N.B., Gitsu, D.V., Popovich, N.S., Sidorov, V.I., Chudinov, S.M.: Sov. Phys. Semicond. **8** (1974) 390 (transl. from Fiz. Tekh. Poluprovodn. **8** (1974) 609).
74G	Gitsu, D.V., Kantser, Ch.T., Stratan, G.I., Cheban, A.G.: Sov. Phys. Semicond. **7** (1974) 1251 (transl. from Fiz. Tekh. Poluprovodn. **7** (1973) 1874).
74R	Roland, G.W., McHugh, J.P., Feichtner, J.D.: J. Electron. Mater **3** (1974) 829.
74Z	Zhitar', V.F., Popovich, N.S., Gitsu, D.V., Radautsan, S.I.: Sov. Phys. Semicond. **8** (1974) 644 (transl. from Fiz. Tekh. Poluprovodn. **8** (1974) 996).
75B	Botgros, I.V., Zbigli, K.R., Stanchu, A.V., Stepanov, G.I., Cheban, A.G., Chumak, G.D.: Inorg. Mater. **11** (1975) 1675 (transl. from Izv. Akad. Nauk SSSR, Neorg. Mater. **11** (1975) 1953).
75S	Stepanov, G.I., Botgros, I.V., Chinik, B.S., Kogalnichanu, N.F., Cheban, A.G.: Sov. Phys. Solid State **17** (1975) 97 (transl. from Fiz. Tverd. Tela **17** (1975) 166).
75V	Varfolomeev, M.B., Sotnikova, M.N., Plyshchev, V.E., Strizhkov, B.V.: Inorg. Mater. **11** (1975) 1209 (transl. from Izv. Akad. Nauk SSSR, Neorg. Mater. **11** (1975) 1416).
76C	Chinik, B.S., Botgros, I.V., Stepanov, G.I.: Sov. Phys. Semicond. **10** (1976) 575 (transl. from Fiz. Tekh. Poluprovodn. **10** (1976) 974).
77B1	Botgros, I.V., Zbigli, K.R., Stanchu, A.V., Stepanov, G.I., Chumak, G.D.: Inorg. Mater. **13** (1977) 975 (transl. from Izv. Akad. Nauk SSSR, Neorg. Mater. **13** (1977) 1202).
77B2	Baidakov, L.A., Funtikov, V.A.: Inorg. Mater. **13** (1977) 749 (transl. from Izv. Akad. Nauk SSSR, Neorg. Mater. **13** (1977) 914).
77G	Guliev, T.N., Rustamov, P.G., Sinechek, V., Mageramov, E.V.: Inorg. Mater. **13** (1977) 517 (transl. from Izv. Akad. Nauk SSSR, Neorg. Mater. **13** (1977) 630).
81V	Valassiades, O., Polychroniadis, E.K., Stoemenos, J., Economou, N.A.: Phys. Status Solidi **(a) 65** (1981) 215.

Physical property	Numerical value	Experimental conditions	Experimental method remarks	Ref.

10.4.5 IV_x–V_y–VI_z compounds

Among the ternary compounds of the IVth, Vth and VIth group of the Periodic System semiconductors of the type V_{12}–IV–VI_{20}, IV–V_2–VI_4 and IV–V_4–VI_7 are of most interest:

V_{12}–IV–VI_{20} compounds: $Bi_{12}SiO_{20}$, $Bi_{12}GeO_{20}$.
IV–V_2–VI_4 compounds: $PbSb_2S_4$, $GeSb_2Te_4$, $SnBi_2Te_4$, $GeBi_2Te_4$.
IV–V_4–VI_7 compounds: $GeBi_4Te_7$, $SnBi_4Te_7$, $GeSb_4Te_7$, $PbBi_4Te_7$.

Furthermore the electro-optical materials $Bi_4Si_3O_{12}$ (eulytine) and $Bi_4Ge_3O_{12}$ (germanoeulytine) are wide-gap semiconductors ($E_g > 4$ eV) [65N, 72D, 78F].

According to [77B] two compounds of the type $mPbS \cdot nSb_2S_3$ ($Pb_5Sb_8S_{17}$ and $Pb_6Sb_{14}S_{27}$) are high-permittivity photoconductors.

$Bi_{12}SiO_{20}$, $Bi_{12}GeO_{20}$

Both materials crystyllize in a cubic structure with space group $T^3 - I23$ [67B1, 67B2]. The lattice constants are $a = 10.10433(5)$ Å at 296 K for $Bi_{12}SiO_{20}$ and 10.1455(8) Å at 298 K for $Bi_{12}GeO_{20}$. The positions of the atoms in the unit cell are very complicated (see e.g. [82B1]). Both materials are optically active [66L, 70F], electro-optic [66L] photoconductors. $Bi_{12}GeO_{20}$ is also known to be piezoelectric [67B1], photo-active [67L] and photo-elastic [69V]. It has unusually good ultrasonic properties (see e.g. [66S, 83G]). $Bi_{12}SiO_{20}$ shows photoelastic behavior, too [78K]. – For further data of elastic constants (also 3rd order), dielectric constants, strain coefficients, electromechanical coupling factors, elastooptic coefficients and refractive index, see Landolt-Börnstein, NS, Vols. III/11 and III/18.

$Bi_{12}SiO_{20}$:

		T [K]		
E_g	3.25 eV	300	from measurements of photo-	73H
	3.40 eV	80	current excitation (see Fig. 1)	
E_c–E_t	0.34 eV	undoped	thermally stimulated photo-	73H
	0.54 eV	sample	conductivity between	
	0.65 eV		80K and 360K	
E_t–E_v	0.26 eV	heavily	for data on impurity centers	73H
	0.31 eV	Al-doped	from photo- and thermo-	
	0.43 eV	sample	luminescence, see [80P]	
E_c–E_t	2.25 eV		luminescence center	73H
E_t–E_v	0.25 eV		recombination center	73H
m_n	14 m_0		photocarrier kinetic response	74L
$\mu_{dr, n}$	0.029(3) cm^2/V s	80	drift mobility from transit time measurements, for temperature dependence, see Fig. 2	73H
l	$8.5 \cdot 10^{-7}$ cm^2/V	80	range of photoelectrons ($\mu_{dr, n} \tau$)	73H
ϱ	$5 \cdot 10^{13}$ Ω cm	p-type		74Z
	$> 10^4$ Ω cm			73H
$\varepsilon(0)$	56			74Z

Wavenumbers of normal modes have been determined by [72V] from Raman spectra and [79W] from reflectivity (see also [80I, 82B2]):

data from reflectivity [79W] (in cm^{-1}):

TO-modes: 89, 99, 107, 115, 136, 175, 195, 208, 237, 288, 314, 353, 462, 531, 579, 609
LO-modes: 91, 101, 112, 118, 168, 185, 196, 212, 257, 289, 351, 374, 506, 557, 591, 615

data from Raman spectra [72V] (in cm^{-1}):

TO-modes: 98.8, 114.4, 209.0, 238.0, 827.4
LO-modes: 53.5, 100.7, 112.4, 167.0, 180.7, 185.0, 213.0, 841.0
LO + TO-modes: 44.4, 50.6, 58.0, 89.2, 105.7, 135.5, 352.0, 509.1

For optical and Raman spectra, see Figs. 1, 3⋯8. For elastic moduli, see [79A].

Physical property	Numerical value	Experimental conditions	Experimental method remarks	Ref.
$Bi_{12}GeO_{20}$:				
E_g	3.25 eV		photoconductivity	71L
E_c–E_t	1.575 eV		trapping level	71L
$\mu_{dr,n}$	$4.51 \cdot 10^{-3}$ cm^2/V s		effective drift mobility in transit time measurements	71L
	$<10^{-3}$ cm^2/V s			68D
	$2.4 \cdot 10^{-2}$ cm^2/V s			80K
ϱ	$8 \cdot 10^{10}$ Ω cm	p-type		67L, 71A
$\varepsilon(0)$	40		see [80L] for temperature dependence	71A
d_{th}	9.222(2) g/cm^3			67A

Wavenumbers of normal modes have been determined by [72V] from Raman spectra and [79W] from reflectivity (see also [80I, 82B2]:

data from reflectivity [79W] (in cm^{-1}):

TO-modes: 97, 105, 123, 130, 177, 190, 205, 232, 271, 303, 356, 459, 526, 571, 600, 682
LO-modes: 100, 114, 129, 154, 179, 193, 209, 254, 281, 355, 372, 497, 555, 578, 612, 894

data from Raman spectra [72V] (in cm^{-1}):

TO-modes: 207.0
LO-modes: 48.1, 54.6, 111.2, 153.0, 178.5, 194.4, 208.5, 691.8
LO+TO-modes: 44.6, 52.4, 57.5, 99.0, 105.8, 124.0, 131.2, 305.0, 338.0, 357.5, 488.3, 678.6

For optical and Raman spectra, see Figs. 5, 6, 8. For elastic moduli, see [79A]. Deep levels are discussed in [81G].

$PbSb_2S_4$, $GeSb_2Te_4$, $GeBi_2Te_4$, $SnBi_2Te_4$

$PbSb_2S_4$ (zinkenite) crystallizes in the $F5_6$ orthorhombic lattice with $a=12.29$ Å, $b=13.6$ Å, $c=8.66$ Å [74D1]. From photoconductivity spectra (Fig. 9) values of 1.13 eV ($T=300$ K) and 1.38 eV ($T=77$ K) for the energy gap have been deduced. The electrical conductivity at RT of an n-type sample was $\sigma = 1.45 \cdot 10^{-7}$ Ω^{-1} cm^{-1} [74D1, 74D2].

$GeSb_2Te_4$ crystallizes in the hexagonal D_{3d}^5—R$\bar{3}$m lattice with $a=4.21$ Å and $c=40.6$ Å. The samples studied [72F] were strongly degenerate n-type with $\sigma_\perp = 4.3 \cdot 10^3$ Ω^{-1} cm^{-1} at RT. The electron Hall mobility at 300 K was 30 cm^2/V s (for the temperature dependence, see Fig. 10). From reflectivity measurements near the plasma edge (Fig. 11) a dielectric constant of 39 and a conductivity effective mass of $0.55\,m_0$ were calculated.

$GeBi_2Te_4$ has the same structure as $GeSb_2Te_4$ with $a=4.28$ Å, $c=39.2$ Å. The optical energy gap is reported to be $E_g = 0.23$ eV at RT. Experimental data on transport parameters can be explained with a two-conduction-band model with $m_{n1} = 0.24\,m_0$, $m_{n2} = 0.88\,m_0$ and E_{c2}–$E_{c1} = 0.13(6)$ eV [79T].

$SnBi_2Te_4$ has the same structure as $GeSb_2Te_4$. Some preliminary results are reported in [74Z].

$GeBi_4Te_7$, $SnBi_4Te_7$, $GeSb_4Te_7$, $PbBi_4Te_7$

These compounds crystallize in a layer structure with space group D_{3d}^3—P$\bar{3}$m1. The lattice parameters are:

Substance	a [Å]	c [Å]	Ref.
$GeSb_4Te_7$	4.21	23.65	74F
$GeBi_4Te_7$	4.36	24.11	76K
$PbBi_4Te_7$	4.42	23.6	71F

For **$SnBi_4Te_7$** only a few preliminary results have been reported in [74Z].

GeSb$_4$Te$_7$ [74F]:

The samples studied were strongly degenerate p-type with $\sigma_\perp = 3.42 \cdot 10^3\,\Omega^{-1}\,cm^{-1}$ and $\sigma_\parallel = 1.032 \cdot 10^3\,\Omega^{-1}\,cm^{-1}$ at room temperature. The transport measurements could be described using a six-valley model for the valence band. Analysis of transport and optical measurements yielded a Hall mobility at RT of 39 cm^2/V s, a dielectric constant of $\varepsilon(0) = 39$ and a conductivity effective mass of 0.46 m_0.

GeBi$_4$Te$_7$ [76K]:

The samples studied were strongly degenerate n-type with $\sigma_\perp = 1.907 \cdot 10^3\,\Omega^{-1}\,cm^{-1}$ and $\sigma_\parallel = 1.017 \cdot 10^3\,\Omega^{-1}\,cm^{-1}$ at room temperature. The transport measurements could be described using a simple three-valley model for the conduction band. Analysis of reflectivity measurements yielded a dielectric constant $\varepsilon(0) = 35$ and a conductivity effective mass of 0.27 m_0.

PbBi$_4$Te$_7$ [71F]:

The samples studied were strongly degenerate n-type with $\sigma_\perp \approx 10^3\,\Omega^{-1}\,cm^{-1}$. The Hall mobility $\mu_\perp$ was of the order of 15 cm^2/V s at room temperature (Fig. 12). From the reflectivity near the plasma edge (Fig. 13) a dielectric constant of $\varepsilon(0) = 44$ and an effective carrier mass of 0.29 m_0 were calculated.

References for 10.4.5

65N   Nitsche, R.: J. Appl. Phys. **36** (1965) 2358.
66L   Lenzo, P.V., Spencer, E.G., Ballman, A.A.: Appl. Opt. **5** (1966) 1688.
66S   Spencer, E.G., Lenzo, P.V., Ballman, A.A.: Appl. Phys. Lett. **9** (1966) 290.
67A   Abrahams, S.C., Jamieson, P.B., Bernstein, J.L.: J. Chem. Phys. **47** (1967) 4034.
67B1   Ballman, A.A.: J. Crystal Growth **1** (1967) 37.
67B2   Bernstein, J.L.: J. Crystal Growth **1** (1967) 45.
67L   Lenzo, P.V., Spencer, E.G., Ballman, A.A.: Phys. Rev. Lett. **19** (1967) 641.
68D   Douglas, G.G., Zitter, R.N.: J. Appl. Phys. **39** (1968) 2133.
69V   Venturini, E.L., Spencer, E.G., Ballman, A.A.: J. Appl. Phys. **40** (1969) 1622.
70F   Feldmann, A., Brower, W.S., Horowitz, D.: Appl. Phys. Lett. **16** (1970) 201.
71A   Aldrich, R.E., Hou, S.L., Harvill, M.L.: J. Appl. Phys. **42** (1971) 493.
71F   Frumar, M., Horák, J.: Phys. Status Solidi (**a**) **6** (1971) K133.
71L   Lenzo, P.V.: J. Appl. Phys. **43** (1971) 1107.
72D   Dickinson, S.K., Hilton, R.M., Lipson, H.G.: Mater. Res. Bull. **7** (1972) 181.
72F   Frumar, M., Tichý, L., Horák, J., Klikorka, J.: Mater Res. Bull. **7** (1972) 1075.
72V   Venugopalan, S., Ramdas, A.K.: Phys. Rev. **B 5** (1972) 4065.
73H   Hou, S.L., Lauer, R.B., Aldrich, R.E.: J. Appl. Phys. **44** (1973) 2652.
74D1   Dmytriv, A.Yu., Koval'skii, P.N., Makarenko, V.V.: Sov. Phys. Semicond. **7** (1974) 1097 (transl. from Fiz. Tekh. Poluprovodn. **7** (1974) 1641).
74D2   Dmytriv, A.Yu., Koval'skii, P.N., Makarenko, V.V.: Sov. Phys. Semicond. **7** (1974) 1493 (transl. from Fiz. Tekh. Poluprovodn. **7** (1974) 2241).
74F   Frumar, M., Tichý, L., Matyás, M., Zelizko, J.: Phys. Status Solidi (**a**) **22** (1974) 535.
74L   Lauer, R.B.: J. Appl. Phys. **45** (1974) 1794.
74Z   Zhukova, T.B., Kutasov, V.A., Parfen'eva, L.S., Smirnov, I.A.: Inorg. Mater. **10** (1974) 1903 (transl. from Izv. Akad. Nauk SSSR, Neorg. Mater. **10** (1974) 2221).
76K   Klikorka, J., Fumar, M., Tichý, L., Panek, L., Tichá, H.: J. Phys. Chem. Solids **37** (1976) 477.
77B   Brillingas, A., Grigas, I., Dmytriv, A.Yu.: Sov. Phys. Solid State **19** (1977) 2013 (transl. from Fiz. Tverd. Tela **19** (1977) 3445).
78F   Flerova, S.A., Bochkova, T.M.: Sov. Phys. Solid State **20** (1978) 703 (transl. from Fiz. Tverd. Tela **20** (1978) 1222).
78K   Kleszczewski, Z.: Arch. Acoust. **3** (1978) 175.
79A   Antonenko, A.M., Volnyanskii, M.D., Kudzin, A.Yu., Kuchukov, E.G.: Sov. Phys. Solid State **21** (1979) 1418 (transl. from Fiz. Tverd. Tela **21** (1979) 2461).
79T   Tichy, L., Frumar, M., Klikorka, J.: Phys. Status Solidi (**a**) **56** (1979) 323.
79W   Wojdowski, W., Lukasiewicz, T., Nazarewicz, W., Zmija, J.: Phys. Status Solidi (**b**) **94** (1979) 649.
80E   Efendiev, Sh.M., Mamedov, A.M., Bagiev, V.E., Elvazonva, G.M.: Sov. Phys. Solid State **22** (1980) 2169 (transl. from Fiz. Tverd. Tela **22** (1980) 3705).

80I	Imaino, W., Ramdas, A.K., Rodriguez, S.: Phys. Rev. B **22** (1980) 5679.
80K	Kostyuk, B.Kh., Kudzin, A.Yu., Sokolyanskii, G.Kh.: Sov. Phys. Solid State **22** (1980) 1429 (transl. from Fiz. Tverd. Tela **22** (1980) 2454).
80L	Link, J., Fontanella, J., Andeen, C.G.: J. Appl. Phys. **51** (1980) 4352.
80P	Panchenko, T.V., Kudzin, A.Yu., Truseeva, N.A.: Sov. Phys. Solid State **22** (1980) 1077 (transl. from Fiz. Tverd. Tela **22** (1980) 1851).
81G	Gudaev, O.A., Detinenko, V.A., Malinowskii, V.K.: Sov. Phys. Solid State **23** (1981) 109 (transl. from Fiz. Tverd. Tela **23** (1981) 195).
82B1	Babonas, G.A., Zhogova, E.A., Zaretskii, Yu.G., Kurbatov, G.A., Ukhanov, Yu.I., Shmartsev, Yu.V.: Sov. Phys. Solid State **24** (1982) 921 (transl. from Fiz. Tverd. Tela **24** (1982) 1612).
82B2	Babonas, G.A., Zaretskii, Yu.G., Kurbatov, G.A., Ukhanov, Yu.I.: Sov. Phys. Solid State **24** (1982) 354 (transl. from Fiz. Tverd. Tela **24** (1982) 626).
83G	Grewal, P.K., Lea, M.J.: J. Phys. C: Solid State Phys. **16** (1983) 247.

10.4.6 V–VI–VII compounds*)

AsSBr, SbSI, SbSBr, SbSeCl, SbSeBr, SbSeI, SbTeI, BiOCl, BiOBr, BiOI, BiSCl, BiSBr, BiSI, BiSeCl, BiSeBr, BiSeI, BiTeBr, BiTeI.

10.4.6.0 Structure

V–VI–VII compounds occurs as ternary compounds in the pseudobinary phase diagrams of the $(V–VII_3)–(V_2–VI_3)$ system (Fig. 1). Several lattice structures exist:

(a) D_{2h}^{16}—Pnam

Most V–VI–VII semiconductors crystallize in this structure. The unit cell contains four formula units. The atoms are arranged in chains along the c axis (Fig. 2).

lattice parameters and densities:

Substance	a [Å]	b [Å]	c [Å]	d_X [g/cm^3]	Ref.
SbSI	8.52	10.13	4.10	5.33	67K
SbSBr	8.20	9.70	3.95	4.94	50D1
SbSeI	8.65	10.38	4.12	5.88	50D2
SbSeBr	8.30	10.20	3.95	5.57	50D2
SbTeI	9.18	10.8	4.23	5.96	51D
BiSCl	7.70	9.87	4.02	6.04	50D1
BiSBr	8.02	9.70	4.01	6.83	50D1
BiSI	8.46	10.15	4.14	6.87	50D1
BiSeCl	12.37	18.10	4.08	7.05	50D2
BiSeBr	8.18	10.47	4.11	6.94	50D2
BiSeI	8.71	10.54	4.19	7.16	50D2

All values cited are measured at room temperature except SbSI (35°C). Besides of the substances cited, AsSBr is mentioned in the literature to have the same structure.

Four of the substances (**SbSI, SbSBr, BiSI, BiSBr**) become ferroelectric below a Curie temperature Θ_C (first order phase transition). The structure of the ferroelectric phase is C_{2v}^9—Pna2$_1$. The phase transition is displacive (see e.g. [67K]). The V- and VI-atoms are shifted along the c axis relative to the VII-atoms. Data for SbSI are: shift of Sb-atoms relative to I-atoms: 0.20 Å according to [67K], 0.103 Å according to [76I]; shift of S-atoms: 0.05 Å [67K], 0.019 Å [76I]. The lattice parameters a, b, c as well as the positions of the atoms perpendicular to the c axis remain unchanged in the phase transition.

---

*) Many of the substances presented in this section have ferroelectric properties. We restrict the following presentation of data to those properties which are important in connection with their semiconducting behavior. For more data, especially about structure parameters, ferroelectric properties and magnetic resonance phenomena we refer to Landolt-Börnstein, NS, Vol. III/16b, section II/14; data on dielectric constants, piezoelectric strain coefficients, electromechanical coupling factors, pyroelectric and electrooptic coefficients are given in great detail in Landolt-Börnstein, NS, Vols. III/11 and III/18.

Physical property	Numerical value	Experimental conditions	Experimental method remarks	Ref.

SbSI is reported to undergo another (second order ?) phase transition to a C_2^2—phase at a lower temperature Θ_C' (see e.g. [75P, 77B1] and Fig. 3). Similar second order phase transitions occur in SbSBr, SbSeI, BiSI, BiSBr, BiSeI according to [69P2].

(b) D_{4h}^7—P4/nmm

The BiO-halogen compounds crystallize in this structure. Sheets of oxygen alternate with sheets of bismuth. Double layers of halogen atoms lie between the metal and oxygen sheets [68B2]. Two formula units are contained in the unit cell of this laminar tetragonal structure.

Substance	a [Å]	c [Å]	lattice stable below
BiOCl	3.883	7.347	575 °C
BiOBr	3.915	8.076	560 °C
BiOI	3.984	9.128	300 °C

(c) D_{3d}^3—P$\bar{3}$m1

BiTeBr and BiTeI crystallize in this hexagonal layered structure [51D].

Substance	a [Å]	c [Å]	Ref.	d_X [g/cm³]	Ref.
BiTeBr	4.23(1)	6.48(1)	68H1	6.90	51D
BiTeI	4.30(1)	6.80(1)	62H2	7.01	

According to [50D2] SbSeCl crystallizes in a D_{2h}^1—Pmmm lattice.

10.4.6.1 Physical properties

AsSBr:

E_g	2.5 eV		temperature dependence of conductivity (Fig. 4)	65A
	2.67 eV		absorption edge	
ϱ	$> 10^{12}\ \Omega\,\text{cm}$			65A
$\mu_{H,n}$	$< 10^{-1}\ \text{cm}^2/\text{V s}$			65A

SbSI:

SbSI is by far the most interesting and most investigated V–VI–VII compound being a photoconductor and having a paraelectric – ferroelectric phase transition near room temperature [62F].

phase transition:

Θ_C	19···23 °C		$T > \Theta_C$: paraelectric phase, $T < \Theta_C$: ferroelectric phase. Θ_C has been found to vary between 19 °C and 23 °C depending on various material parameters. The value most quoted is $\Theta_C = 22$ °C. For the variation of Θ_C in the systems SbSI–SbSeI and SbSI–SbSBr, see [64N], in SbSI–BiSI, see [78T].	
	15.70 °C			78I1, 78I2
$d\Theta_C/dp$	$-37.0(20)$ K/kbar		pressure dependence of dielectric constant	68S1
	-40 K/kbar		pressure effects in Raman spectra	72T
	$-36.5(20)$ K/kbar	$p < 2$ kbar		72S4
$d\Theta_C/dE$	2.2 K/kV cm^{-1}		shift in electric field	69P2

(continued)

Physical property	Numerical value	Experimental conditions	Experimental method remarks	Ref.
phase transition (continued)				
Θ'_C	$-40\,°C$		(second order) phase transition to a C_2^2-lattice (see sect. 10.4.7.0)	77B1
	$-36.61\,°C$			78I1, 78I2

Electronic properties

Brillouin zone: Fig. 5, band structure: Fig. 6 (paraelectric phase), Fig. 7 (ferroelectric phase)

The calculated band structures show an indirect gap in both phases: $Z_{1c}-U_{5,6v}$ in the paraelectric phase, $Z_{1c}-R_{3,4v}$ in the ferroelectric phase. The lowest direct transition occurs at U: $U_{7,8c}-U_{5,6v}$ for $E\|c$, $U_{7,8c}-U_{1,2v}$ for $E\perp c$ (paraelectric phase), $U_{3,4c}-U_{3,4v}$ for $E\|c$, $U_{3,4c}-U_{1,2v}$ for $E\perp c$ (ferroelectric phase). In contrast to [73N] the band structure presented in [74F] shows a direct gap at Γ.

energy gap (in eV):

$E_{g,\,dir}$	1.88	$E\|c$, $T=25\,°C$	absorption edge, Fig. 8	63H
	1.95	$E\perp c$	(direct transitions, for indirect	
	1.95	$E\|c$	transitions, see Fig. 9)	68C
	1.97	$E\perp c$		
	2.03		thermoabsorption (Fig. 10c)	81A, 82G
$\Delta E_{g,\,dir}$	0.02	$25\,°C$	shift at Θ_C going from the paraelectric phase to the ferroelectric phase	63H

temperature dependence of energy gap (in eV/K):

dE_g/dT	$-2.51\cdot10^{-3}$	$T<\Theta_C$	thermoabsorption	81A,
	$-8.83\cdot10^{-4}$	$T>\Theta_C$		82G
	$-2.2(2)\cdot10^{-3}$	$T<\Theta_C$	absorption	63H
	$-0.9(2)\cdot10^{-3}$	$T>\Theta_C$	for temperature dependence of E_g, see also Figs. 10a, b and c	
	$-2.6\cdot10^{-2}$	$T\approx\Theta_C$	photoconductivity, absorption	66N,
	$-20\cdot10^{-4}$	$T<\Theta_C$		66F
	$-12\cdot10^{-4}$	$T>\Theta_C$		
	$-1.8\cdot10^{-3}$	$T<\Theta_C$	absorption (Fig. 11)	73Z
	$-7\cdot10^{-4}$	$T>\Theta_C$		

pressure dependence of energy gap:

dE_g/dp	$-1.0(3)\cdot10^{-4}\,eV/atm$	$p<0.5$ atm, $T=15\,°C$		65G
	$10^{-6}\,eV/atm$	$p>0.5$ atm		
	$-5(2)\cdot10^{-6}\,eV/atm$	$T>\Theta_C$		66F

For the dependence of the absorption edge on an electric field, see Figs. 12···14; see also Fig. 8.

higher transitions (energy of critical points in reflection spectra) (in eV):

		T [K]		
E	2.5	5, $E\|c$	wavelength modulated spectroscopy	74F
	2.7		(Fig. 15)	
	3.08		(for the temperature dependence	
	3.8		of transition energies near the	
	4.2		phase transitions, see [81S])	
	2.5	5, $E\perp c$		
	3.0			
	3.37			
	3.88			

(continued)

Physical property	Numerical value	Experimental conditions	Experimental method remarks	Ref.
higher transitions (continued)				
E	2.49	110, $E \parallel c$		74F
	2.63			
	2.78			
	2.49	110, $E \perp c$		
	2.58			
	2.92			
	2.37 (2.42)	110, $E \parallel c$	temperature modulated spectro-	76E
	2.46 (2.50)		scopy (Fig. 15). Values obtained	
	2.50 (2.63)		from Δc_1, values in brackets	
	2.61 (2.72)		from $\Delta \varepsilon_2$.	
	2.71			
	2.49 (2.48)	110, $E \perp c$		
	2.58 (2.58)			
	2.91 (2.94)			
	2.32	90, $E \parallel c$	reflectance (Fig. 16)	71B3
	2.42			
	2.49			
	2.63			
	2.82			
	2.45	90, $E \perp c$		
	2.96			
	2.086	286, $E \parallel c$	electroreflectance (Fig. 17)	71G
	2.305			
	2.597			
	1.911	286, $E \perp c$		
	2.131			
	2.435			
	2.737			
	2.060	296, $E \parallel c$		
	2.279			
	2.547			
	1.893	296, $E \perp c$		
	2.105			
	2.406			
	2.677			
	2.044	306, $E \parallel c$		
	2.258			
	2.510			
	1.884	306, $E \perp c$		
	2.086			
	2.383			
	2.627			

Lattice properties

Brillouin zone: Fig. 5, phonon dispersion relations: Fig. 18

The vibrational spectrum of SbSI is very complicated. The unit cell of the lattice contains 4 formula units. Thus there are 3 acoustic and 33 optical branches. For the analysis of phonon spectra often a simplified lattice is chosen by neglecting the interaction between the double chains in the unit cell. This leads to C_{2h}^2–$P2_1/m$ symmetry in the paraelectric phase and C_2^2–$P2_1$ symmetry in the ferroelectric phase with only two formula units in the unit cell. The optical activity of the optical modes in this description is:

Physical property	Numerical value	Experimental conditions	Experimental method remarks	Ref.
paraelectric phase:	infrared active: 2 A_u-modes ($E\|c$),		4 B_u-modes ($E\perp c$)	
	Raman active: 6 A_g-modes,		3 B_g-modes	
ferroelectric phase:	infrared active: 8A-modes ($E\|c$),		7 B-modes ($E\perp c$)	
	these modes are also Raman active.			

For a detailed discussion, see [71A] and [71B1].

wavenumbers of infrared active modes (in cm^{-1}) ($k \simeq 0$): (see also Fig. 19):

		T [K]		
$\bar{\nu}_T$	9	298 K, $E \| c$	designation according to	71A
$\bar{\nu}_L$	111	(paraelectric	simplified C_{2h}^2 lattice	
$\bar{\nu}_T$	179(10)	phase)		
$\bar{\nu}_L$	261(10)			
$\bar{\nu}_T$	78	298, $E \perp c$	designation according to	71A
$\bar{\nu}_L$	82	(paraelectric	simplified C_{2h}^2 lattice	
$\bar{\nu}_T$	120	phase)		
$\bar{\nu}_L$	124			
$\bar{\nu}_T$	270			
$\bar{\nu}_L$	276			
$\bar{\nu}_T$	327			
$\bar{\nu}_L$	332			
$\bar{\nu}_T$	16	275, $E \| c$	designation according to	71A
$\bar{\nu}_T$	60	(ferroelectric	simplified C_2^2 lattice	
$\bar{\nu}_L$	100	phase)	(for further temperatures and	
$\bar{\nu}_T$	108		figures, see [71A])	
$\bar{\nu}_L$	115			
$\bar{\nu}_T$	138			
$\bar{\nu}_L$	140			
$\bar{\nu}_T$	178(10)			
$\bar{\nu}_L$	217(10)			
$\bar{\nu}_T$	247			
$\bar{\nu}_L$	271			
$\bar{\nu}_T$	80	80, $E \perp c$	designation according to	71A
$\bar{\nu}_L$	85	(ferroelectric	simplified C_2^2 lattice	
$\bar{\nu}_T$	124	phase)		
$\bar{\nu}_L$	128			
$\bar{\nu}_T$	245(10)			
$\bar{\nu}_L$	250(10)			
$\bar{\nu}_T$	270			
$\bar{\nu}_L$	272			
$\bar{\nu}_T$	≈ 320			
$\bar{\nu}_T$	332			
$\bar{\nu}_L$	334			

wavenumbers of Raman active modes (in cm^{-1}) (Figs. 20···23):

		T [K]		
$\bar{\nu}(A_g)$	51	299	A_g-symmetry in simplified C_{2h}^2	71A
	66	(paraelectric	lattice	
	107	phase)		
	137			
	149			
	329			

(continued)

Physical property	Numerical value	Experimental conditions	Experimental method remarks	Ref.
wavenumbers of Raman active modes (continued)				
$\bar{\nu}(B_g)$	37		B_g-symmetry in simplified C_{2h}^2	71A
	212		lattice	
	239			
	soft mode	250 (ferroelectric phase)	for the temperature dependence of the soft mode and its interaction with low lying "hard modes", see Fig. 23	72T
$\bar{\nu}(A)$	54	250	A-symmetry in simplified C_2^2	71A
	70	(ferroelectric	lattice	
	109	phase)		
	139			
	153			
	248			
	320			
$\bar{\nu}(B)$	39	250	B-symmetry in simplified C_2^2	71A
	59	(ferroelectric	lattice	
	119	phase)		
	212			
	240			
	330			

For lower temperatures the simplified space group symmetry selection rules break down and the full group symmetry has to be taken into account. For data, see [71A] and Fig. 21. Other infrared and Raman data (partly deviating from the values given above) have been published in [71B1, 71C, 69R, 69P1, 68B1]. For the phonon spectrum near the phase transition, see [82A].

phonon energies from other experiments:

$\hbar\omega_{ph}$	0.09 eV	$T = 24\,°C$	phonon participating in indirect	68G1
	0.14 eV	$-164\,°C$	transitions as obtained from Fig. 10	
$\bar{\nu}_{ph}$	210 cm^{-1}		phonon dominating in the electron-	73Z
	220 cm^{-1}		phonon interaction in Urbach tails of the absorption edge	70K

elastic moduli (in 10^8 N cm^{-2}):

c_{11}	3.09(9)	paraelectric		70S
c_{22}	3.27(4)	phase,		
c_{33}	4.95(6)	$T = 22\,°C$		
c_{44}	2.21(3)			
c_{55}	0.92(3)			
c_{66}	0.60(2)			
c_{12}	0.96(9)			
c_{13}	0.93(30)			
c_{23}	1.58(12)			
c_{11}	3.06(9)	ferroelectric		70S
c_{22}	3.14(4)	phase,		
c_{33}	5.18(6)	$T \approx 12\,°C$		
c_{44}	2.24(3)			
c_{55}	0.99(3)			
c_{66}	0.59(2)			
c_{12}	0.85(9)			
c_{13}	0.97(30)			
c_{23}	1.44(12)			

Physical property	Numerical value	Experimental conditions	Experimental method remarks	Ref.

Optical properties

Optical spectra are shown in Figs. 8···17, 19. Values of the absorption edge are often not precise because the line shape of the absorption edge shows Urbach tails (strong phonon-electron interaction) [70K, 73Z]. The absorption edge shifts to shorter wavelengths in electric fields (electro-optical effect [62K]), Figs. 12···14.

refractive index (at $\lambda = 633$ nm):

n_a	2.87	paraelectric	for temperature and wavelength	70S
n_b	3.63	phase,	dependence, see Figs. 24, 25	
n_c	4.55	$T = 22\,°C$		
n_a	2.87	ferroelectric		
n_b	3.57	phase,		
n_c	4.44	$T \approx 12\,°C$		

dielectric constant:

$\varepsilon_{\parallel c}$	$6.2 \cdot 10^4$	at Θ_C, $f = 1$ kHz	temperature dependence, see Figs. 26, 27. Above Θ_C ε obeys a Curie-Weiss law with $\varepsilon = C/(T - \Theta_{pe})$, $C = 2.33 \cdot 10^5\,°C$, $\Theta_{pe} = 16.0\,°C$ For the dependence of ε on pressure, see Fig. 28 and [68S1, 69G].	64M

Transport properties

Electrical conduction in SbSI is dominated by contact phenomena due to the presence of strong field regions near the contacts [73A] and space charge limited currents.

Photoconductivity is generated by excitation of electrons from deep trapping levels.

ϱ	$10^8 \cdots 10^9\ \Omega$ cm	along c axis		68C
$\mu_{H,n}$	$50 \cdots 100$ cm^2/V s		mobility of photoexcited carriers measured with the photo Hall effect in a Li doped sample	67A
E_A	$0.52 \cdots 0.58$ eV	paraelectric	E_A: thermal activation energy of	64S
	0.62 eV	phase	electrical conductivity σ_{33}	66N
	0.38 eV		(see Fig. 29)	77I
	$0.98 \cdots 1.2$ eV	ferroelectric		64S
	0.83 eV	phase		66N
$E_t - E_v$	0.40 eV		energy of trapping level above the valence band; dominant trap for holes measured from thermally stimulated currents	67A
	0.21 eV			73S,
	0.41 eV			72K
	0.43 eV			
	0.38 eV		value in paraelectric phase (higher value in ferroelectric phase)	77I
$E_c - E_t$	$1.9 \cdots 2.0$ eV		energy of trapping level below the conduction band from maximum of photocurrent, Fig. 30 (similar values follow from the peak in short-circuit photo-voltaic current)	73I

For details on photoconduction and photovoltaic effect in SbSI, see e.g. [60N, 67A, 71B2, 71B4, 71I, 72B2, 72B3], for space charge limited currents especially [69P3].

Physical property	Numerical value	Experimental conditions	Experimental method remarks	Ref.
Further properties				
T_m	$\approx 400\,°C$			64M
ΔH_c	$58\,cal/mol$		transition heat at Θ_C	68S
ΔS_c	$0.18\,cal/mol\,K$		entropy change at Θ_C	68S

magnetic susceptibility: Fig. 31;
SbSI shows an electromechanical effect (elongation of the crystal along the c axis in an electric field [62K]), see Fig. 32; see also [77B2]
heat capacity: Fig. 33, thermal conductivity: Fig. 34

SbSBr:				
Θ_C	$-180\,°C$		ferroelectric-paraelectric phase transition	64N
$d\Theta_C/dE$	$1.1\,K/kV\,cm^{-1}$		shift in electric field	69P2
Θ'_C	$-95\,°C$		second order phase transition	69P2
E_g	$2.26\,eV$	$E\perp c$, RT		68C
	$2.20\,eV$	$E\parallel c$		
	$2.31\,eV$	$E\perp c$, $T=19\,°C$	value at absorption coefficient	68O
	$2.28\,eV$	$E\parallel c$	of $500\,cm^{-1}$	
dE_g/dT	$-8\cdot10^{-4}\,eV/K$	$20\,°C$		68C
ϱ	$10^6\,\Omega\,cm$	along c axis		68C

absorption spectra: Figs. 35, 36, Raman spectra: Fig. 37, magnetic susceptibility: Fig. 31

n_a	2.64	$\lambda=570\,nm$	refractive index along crystal	68O
n_b	3.13		axes	
n_c	4.01			

SbSeBr:				
E_g	$1.92\,eV$	$E\perp c$		68C
	$1.88\,eV$	$E\parallel c$		
dE_g/dT	$-7.2\cdot10^{-4}\,eV/K$			68C

photocurrent vs. wavelength: Fig. 38

SbSeI:				
Θ'_C	$-50\,°C$		second order phase transition	69P2
E_g	$1.68\,eV$	$E\perp c$		68C
	$1.66\,eV$	$E\parallel c$		
dE_g/dT	$-8\cdot10^{-4}\,eV/K$			68C
$\varrho_{\parallel c}$	$10^7\,\Omega\,cm$			68C

magnetic susceptibility: Fig. 31

SbTeI:				
E_g	$1.28\,eV$	$E\perp c$		68C
	$1.25\,eV$	$E\parallel c$		
$\varrho_{\parallel c}$	$10^4\,\Omega\,cm$			68C

BiOCl:				
$E_{g,\,ind}$	$3.455\,eV$	RT	optical absorption	72S3
$E_{g,\,dir}$	$3.50\,eV$	RT	optical absorption	72S3
dE_g/dT	$-6.3\cdot10^{-4}\,eV/K$	$T=90\cdots600\,K$		72S3

(continued)

Physical property	Numerical value	Experimental conditions	Experimental method remarks	Ref.
BiOCl (continued)				
$\hbar\omega_{TO}(\Gamma)$	35.3 meV		infrared bands, Fig. 39	73P
	65.4 meV			
	(16.6 meV)			
$\hbar\omega$	20 meV		phonons participating in indirect	72S3
	25 meV		transitions	
	34 meV			
	52 meV			

luminescence spectra: Fig. 40, trapping levels, thermally stimulated currents: [71K], photoemission: [72B1]

BiOBr:				
$E_{g, ind}$	2.924 eV	$T = 293$ K	absorption, Fig. 42	72S3
$E_{g, dir}$	3.00 eV		absorption, Fig. 41	72S3
dE_g/dT	$-7.4 \cdot 10^{-4}$ eV/K	$90 \cdots 600$ K		72S3
dE_g/dp	$4.3 \cdot 10^{-5}$ eV/atm			72S1
$\hbar\omega_{TO}(\Gamma)$	32.8 meV		infrared bands, Fig. 39	73P
	64.4 meV			
	(14.6 meV)			
$\hbar\omega$	11 meV		phonons participating in indirect	72S3
	15 meV		transitions	
	25 meV			
	35 meV			

luminescence spectra: Fig. 40, trapping levels, thermally stimulated currents: [71K], photoemission: [72B1]

BiOI:				
$E_{g, ind}$	1.890 eV	RT	absorption edge, Fig. 44	72S3
$E_{g, dir}$	1.94 eV	RT	absorption edge, Fig. 43	72S3
dE_g/dT	$-8.1 \cdot 10^{-4}$ eV/K	$T = 90 \cdots 600$ K		72S3
	$-5.6 \cdot 10^{-4}$ eV/K	$T < 300$ K	phase transition at 300 K?	73B1
	$-8.2 \cdot 10^{-4}$ eV/K	$T > 300$ K		
dE_g/dp	10^{-5} eV/atm		shift of absorption edge	72S1
$\hbar\omega_{TO}(\Gamma)$	60.4 meV		infrared bands, Fig. 39	73P
	30.8 meV			
	12.8 meV			
	70.6 meV			
$\hbar\omega$	17 meV		phonons participating in indirect	72S3
	21 meV		transitions	
	25 meV			
	31 meV			
	51 meV			
	70 meV			

luminescence: Fig. 40, trapping levels, thermally stimulated currents: [71K], photoemission: [72B1]

BiSCl:				
E_g	1.93 eV	$E \perp c$		68C
	1.89 eV	$E \parallel c$		
ϱ	$10^3 \cdots 10^4$ Ω cm		see also [60N, 64N]	68C

magnetic susceptibility: Fig. 31

Physical property	Numerical value	Experimental conditions	Experimental method remarks	Ref.
BiSBr:				
Θ_C	$-170\,°C$		ferroelectric-paraelectric phase transition	64N
$d\Theta_C/dE$	$1.6\,K/kV\,cm^{-1}$		shift in electric field	69P2
Θ'_C	$-140\,°C$		second order phase transition	69P2
E_g	$1.97\,eV$	$E \perp c$		68C,
	$1.95\,eV$	$E \parallel c$		68H1
dE_g/dT	$-7.6 \cdot 10^{-4}\,eV/K$			68C
ϱ	$10^3 \cdots 10^4\,\Omega\,cm$	along c axis, n-type sample		68C
BiSI:				
Θ_C	$-160\,°C$		ferroelectric-paraelectric phase transition	64N
$d\Theta_C/dE$	$3.0\,K/kV\,cm^{-1}$		shift in electric field	69P2
Θ'_C	$-40\,°C$		second order phase transition	69P2
E_g	$1.58\,eV$	$E \perp c$		68C
	$1.56\,eV$	$E \parallel c$		
dE_g/dT	$-7 \cdot 10^{-4}\,eV/K$			68C
ϱ	$10^7\,\Omega\,cm$			68C
magnetic susceptibility: Fig. 31				
BiSeBr:				
E_g	$1.54\,eV$	$E \perp c$	see Fig. 45	68H1
	$1.50\,eV$	$E \parallel c$		
magnetic susceptibility: Fig. 31				
BiSeI:				
Θ'_C	$-140\,°C$		second order phase transition	69P2
E_g	$1.32\,eV$	$E \perp c$		68C
	$1.3\,eV$	$E \parallel c$		
dE_g/dT	$-6.5 \cdot 10^{-4}\,eV/K$			68C
ϱ	$10^2 \cdots 10^3\,\Omega\,cm$	along c axis		68C
magnetic susceptibility: Fig. 31, optical spectra, photoconductivity: Figs. 46 ··· 48				

Physical property	Numerical value	$T\,[K]$	Experimental method remarks	Ref.
BiTeBr:				
E_g	$0.55\,eV$	$E \parallel c$		68C
$E_{g,\,ind}$	$0.472\,eV$	295	absorption edge, Fig. 50	73B2
	$0.515\,eV$	5		
$E_{g,\,dir}$	$0.501\,eV$	295	absorption edge, Fig. 49	73B2
	$0.572\,eV$	5	(see also Fig.51)	
	$0.59\,eV$		absorption edge	68H1
$dE_{g,\,ind}/dT$	$-1.85 \cdot 10^{-4}\,eV/K$	$T > 100$		73B2
$dE_{g,\,dir}/dT$	$-2.93 \cdot 10^{-4}\,eV/K$	$T > 100$		73B2
$\hbar\omega$	$14\,meV$		phonons participating in indirect transitions, Fig. 50	73B2
	$29\,meV$			
	$44\,meV$			
d_{exp}	$6.65\,g/cm^3$			51D
n-type conduction, some conclusions about band structure: [73B2], conductivity and Hall coefficient: Fig. 52				

Physical property	Numerical value	Experimental conditions	Experimental method remarks	Ref.
BiTeI:				
		T [K]		
E_g	0.39 eV	295	absorption, reflection	70H1
	0.46 eV	$E \parallel c$	direct transition (?)	68C
$E_{g,\,ind}$	0.448 eV	295	position of the indirect absorption	68C
	0.479 eV	5	edge (gap including energy of the participating phonon), Fig. 57	
$\mu_H^{\perp}$	280···445 cm²/V s	degenerate sample, RT	conductivity, Hall effect (see also Figs. 53, 54)	68H2
m_n	0.20···0.25 m_0	$n = 3.5···$ $7.6 \cdot 10^{19}$ cm^{-3}		70H1
$\varepsilon(0)$	14.5(15)		absorption of free carriers	70H1
$\varepsilon(\infty)$	19(2)	$E \perp c$		80L
d_{exp}	6.91 g/cm³			51D

magnetic susceptibility: Fig. 31, optical spectra: Figs. 55···57

10.4.6.2 References for 10.4.6

50D1   Dönges, E.: Z. Anorg. Allg. Chem. **263** (1950) 112.
50D2   Dönges, E.: Z. Anorg. Allg. Chem. **263** (1950) 280.
51D   Dönges, E.: Z. Anorg. Allg. Chem. **265** (1951) 56.
60N   Nitsche, R., Merz, W.J.: J. Phys. Chem. Solids **13** (1960) 154.
62F   Fatuzzo, E., Harbeke, G., Merz, W.J., Nitsche, R., Roetschi, R., Ruppel, W.: Phys. Rev. **127** (1962) 2036.
62K   Kern, R.: J. Phys. Chem. Solids **23** (1962) 249.
63H   Harbeke, G.: J. Phys. Chem. Solids **24** (1963) 957.
64M   Mori, T., Tamura, H.: J. Phys. Soc. Jpn **19** (1964) 1247.
64N   Nitsche, R., Roetschi, H., Wild, P.: Appl. Phys. Lett. **4** (1964) 210.
64S   Sasaki, Y.: Jpn. J. Appl. Phys. **3** (1964) 558.
65A   Antonini, J.F., Brun, R.: Nuovo Cimento **35** (1965) 956.
65G   Gulyamov, K., Lyakhovitskaya, V.A., Tikhomirova, N.A., Fridkin, V.M.: Sov. Phys. Dokl. **10** (1965) 331 (transl. from Dokl. Akad. Nauk SSSR **161** (1965) 1060).
65M   Mori, T., Tamura, H., Sawaguchi, E.: J. Phys. Soc. Jpn **20** (1965) 281.
66F   Fridkin, V.M., Gulyamov, K., Lyakhovitskaya, V.A., Nosov, V.N., Tikhomirova, N.A.: Sov. Phys. Solid State **8** (1966) 1510 (transl. from Fiz. Tverd. Tela **8** (1966) 1907).
66N   Nosov, V.N., Fridkin, V.M.: Sov. Phys. Solid State **8** (1966) 113 (transl. from Fiz. Tverd. Tela **8** (1966) 148).
66T   Tatsukazi, I., Itoh, K., Ueda, S., Shindo, Y.: Phys. Rev. Lett. **17** (1966) 198.
67A   Alekseeva, V.G., Landsberg, E.G.: Sov. Phys. Solid State **8** (1967) 2518 (transl. from Fiz. Tverd. Tela **8** (1966) 3138).
67J   Johannes, R., Haas, W.: Appl. Opt. **6** (1967) 1059.
67K   Kikuchi, A., Oka, Y., Sawaguchi, E.: J. Phys. Soc. Jpn. **23** (1967) 337.
67O   Ohi, K., Arizumi, O.: J. Phys. Soc. Jpn. **22** (1967) 1307.
67U   Ueda, S., Tatsuzaki, I., Shindo, Y.: Phys. Rev. Lett. **18** (1967) 453.
68B1   Blinc, R., Mali, M., Novak, A.: Solid State Commun. **6** (1968) 327.
68B2   Bonnaire, R.: C.R. Acad. Sci. Paris **266** (1968) 1415.
68C   Chepur, D.V., Bercha, D.M., Turyanitsa, I.D., Slivka, V.Yu.: Phys. Status Solidi **30** (1968) 461.
68G1   Gerzanich, E.I.: Sov. Phys. Solid State **9** (1968) 2358 (transl. from Fiz. Tverd. Tela **8** (1967) 2995).
68G2   Gerzanich, E., Brygalov, I.A., Rakcheev, A.D., Lyakhovitskaya, V.A.: Sov. Phys. Crystallogr. **13** (1969) 776 (transl. from Kristallografiya **13** (1968) 898).
68H1   Horak, J., Rodot, H.: C.R. Acad. Sci. Paris **267** (1968) 1427.

68H2   Horak, J., Rodot, H.: C.R.Acad. Sci. Paris **267** (1968) 363.
68H3   Horak, J., Turjanica, I.D., Klazar, J., Kozalova, M.: Krist. Tech. **3** (1968) 241.
68O   Ohi, K.: J. Phys. Soc. Jpn. **25** (1968) 1369.
68S1   Samara, G.A.: Phys. Lett. **27A** (1968) 232.
68S2   Steigmeier, E.F., Merz, W.J.: Helv. Phys. Acta **41** (1968) 1206
69G   Gerzanich, E.I., Fridkin, V.M.: Sov. Phys. Solid State **10** (1969) 2452 (transl. from Fiz. Tverd. Tela **10** (1968) 3111).
69P1   Petzelt, J.: Phys. Status Solidi **36** (1969) 321.
69P2   Pikka, T.A., Fridkin, V.M.: Sov. Phys. Solid State **10** (1969) 2668 (transl. from Fiz. Tverd. Tela **10** (1968) 3378).
69P3   Popov, Yu.M., Nesterenko, P.S.: Sov. Phys. Semicond. **2** (1969) 1151 (transl. from Fiz. Tekh. Poluprovodn. **2** (1968) 1373).
69R   Riede, V.: Phys. Lett. **29A** (1969) 715.
70H1   Horak, J.: J. Phys. (Paris) **31** (1970) 121.
70H2   Harbeke, G., Steigmeier, E.F., Wehner, R.K : Solid State Commun. **8** (1970) 1765.
70K   Kamimura, H., Shapiro, S.M., Balkanski, M.: Phys. Lett. **33A** (1970) 277.
70O   Ohi, K.: J. Phys. Soc. Jpn. Suppl. **28** (1970) 84.
70P   Perry, C.H., Agrawal, D.K.: Solid State Commun. **8** (1970) 225.
70S   Sandercock, J.R.: Opt. Commun. **3** (1970) 73.
71A   Agrawal, D.K., Perry, C.H.: Phys. Rev. **B4** (1971) 1893.
71B1   Balkanski, M., Teng, M.K., Shapiro, S.M., Ziolkiewicz, M.K.: Phys. Status Solidi (b) **44** (1971) 355.
71B2   Bezdetnyi, N.M., Zeinally, A.Kh., Lebedeva, N.N., Sheinkman, M.K.: Sov. Phys. Semicond. **5** (1971) 904 (transl. from Fiz. Tekh. Poluprovodn. **5** (1971) 1016).
71B3   Bercha, D.M., Slivka, V.Yu., Syrbu, N.N., Turyanitsa, I.S., Chepur, D.V.: Sov. Phys. Solid State **13** (1971) 217 (transl. from Fiz. Tverd. Tela **13** (1971) 276).
71B4   Bezdetnyi, N.M., Zeinally, A.Kh., Lebedeva, N.N., Sheinkman, M.K.: Sov. Phys. Solid State **12** (1971) 1990 (transl. from Fiz. Tverd. Tela **12** (1970) 2480).
71C   Chisler, E.V., Savatinova, I.T., Fridkin, V.M.: Sov. Phys. Solid State **12** (1971) 2327 (transl. from Fiz. Tverd. Tela **12** (1970) 2882).
71G   Golik, L.L., Elinson, M.I.: Sov. Phys. Solid State **12** (1971) 2338 (transl. from Fiz. Tverd. Tela **12** (1970) 2895).
71K   Kopinets, I.F., Rubish, I.D., Shtilikha, I.V., Chepur, D.V.: Sov. Phys. Semicond. **5** (1971) 649 (transl. from Fiz. Tekh. Poluprovodn. **5** (1971) 740).
71I   Irie, K.: J. Phys. Soc. Jpn. **30** (1971) 1506.
72B1   Bentsa, V.M., Shtilikha, M.V., Chepur, D.V., Kampi, Yu.Yu., Matyashevskii, V.V.: Sov. Phys. Solid State **14** (1972) 671 (transl. from Fiz. Tverd Tela **14** (1972) 787).
72B2   Bezdetnyi, N.M., Gorbatov, G.Z., Zeinally, A.Kh., Lebedeva, N.N., Sheinkman, M.K.: Sov. Phys. Solid State **14** (1972) 477 (transl. from Fiz. Tverd. Tela **14** (1972) 574).
72B3   Bezdetnyi, N.M., Gorbatov, G.Z., Zeinally, A.Kh., Lebedeva, N.N.: Sov. Phys. Semicond. **6** (1972) 1047 (transl. from Fiz. Tekh. Poluprovodn. **6** (1972) 1189).
72K   Kikneshi, A.A., Semak, D.G.: Sov. Phys. Semicond. **6** (1972) 449 (transl. from Fiz. Tekh. Poluprovodn. **6** (1972) 526).
72S1   Shtilikha, M.V., Chepur, D.V.: Sov. Phys. Semicond. **6** (1972) 1019 (transl. from Fiz. Tekh. Poluprovodn. **6** (1972) 1162).
72S2   Shtilikha, M.V., Chepur, D.V.: Sov. Phys. Semicond. **6** (1972) 962 (transl. from Fiz. Tekh. Poluprovodn. **6** (1972) 1108).
72S3   Shtilikha, M.V., Chepur, D.V.: Sov. Phys. Semicond. **6** (1972) 389 (transl. from Fiz. Tekh. Poluprovodn. **6** (1972) 451).
72S4   Syrkin, L.N., Polandov, I.N., Kachalov, N.P., Gamynin, E.V.: Sov. Phys. Solid State **14** (1972) 517 (transl. from Fiz. Tverd. Tela **14** (1972) 610).
72S5   Sheinkman, M.K., Krolovets, N.M., Savchenko, E.A., Tatarenko, L.N.: Sov. Phys. Solid State **14** (1972) 253 (transl. from Fiz. Tverd. Tela **14** (1972) 302).
72T   Teng, M.K., Balkanski, M., Massot, M.: Phys. Rev. **B5** (1972) 1031.
73A   Artobolevskaya, E.S., Chenskii, E.V., Gvozdover, R.S., Petrov, V.I.: Sov. Phys. Solid State **14** (1973) 1935 (transl. from Fiz. Tverd. Tela **14** (1972) 2236).
73B1   Berezhnoi, A.A., Mamedov, A.M.: Sov. Phys. Solid State **15** (1973) 1023 (transl. from Fiz. Tverd. Tela **15** (1973) 1525).

73B2	Borets, A.N., Puga, G.D., Chepur, D.V.: Sov. Phys. Solid State **15** (1973) 1255 (transl. from Fiz. Tverd. Tela **15** (1973) 1884).
73I	Irie, K.: J. Phys. Soc. Jpn. **34** (1973) 1530.
73N	Nako, K., Balkanski, M.: Phys. Rev. **B8** (1973) 5759.
73P	Puga, G.D., Borets, A.N., Bercha, D.M., Shtilikha, M.V.: Sov. Phys. Solid State **14** (1973) 1830 (transl. from Fiz. Tverd. Tela **14** (1972) 2125).
73S	Semak, D.G., Kikineshi, A.A., Chepur, D.V.: Phys. Status Solidi (**a**) **15** (1973) 533.
73Z	Zeinally, A.Kh., Mamedoc, A.M., Efendiev, Sh.M.: Sov. Phys. Semicond. **7** (1973) 271 (transl. from Fiz. Tekh. Poluprovodn. **7** (1973) 383).
74B	Bercha, D.M., Zayachkovskii, M.P., Kolosyuk, V.N., Slivka, V.Yu., Suslikov, L.M.: Sov. Phys. Semicond. **8** (1974) 721 (transl. from Fiz. Tekh. Poluprovodn. **8** (1973) 1106).
74F	Fong, C.Y., Petroff, Y., Kohn, S., Shen, Y.R.: Solid State Commun. **14** (1974) 681.
74P	Puga, G.D., Borets, A.N., Chepur, D.V.: Sov. Phys. Semicond. **8** (1974) 748 (transl. From Fiz. Tekh. Poluprovodn. **8** (1974) 1151).
75P	Peercy, P.S.: Phys. Rev. Lett. **35** (1975) 1581.
76B	Bercha, D.M., Zayachkovskii, M.P., Zayachkovskaya, N.F., Maslyuk, V.T.: Sov. Phys. Solid State **17** (1976) 1393.
76E	Efendief, S.M., Zeynalli, A.H., Grandolfo, M., Mariutti, G.: Solid State Commun. **18** (1976) 167.
76I	Itoh, K., Matsunaga, H., Nakamura, E.: J. Phys. Soc. Jpn. **41** (1976) 1679.
76M	Matyas, M., Horak, J.: Phys. Status Solidi (**a**) **36** (1976) K137.
77B1	Bercha, D.M., Baletskii, D.Yu., Nebola, I.I.: Sov. Phys. Solid State **18** (1977) 2033 (transl. from Fiz. Tverd. Tela **18** (1976) 3494).
77B2	Belyaev, L.M., Grekov, A.A., Zaks, P.L., Lyakhovitskaya, V.A., Popovkin, B.A., Protsenko, N.P., Spitsyna, V.D., Syrkin, L.N., Tatarenko, L.N., Feoktisova, N.N.: Sov. Phys. Acoust. **23** (1977) 463 (transl. from Akust. Zh. **23** (1977) 810).
77I	Irie, K.: Phys. Status Solidi (**a**) **39** (1977) 313.
77P	Pierrefeu, A., Steigmeier, E.F., Dorner, B.: Phys. Status Solidi (**b**) **80** (1977) 167.
78D	Nguyen Tat Dich, Lostak, P., Horak, J.: Czech. J. Phys. **B28** (1978) 1297.
78I1	Inushima, T., Uchinokura, K., Matsuura, E.: Solid State Commun. **26** (1978) 29.
78I2	Inushima, T., Uchinokura, K., Matsuura, E.: J. Phys. Soc. Jpn. **44** (1978) 1656.
78T	Teng, M.K., Massot, M., Balkanski, M., Ziolkiewicz, S.: Phys. Rev. **B17** (1978) 3895.
80G	Gommonai, A.V., Koperles, B.M., Groshik, I.I., Gurzan, M.I.: Sov. Phys. Solid State **22** (1980) 546 (transl. from Fiz. Tverd. Tela **22** (1980) 930).
80L	Lostak, P., Horak, J., Vasko, A., Nguyen Tat Dich: Phys. Status Solidi (**a**) **59** (1980) 311.
80M	Matyas, M., Horak, J., Klubickova, B.: Phys. Status Solidi (**a**) **61** (1980) 419.
81A	Angelini, V., Casalboni, M., Grandolfo, M., Somma, F., Vecchia, P., Effendiev, Sh.M.: Ferroelectrics **34** (1981) 231.
81S	Sobolev, V.V., Turyshev, M.V., Lyakhovitskaya, V.A.: Sov. Phys. Solid State **23** (1981) 1442 (transl. from Fiz. Tverd. Tela **23** (1981) 2463).
82A	Afanas'eva, N.I., Buriakov, V.M., Vinogradov, E.A., Goncharov, A.F., Zhizhin, G.N.: Sov. Phys. Solid State **24** (1982) 116 (transl. from Fiz. Tverd. Tela **24** (1982) 211).
82G	Grandolfo, M., Somma, F., Vecchia, P., Effendiev, Sh.M.: Ferroelectrics **42** (1982) 203.
82I	Inushima, T., Uchinokura, K., Sasahara, K., Matsuura, E.:. Phys. Rev. **B26** (1982) 2525.

Physical property	Numerical value	Experimental conditions	Experimental method remarks	Ref.

10.4.7 Other ternary compounds

(a) Ternary compounds in pseudobinary systems of the type $(I_2–VI)_m(III_2–VI_3)_n$

The admixture of $I_2–VI$ compounds to the tetrahedrally coordinated $\square–III_2–VI_3$ compounds (section 10.1.1) may formally be considered as a gradual filling of the cationic vacant sites with monovalent atoms, eventually reaching the tetrahedral $I–III–VI_2$ compounds (section 10.1.2). In these $(I_2–VI)_m(III_2–V_3)_n$ phases several ordered ternary compounds have been found.

Phase diagrams of such pseudobinary systems can be found in the following references:

$Cu_2Se–Ga_2Se_3$ [67P2], $Cu_2Te–Ga_2Te_3$ [73C, 67P2], $Cu_2S–In_2S_3$ [80B], $Cu_2Se–In_2Se_3$ [67P1], $Cu_2Te–In_2Te_3$ [67P1], $Ag_2S–Ga_2S_3$ [76B], $Ag_2Se–Ga_2Se_3$ [67P2], $Ag_2Te–Ga_2Te_3$ [67P2], $Ag_2Se–In_2Se_3$ [67P1], $Ag_2Te–In_2Te_3$ [67P1, 67C1].

Many of the compounds reviewed in the following have not been found in these phase diagrams. For $m = n = 1$ ($I–III–VI_2$ compounds) we refer to section 10.1.2.

$m = 3$, $n = 5$: $I_3–III_5–VI_9$ compounds [80T1, 80T2]

These compounds are isostructural derivatives of In_2Se_3, the only difference being that the tetrahedral In cations are replaced by monovalent Cu (or Ag) atoms, the overall valence balance of the compound remaining unchanged.

$Cu_3In_5Se_9$:

E_g	1.10 eV	$T = 300$ K	photoconductivity maximum, Fig. 1	
	1.18 eV	77 K		
	0.96 eV	0 K	extrapolated from conductivity, Fig. 2	
			mobility of holes: Fig. 3	
			absorption spectrum: Fig. 4	
a	8.47 Å			
c	17.41 Å			
T_m	1025 °C			
d	5.568 g/cm³			

$Cu_3In_5Te_9$: No data except: $a = 8.78$ Å, $c = 18.66$ Å.

$Cu_3Ga_5Se_9$:

E_g	1.74 eV	$T = 300$ K	reflectivity	
a	8.01 Å			
c	16.46 Å			
T_m	1100 °C			
d	5.330 g/cm³			

$Ag_3In_5Se_9$:

E_g	1.22 eV	$T = 300$ K	photoconductivity	
T_m	825 °C			
d	5.668 g/cm³			

$Ag_3Ge_5Se_9$:

E_g	1.92 eV	$T = 300$ K	reflectivity	
T_m	884 °C			
d	6.720 g/cm³			

Physical property	Numerical value	Experimental conditions	Experimental method remarks	Ref.

m = 1, n = 2: I_2–III_4–VI_7 compounds (sphalerite structure)

$Cu_2Ga_4Te_7$ [73C]:

$E_{g,th}$	1.08 eV		conductivity, Fig. 5	
E_g	1.04 eV	$T = 300$ K	absorption	
$\mu_{H,p}$	120 cm^2/V s	300 K, $p = 10^{18}$ cm^{-3}	see Fig. 6; for other transport parameters, see Figs. 7, 8	
a	5.93 Å			
T_m	874 °C			
d	5.93 g/cm^3		X-ray	
	5.84 g/cm^3		experimental	

$Cu_2In_4Te_7$ [72C]:

$E_{g,th}$	1.10 eV		conductivity, Fig. 9	
$\mu_{H,p}$	0.27 cm^2/V s	$p = 7.3$ $\cdot 10^{15}$ cm^{-3}, $T = 300$ K	see Fig. 10; for other transport parameters, see Figs. 11, 12	
a	6.16 Å			
T_m	795(5) °C			
d	6.02 g/cm^3		X-ray	
	5.93 g/cm^3		experimental	

An ordered cubic phase close to $Cu_2In_4Se_7$ has been reported in [77L2].

m = 1, n = 3: I–III_3–VI_5 compounds (sphalerite structure)

$CuIn_3Te_5$ [71C]:

$E_{g,th}$	1.20 eV		conductivity, Fig. 13	
$\mu_{H,p}$	0.1 cm^2/V s	$p = 3 \cdot 10^{15}$ cm^{-3}, $T = 300$ K	for other transport parameters, see Figs. 14, 15	
a	6.16 Å			
T_m	772(5) °C			
d	5.94 g/cm^3		experimental	

$AgIn_3Te_5$ with $E_g = 1.1(1)$ eV, $a = 6.2476$ Å has been reported in [64O].

m = 1, n = 5: I–III_5–VI_8 compounds (spinel type)

$AgIn_5S_8$ [77P]:

E_g	1.7 eV	$T = 300$ K	optical gap, direct transition, Fig. 16	
$\mu_{H,n}$	4 cm^2/V s	$n = 1.4$ $\cdot 10^{18}$ cm^{-3}, $T = 300$ K	Fig. 17	
a	10.822 Å			
T_m	1075(10) °C			
d	4.85 g/cm^3		experimental	

Other compounds of this type cited in the literature are $AgIn_5Se_8$ [67P1], $AgIn_5Te_8$ [67P1], $CuIn_5S_8$ [80B].

m = 1, n = 9: I–III_9–VI_{14} compounds

Only $AgIn_9Te_{14}$ is cited as semiconductor with $E_{g,th} \approx 1.50$ eV, $\mu_{H,n} = 40$ cm^2/V s at 600 K [67C1], see Figs. 18, 19.

m = 1, n = 10: I_2–III_{20}–VI_{31} compounds

$Ag_2Ga_{20}S_{31}$ has sphalerite structure according to [76B]. No physical properties are known.
Of this system also Ag_9GaS_6 (bcc structure and T_d^2— F$\bar{4}$3m modifications) is cited in [76B].

(b) Some other semiconducting ternary compounds cited in the literature
(All data at room temperature)

Cd_2SnO_4: This compound has been prepared as crystalline powder or (amorphous) thin film only [77L1, 77S, 72N]. An optical gap of $E_g = 2.06$ eV and a carrier mobility of $\mu_H = 35$ cm^2/V s have been reported. Space group: D_{2h}^9—Pbam, $a = 5.5674(5)$ Å, $b = 9.8871(9)$ Å, $c = 3.1923(4)$ Å.

$CdSnO_3$: According to [77S] $CdSnO_3$ is a semiconductor with $E_g = 0.3$ eV. Structure: ilmenite (C_{3i}^2—R$\bar{3}$), $a = 5.4530(5)$ Å, $c = 14.960(3)$ Å.

Li_3CuO_3: n-type semiconductor with $E_g = 0.88$ eV [75M]

Hg_3PS_3, Hg_3PS_4: Semiconductors with E_g around 2 eV [68O]

InOF: Degenerate semiconductor with negative thermoelectric power [67C2]

$BaCu_4S_3$: Degenerate semiconductor at high temperature [76E]

$Cd_4(P, As)_2(Cl, Br, I)_3$: Cubic semiconductors, E_g about 1.8···2.3 eV [63S]

References for 10.4.7

63S   Suchov, L., Stemple, N.R.: J. Electrochem. Soc. **110** (1963) 766.

64O   O'Kane, D.F., Mason, D.R.: J. Electrochem. Soc. **114** (1964) 546.

67P1   Palatnik, L.S., Rogacheva, E.I.: Sov. Phys. Dokl. **12** (1967) 503.

67P2   Palatnik, L.S., Belova, E.K.: Inorg. Mater. **3** (1967) 1914 (transl. from Izv. Akad. Nauk SSSR Neorg. Mater. **3** (1967) 2194).

67C1   Chiang, P.W., O'Kane, D.F., Mason, D.R.: J. Electrochem. Soc. **114** (1967) 759.

67C2   Chamberland, B.L., Babcock, K.R.: Mater. Res. Bull. **2** (1967) 661.

68O   Olekseyuk, I.D., Golovei, I.D.: Inorg. Mater. **4** (1968) 1462 (transl. from Izv. Akad. Nauk SSSR Neorg. Mater. **4** (1968) 1676).

71C   Congiu, A., Garbato, L., Serci, S.: Phys. Status Solidi (a) **5** (1971) K15.

72C   Congiu, A., Garbato, L., Manca, P., Serci, S.: J. Electrochem. Soc. **119** (1972) 280.

72N   Nozik, A.J.: Phys. Rev. **B6** (1972) 453.

73C   Congiu, A., Garbato, L., Manca, P.: Mater. Res. Bull. **8** (1973) 293.

75M   Migeon, H.N., Courtois, A., Zanne, M., Gleitzer, Ch., Aubry, J.: Rev. Chim. Miner. **12** (1975) 203.

76B   Brandt, G., Krämer, V.: Mater. Res. Bull. **12** (1976) 1381.

76E   Eliezer, Z., Steinfink, H.: Mater. Res. Bull. **11** (1976) 385.

77L1   Lloyd, P.: Thin Solid Films **41** (1977) 113.

77L2   Lesueur, R., Djega-Mariadassou, C., Charpin, P., Albany, J.H.: 3d Conf. on Ternary Compounds, Edinburgh 1977, Conference Series No 35, The Institute of Physics: London **1977**, p. 15.

77P   Paorici, C., Zanotti, L., Romeo, N., Sberveglieri, G., Tarricone, L.: Mater. Res. Bull. **12** (1977) 1207.

77S   Shannon, R.D., Gillson, J.L., Bouchard, R.J.: J. Phys. Chem. Solids **38** (1977) 877.

80B   Binsma, J.J.M., Giling, L.J., Bloem, J.: J. Cryst. Growth **50** (1980) 429.

80T1   Tagirov, V.I., Gakhramanov, A.G., Guseinov, A.G., Aliev, F.M., Guseinov, G.G.: Sov. Phys. Crystallogr. **25** (1980) 237 (transl. from Kristallografiya **25** (1980) 411).

80T2   Tagirov, V.I., Gakhramanov, A.G., Guseinov, A.G., Aliev, F.M.: Sov. Phys. Semicond. **14** (1980) 831 (transl. from Fiz. Tekh. Poluprovodn. **14** (1980) 1403).

Figures

B. Physical data of semiconductors VI

10 Ternary compounds

10.1 Tetrahedrally bonded ternary and quasi-binary compounds

10.1.0 Introduction, general remarks on structure and properties (For tables, see p. 9ff.)

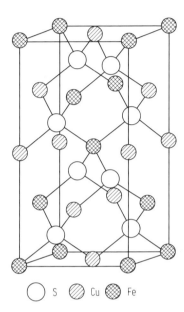

Fig. 1. Structure of chalcopyrite [75S].

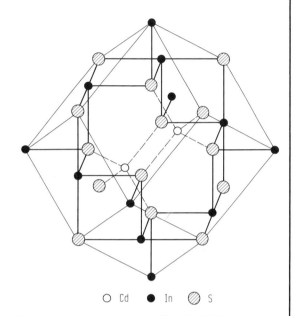

Fig. 3. Structure of spinel (e.g. $CdIn_2S_4$) [65W].

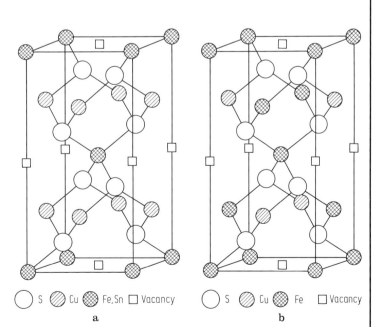

Fig. 2. a) Structure of defect stannite [81M]. b) Structure of defect chalcopyrite or thiogallate [81M].

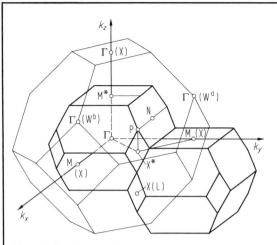

Fig. 4. Brillouin zone of zincblende (narrow lines), two Brillouin zones of chalcopyrite (broad lines) [81M]. M*, X*: Symmetry points M and X are sometimes designated T (or Z) and N, respectively, in the tables and figures.

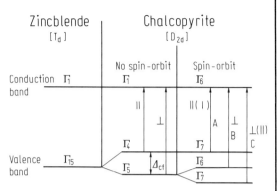

Fig. 5. Selection rules for the transitions in the energy gap of chalcopyrite [81M].

10.1.1 III$_2$–VI$_3$ compounds (For tables, see p. 12ff.)

α– Ga$_2$S$_3$

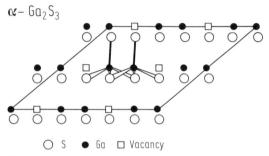

○ S   ● Ga   □ Vacancy

Fig. 1. α-Ga$_2$S$_3$. Projection of the structure of α-Ga$_2$S$_3$ on the plane perpendicular to the c-axis. [76C].

β– In$_2$S$_3$

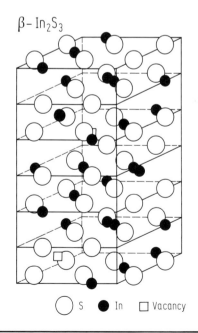

○ S   ● In   □ Vacancy

α– In$_2$Te$_3$ – II

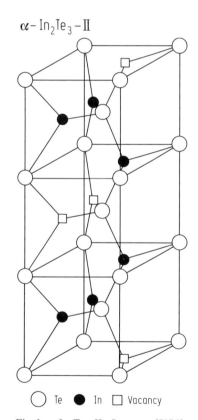

○ Te   ● In   □ Vacancy

Fig. 3. α-In$_2$Te$_3$–II. Structure [78B2].

◄ Fig. 2. β-In$_2$S$_3$. Primitive unit cell [70L].

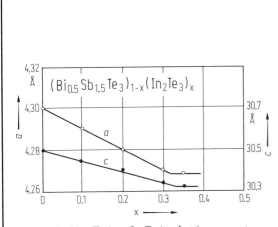

Fig. 4. $(Bi_{0.5}Sb_{1.5}Te_3)_{1-x}(In_2Te_3)_x$. Lattice parameters vs. composition [73A]. For $x=1$: α-phase.

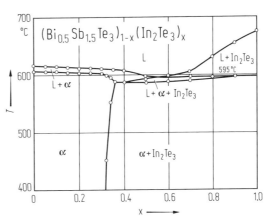

Fig. 5. $(Bi_{0.5}Sb_{1.5}Te_3)_{1-x}(In_2Te_3)_x$. Phase diagram [73A]. α: solid solution. For $x=1$: α-phase.

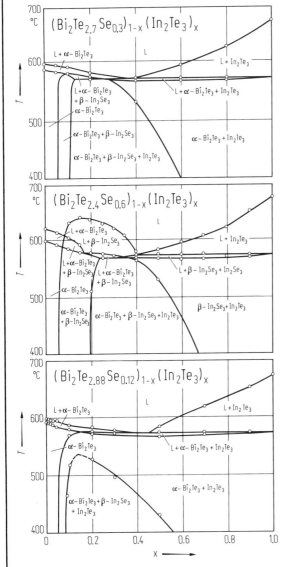

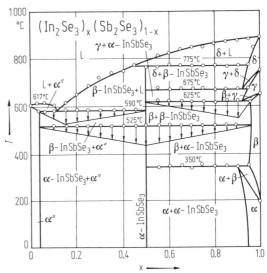

Fig. 7. $(In_2Se_3)_x(Sb_2Se_3)_{1-x}$. Phase diagram [77G]. α, β, γ, δ stand for α-, β-, γ-, δ-In_2Se_3, α'': Sb_2Se_3 with some Sb replaced by In.

◄

Fig. 6. $(Bi_2Te_{2.88}Se_{0.12})_{1-x}(In_2Te_3)_x$, $(Bi_2Te_{2.7}Se_{0.3})_{1-x}(In_2Te_3)_x$, $(Bi_2Te_{2.4}Se_{0.6})_{1-x}(In_2Te_3)_x$. Phase diagram [74A2]. α-Bi_2Te_3 implies a Bi_2Te_3 based solid solution.

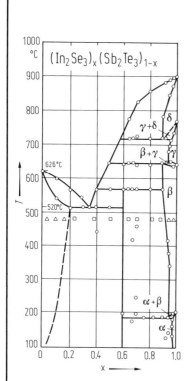

Fig. 8. (In$_2$Se$_3$)$_x$(Sb$_2$Te$_3$)$_{1-x}$. Phase diagram [72B1]. ○ experimental points, △ one phase, □ two phases; see caption of Fig. 7.

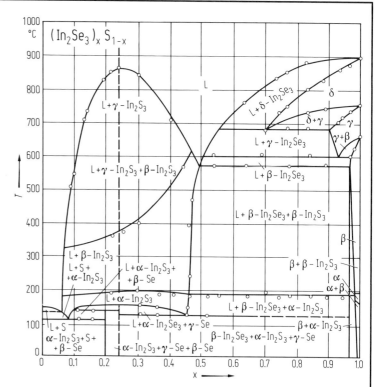

Fig. 10. (In$_2$Se$_3$)$_x$S$_{1-x}$. Phase diagram [73R2]. α, β, γ, δ: see Fig. 7.

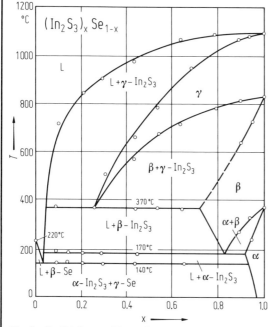

Fig. 9. (In$_2$S$_3$)$_x$Se$_{1-x}$. Phase diagram [73R2]. α, β, γ stand for α-, β-, γ-In$_2$S$_3$.

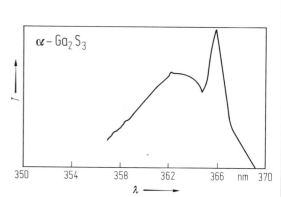

Fig. 11. α-Ga$_2$S$_3$. Transmission vs. wavelength at T = 1.6 K [77M2]. Unpolarized light.

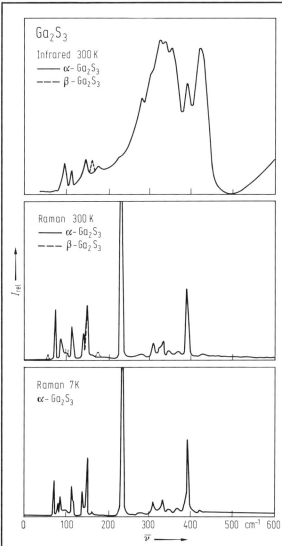

Fig. 12. Ga$_2$S$_3$. Infrared and Raman spectra (relative intensity vs. wavenumber) [78L1].

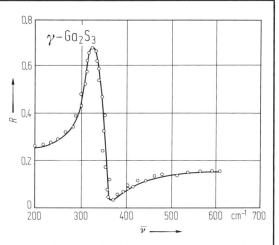

Fig. 13. γ-Ga$_2$S$_3$. Reflectivity vs. wavenumber. Experimental points and calculated curve [73K].

For Fig. 14, see next page.

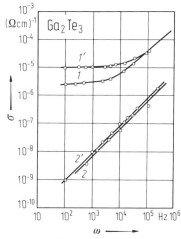

Fig. 15. Ga$_2$Te$_3$. Conductivity vs. frequency for two samples: 1,1′: $T = 300$ K; 2,2′: $T = 77$ K [78B2]. Orientation dependence not given.

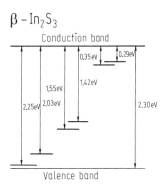

Fig. 16. β-In$_2$S$_3$. Local levels in the band gap obtained by optical measurements (schematic diagram for $T = 0$ K) [67N].

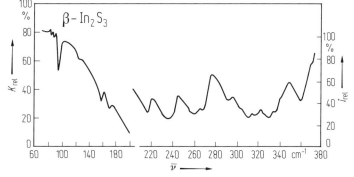

Fig. 17. β-In$_2$S$_3$. Infrared absorption vs. wavenumber [70L]. Unpolarized light. Note change of scale at 200 cm^{-1}.

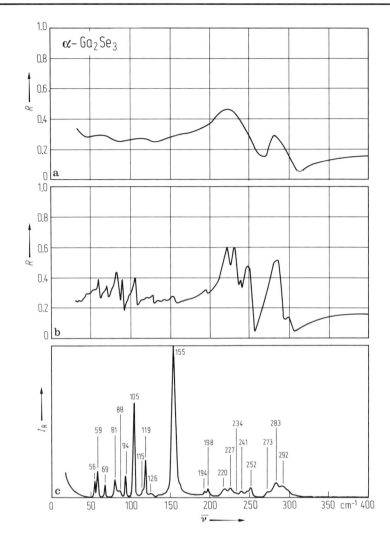

Fig. 14. α-Ga$_2$Se$_3$. Infrared reflectivity and Raman intensity vs. wavenumber [73F2]. (a) disordered α-Ga$_2$Se$_3$; IR, (b) ordered α-Ga$_2$Se$_3$; IR, (c) α-Ga$_2$Se$_3$; Raman.

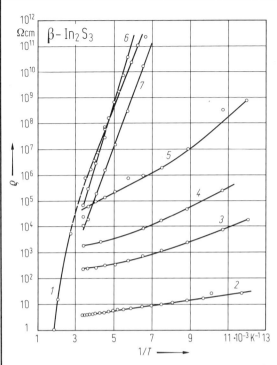

Fig. 18. β-In$_2$Se$_3$. Resistivity vs. reciprocal temperature for five undoped samples (1···5), one Cd-doped (6) and one P-doped sample (7) [65R]. Orientation not specified.

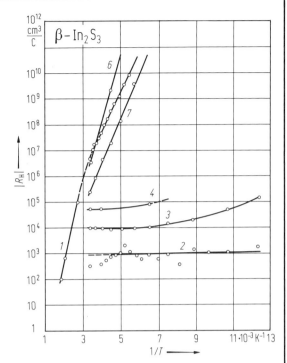

Fig. 19. β-In$_2$S$_3$. Hall coefficient vs. reciprocal temperature for samples 1···4, 6, 7 of Fig. 18 [65R]. Orientation not specified.

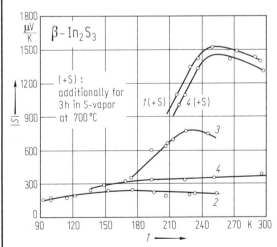

Fig. 20. β-In$_2$S$_3$. Thermoelectric power vs. temperature for two undoped samples (1, 2), one Cd-doped (3) and one P-doped sample (4). [65R]. Orientation not specified.

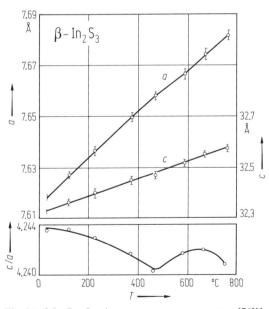

Fig. 21. β-In$_2$S$_3$. Lattice parameters vs. temperature [76K].

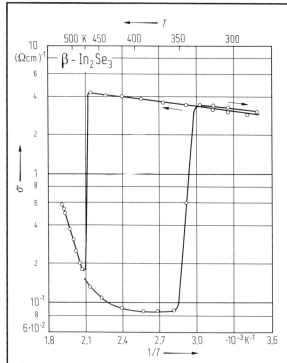

Fig. 22. β-In$_2$Se$_3$. Electrical conductivity vs. (reciprocal) temperature [71B]. σ⊥c.

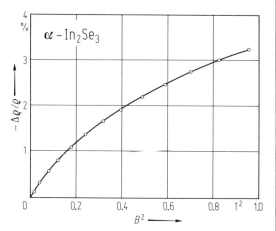

Fig. 23. α-In$_2$Se$_3$. Negative transverse magnetoresistance (⊥c) vs. magnetic field squared at $T = 5$ K [74R].

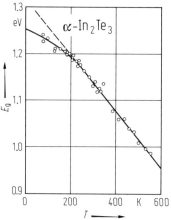

Fig. 25. α-In$_2$Te$_3$. Energy gap vs. temperature [75H].

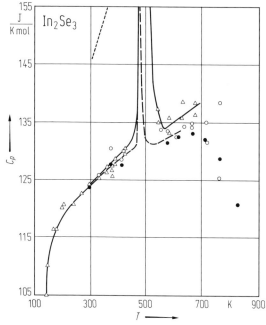

Fig. 24. In$_2$Se$_3$. Heat capacity vs. temperature (data from several samples, dotted line from other literature) [76M].

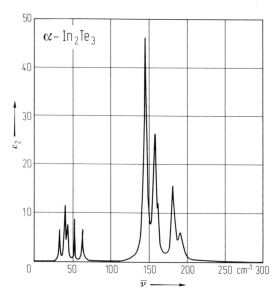

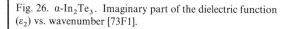

Fig. 26. α-In$_2$Te$_3$. Imaginary part of the dielectric function (ε_2) vs. wavenumber [73F1].

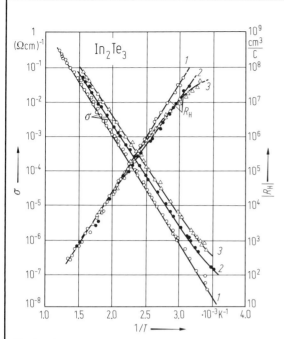

Fig. 27. In$_2$Te$_3$. Conductivity and Hall coefficient vs. reciprocal temperature. Degree of ordering decreases in the sequence *1, 2, 3*, i.e. disorder can be quenched [60Z2].

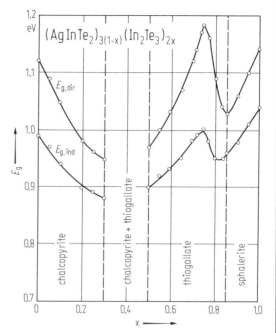

Fig. 29. (AgInTe$_2$)$_{3(1-x)}$(In$_2$Te$_3$)$_{2x}$. Direct and indirect gaps vs. composition at $T = 300$ K [75M]. (For x = 1: β-In$_2$Te$_3$.)

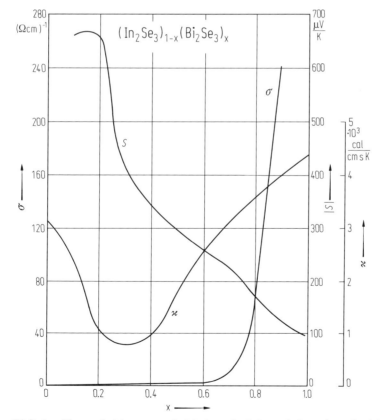

Fig. 28. (In$_2$Se$_3$)$_{1-x}$(Bi$_2$Se$_3$)$_x$. Thermoelectric power, electrical conductivity and thermal conductivity vs. composition at RT [70G]. (For x = 0: In$_2$Se$_3$ phase unclear.)

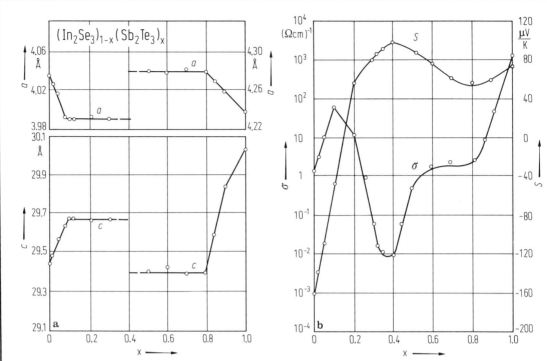

Fig. 30. $(In_2Se_3)_{1-x}(Sb_2Te_3)_x$. (a) Lattice parameters, (b) conductivity and thermoelectric power at RT vs. composition. Orientations not specified [73R2]. (For $x=0$: β-In_2Se_3.)

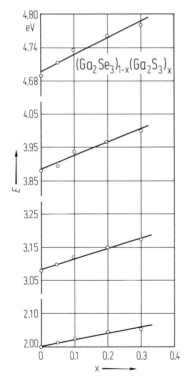

Fig. 31. $(Ga_2Se_3)_{1-x}(Ga_2S_3)_x$. Peaks in optical reflection vs. composition [72B2]. (For $x=0$: α-Ga_2Se_3.)

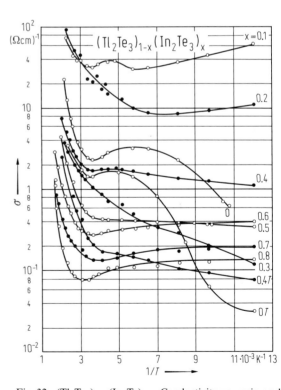

Fig. 32. $(Tl_2Te_3)_{1-x}(In_2Te)_{3x}$. Conductivity vs. reciprocal temperature. T: tempered sample [71N].

10.1.2 I–III–VI$_2$ compounds (For tables, see p. 26 ff.)

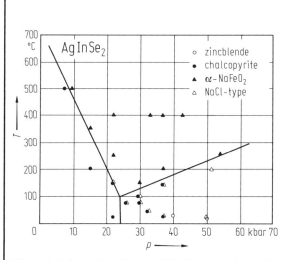

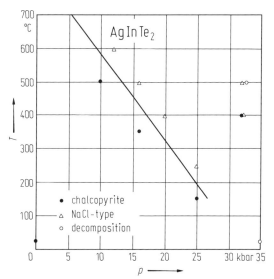

Fig. 1. AgInSe$_2$. $T-p$ diagram [77J]. From powder at RT and 1 bar after $T-p$ treatment.

Fig. 2. AgInTe$_2$. $T-p$ diagram [77J]. Cf. Fig. 1.

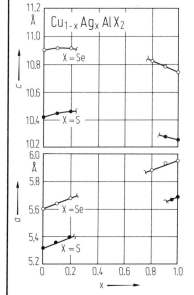

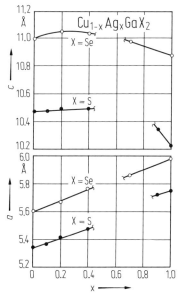

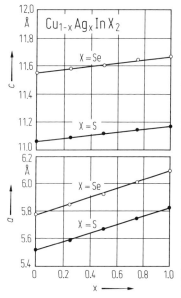

Fig. 3. Cu$_{1-x}$Ag$_x$AlX$_2$ (X = S,Se). Lattice parameters vs. composition at RT [75S1].

Fig. 4. Cu$_{1-x}$Ag$_x$GaX$_2$ (X = S,Se). Lattice parameters vs. composition at RT [75S1].

Fig. 5. Cu$_{1-x}$Ag$_x$InX$_2$ (X = S,Se). Lattice parameters vs. composition at RT [75S1].

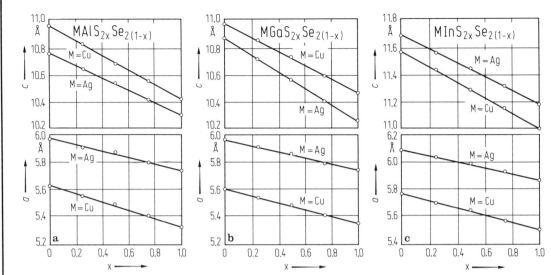

Fig. 6. (a) MAlS$_{2x}$Se$_{2(1-x)}$, (b) MGaS$_{2x}$Se$_{2(1-x)}$, (c) MInS$_{2x}$Se$_{2(1-x)}$ (M = Cu, Ag). Lattice parameters vs. composition at RT [73R].

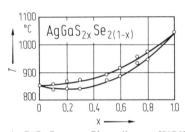

Fig. 7. AgGaS$_{2x}$Se$_{2(1-x)}$. Phase diagram [79B2].

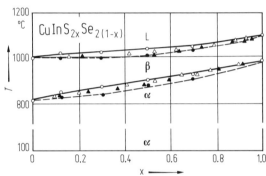

Fig. 8. CuInS$_{2x}$Se$_{2(1-x)}$. Phase diagram for several different samples (α = chalcopyrite, β = disordered zincblende). Lines delimit range of values [77T5].

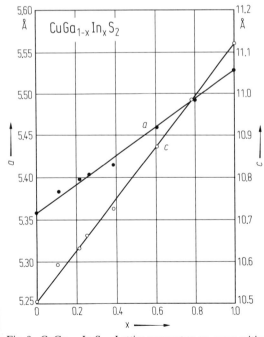

Fig. 9. CuGa$_{1-x}$In$_x$S$_2$. Lattice parameters vs. composition at RT [75S1].

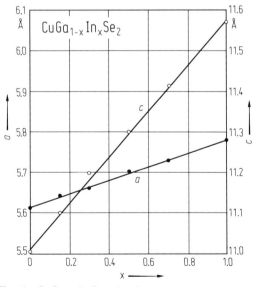

Fig. 10. CuGa$_{1-x}$In$_x$Se$_2$. Lattice parameters vs. composition [79P].

10.1.2 I–III–VI₂ compounds

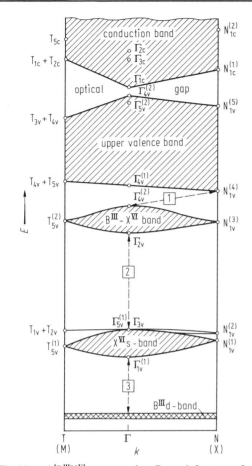

Fig. 10α. A^{I}B^{III}X$_2^{VI}$ compounds. General features of the band structure. Shaded and cross-hatched areas denote major subbands, and boxed numbers indicate the three internal gaps [83J]. Symmetry points T and N correspond to M and X, respectively, in the Brillouin zone of Fig. 4 of section 10.1.0.

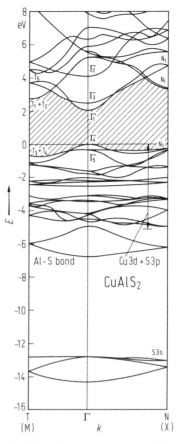

Fig. 10β. CuAlS₂. Self-consistent electronic band structure (compare Fig. 10α) [83J]. Note: all optical band gaps of the CuBX₂ semiconductors are found to be 1···1.5 eV too small relative to experiment. For discussion, see [83J]. Symmetry points T and N correspond to M and X, respectively, in the Brillouin zone of Fig. 4 of section 10.1.0.

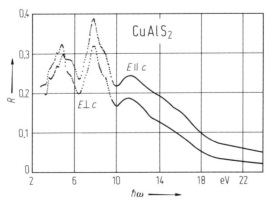

Fig. 11. CuAlS₂. Reflectivity vs. photon energy at RT [77R].

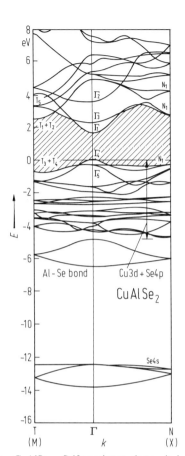

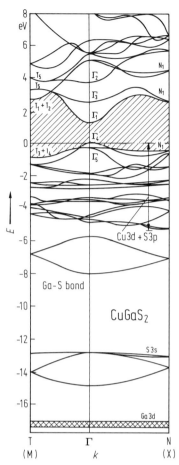

Fig. 11α. CuAlSe$_2$. Self-consistent electronic band structure (compare Fig. 10α) [83J]. See caption of Fig. 10β.

Fig. 11β. CuGaS$_2$. Self-consistent electronic band structure (compare Fig. 10α) [83J]. See caption of Fig. 10β.

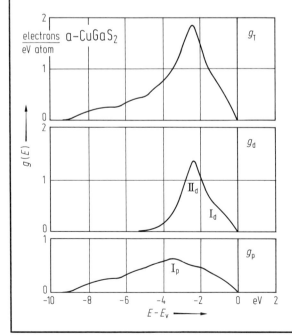

◄

Fig. 12. CuGaS$_2$-amorphous. Partial densities of valence states (p- and d-states), total density of valence states: $g_T = g_p + g_d$. I_p, I_d, II_d denote features in g_T [76B1].

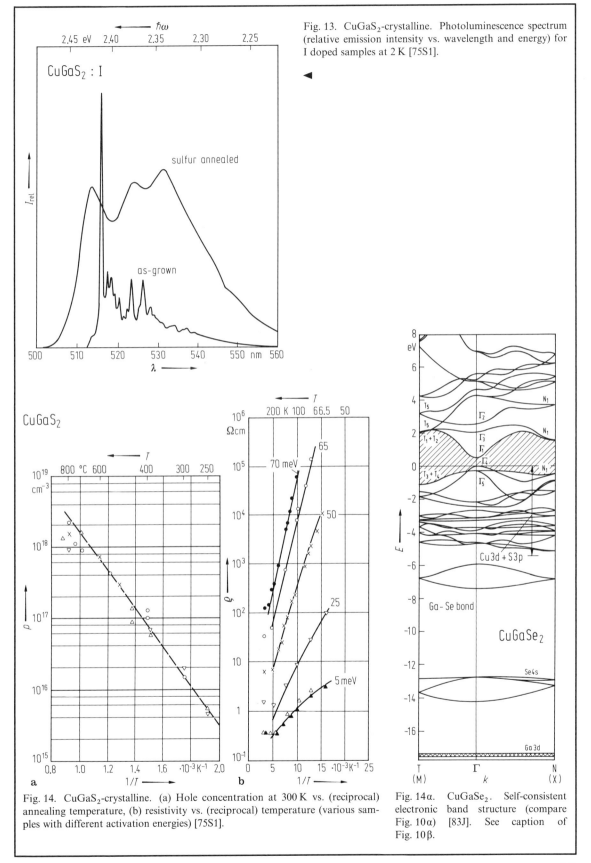

Fig. 13. CuGaS$_2$-crystalline. Photoluminescence spectrum (relative emission intensity vs. wavelength and energy) for I doped samples at 2 K [75S1].

Fig. 14. CuGaS$_2$-crystalline. (a) Hole concentration at 300 K vs. (reciprocal) annealing temperature, (b) resistivity vs. (reciprocal) temperature (various samples with different activation energies) [75S1].

Fig. 14α. CuGaSe$_2$. Self-consistent electronic band structure (compare Fig. 10α) [83J]. See caption of Fig. 10β.

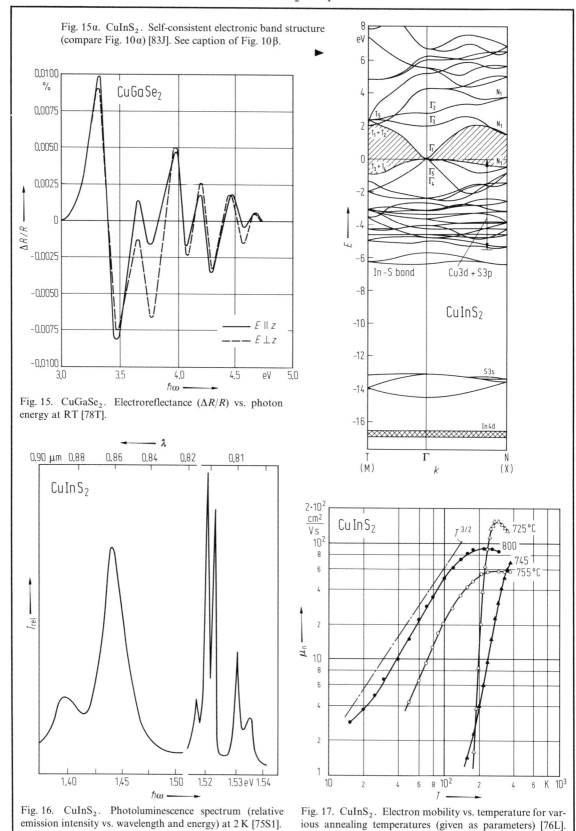

Fig. 15α. CuInS$_2$. Self-consistent electronic band structure (compare Fig. 10α) [83J]. See caption of Fig. 10β.

Fig. 15. CuGaSe$_2$. Electroreflectance ($\Delta R/R$) vs. photon energy at RT [78T].

Fig. 16. CuInS$_2$. Photoluminescence spectrum (relative emission intensity vs. wavelength and energy) at 2 K [75S1].

Fig. 17. CuInS$_2$. Electron mobility vs. temperature for various annealing temperatures (given as parameters) [76L]. Annealing atmosphere: In; single crystalline samples.

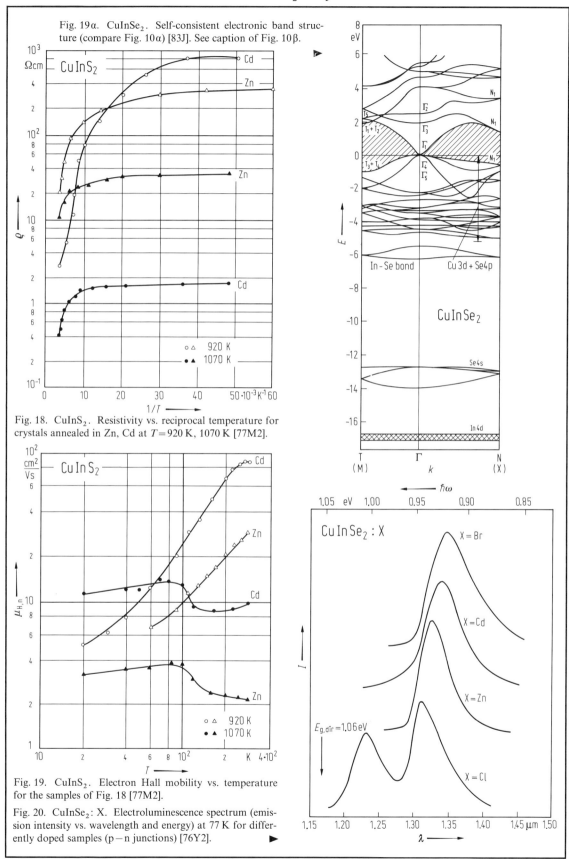

Fig. 19α. CuInSe$_2$. Self-consistent electronic band structure (compare Fig. 10α) [83J]. See caption of Fig. 10β.

Fig. 18. CuInS$_2$. Resistivity vs. reciprocal temperature for crystals annealed in Zn, Cd at $T=920$ K, 1070 K [77M2].

Fig. 19. CuInS$_2$. Electron Hall mobility vs. temperature for the samples of Fig. 18 [77M2].

Fig. 20. CuInSe$_2$: X. Electroluminescence spectrum (emission intensity vs. wavelength and energy) at 77 K for differently doped samples (p–n junctions) [76Y2].

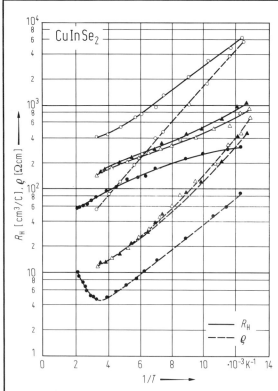

Fig. 21. CuInSe$_2$. Resistivity and Hall coefficient vs. reciprocal temperature for four p-type samples [79I].

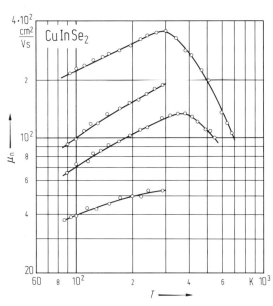

Fig. 22. CuInSe$_2$. Electron mobility vs. temperature for four n-type samples [78N2].

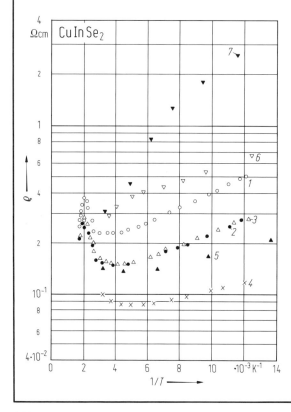

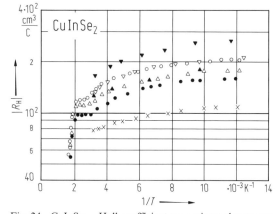

Fig. 24. CuInSe$_2$. Hall coefficient vs. reciprocal temperature for the samples of Fig. 23 [79I].

◄

Fig. 23. CuInSe$_2$. Resistivity vs. reciprocal temperature for seven n-type samples [79I].

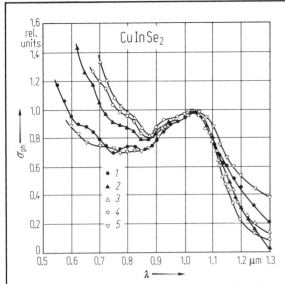

Fig. 25. CuInSe$_2$. Photoconductivity vs. wavelength for $T = 100$ K (1), 125 K (2), 152 K (3), 178 K (4), 192 K (5) [80A2]. Cf. Fig. 41.

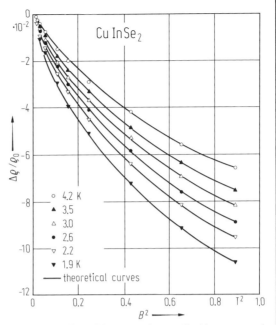

Fig. 26. CuInSe$_2$. Magnetoresistance ($\Delta\varrho/\varrho$) vs. magnetic field squared at different temperatures for sample 6 of Fig. 23 [79I].

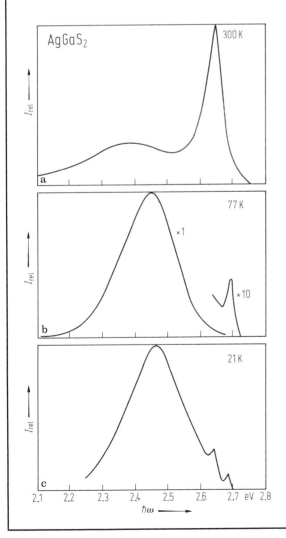

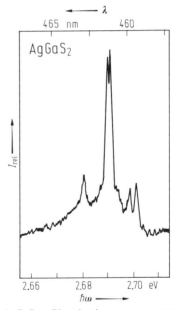

Fig. 27. AgGaS$_2$. Photoluminescence spectrum (relative emission intensity vs. wavelength and energy) at 2 K [75S1].

◀

Fig. 28. AgGaS$_2$. Cathodoluminescence spectrum (relative emission intensity vs. photon energy) at different temperatures [75S1].

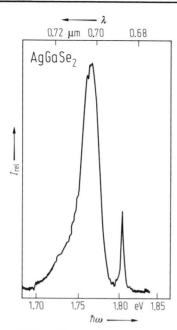

Fig. 29. AgGaSe$_2$. Photoluminescence spectrum (relative emission intensity vs. wavelength and energy) at 2 K [75S1].

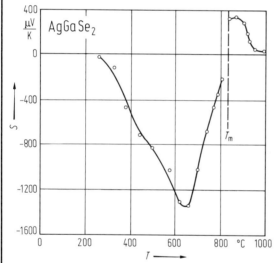

Fig. 31. AgGaSe$_2$. Thermoelectric power vs. temperature [75S1].

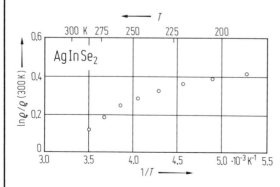

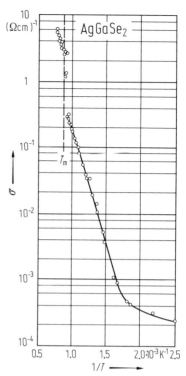

Fig. 30. AgGaSe$_2$. Electrical conductivity vs. reciprocal temperature for n-type sample [75S1].

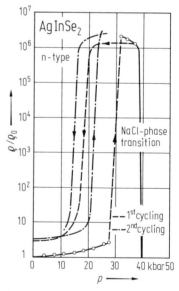

Fig. 32. AgInSe$_2$. Resistivity vs. pressure at 300 K [78J].

◀

Fig. 33. AgInSe$_2$. Resistivity vs. (reciprocal) temperature at a pressure chosen for maximum ϱ [78J].

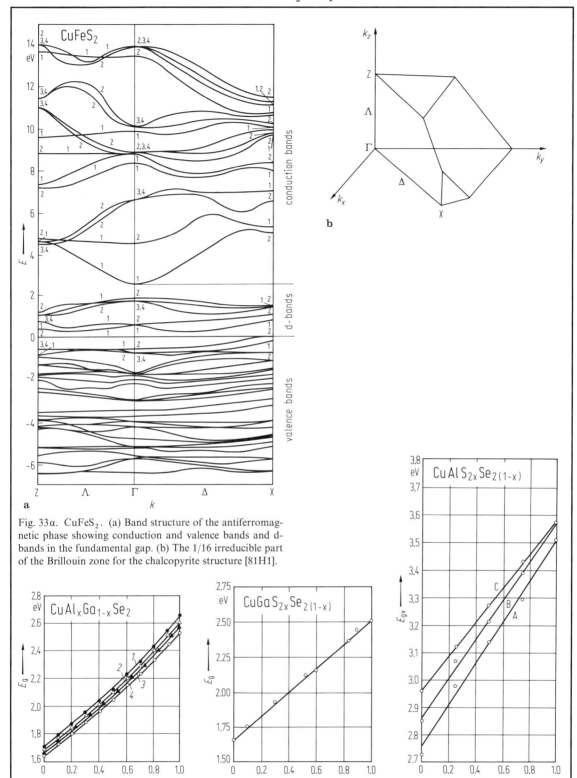

Fig. 33α. CuFeS$_2$. (a) Band structure of the antiferromagnetic phase showing conduction and valence bands and d-bands in the fundamental gap. (b) The 1/16 irreducible part of the Brillouin zone for the chalcopyrite structure [81H1].

Fig. 34α. CuAl$_x$Ga$_{1-x}$Se$_2$. Compositional dependence of the optical band gap. *1, 2*: $T = 77$ K; *3, 4*: $T = 293$ K; *1, 3*: $E\perp c$; *2, 4*: $E \parallel c$ [83B].

Fig. 35. CuGaS$_{2x}$Se$_{2(1-x)}$. Energy gap vs. composition at 77 K from unpolarized absorption [77B1].

Fig. 34. CuAlS$_{2x}$Se$_{2(1-x)}$. Exciton energies vs. composition at 20 K [77Y].

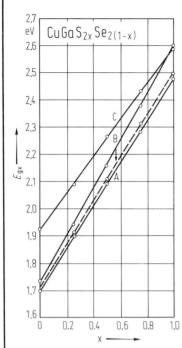

Fig. 36. CuGaS$_{2x}$Se$_{2(1-x)}$. Exciton energies vs. composition at 20 K; dashed line: first excited state of A-exciton [77Y].

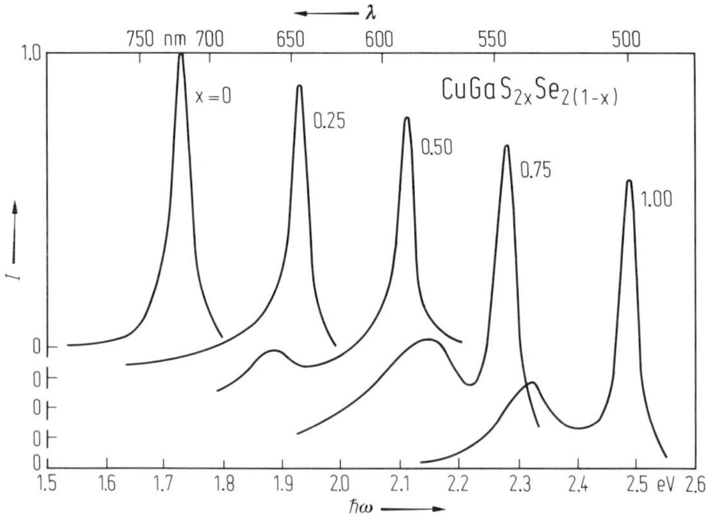

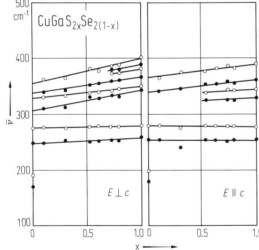

Fig. 37. CuGaS$_{2x}$Se$_{2(1-x)}$. Photoluminescence spectrum (normalized intensity vs. photon energy and wavelength) for various compositions at 77 K [77T1].

Fig. 38. CuGaS$_{2x}$Se$_{2(1-x)}$. Infrared phonon wavenumbers vs. composition at RT (○ LO, ● TO) [77B2].

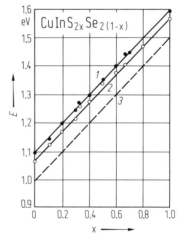

Fig. 39. CuInS$_{2x}$Se$_{2(1-x)}$. Composition dependence of (1) energy gap at 77 K, (2) energy gap at 300 K (from photoemf), (3) luminescence maximum at $T = 77$ K and 300 K [79B3].

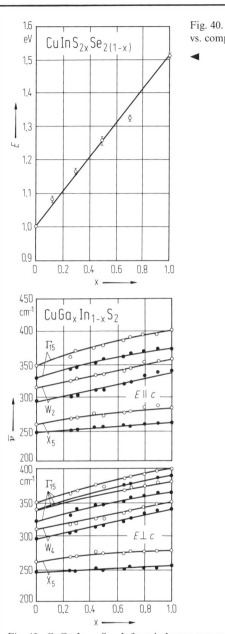

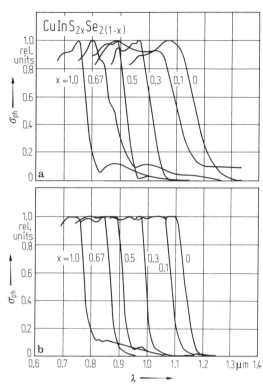

Fig. 40. CuInS$_{2x}$Se$_{2(1-x)}$. Cathodoluminescence maximum vs. composition at 90 K [75D].

Fig. 41. CuInS$_{2x}$Se$_{2(1-x)}$. Photoconductivity vs. wavelength (a) $T = 300$ K, (b) $T = 77$ K [79B3].

Fig. 42. CuGa$_x$In$_{1-x}$S$_2$. Infrared phonon wavenumbers vs. composition at RT (○ LO, ● TO) [78B2].

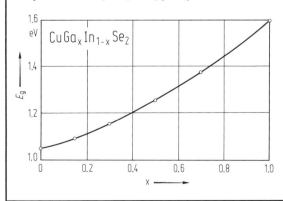

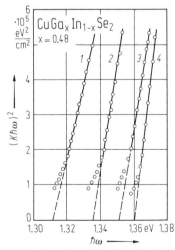

Fig. 44. CuGa$_x$In$_{1-x}$Se$_2$. $(Kh\omega)^2$ vs. $h\omega$ for x = 0.48 [79H2]. (1) $T = 296$ K, (2) $T = 199$ K, (3) $T = 100$ K, (4) $T = 22$ K (polycrystalline samples).

Fig. 43. CuGa$_x$In$_{1-x}$Se$_2$. Energy gap vs. composition at 300 K from unpolarized absorption [79P].

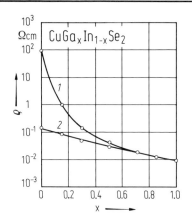

Fig. 45. CuGa$_x$In$_{1-x}$Se$_2$. Resistivity vs. composition at 300 K [79P]. (*1*) as grown samples, (*2*) Se annealed samples. (No anisotropy reported.)

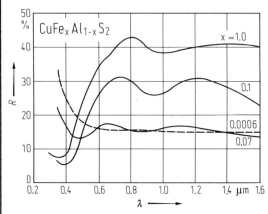

Fig. 46. CuFe$_x$Al$_{1-x}$S$_2$. Reflectivity (unpolarized) vs. wavelength at RT for samples of different composition [77T6].

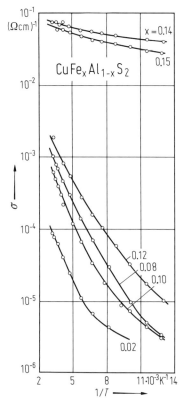

Fig. 48. CuFe$_x$Al$_{1-x}$S$_2$. Conductivity vs. reciprocal temperature for samples of different composition [77T6]. (No anisotropy reported.)

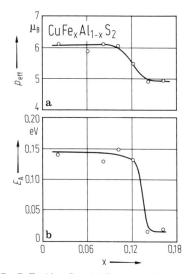

Fig. 47. CuFe$_x$Al$_{1-x}$S$_2$. (a) Paramagnetic moment vs. Fe concentration (at RT), (b) activation energy of conductivity vs. Fe concentration ($T = 70\cdots300$ K) [77T6].

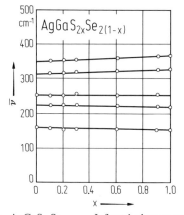

Fig. 49. AgGaS$_{2x}$Se$_{2(1-x)}$. Infrared phonon wavenumbers vs. composition at RT [79B2].

10.1.3 II–IV–V$_2$ compounds (For tables, see p. 68 ff.)

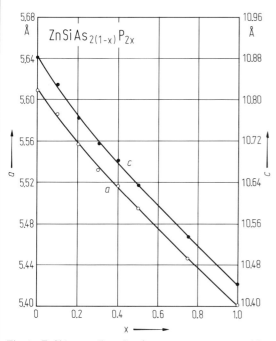

Fig. 1. ZnSiAs$_{2(1-x)}$P$_{2x}$. Lattice constants vs. composition at $T=300$ K [75S1].

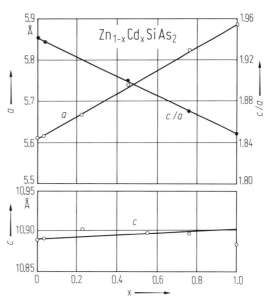

Fig. 2. Zn$_{1-x}$Cd$_x$SiAs$_2$. Lattice constants vs. composition at $T=300$ K [76J].

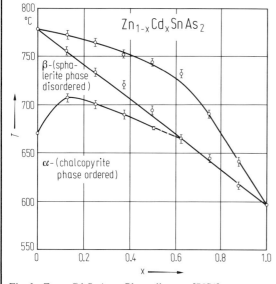

Fig. 3. Zn$_{1-x}$Cd$_x$SnAs$_2$. Phase diagram [75S1].

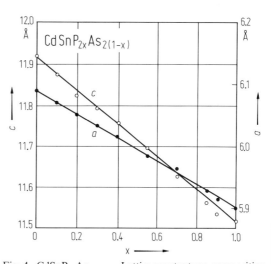

Fig. 4. CdSnP$_{2x}$As$_{2(1-x)}$. Lattice constants vs. composition at $T=300$ K [77G].

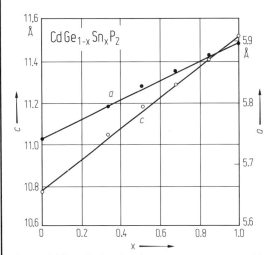

Fig. 5. CdGe$_{1-x}$Sn$_x$P$_2$. Lattice constants vs. composition at $T = 300$ K [77G].

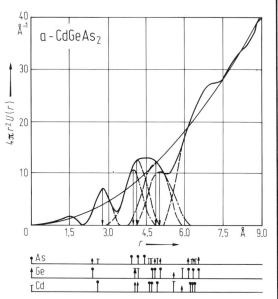

Fig. 6. CdGeAs$_2$-amorphous. Radial distribution function (RDF) (thick line), decomposition (dotted lines), for a uniform sphere (thin line), r: radial distance. The arrows etc. below show the atomic positions in the crystal [76P2].

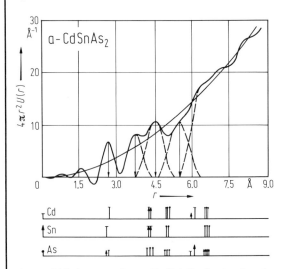

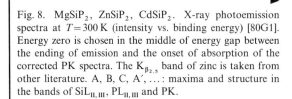

Fig. 7. CdSnAs$_2$-amorphous. Radial distribution function [76P2]. For explanations, see Fig. 6.

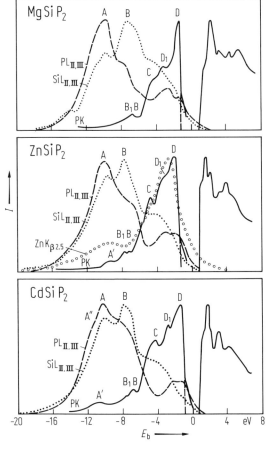

Fig. 8. MgSiP$_2$, ZnSiP$_2$, CdSiP$_2$. X-ray photoemission spectra at $T = 300$ K (intensity vs. binding energy) [80G1]. Energy zero is chosen in the middle of energy gap between the ending of emission and the onset of absorption of the corrected PK spectra. The K$_{\beta_{2.5}}$ band of zinc is taken from other literature. A, B, C, A′, ...: maxima and structure in the bands of SiL$_{II, III}$, PL$_{II, III}$ and PK.

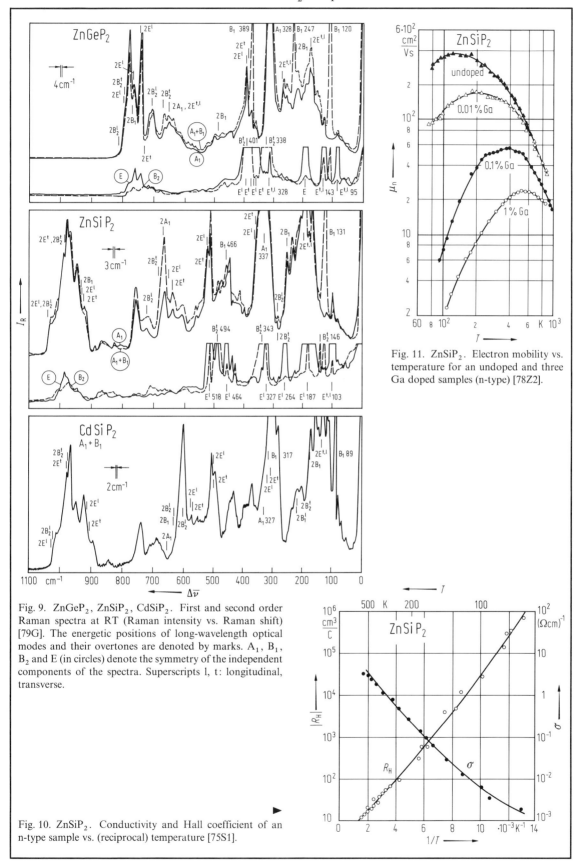

Fig. 11. ZnSiP$_2$. Electron mobility vs. temperature for an undoped and three Ga doped samples (n-type) [78Z2].

Fig. 9. ZnGeP$_2$, ZnSiP$_2$, CdSiP$_2$. First and second order Raman spectra at RT (Raman intensity vs. Raman shift) [79G]. The energetic positions of long-wavelength optical modes and their overtones are denoted by marks. A$_1$, B$_1$, B$_2$ and E (in circles) denote the symmetry of the independent components of the spectra. Superscripts l, t: longitudinal, transverse.

Fig. 10. ZnSiP$_2$. Conductivity and Hall coefficient of an n-type sample vs. (reciprocal) temperature [75S1].

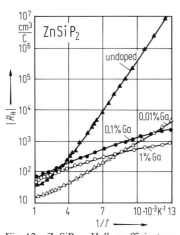

Fig. 12. ZnSiP$_2$. Hall coefficient vs. reciprocal temperature for the n-type samples of Fig. 11 [78Z1].

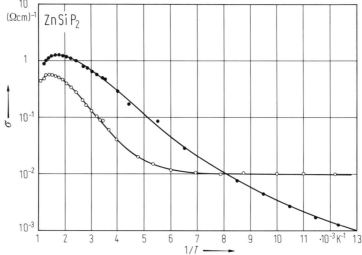

Fig. 13. ZnSiP$_2$. Hall coefficient vs. reciprocal temperature for a sample with high donor concentration (● hopping, ○ impurity band conduction) [76S].

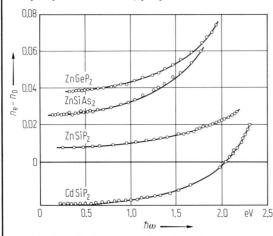

Fig. 14. ZnSiP$_2$. Conductivity vs. reciprocal temperature for a sample with high donor concentration (● hopping, ○ impurity band conduction) [76S].

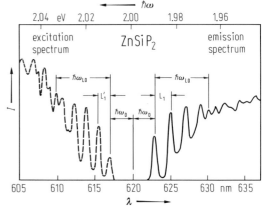

Fig. 15. ZnGeP$_2$, ZnSiAs$_2$, ZnSiP$_2$, CdSiP$_2$. Birefringence ($n_e - n_o$) vs. photon energy at $T = 300$ K [77A1].

Fig. 16. ZnSiP$_2$. Photoluminescence spectrum (emission and excitation intensity vs. wavelength (photon energy)) at $T = 1.8$ K. L$_1$ = local mode sideband [75S1].

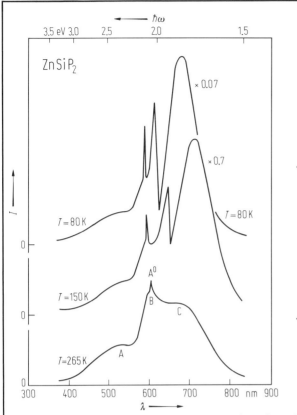

Fig. 17. ZnSiP$_2$. Cathodoluminescence spectrum (intensity vs. wavelength (photon energy)) at $T = 80$ K, 150 K, 265 K [76B1].

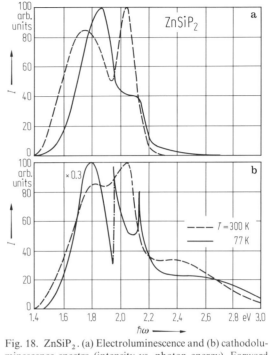

Fig. 18. ZnSiP$_2$. (a) Electroluminescence and (b) cathodoluminescence spectra (intensity vs. photon energy). Forward biased Au—ZnSiP$_2$ diode [77S1].

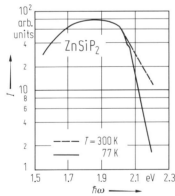

Fig. 19. ZnSiP$_2$. Electroluminescence spectrum (intensity vs. photon energy at $T = 300$ K; unpolarized light). Reverse biased Au—ZnSiP$_2$ diode. [77S1].

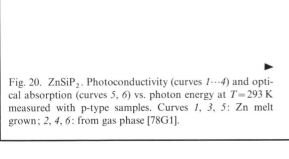

Fig. 20. ZnSiP$_2$. Photoconductivity (curves $1 \cdots 4$) and optical absorption (curves 5, 6) vs. photon energy at $T = 293$ K measured with p-type samples. Curves 1, 3, 5: Zn melt grown; 2, 4, 6: from gas phase [78G1].

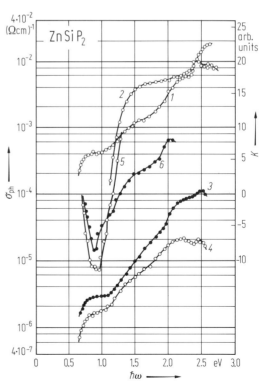

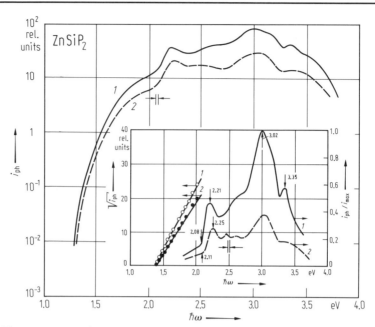

Fig. 21. ZnSiP$_2$. Photocurrent vs. photon energy at $T = 300$ K for an n-type sample; curve *1*: $E \perp c$, *2*: $E \parallel c$. Insert shows $i_{ph}^{1/2}$ and i_{ph}/i_{max} vs. $\hbar\omega$. [77M1].

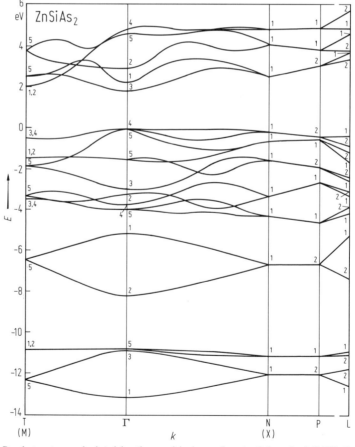

Fig. 21α. ZnSiAs$_2$. Band structure calculated by the empirical pseudopotential method [81H]. Symmetry points T and N correspond to M and X, respectively, in the Brillouin zone of Fig. 4 in 10.1.0.

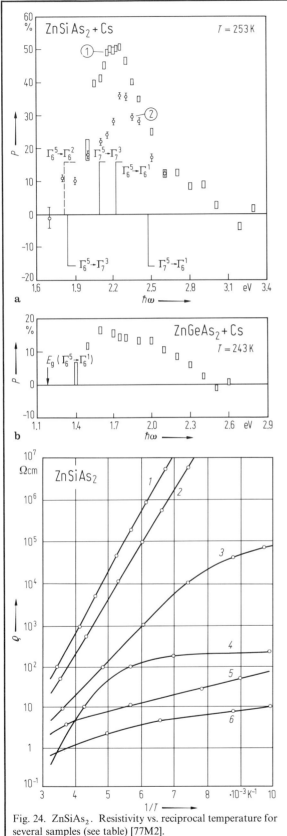

Fig. 22. ZnSiAs$_2$ (a), ZnGeAs$_2$ (b). Photoelectron spin polarization P by irradiation with circularly polarized light vs. photon energy. Rectangles: one cesiation, ⚬⚬ two cesiations (2nd cesiation 19 h later, cesiations at $T = 250$ K, $p = 6 \cdot 10^{-10}$ Torr). Vertical lines indicate onset of transitions at Γ (lines pointing up: α spins, others: β spins). Onset of transition $\Gamma_6^5 - \Gamma_6^2$ (dashed line) is uncertain [79Z]. $\Gamma_6^5 \ldots$ etc.: upper index indicates the single group representation from which the state originates.

◀

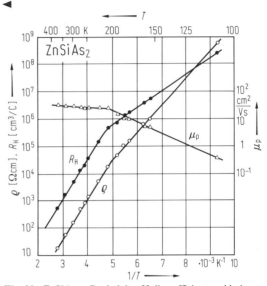

Fig. 23. ZnSiAs$_2$. Resistivity, Hall coefficient and hole mobility vs. (reciprocal) temperature for a p-type sample [75S1].

Fig. 24. ZnSiAs$_2$. Resistivity vs. reciprocal temperature for several samples (see table) [77M2].

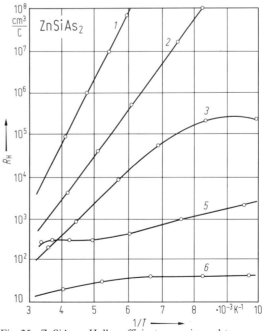

Fig. 25. ZnSiAs$_2$. Hall coefficient vs. reciprocal temperature for the samples of Fig. 24 [77M2].

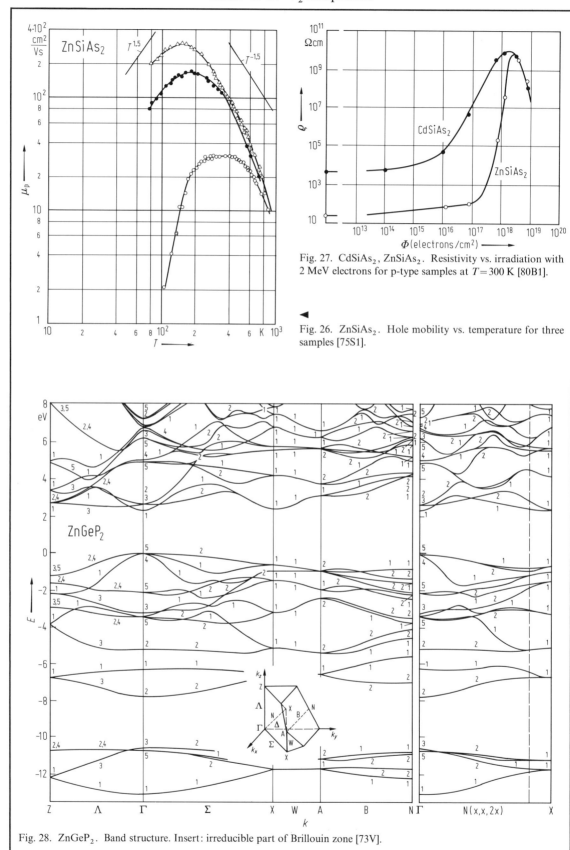

Fig. 27. CdSiAs$_2$, ZnSiAs$_2$. Resistivity vs. irradiation with 2 MeV electrons for p-type samples at $T = 300$ K [80B1].

◄

Fig. 26. ZnSiAs$_2$. Hole mobility vs. temperature for three samples [75S1].

Fig. 28. ZnGeP$_2$. Band structure. Insert: irreducible part of Brillouin zone [73V].

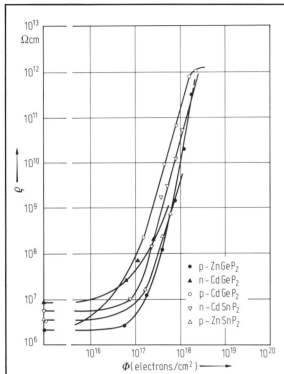

Fig. 29. ZnGeP$_2$, CdGeP$_2$, CdSnP$_2$, ZnSnP$_2$. Resistivity vs. irradiation with 2 MeV electrons [78B3].

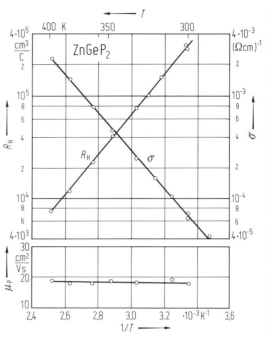

Fig. 31. ZnGeP$_2$. Conductivity, Hall coefficient and hole mobility vs. (reciprocal) temperature for a p-type sample with $p \approx 10^{13}$ cm^{-3} [75S1].

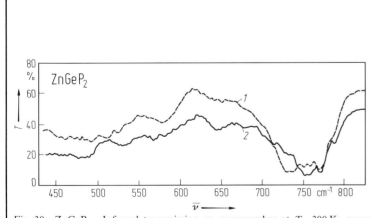

Fig. 30. ZnGeP$_2$. Infrared transmission vs. wavenumber at $T = 300$ K; curve 1: $E \parallel c$, 2: $E \perp c$ [75G1].

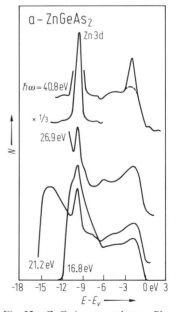

Fig. 32. ZnGeAs$_2$-amorphous. Photoemission spectra (electron counts N vs. binding energy) at $T = 300$ K for various excitation energies [76B2].

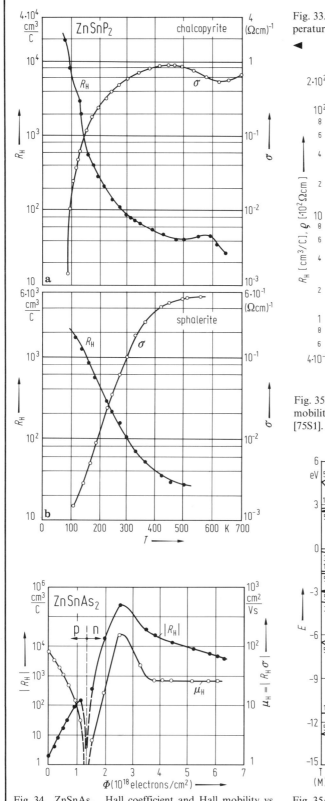

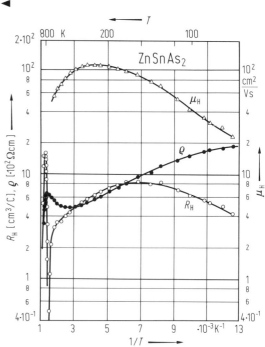

Fig. 33. ZnSnP$_2$. Conductivity and Hall coefficient vs. temperature for p-type samples [75S1].

Fig. 35. ZnSnAs$_2$. Resistivity, Hall coefficient and Hall mobility vs. (reciprocal) temperature for a p-type sample [75S1].

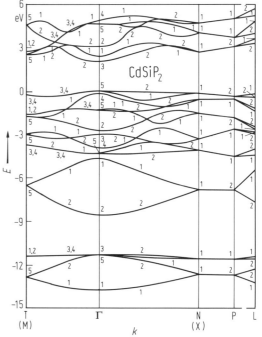

Fig. 34. ZnSnAs$_2$. Hall coefficient and Hall mobility vs. irradiation with 2 MeV electrons. Note the change of type at $\Phi = 1.3 \cdot 10^{18}$ electrons cm^{-2} [76B4].

Fig. 35α. CdSiP$_2$. Band structure calculated with an empirical pseudopotential method [79C]. For symmetry points T and N, cf. caption of Fig. 21α.

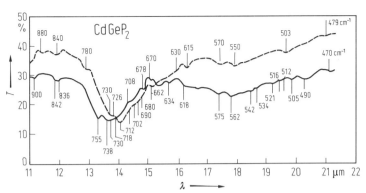

Fig. 36. CdGeP$_2$. Infrared transmission vs. wavelength at $T = 300$ K. Solid line $E \parallel c$, dashed line $E \perp c$ [75G5].

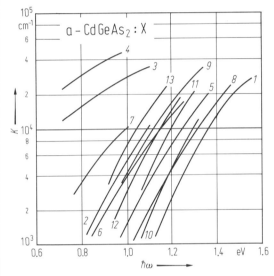

Fig. 37. CdGeAs$_2$: X-(amorphous). Absorption coefficient vs. photon energy for various samples: curve 1: nominally pure, 2: 0.42 at% Ni, 3: 1.3 at% Ni, 4: 3.3 at% Ni, 5: 6.3 at% Cu, 6: 20 at% Te, 7: 57 at% Te, 8: 2.2 at% Cd, 9: 32 at% Cd, 10: 2.7 at% Ge, 11: 5.7 at% Ge, 12: 4.7 at% As, 13: 11.2 at% As [79O].

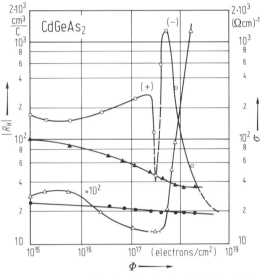

Fig. 38. CdGeAs$_2$. Conductivity (○, ●) and Hall coefficient (△, ▲) vs. irradiation with 2 MeV electrons at $T = 300$ K [78B5]. (●, ▲) initially n-type sample, (○, △) initially p-type sample. Note the change of type at $\Phi \approx 3 \cdot 10^{17}$ electrons cm^{-2} (p→n).

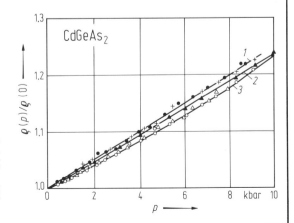

Fig. 39. CdGeAs$_2$. Resistivity vs. pressure at $T = 300$ K [75K3]. $+ 4 \cdot 10^{17}$ cm^{-3} (curve 1) Te doped, ▲ $6 \cdot 10^{17}$ cm^{-3} (curve 2) Te doped, ○ $7 \cdot 10^{17}$ cm^{-3} (curve 3) Te doped, ● $1.7 \cdot 10^{18}$ cm^{-3} In doped, △ $2.4 \cdot 10^{18}$ cm^{-3} In doped. All samples n-type.

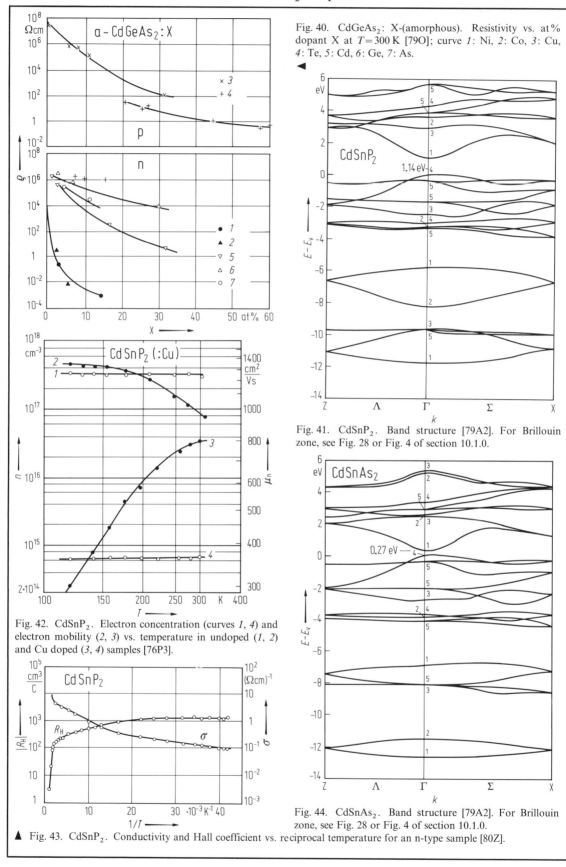

Fig. 40. CdGeAs₂: X-(amorphous). Resistivity vs. at%
dopant X at $T = 300$ K [79O]; curve 1: Ni, 2: Co, 3: Cu,
4: Te, 5: Cd, 6: Ge, 7: As.

Fig. 41. CdSnP₂. Band structure [79A2]. For Brillouin
zone, see Fig. 28 or Fig. 4 of section 10.1.0.

Fig. 42. CdSnP₂. Electron concentration (curves 1, 4) and
electron mobility (2, 3) vs. temperature in undoped (1, 2)
and Cu doped (3, 4) samples [76P3].

Fig. 44. CdSnAs₂. Band structure [79A2]. For Brillouin
zone, see Fig. 28 or Fig. 4 of section 10.1.0.

▲ Fig. 43. CdSnP₂. Conductivity and Hall coefficient vs. reciprocal temperature for an n-type sample [80Z].

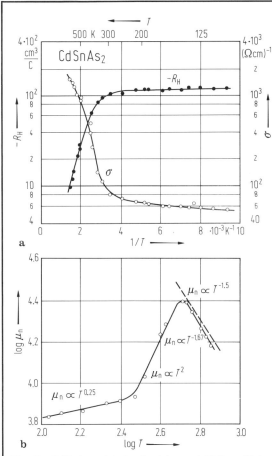

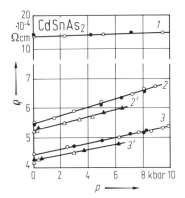

Fig. 45. CdSnAs$_2$. (a) Conductivity and Hall coefficient vs. (reciprocal) temperature, (b) log electron mobility vs. log T for an n-type sample with $n = 6 \cdot 10^{16}$ cm^{-3}. T in K. μ in cm^2 V^{-1} s^{-1} [75S1].

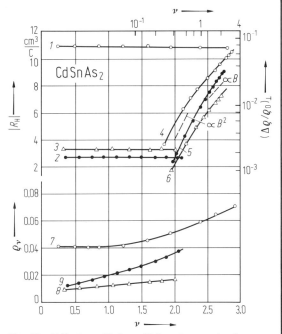

Fig. 46. CdSnAs$_2$. Hall coefficient (curves $1\cdots3$), transverse magnetoresistance (curves $4\cdots6$) and parameter $\varrho_v = (\Delta\varrho/\varrho_o)\,[v^2/(1+v^2)]$ (curves $7\cdots9$) vs. $v = R_H \sigma B$ for three n-type samples ($n = 10^{18}$ cm^{-3}) (○, △, ●). Upper scale: curves $4\cdots6$, lower scale: curves $1\cdots3$, $7\cdots9$ [75D2].

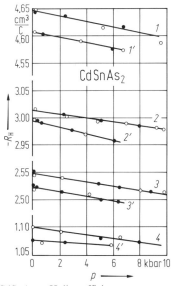

Fig. 47. CdSnAs$_2$. Resistivity vs. pressure at $T = 298$K (curves $1\cdots3$) and 77 K ($2'$, $3'$). Sample 1: polycrystalline, samples $2, 3$: single crystals (n-type). (Full symbols: pressure decreasing, open symbols: pressure increasing). [76D2].

Fig. 48. CdSnAs$_2$. Hall coefficient vs. pressure for n-type samples at $T = 298$ K (curves $1\cdots4$) and 77 K ($1'\cdots4'$). All samples are single crystals. Samples 2 and 3 are the same as in Fig. 47. Symbols as in Fig. 47 [76D2].

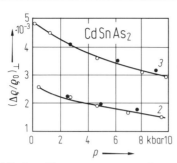

Fig. 49. CdSnAs$_2$. Transverse magnetoresistance vs. pressure at $T = 298$ K for two n-type samples. Same samples as in Fig. 47. Symbols as in Fig. 47 [76D2].

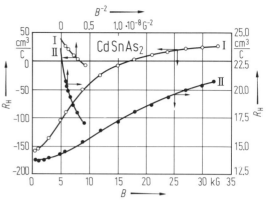

Fig. 50. CdSnAs$_2$. Hall coefficient vs. magnetic field B and vs. B^{-2} (o left hand ordinate, ● right hand ordinate, — theory) for two p-type samples with $p = 1.79 \cdot 10^{17}$ cm^{-3} (sample I) and $p = 2.79 \cdot 10^{17}$ cm^{-3} (sample II) [75D2].

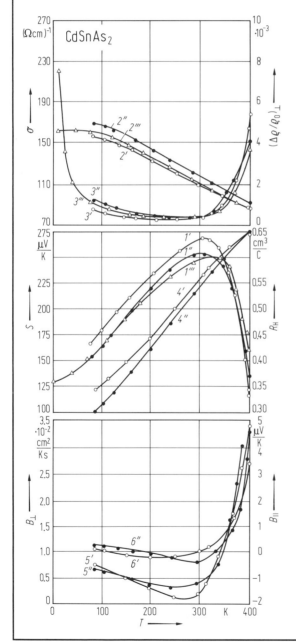

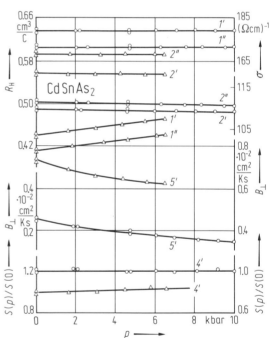

Fig. 52. CdSnAs$_2$. Pressure dependence of the same parameters as in Fig. 51 for two p-type samples with $p \approx 10^{19}$ cm^{-3} ($T = 300$ K) at 85 K (△) and 305 K (o) [80D3].

Fig. 51. CdSnAs$_2$. Hall coefficient ($1'\cdots1'''$), conductivity ($2'\cdots2'''$), transverse magnetoresistance ($3'\cdots3'''$), thermoelectric power ($4'\cdots4''$), transverse ($5'\cdots5''$) and longitudinal ($6'\cdots6''$) Nernst-Ettingshausen coefficient for three p-type samples with $p \approx 10^{19}$ cm^{-3} ($T = 300$ K) vs. temperature [80D3]. $B_\parallel = S_H - S$. For definition of $B_\perp$, see original paper.

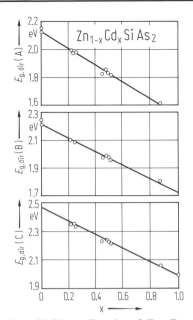

Fig. 53. $Zn_{1-x}Cd_xSiAs_2$. Energies of $\Gamma_{15}-\Gamma_1$ transitions (A, B, C bands) vs. composition [76J].

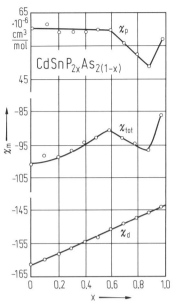

Fig. 54. $CdSnP_{2x}As_{2(1-x)}$. Magnetic susceptibility vs. composition. χ in CGS-emu, χ_d: diamagnetic part, χ_{tot}: total susceptibility, χ_p: paramagnetic part [74T1].

10.1.4 I₂–IV–VI₃ compounds (For tables, see p. 111 ff.)

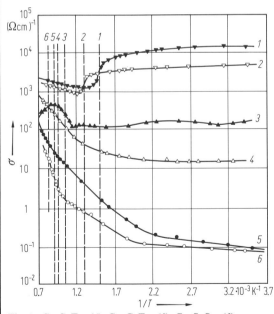

Fig. 1. Cu_2SnTe_3 (*1*), Cu_2GeTe_3 (*2*), Cu_2SnSe_3 (*3*), Cu_2GeSe_3 (*4*), Cu_2SnS_3 (*5*), Cu_2GeS_3 (*6*). Conductivity vs. reciprocal temperature (the perpendicular lines are melting temperatures) [70A1]. (*1, 2*): metallic solid, semiconducting liquid, (*3···6*): intrinsic semiconductor.

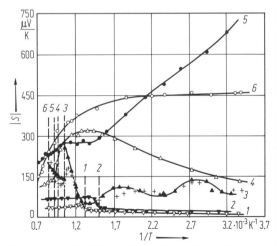

Fig. 2. Cu_2SnTe_3 (*1*), Cu_2GeTe_3 (*2*), Cu_2SnSe_3 (*3*), Cu_2GeSe_3 (*4*), Cu_2SnS_3 (*5*), Cu_2GeS_3 (*6*). Thermoelectric power vs. reciprocal temperature (the perpendicular lines are melting temperatures) [70A1]. See also Fig. 1.

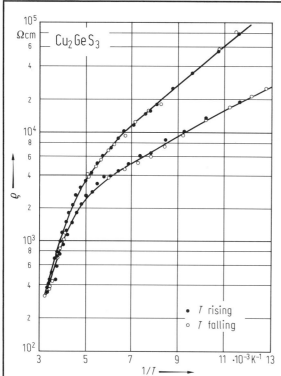

Fig. 3. Cu$_2$GeS$_3$. Resistivity vs. reciprocal temperature [74K]. n-type sample.

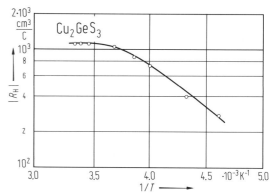

Fig. 4. Cu$_2$GeS$_3$. Hall coefficient vs. reciprocal temperature [74K]. n-type sample.

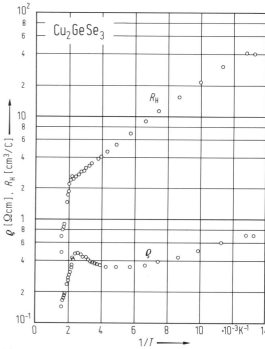

Fig. 6. Cu$_2$GeSe$_3$. Resistivity and Hall coefficient vs. reciprocal temperature; after annealing at $T = 970$ K [71E]. p-type material.

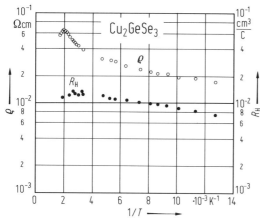

Fig. 5. Cu$_2$GeSe$_3$. Resistivity and Hall coefficient vs. reciprocal temperature; as grown material, p-type ($p = 6 \cdot 10^{20}$ cm^{-3}) [71E].

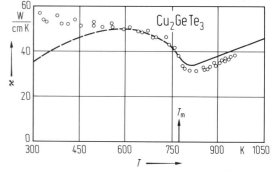

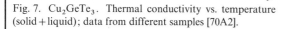

Fig. 7. Cu$_2$GeTe$_3$. Thermal conductivity vs. temperature (solid + liquid); data from different samples [70A2].

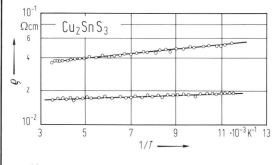

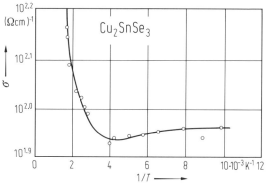

Fig. 10. Cu$_2$SnSe$_3$. Conductivity vs. reciprocal temperature [69B]. p-type sample. See also Fig. 1.

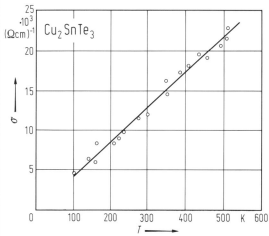

Fig. 12. Cu$_2$SnTe$_3$. Conductivity vs. temperature [69B].

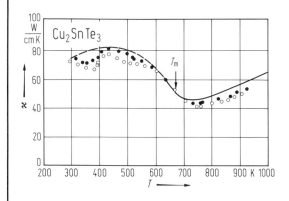

Fig. 8. Cu$_2$SnS$_3$. Resistivity vs. reciprocal temperature for two n-type samples (monoclinic form) [74K].

◀

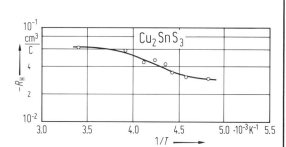

Fig. 9. Cu$_2$SnS$_3$. Hall coefficient vs. reciprocal temperature for an n-type sample (monoclinic form) [74K].

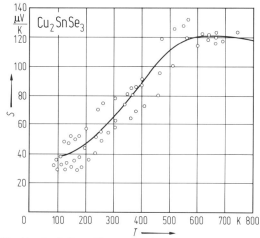

Fig. 11. Cu$_2$SnSe$_3$. Thermoelectric power vs. temperature [69B]. p-type sample.

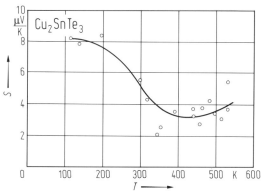

Fig. 13. Cu$_2$SnTe$_3$. Thermoelectric power vs. temperature [69B].

◀

Fig. 14. Cu$_2$SnTe$_3$. Thermal conductivity vs. temperature (solid + liquid); data from different samples [70A2].

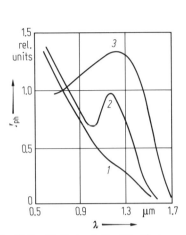

Fig. 15. Ag$_2$GeSe$_3$ (*1, 2*), Ag$_2$SnSe$_3$ (*3*). Photoconductivity vs. wavelength at (*1*) 293 K, (*2*) 77 K, (*3*) 293 K [64K].

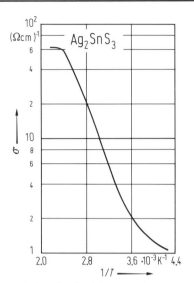

Fig. 16. Ag$_2$SnS$_3$. Conductivity vs. reciprocal temperature [64K].

10.1.5 I$_3$–V–VI$_4$ compounds (For tables, see p. 118 ff.)

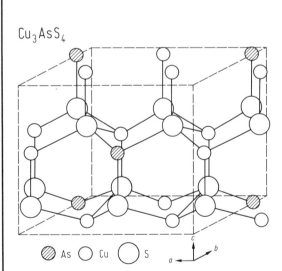

Cu$_3$AsS$_4$

Fig. 1. Cu$_3$AsS$_4$. Structure of enargite [70A1].

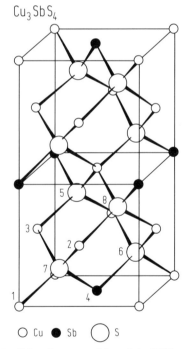

Cu$_3$SbS$_4$

Fig. 2. Cu$_3$SbS$_4$. Structure of famatinite [70P].

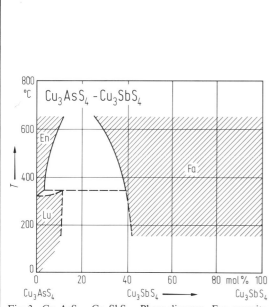

Fig. 3. Cu$_3$AsS$_4$–Cu$_3$SbS$_4$. Phase diagram. En: enargite, Lu: Luzonite, Fa: Famatinite [69T].

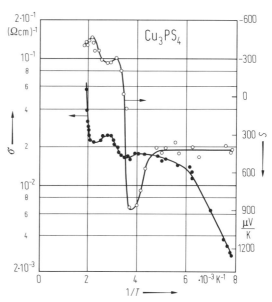

Fig. 5. Cu$_3$PS$_4$. Conductivity and thermoelectric power vs. reciprocal temperature [75H].

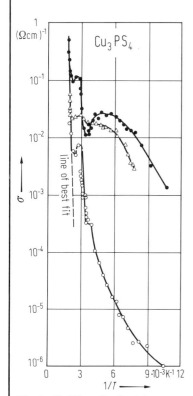

Fig. 4. Cu$_3$PS$_4$. Conductivity vs. reciprocal temperature (several samples) [75H].

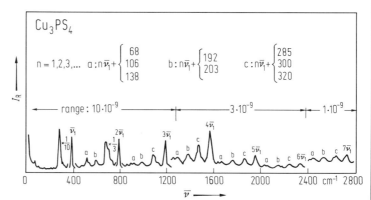

Fig. 6. Cu$_3$PS$_4$. Raman intensity vs. wavenumber [78T]. $T = 77$ K, pump wavelength: $\lambda = 476$ nm. Intensity (arb. units) rescaled as marked (range: etc.). $\bar{\nu}_1$ is the wavenumber of the symmetric stretching mode of PS$_4$: $\bar{\nu} = 392$ cm^{-1}. Several progressions of overtones marked (see table in figure).

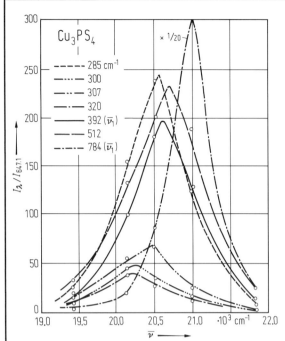

Fig. 7. Cu_3PS_4. Raman intensity vs. pump wavenumber for various phonons (see table upper left) [78T]. Intensity normalized to excitation with $\lambda = 647.1$ nm.

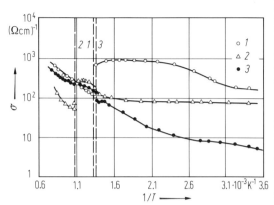

Fig. 8. Cu_3AsSe_4 (1), Cu_3AsS_4 (2), Cu_3SbSe_4 (3). Conductivity vs. reciprocal temperature (solid + liquid) (the perpendicular lines are melting temperatures) [70A2].

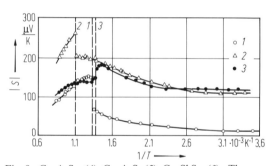

Fig. 9. Cu_3AsSe_4 (1), Cu_3AsS_4 (2), Cu_3SbSe_4 (3). Thermoelectric power vs. reciprocal temperature (solid + liquid) (the perpendicular lines are melting temperatures) [70A2].

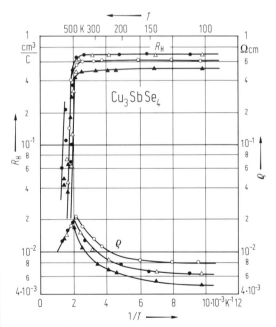

Fig. 10. Cu_3SbSe_4. Resistivity, Hall coefficient vs. (reciprocal) temperature (various samples) [69N]. Compare Fig. 8.

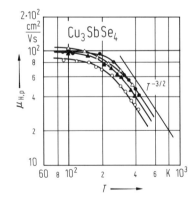

Fig. 11. Cu_3SbSe_4. Hall mobility of holes vs. temperature (various samples) [69N].

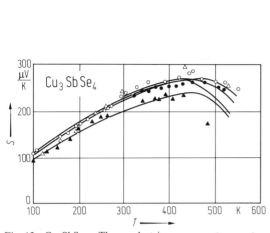

Fig. 12. Cu$_3$SbSe$_4$. Thermoelectric power vs. temperature (various samples); solid lines: theory [69N]. Compare Fig. 9.

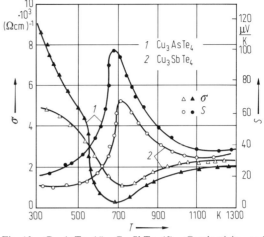

Fig. 13. Cu$_3$AsTe$_4$ (*1*), Cu$_3$SbTe$_4$ (*2*). Conductivity and thermoelectric power vs. temperature [77G].

10.1.6 II–III$_2$–VI$_4$ compounds (For tables, see p. 124 ff.)

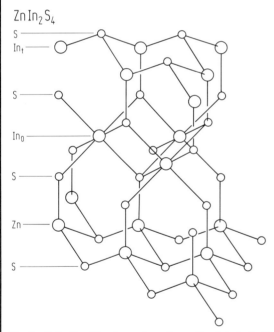
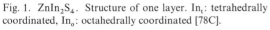

Fig. 1. ZnIn$_2$S$_4$. Structure of one layer. In$_t$: tetrahedrally coordinated, In$_o$: octahedrally coordinated [78C].

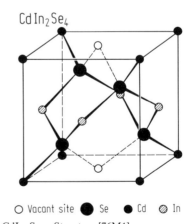

○ Vacant site ● Se ● Cd ⊘ In

Fig. 2. CdIn$_2$Se$_4$. Structure [76M1].

373

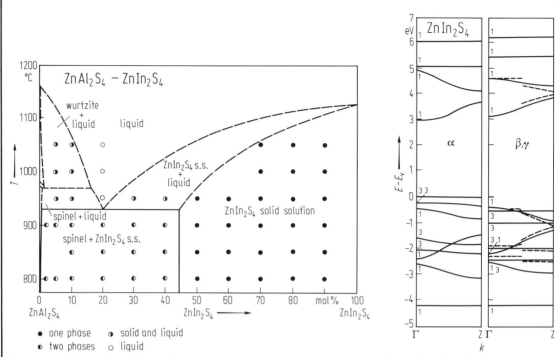

Fig. 3. ZnAl₂S₄—ZnIn₂S₄. Phase diagram; s.s.: solid solution [78B].

Fig. 5. ZnIn₂S₄. Band structure of α, β, and γ polytypes. Dashed lines: γ-phase, unfolded into BZ of β [79A2]. Γ−Z≡k∥c.

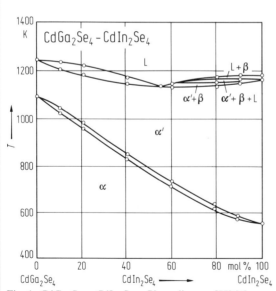

Fig. 4. CdGa₂Se₄—CdIn₂Se₄. Phase diagram [77M1].

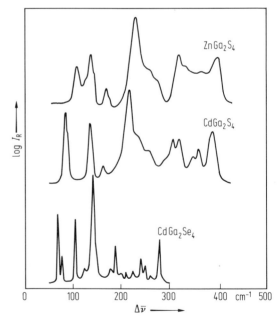

Fig. 4α. ZnGa₂S₄, CdGa₂S₄, CdGa₂Se₄. Raman spectra (Raman intensity vs. Raman shift) [83L1].

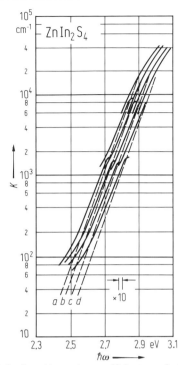

Fig. 6. ZnIn$_2$S$_4$. Absorption coefficient vs. photon energy ($E \perp c$) at (a) 77 K, (b) 180 K, (c) 250 K, (d) 300 K. The dashed curves represent the absorption coefficient after the correction for reflection losses on the measured absorption [73B].

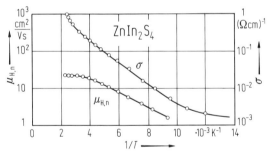

Fig. 8. ZnIn$_2$S$_4$. Conductivity and Hall mobility vs. reciprocal temperature [76G1]. $E \perp c$; n-type sample.

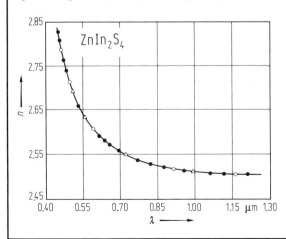

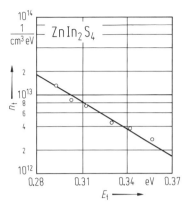

Fig. 7. ZnIn$_2$S$_4$. Trap density vs. trap depth from photoconductivity relaxation time measurements [76H].

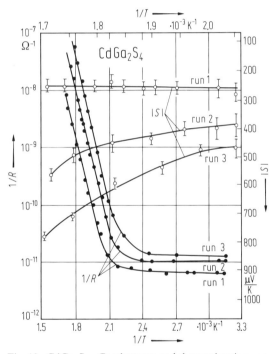

Fig. 10. CdGa$_2$S$_4$. Conductance and thermoelectric power vs. reciprocal temperature for 3 runs. n-type samples, $E \parallel c$. [77K3]. (Upper abscissa is for S, lower abscissa is for $1/R$.)

◀ Fig. 9. ZnIn$_2$S$_4$. Refractive index vs. wavelength for three samples [74A1]. $E \perp c$, $T = 300$ K.

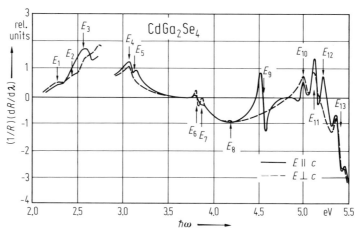

Fig. 10a. CdGa$_2$Se$_4$. Wavelength modulated reflection spectrum (derivative of reflectance over reflectance vs. photon energy) for two polarization directions at $T = 300$ K [81K].

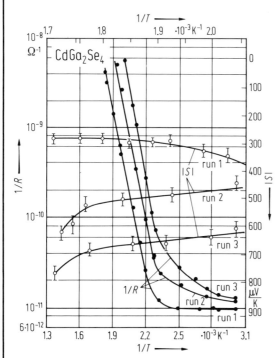

Fig. 11. CdGa$_2$Se$_4$. Conductance and thermoelectric power vs. reciprocal temperature for 3 runs. n-type samples, $E \parallel$ (112). [77K3]. (Upper abscissa is for S, lower abscissa is for $1/R$.)

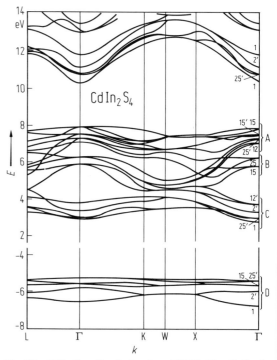

Fig. 12. CdIn$_2$S$_4$. Band structure [77B1]. For Brillouin zone, see e.g. Vol. III/17a, p. 349.

10.1.6 II–III$_2$–VI$_4$ compounds

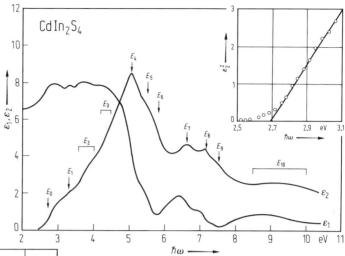

Fig. 13. CdIn$_2$S$_4$. Real (ε_1) and imaginary (ε_2) part of dielectric constant vs. photon energy. Insert: ε_2^2 vs. photon energy. Critical points in ε_2 are designated by E_i [78G1].

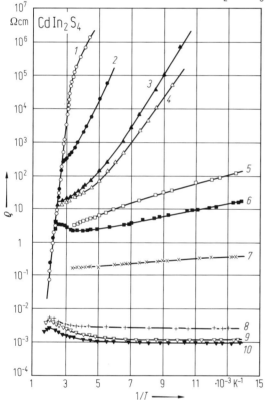

Fig. 14. CdIn$_2$S$_4$. Resistivity vs. reciprocal temperature for various samples:

Fig. 15. CdIn$_2$S$_4$. Hall coefficient vs. reciprocal temperature for the samples of Fig. 14 [76E].

Sample	n at $T = 125$ K cm^{-3}	N_I cm^{-3}	μ_H at $T = 125$ K cm^{-3} V^{-1} s^{-1}
3	$2.7 \cdot 10^{13}$	$8.3 \cdot 10^{18}$	25
4	$1.2 \cdot 10^{14}$	$1.2 \cdot 10^{18}$	20
5	$1.5 \cdot 10^{16}$	$1.5 \cdot 10^{18}$	33
6	$4.6 \cdot 10^{16}$	$1.1 \cdot 10^{18}$	56
7	$4.8 \cdot 10^{17}$	$1.4 \cdot 10^{18}$	100
9	$1.6 \cdot 10^{19}$	$3.9 \cdot 10^{19}$	320
10	$1.3 \cdot 10^{19}$	$3.1 \cdot 10^{19}$	400

Samples 9, 10: from stoichiometric melt, samples 1···8: with S excess. N_I: doping concentration, n: number of charge carriers. Samples 1, 2, 8 are not explained [76E].

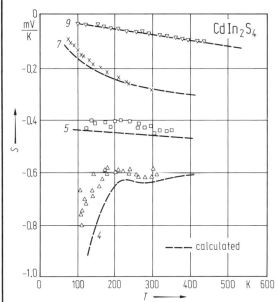

Fig. 16. CdIn₂S₄. Seebeck coefficient vs. temperature for some of the samples of Fig. 14. The dashed lines are calculated [76E].

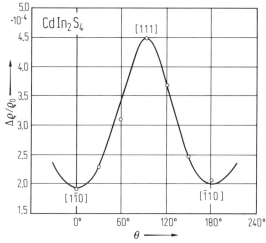

Fig. 18. CdIn₂S₄. Magnetoresistance $\Delta\varrho/\varrho$ vs. angle θ to the [1$\bar{1}$0] axis; $B = 8.5$ kG; $T = 290$ K [73A].

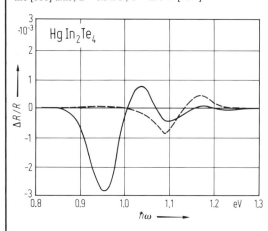

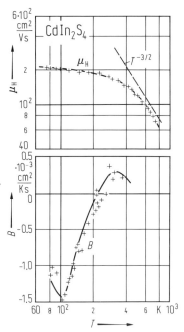

Fig. 17. CdIn₂S₄. Nernst coefficient and Hall mobility vs. temperature for sample 8 of Fig. 14 [76E].

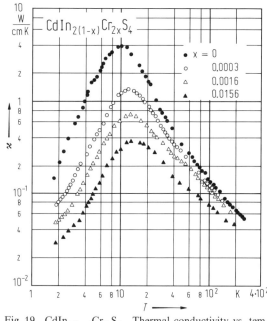

Fig. 19. CdIn₂₍₁₋ₓ₎Cr₂ₓS₄. Thermal conductivity vs. temperature for various Cr-concentrations [77E1].

Fig. 20. HgIn₂Te₄. Electroreflectance $\Delta R/R$ vs. photon energy. Full line: $E \perp c$, dashed line: $E \parallel c$. [77M2].

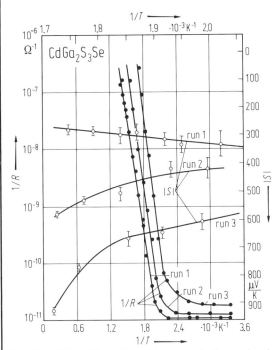

Fig. 21. CdGa₂S₃Se. Conductance and thermoelectric power vs. reciprocal temperature for 3 runs. n-type samples, $E\|c$. [77K3]. (Upper abscissa is for S, lower abscissa is for $1/R$.)

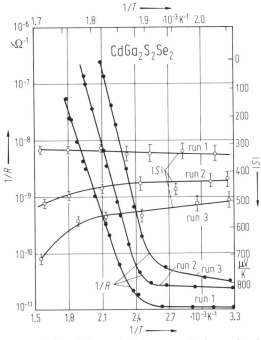

Fig. 22. CdGa₂S₂Se₂. Conductance and thermoelectric power vs. reciprocal temperature for 3 runs. n-type samples, $E\| (112)$. [77K3]. (Upper abscissa is for S, lower abscissa is for $1/R$.)

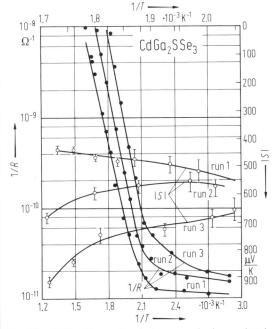

Fig. 23. CdGa₂SSe₃. Conductance and thermoelectric power vs. reciprocal temperature for 3 runs. n-type samples, $E\| (112)$. [77K3]. (Upper abscissa is for S, lower abscissa is for $1/R$.)

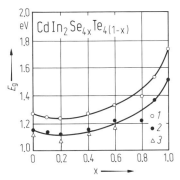

Fig. 24. CdIn₂Se₄ₓTe₄₍₁₋ₓ₎. Energy gap vs. composition at $T = 293$ K; curve 1: $E_{g,\text{dir}}$, 2: $E_{g,\text{ind}}$, 3: $E_{g,\text{th}}$ [72K3].

10.1.7 Other ordered vacancy compounds (For tables, see p. 150ff.)

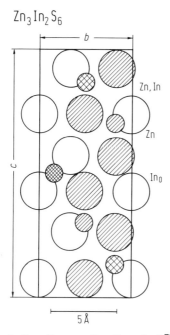

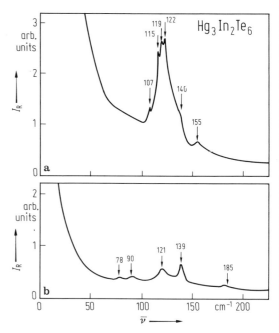

Fig. 1. $Zn_3In_2S_6$. Cross section through $(2\bar{1}\bar{1}0)$ plane. Atoms in plane are shaded [67D].

Fig. 2. $Hg_3In_2Te_6$. Raman intensity vs. wavenumber. (a) $E_i \parallel E_s$; (b) $E_i \perp E_s$ [77M].

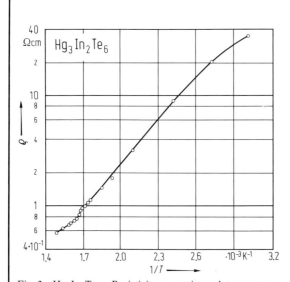

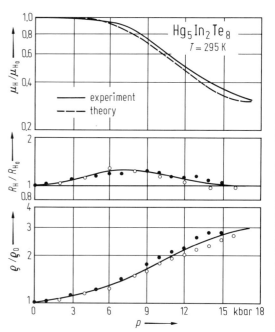

Fig. 3. $Hg_3In_2Te_6$. Resistivity vs. reciprocal temperature [70M].

Fig. 4. $Hg_5In_2Te_8$. Normalized Hall mobility (μ_H/μ_{H_0}), Hall coefficient (R_H/R_{H_0}), and resistivity (ϱ/ϱ_0) vs. pressure [72P]. Sample with $\mu = 450\ cm^2\ V^{-1}\ s^{-1}$ at $p = 0$, $T = 300\ K$ $(\varrho_0, R_{H_0}, \mu_{H_0}$ not quoted).

10.1.8 Quaternary compounds (For tables, see p. 153 ff.)

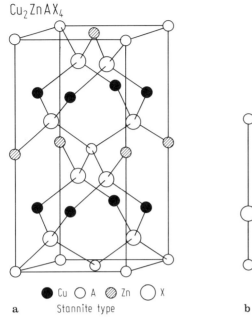

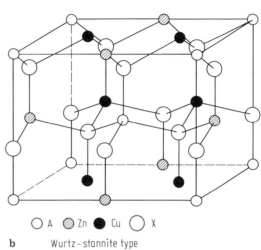

Fig. 1. Cu_2ZnAX_4. (a) Stannite. Structure [77S]. (b) Wurtz-stannite. Structure [77S].

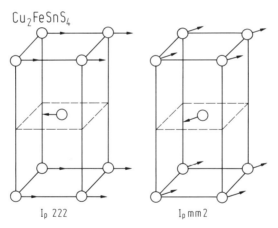

Fig. 2. Cu_2FeSnS_4. Two possibilities of magnetic structures [72G].

10.2 Ternary transition-metal compounds

10.2.1 Chalcogenides MeM$_2$S$_4$ (Me = Mn, Fe, Co, Ni, Cd; M = In, Ga, Sb, Sn, Rh) (For tables, see p. 157ff.)

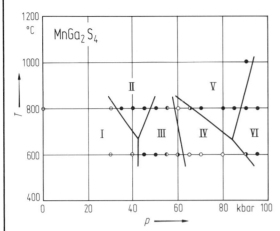

Fig. 1. MnGa$_2$S$_4$. Phase diagram [71Y].

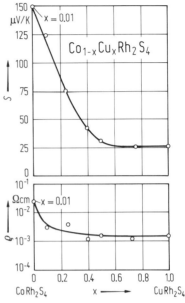

Fig. 2. Co$_{1-x}$Cu$_x$Rh$_2$S$_4$. Seebeck coefficient at room temperature and resistivity ($T = 80$ K) vs. composition [69L].

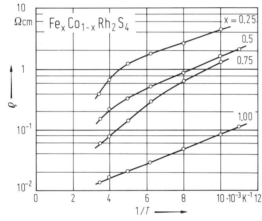

Fig. 3. Co$_{1-x}$Fe$_x$Rh$_2$S$_4$. Resistivity vs. reciprocal temperature [76K].

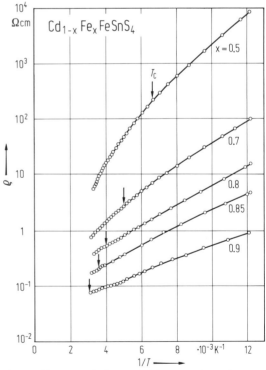

Fig. 4. Cd$_{1-x}$Fe$_x$FeSnS$_4$. Resistivity vs. reciprocal temperature (hot pressed samples) [73H].

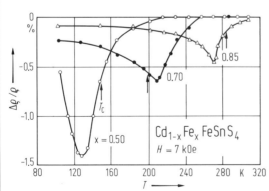

Fig. 5. Cd$_{1-x}$Fe$_x$FeSnS$_4$. Magnetoresistance vs. temperature [73H].

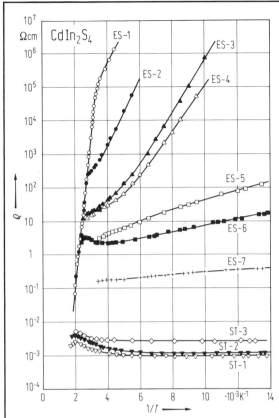

Fig. 7. CdIn$_2$S$_4$. Temperature dependence of the electrical resistivity for different samples:

Sample	N_I cm^{-3}	n at $T=125$ K cm^{-3}	μ_H at $T=120$ K cm^2 V^{-1} s^{-1}
ST-1	$3.1 \cdot 10^{19}$	$1.3 \cdot 10^{19}$	400
ST-2	$3.9 \cdot 10^{19}$	$1.6 \cdot 10^{19}$	320
ES-3	$8.3 \cdot 10^{18}$	$2.7 \cdot 10^{13}$	25
ES-4	$1.2 \cdot 10^{18}$	$1.2 \cdot 10^{14}$	20
ES-5	$1.5 \cdot 10^{18}$	$1.5 \cdot 10^{16}$	33
ES-6	$1.1 \cdot 10^{18}$	$4.6 \cdot 10^{16}$	56
ES-7	$1.4 \cdot 10^{18}$	$4.8 \cdot 10^{17}$	100

ST = samples from stoichiometric melt; ES = samples from melt with S excess; N_I = Doping concentration; n = Number of charge carriers. ES-1, ES-2, ES-8 and ST-3 are not explained in the original paper [76E].

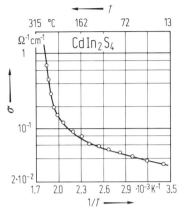

Fig. 6. CdIn$_2$S$_4$. Electrical conductivity vs. (reciprocal) temperature [73A].

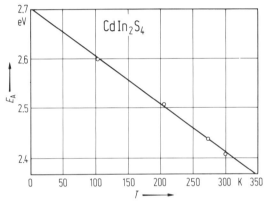

Fig. 8. CdIn$_2$S$_4$. Activation energy vs. temperature [73N].

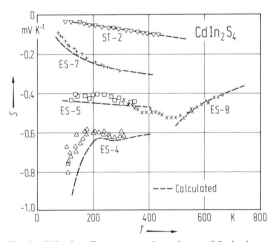

Fig. 9. CdIn$_2$S$_4$. Temperature dependence of Seebeck coefficient for different samples [76E]. The dashed lines are calculated; for samples numbers, see Fig. 7.

10.2.2 Chalcogenides MCr$_2$S$_4$ (M = Cd, Fe, Co, Cu, Hg, Zn, Mn, V, Ba) (For tables, see p. 163 ff.)

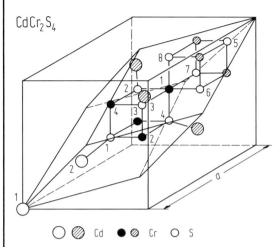

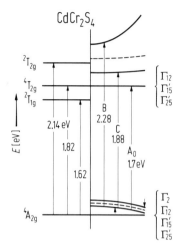

Fig. 1. CdCr$_2$S$_4$. Crystal structure of spinel. The hatched circles are outside the primitive unit cell [80K].

Fig. 3. CdCr$_2$S$_4$. Band structure model. The model shows the position of the different levels at $T = 4.2$ K, $T \ll T_C$. The left hand side of the CdCr$_2$S$_4$ scheme gives the single-ion many-electron states, the right hand side gives the itinerant one-electron states. The position of the red-shifting bands at RT is indicated by dashed lines [70H]. For temperature dependence of bands A$_0$, B, C, see Fig. 6.

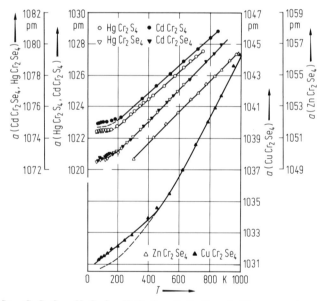

Fig. 2. CdCr$_2$S$_4$, CdCr$_2$Se$_4$, CuCr$_2$Se$_4$, HgCr$_2$S$_4$, HgCr$_2$Se$_4$, ZnCr$_2$Se$_4$. Lattice parameter vs. temperature [74S]. The dashed line is a normal thermal expansion curve deduced from the Grüneisen and Debye theory.

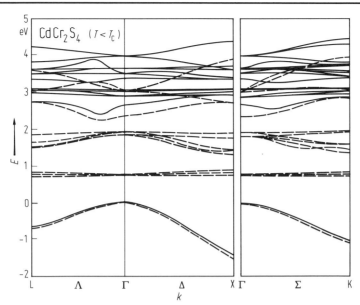

Fig. 4. CdCr$_2$S$_4$. Calculated band structure on the symmetry lines $\Gamma-\Lambda-L$, $\Gamma-\Delta-X$ and $\Gamma-\Sigma-K$ ($T < T_C$). The broken and solid curves represent the energy band for the majority and minority spin, respectively. The valence bands, except for the top band, are omitted [80K].

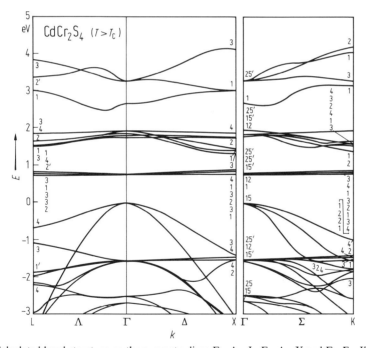

Fig. 5. CdCr$_2$S$_4$. Calculated band structure on the symmetry lines $\Gamma-\Lambda-L$, $\Gamma-\Delta-X$ and $\Gamma-\Sigma-K$ ($T > T_C$) [80K].

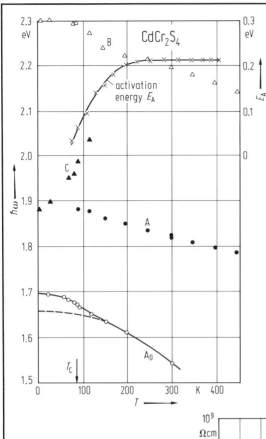

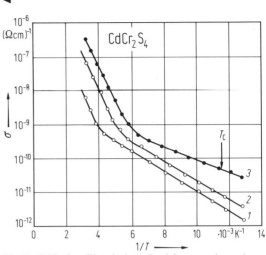

Fig. 6. CdCr$_2$S$_4$. Activation energy vs. temperature obtained from resistivity measurements on a single crystal [71T]. A, B, and C are due to [69B] and A$_0$ is due to [70H] obtained by optical absorption measurements. For identification of transition A$_0$, B, C, see Fig. 3.

Fig. 7. CdCr$_2$S$_4$. Electrical conductivity vs. reciprocal temperature. Curve 1: undoped before annealing; 2: undoped after annealing in vacuum; 3: doped with 5 at % Gd. Annealing condition: 200 h at 550°C in 1.33·10^{-2} Pa vacuum [77N].

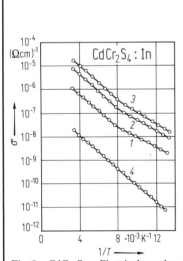

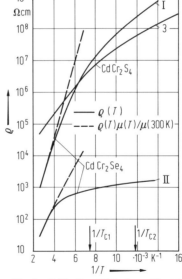

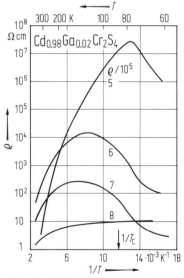

Fig. 8. CdCr$_2$S$_4$. Electrical conductivity vs. reciprocal temperature of single crystals doped with 0.2 (curve 1), 0.3 (2) and 0.6 (3) at % In and undoped (4) [78R].

Fig. 9. CdCr$_2$S$_4$, CdCr$_2$Se$_4$. Electrical resistivity of undoped CdCr$_2$Se$_4$ (samples I and II), and of undoped CdCr$_2$S$_4$ (sample 3), vs. reciprocal temperature (all samples polycrystals). T_{C1} and T_{C2} are the Curie temperatures for CdCr$_2$Se$_4$ and CdCr$_2$S$_4$, respectively. The dashed lines are inversely proportional to the electron concentration [73L]. For sample numbers, see respective table.

Fig. 10. CdCr$_2$S$_4$. Electrical resistivity of 2 at% Ga doped polycrystalline hot pressed samples (samples 5, 6, 7 and 8), vs. (reciprocal) temperature [73L]. Values of sample 5 have to be multiplied by 10^5. For sample numbers, see respective table.

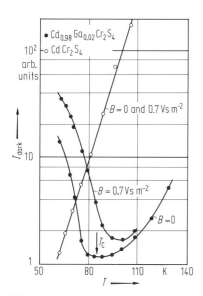

Fig. 11. CdCr$_2$S$_4$. Dark conduction vs. temperature for a Ga doped and an undoped polycrystalline sample [74L1].

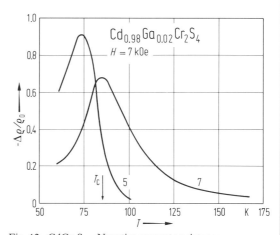

Fig. 12. CdCr$_2$S$_4$. Negative magnetoresistance

$$-\Delta\varrho/\varrho_0 = -(\varrho(H=7\text{kOe}) - \varrho(H=0))/\varrho(H=0)$$

for 2 at% Ga doped polycrystalline samples vs. temperature [73L]. For sample numbers, see respective table.

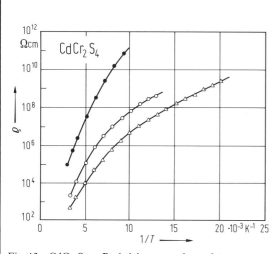

Fig. 13. CdCr$_2$S$_4$. Resistivity vs. reciprocal temperature for three thin films [77G1].

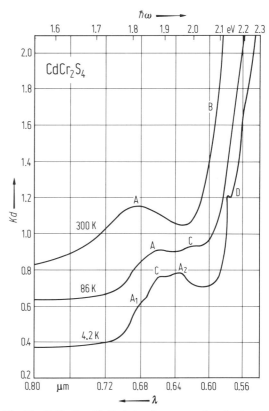

Fig. 14. CdCr$_2$S$_4$. Optical density vs. wavelength at various temperatures ($7 \cdot 10^{-5}$ cm thick single crystal film). Characteristic optical transitions are indicated. The spectra are displaced by 0.2 optical density for clarity. – Peak A and edge B have similar blue shifts with decreasing temperature. C appears only in the range of T_c and below and has the red shift characteristic to some magnetic semiconductors (cf. Fig. 6) [69B].

387

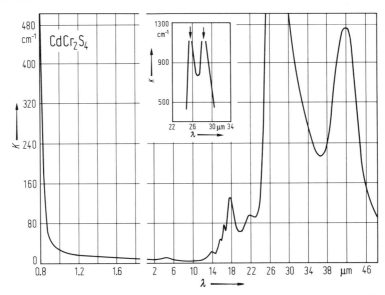

Fig. 15. CdCr$_2$S$_4$. Absorption coefficient ($T = 300$ K) of hot pressed samples vs. wavelength (the inset shows the resolved reststrahl peaks at 26 and 28.8 μm) [71M]. K includes all losses except reflective losses.

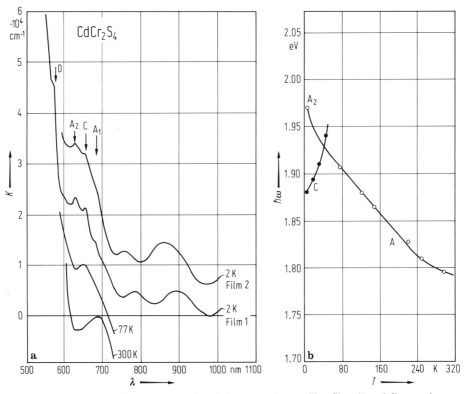

Fig. 16. CdCr$_2$S$_4$. Absorption coefficient vs. wavelength for two polycrystalline films (1 and 2) at various temperatures where the scale of the ordinate is for the spectrum of film 1 at 2 K and the other spectra are displaced. (b) Absorption peaks A and C for another film vs. temperature [74T].

Fig. 17. CdCr$_2$S$_4$. Optical density vs. wavelength in a polycrystalline CdCr$_2$S$_4$ film at T(K): curve *1*: 16, *2*: 80, *3*: 293. Film 0.5 μm thick. The spectra for 293 K and 80 K are shifted along the abscissa by 0.8 and 0.3, respectively [76G3]. Peaks A$_1$ and D are assigned to crystal field transitions of Cr^{3+}. Edge B is assigned to electron transitions from the valence band to the conduction band. ▶

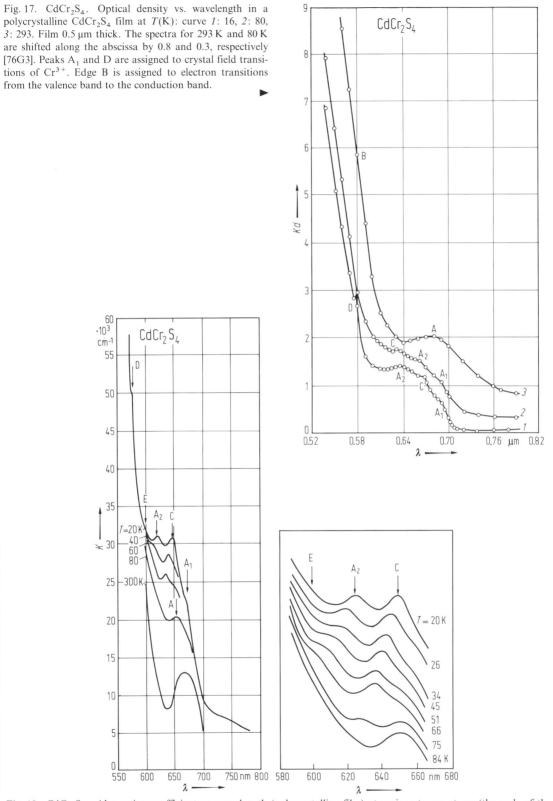

Fig. 18. CdCr$_2$S$_4$. Absorption coefficient vs. wavelength (polycrystalline film) at various temperatures (the scale of the ordinate is for the spectrum at 20 K and the other spectra are displaced) [78K].

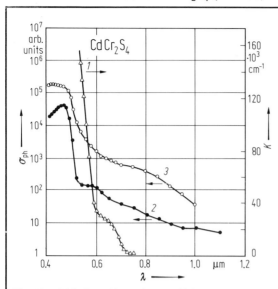

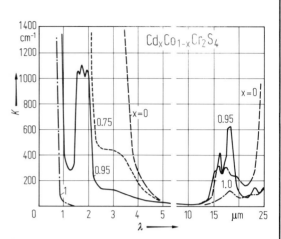

Fig. 19. CdCr$_2$S$_4$. Absorption coefficient (curve *1*) and photoconductivity (*2, 3*) vs. wavelength for films at the following temperatures: *1, 2*: 77 K; *3*: 293 K [77G1].

Fig. 22. Cd$_x$Co$_{1-x}$Cr$_2$S$_4$. Optical absorption coefficient vs. wavelength for polycrystalline samples [73C].

For Fig. 20, see next page.

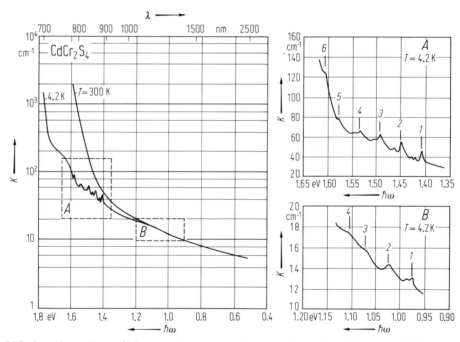

Fig. 21. CdCr$_2$S$_4$. Absorption coefficient vs. photon energy (hot pressed samples). Figs. A and B show the prominent phonon assisted transitions on an expanded scale, at 4.2 K [71M].

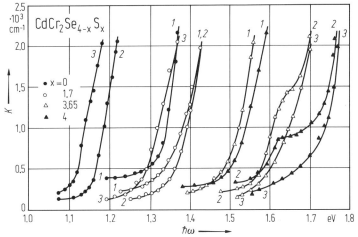

Fig. 20. CdCr$_2$Se$_{4-x}$S$_x$. Absorption coefficient vs. photon energy for various compositions at different temperatures. Curve 1: 300 K; 2: 80 K; 3: 20 K [76K2]. ▶

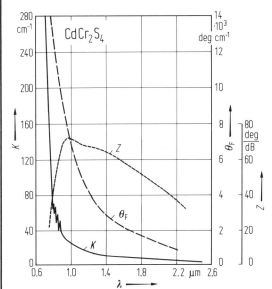

Fig. 23. CdCr$_2$S$_4$. Optical absorption coefficient, Faraday rotation θ_F and figure of merit Z vs. wavelength (hot pressed samples, $T = 4.2$ K, $H = 6$ kOe) [71A1]. Z is defined as the rotation per unit thickness divided by the attenuation in decibels per unit thickness at saturating magnetic field.

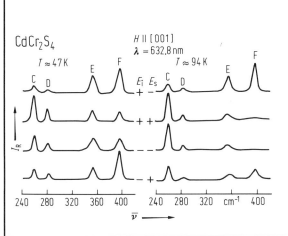

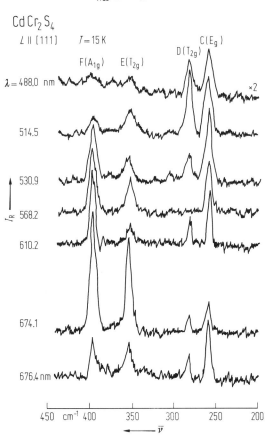

Fig. 25. CdCr$_2$S$_4$. Raman spectra (intensity vs. wavenumber) with various excitation wavelengths of light at T = 15 K, single crystal sample. L is the electric vector of incident and scattered light, λ is the wavelength of excitation light [76K3].

◀

Fig. 24. CdCr$_2$S$_4$. Stokes component of the Raman spectrum (intensity vs. wavenumber) with various circular-polarization configurations at T = 47 and 94 K (applied field: 2.2 kOe) [80K]. +, −: right and left hand circularly polarized light. See also [70S].

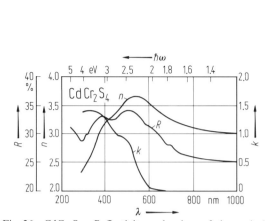

Fig. 26. CdCr$_2$S$_4$. Reflectivity and values of the optical constants n and k vs. wavelength ($T = 80$ K); hot pressed samples [71W].

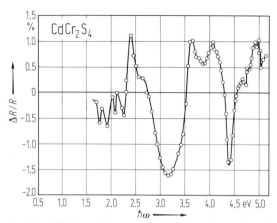

Fig. 28. CdCr$_2$S$_4$. Magnetoreflectance (rel. change in reflectance vs. photon energy) at liquid helium temperatures [71A2].

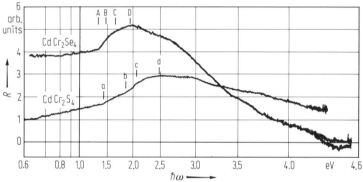

Fig. 27. CdCr$_2$S$_4$, CdCr$_2$Se$_4$. Reflectivity vs. photon energy (room temperature) [71F]. Four optical transitions are indicated for both (single crystal) substances.

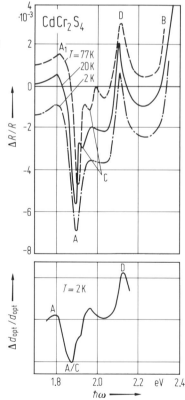

Fig. 29. CdCr$_2$S$_4$. Polarization-modulated magnetoreflectance (rel. change in reflectance vs. photon energy) for three temperatures below T_c (hot pressed sample). The lower part shows the differential optical density spectrum at $T = 2$ K [73P1]. B represents the absorption edge. A$_1$ and D are crystal field transitions of Cr^{3+} and are assigned to ^{4}A$_{2g} \rightarrow {}^2$T$_{2g}$ (D) and ^{4}A$_{2g} \rightarrow {}^4$T$_{2g}$ (A$_1$). The origin of the two red shifting structures A and C is not established in a clear manner.

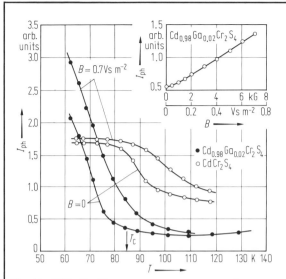

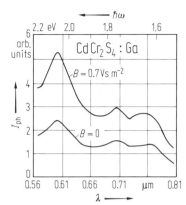

Fig. 31. CdCr$_2$S$_4$. Photocurrent vs. wavelength for a 2 at% Ga doped polycrystalline sample [74L1]. For undoped CdCr$_2$S$_4$, cf. Fig. 30.

Fig. 30. CdCr$_2$S$_4$. Photocurrent vs. temperature for a pure and a 2 at% Ga doped polycrystalline sample for excitation at $\lambda = 0.6\,\mu$m (2.1 eV) [74L1]. The inset shows the magnetic field dependence of the photocurrent for Cd$_{0.98}$Ga$_{0.02}$Cr$_2$S$_4$ at 77 K.

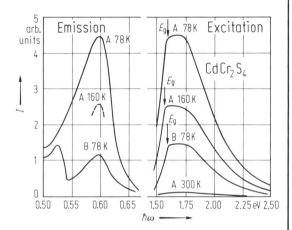

Fig. 33. CdCr$_2$S$_4$. Photoluminescence emission and excitation spectra (intensity vs. photon energy) at various temperatures [73O]. A refers to a 0.1 at% In doped sample, and B refers to a non-doped-vacuum-annealed sample (both are single crystals). The arrows indicate the positions of absorption edges, E_g, at the temperatures concerned.

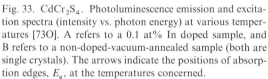

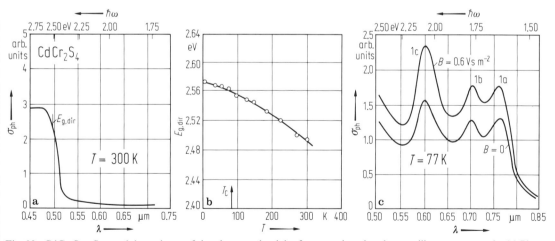

Fig. 32. CdCr$_2$S$_4$. Spectral dependence of the photoconductivity for an undoped, polycrystalline n-type sample. (a) Photoconductivity vs. wavelength. Sample resistivity $\varrho = 6 \cdot 10^4\ \Omega$ cm at RT and activation energy $E_A = 0.22$ eV. (b) Temperature variation of the direct edge, $E_{g,dir}$. (c) Photoconductivity vs. wavelength. Sample resistivity $\varrho = 6 \cdot 10^{10}\ \Omega$ cm at RT and activation energy $E_A = 0.43$ eV. The magnetic flux density of the applied transverse magnetic field was $B = 0.6$ Vs m^{-2} [72L].

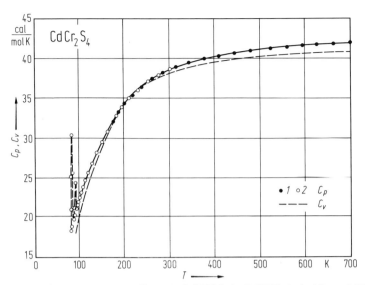

Fig. 34. CdCr$_2$S$_4$. Heat capacity vs. temperature. Curve _1_: C_p [77K1], _2_: C_p [75B], dashed line: C_v [77K1].

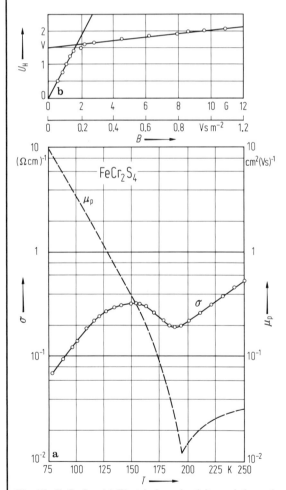

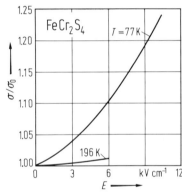

Fig. 35. FeCr$_2$S$_4$. Relative conductivity vs. electrical field for a single crystal sample [73N]. The values were measured 30 ns after the voltage pulse had been applied at $T = 77$ K and 196 K. The repetition rate of the voltage pulse was 20 Hz. See also [74N].

For Fig. 37, see next page.

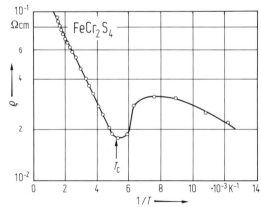

Fig. 36. FeCr$_2$S$_4$. (a) Electrical conductivity and theoretical hole mobility vs. temperature for a single crystal. (b) Hall voltage vs. magnetic field [74N].

Fig. 38. FeCr$_2$S$_4$. Resistivity vs. reciprocal temperature [68G].

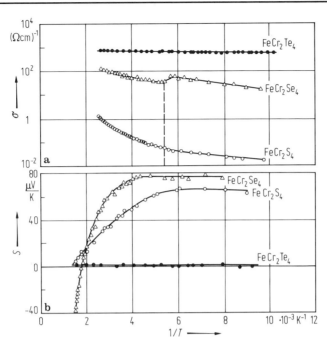

Fig. 37. FeCr$_2$S$_4$, FeCr$_2$Se$_4$, FeCr$_2$Te$_4$. Electrical conductivity (a) and Seebeck coefficient (b) vs. reciprocal temperature [75V]. Orientation not specified.

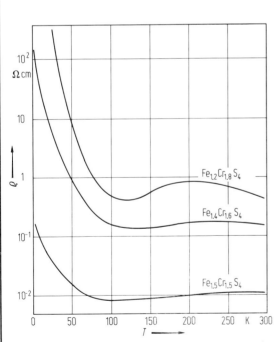

Fig. 39. Fe$_{1+x}$Cr$_{2-x}$S$_4$. Resistivity vs. temperature for three compositions [70R].

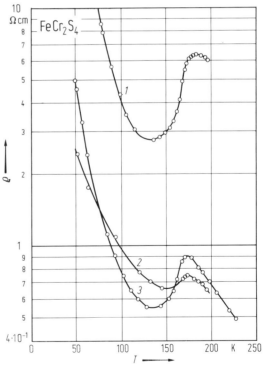

Fig. 40. FeCr$_2$S$_4$. Resistivity vs. temperature [72G]. Curve 1: S annealed, 2: Cd doped, 3: In doped; single crystal samples.

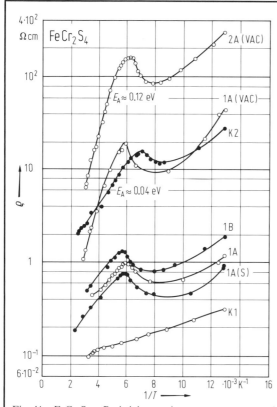

Fig. 41. FeCr$_2$S$_4$. Resistivity vs. inverse temperature of various single crystal samples [73W2]. 1A, 1B: as-grown; 1A(VAC): vacuum annealed at $T = 575\,°C$ for $t = 68\,h$; 1A(S): S annealed at $T = 700\,°C$ for $t = 72\,h$; 2A(VAC): vacuum annealed at $T = 575\,°C$ for $t = 110\,h$; K1: Cu doped ≈ 9 at%; K2: Zn doped ≈ 3 at%.

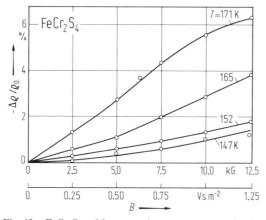

Fig. 43. FeCr$_2$S$_4$. Magnetoresistance vs. magnetic flux density at different temperatures between 147 K and 171 K [68G].

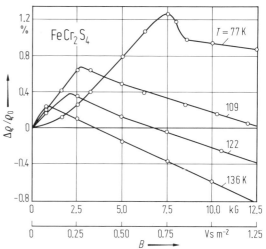

Fig. 42. FeCr$_2$S$_4$. Magnetoresistance vs. magnetic flux density, at different temperatures $T < T_C$ between 77 K and 136 K [68G].

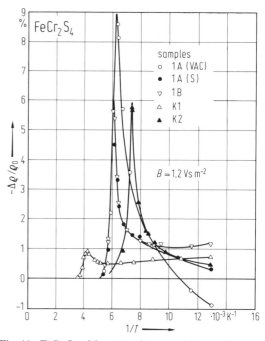

Fig. 44. FeCr$_2$S$_4$. Magnetoresistance vs. inverse temperature of various single crystal samples [73W2]. 1B: as-grown; 1A(VAC): vacuum annealed at $T = 575\,°C$ for $t = 68\,h$; 1A(S): S annealed at $T = 700\,°C$ for $t = 72\,h$; K1: Cu doped ≈ 9 at%; K2: Zn doped ≈ 3 at%. See Fig. 41 for resistance of the same samples.

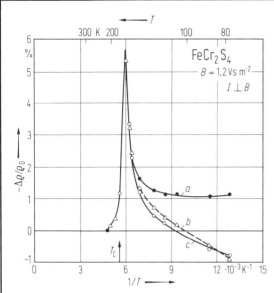

Fig. 45. FeCr$_2$S$_4$. Magnetoresistance vs. (inverse) temperature with the magnetic field rotated in a plane perpendicular to the current direction by an angle (b) 90° and (c) 60° away from the direction of magnetic field (a); single crystal sample [73W1].

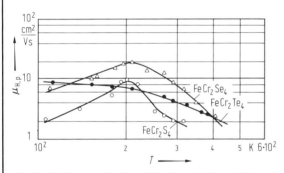

Fig. 47. FeCr$_2$S$_4$, FeCr$_2$Se$_4$, FeCr$_2$Te$_4$. Hall mobilities vs. temperature [75V]. Orientation not specified.

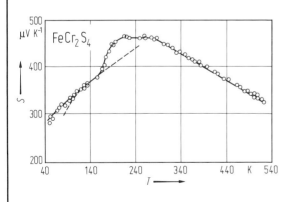

Fig. 48. FeCr$_2$S$_4$. Seebeck coefficient vs. temperature for single crystals [66H2]. The dashed line shows the calculated $S(T)$ curve for an extrinsic non-degenerate semiconductor matching the experimental data between 120 K and 160 K. The additional part above 160 K is supposed to be due to the magnetic nature of FeCr$_2$S$_4$. Cf. Fig. 37.

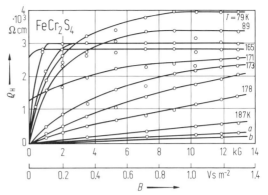

Fig. 46. FeCr$_2$S$_4$. Hall resistivity, $\varrho_H = U_H d/I$, vs. magnetic flux density at different temperatures near $T_C = 177$ K (single crystals) [72G]. Note: curves a and b are supposed to be measured at 191 K and 196 K, respectively.

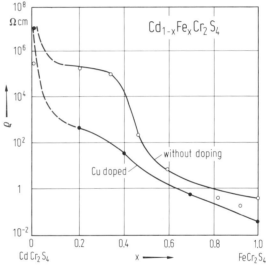

Fig. 49. Cd$_{1-x}$Fe$_x$Cr$_2$S$_4$. Electrical room-temperature resistivity for Cu doped and undoped single crystals vs. composition [78T]. The dashed lines mean that in these regions the number of measuring points is too small; they roughly give the expected behavior.

◄

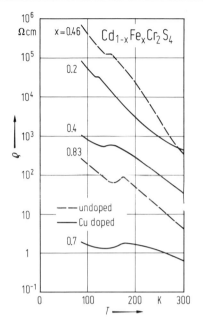

Fig. 50. Cd$_{1-x}$Fe$_x$Cr$_2$S$_4$. Electrical resistivity vs. temperature for Cu doped and undoped single crystals [78T].

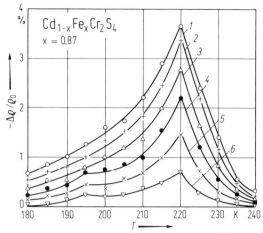

Fig. 51. Cd$_{1-x}$Fe$_x$Cr$_2$S$_4$. Magnetoresistance of a single crystal with x = 0.87 vs. temperature for different magnetic flux densities: curve 1: B = 0.6, 2: B = 0.5, 3: B = 0.4, 4: B = 0.3, 5: B = 0.2, 6: B = 0.1 V s m^{-2} [74B].

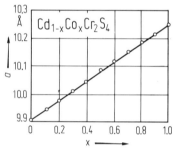

Fig. 53. Cd$_{1-x}$Co$_x$Cr$_2$S$_4$. Lattice parameter vs. composition [75T].

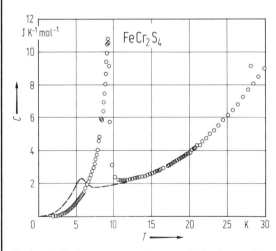

Fig. 52. FeCr$_2$S$_4$. Heat capacity per mol of polycrystalline Fe$_{0.97}$Cr$_2$S$_4$ vs. temperature [75L3]. Results on an earlier sample of composition FeCr$_2$S$_4$ are represented by the dashed curve.

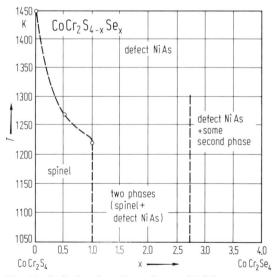

Fig. 54. CoCr$_2$S$_{4-x}$Se$_x$. Phase diagram [73G2].

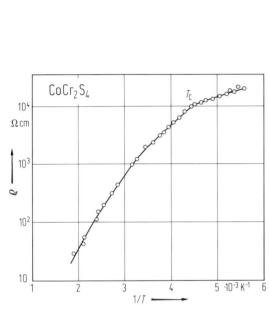

Fig. 55. CoCr$_2$S$_4$. Electrical resistivity vs. inverse temperature for single crystal material [69P].

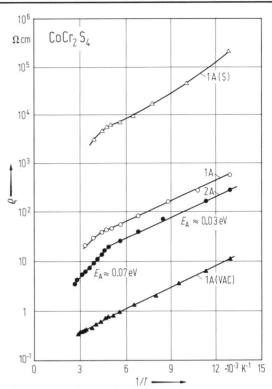

Fig. 56. CoCr$_2$S$_4$. Resistivity vs. reciprocal temperature (1A, 2A: as-grown; 1A(VAC): vacuum annealed at 600°C for 68 h; 1A (S): sulfur annealed at 700°C for 72 h) [73W2].

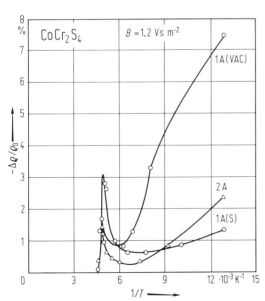

Fig. 57. CoCr$_2$S$_4$. Magnetoresistance vs. reciprocal temperature ($B = 1.2$ Vs m^{-2}) for single crystal material. (2A: as-grown; 1A(VAC): vacuum annealed at 600°C for 68 h; 1A(S): sulfur annealed at 700°C for 72 h) [73W2].

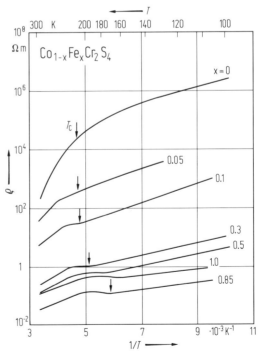

Fig. 58. Co$_{1-x}$Fe$_x$Cr$_2$S$_4$. Resistivity vs. (inverse) temperature, for polycrystalline samples of various compositions [76T]. Arrows indicate Curie temperatures.

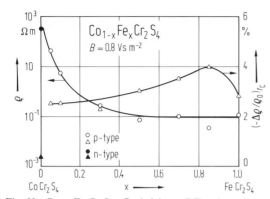

Fig. 59. Co$_{1-x}$Fe$_x$Cr$_2$S$_4$. Resistivity at RT and maximum magnetoresistance $(-\Delta\varrho/\varrho_0)_{T_C}=((\varrho_{B=0.8\,T}-\varrho_{B=0})/\varrho_{B=0})_{T_C}$ vs. composition; polycrystalline sample [76T].

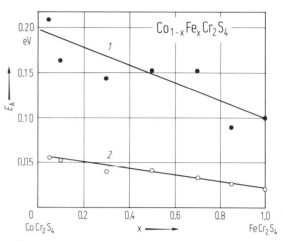

Fig. 60. Co$_{1-x}$Fe$_x$Cr$_2$S$_4$. Thermal activation energy of the electrical conductivity at high *1* and low *2* temperatures with respect to T_C, vs. composition; polycrystalline sample [76T].

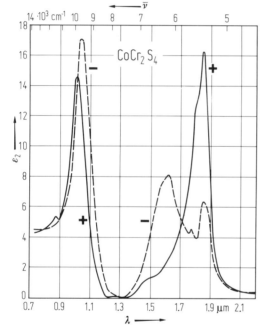

Fig. 61. CoCr$_2$S$_4$. Imaginary part of the dielectric constant vs. wavelength (for right (———) and left (-------) hand circularly polarized light) computed by Kramers-Kronig analysis [73A1, 74A].

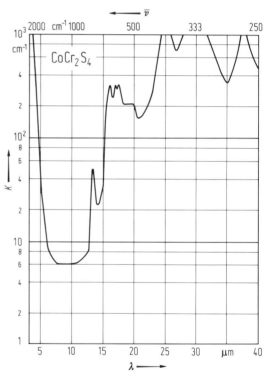

Fig. 62. CoCr$_2$S$_4$. Absorption coefficient vs. wavenumber (wavelength) for a hot pressed sample at room temperature [72C].

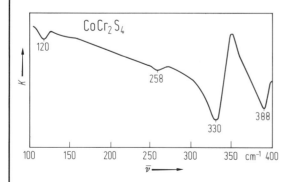

Fig. 63. CoCr$_2$S$_4$. Absorption coefficient vs. wavenumber in the IR region [75L1].

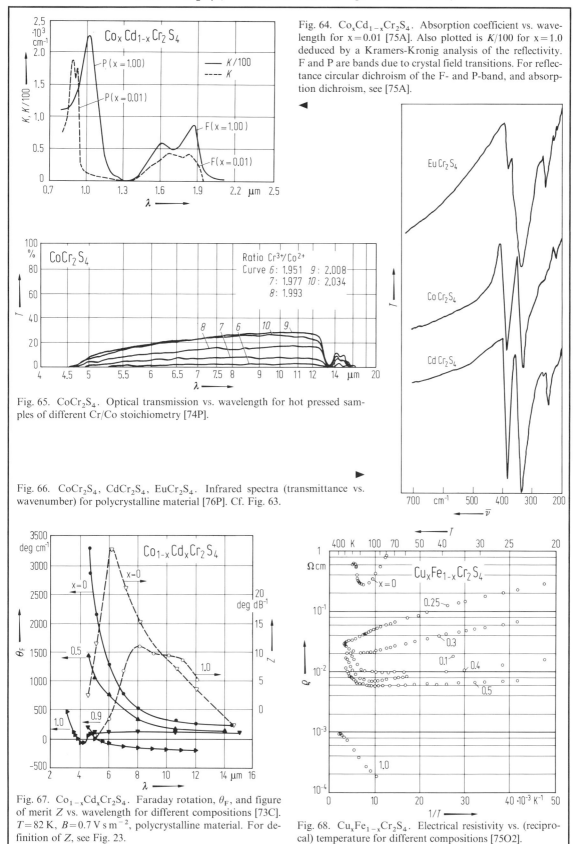

Fig. 64. Co$_x$Cd$_{1-x}$Cr$_2$S$_4$. Absorption coefficient vs. wavelength for x = 0.01 [75A]. Also plotted is K/100 for x = 1.0 deduced by a Kramers-Kronig analysis of the reflectivity. F and P are bands due to crystal field transitions. For reflectance circular dichroism of the F- and P-band, and absorption dichroism, see [75A].

Fig. 65. CoCr$_2$S$_4$. Optical transmission vs. wavelength for hot pressed samples of different Cr/Co stoichiometry [74P].

Fig. 66. CoCr$_2$S$_4$, CdCr$_2$S$_4$, EuCr$_2$S$_4$. Infrared spectra (transmittance vs. wavenumber) for polycrystalline material [76P]. Cf. Fig. 63.

Fig. 67. Co$_{1-x}$Cd$_x$Cr$_2$S$_4$. Faraday rotation, θ_F, and figure of merit Z vs. wavelength for different compositions [73C]. $T = 82$ K, $B = 0.7$ V s m^{-2}, polycrystalline material. For definition of Z, see Fig. 23.

Fig. 68. Cu$_x$Fe$_{1-x}$Cr$_2$S$_4$. Electrical resistivity vs. (reciprocal) temperature for different compositions [75O2].

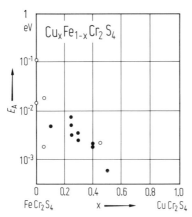

Fig. 69. Cu$_x$Fe$_{1-x}$Cr$_2$S$_4$. Thermal activation energy of the electrical conductivity vs. composition [75A2]. Open circles: from [68H].

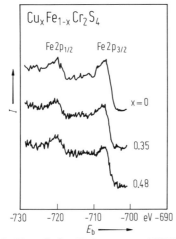

Fig. 71. Cu$_x$Fe$_{1-x}$Cr$_2$S$_4$. Fe 2p spectra of XPS [80A].

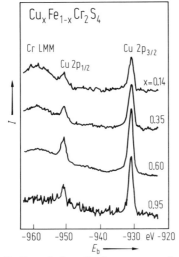

Fig. 72. Cu$_x$Fe$_{1-x}$Cr$_2$S$_4$. Cu 2p spectra and Cr LMM Auger lines of XPS [80A].

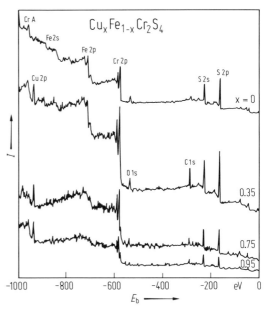

Fig. 70. Cu$_x$Fe$_{1-x}$Cr$_2$S$_4$. XPS spectrum (intensity vs. binding energy) [80A]. (E_b is referred to C 1s peak at −284 eV which appeared due to a hydro-carbon contamination or the polish with diamond file)

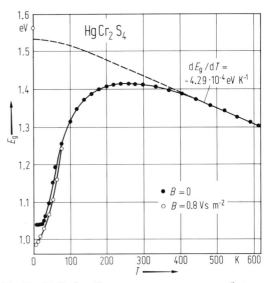

Fig. 73. HgCr$_2$S$_4$. Energy gap vs. temperature (between 4.2 K and 600 K) for $B = 0$ and 0.8 V s m^{-2}. Broken line is the estimated gap for the paramagnetic state [70L].

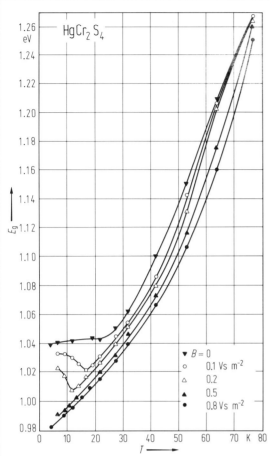

Fig. 74. HgCr₂S₄. Energy gap vs. temperature between 4.2 K and 78 K with the magnetic field as a parameter; single crystal material [70L].

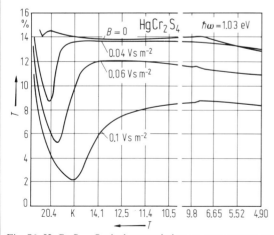

Fig. 76. HgCr₂S₄. Optical transmission vs. temperature at constant energy ($\hbar\omega = 1.03$ eV) with magnetic field as a parameter; single crystal sample [70L].

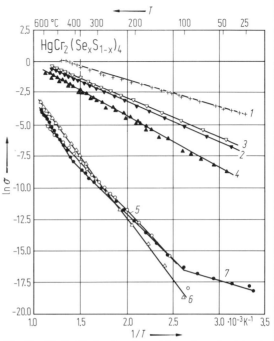

Fig. 75. HgCr₂(SeₓS₁₋ₓ)₄. Logarithm of conductivity (σ in Ω^{-1} cm^{-1}) vs. (reciprocal) temperature (curve 1: x = 1, 2: 0.9375, 3: 0.875, 4: 0.750, 5: 0.125, 6: 0.0625, 7: 0) [75O1]. For activation energies, see original paper.

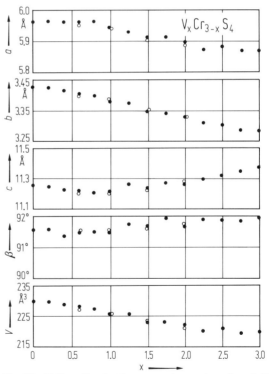

Fig. 77. VₓCr₃₋ₓS₄. Lattice parameters a, b, c, β, and V (cell volume) vs. composition. Solid circles are those of the annealed sample and open circles are those of the quenched sample [81T].

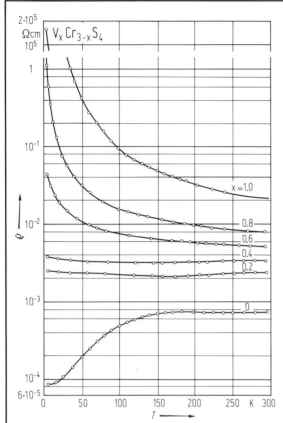

Fig. 78. V$_x$Cr$_{3-x}$S$_4$. Resistivity vs. temperature for sintered material [79T].

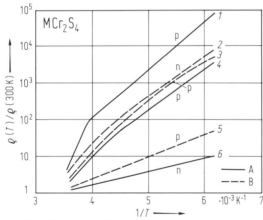

Fig. 80. MCr$_2$S$_4$ (M = Ba, Eu, Sr). Electrical resistivity vs. reciprocal temperature of compounds (single crystal needles) with different generating conditions:

A: 4AX + 2CrCl$_3$ → ACr$_2$X$_4$ + 3ACl$_2$,

B: AX + Cr$_2$X$_3$ → ACr$_2$X$_4$ [71L3].

Activation energy and resistivity at room temperature are given in brackets: curve 1: BaCr$_2$S$_4$ (E_A = 0.25 eV, ϱ = 1.3 · 10^5 Ω cm); 2: EuCr$_2$S$_4$ (E_A = 0.22 eV, ϱ = 5 · 10^3 Ω cm); 3: SrCr$_2$S$_4$ (E_A = 0.35 eV, ϱ = 3 · 10^4 Ω cm); 4: SrCr$_2$S$_4$ (E_A = 0.20 eV, ϱ = 1 · 10^4 Ω cm); 5: BaCr$_2$S$_4$ (E_A = 0.12 eV, ϱ = 2.5 · 10^2 Ω cm); 6: EuCr$_2$S$_4$ (E_A = 0.068 eV, ϱ = 1 · 10^2 Ω cm).

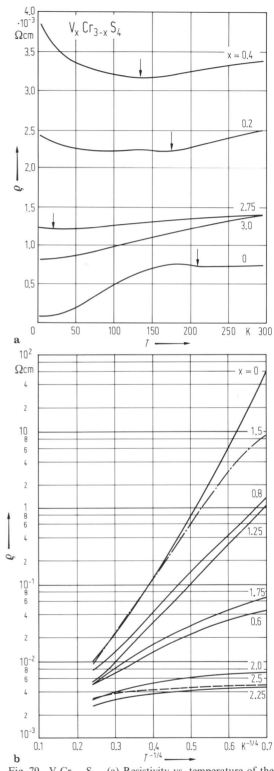

Fig. 79. V$_x$Cr$_{3-x}$S$_4$. (a) Resistivity vs. temperature of the metallic compounds. The arrows indicate the magnetic transition temperature determined from ϱ anomaly. (b) ϱ vs. $T^{-1/4}$ of the semiconductive compounds. Chained line and dashed line are used in order only to avoid confusion [81T].

10.2.3 Chalcogenides MCr₂X₄ (M = Cd, Cu, Hg, Zn, Fe, Ni, V, Ba, Co; X = Se, Te) (For tables, see p. 174 ff.)

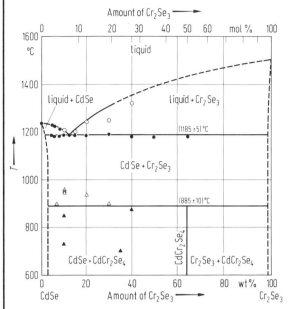

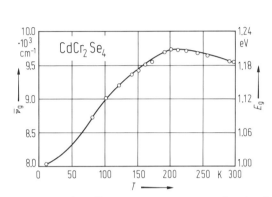

Fig. 2. $CdCr_2Se_4$. Energy gap vs. temperature [70S2]. Cf. Fig. 61.

• DTA heating curves   △ CdSe + Cr₂Se₃ after annealing
○ DTA cooling curves   ▲ CdSe + CdCr₂Se₄ after annealing
(complete or partial transformation)

Fig. 1. $CdCr_2Se_4$. CdSe—Cr₂Se₃ phase diagram [73B].

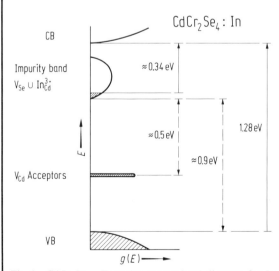

Fig. 3. $CdCr_2Se_4$. Tentative energy level diagram for In doped, n-type material at $T = 80$ K [77T2]. $g(E) =$ density of states; $V_{Cd} = $ Cd vacancy; impurity band is formed by Se vacancies and In^{3+} ions on Cd-lattice sites.

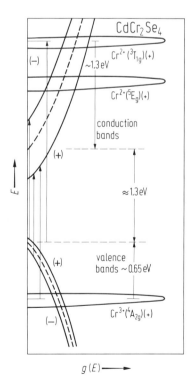

Fig. 4. $CdCr_2Se_4$. Energy level diagram at $T < T_C$ [70H, 75T]. Conduction and valence bands at $T > T_C$ are indicated by dashed lines; $g(E) =$ density of states.

405

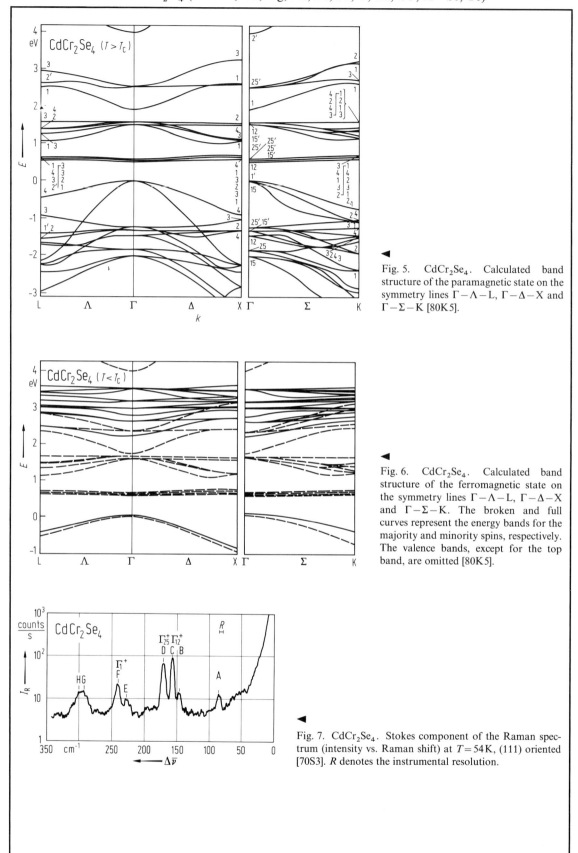

Fig. 5. CdCr$_2$Se$_4$. Calculated band structure of the paramagnetic state on the symmetry lines Γ−Λ−L, Γ−Δ−X and Γ−Σ−K [80K5].

Fig. 6. CdCr$_2$Se$_4$. Calculated band structure of the ferromagnetic state on the symmetry lines Γ−Λ−L, Γ−Δ−X and Γ−Σ−K. The broken and full curves represent the energy bands for the majority and minority spins, respectively. The valence bands, except for the top band, are omitted [80K5].

Fig. 7. CdCr$_2$Se$_4$. Stokes component of the Raman spectrum (intensity vs. Raman shift) at $T = 54$ K, (111) oriented [70S3]. R denotes the instrumental resolution.

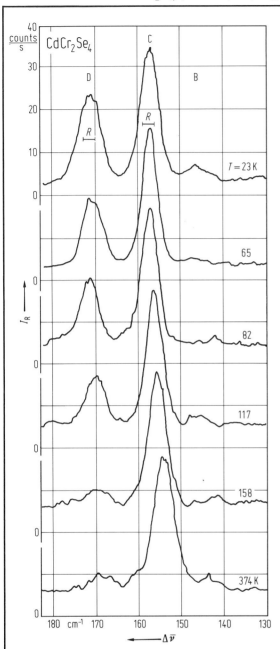

Fig. 8. CdCr$_2$Se$_4$. Stokes component of the Raman spectrum lines B, C, and D (cf. Fig. 7), at various temperatures, below and above the Curie temperature $T_C = 130$ K [70S3]. R denotes the instrumental resolution. Cf. respective table.

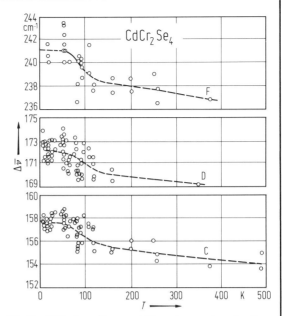

Fig. 9. CdCr$_2$Se$_4$. Raman shift vs. temperature for three of the lines listed in the tables [70S3]. Cf. Fig. 7.

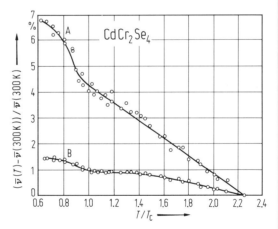

Fig. 10. CdCr$_2$Se$_4$. Infrared active phonon frequencies vs. temperature normalized to the Curie temperature. (A) 75 cm^{-1} phonon; (B) 290 cm^{-1} phonon [71B2]. Polycrystalline sample.

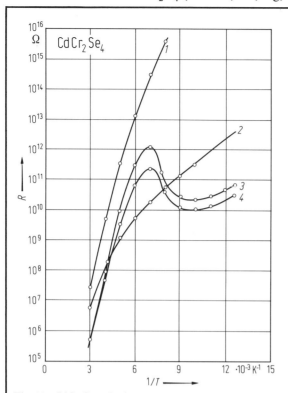

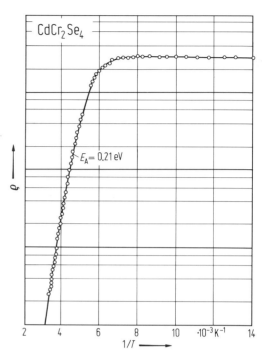

Fig. 11. CdCr$_2$Se$_4$. Resistance vs. reciprocal temperature. Curves *1*, *2*: crystals not annealed; *3*, *4*: annealed in vacuum at 550°C for 5 h [79K1]. See also Fig. 13.

Fig. 12. CdCr$_2$Se$_4$. Resistivity vs. reciprocal temperature for a Se-deficient undoped polycrystalline sample [70A]. E_A is the thermal activation energy.

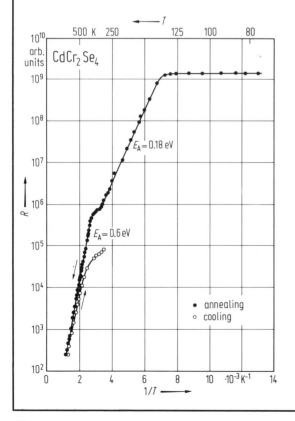

Fig. 13. CdCr$_2$Se$_4$. Resistance vs. (reciprocal) temperature for a single crystal [74P]. Note: The activation energy E_A = 0.18 eV is ascribed to selenium vacancies, the E_A = 0.6 eV level to cadmium vacancies. Selenium vacancies are donors, the measurement of their activation energy requires the conduction to be n-type. The cadmium vacancies, however, are acceptors, the measurement of their activation energy requires a p-type conduction. Since a change of the sign of the thermoelectric voltage at the vicinity of the room temperature has not been reported the interpretation given is questionable. Alternatively the two energies could be ascribed to the first and the second ionization energy of the doubly ionizable cadmium vacancies, if the conduction is p-type in the whole temperature range.

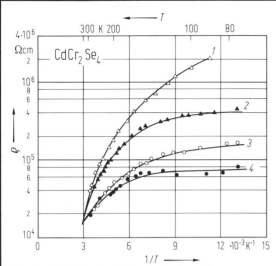

Fig. 14. $CdCr_2Se_4$. Resistivity vs. (reciprocal) temperature; curve *1*: in a static field; *2, 3, 4*: at frequencies 10^5, 10^6, and 10^8 Hz, respectively [82K].

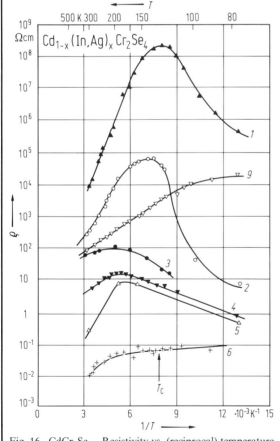

Fig. 16. $CdCr_2Se_4$. Resistivity vs. (reciprocal) temperature for samples doped with In and Ag [82K].
Curve *1*: $Cd_{0.96}In_{0.04}Cr_2Se_4$, *2*: $Cd_{0.994}In_{0.006}Cr_2Se_4$, *3, 4*: $Cd_{0.988}In_{0.012}Cr_2Se_4$, *5*: $Cd_{0.976}In_{0.024}Cr_2Se_4$, *6*: $Cd_{0.91}In_{0.09}Cr_2Se_4$, *9*: $Cd_{0.999}Ag_{0.001}Cr_2Se_4$.

For Fig. 17, see next page.

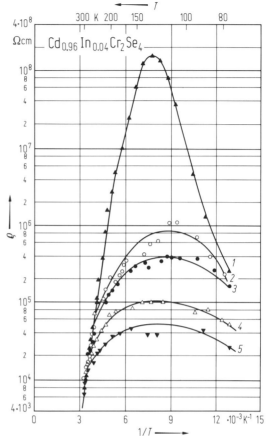

Fig. 15. $Cd_{0.96}In_{0.04}Cr_2Se_4$. Resistivity vs. (reciprocal) temperature; curve *1*: in a static field; *2⋯5*: at frequencies $1.5 \cdot 10^4$, 10^5, 10^6, and 10^8 Hz, respectively [82K].

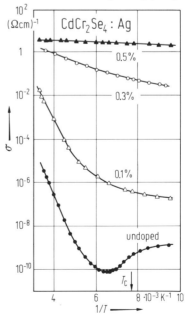

Fig. 18. $CdCr_2Se_4$. Electrical conductivity vs. reciprocal temperature for Ag doped and undoped, p-type single crystals. $T_C = 130$ K [76A].

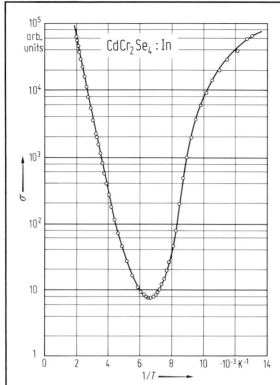

Fig. 17. CdCr$_2$Se$_4$. Electrical conductivity, vs. reciprocal temperature for In doped, n-type single crystals [68W2].

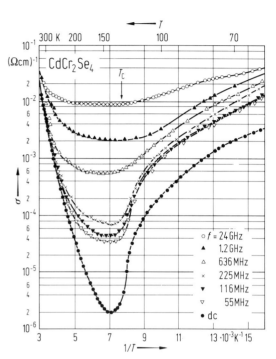

Fig. 19. CdCr$_2$Se$_4$. Electrical conductivity at various frequencies vs. (reciprocal) temperature in the temperature range from 63 K to RT for polycrystalline material [72K1].

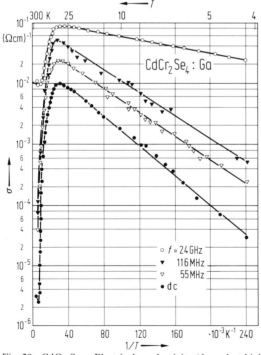

Fig. 20. CdCr$_2$Se$_4$. Electrical conductivity (dc and at high frequencies) for a 0.1 at % Ga doped, hot pressed sample vs. (reciprocal) temperature in the range from $T = 4.2$ K to RT [72K1].

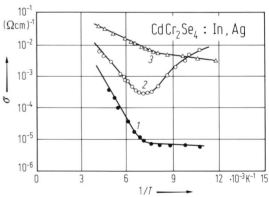

Fig. 21. CdCr$_2$Se$_4$. Conductivity vs. reciprocal temperature. Curve 1: original film, 2: doped with 0.3 at % In, 3: doped with 1.2 at % Ag [82P].

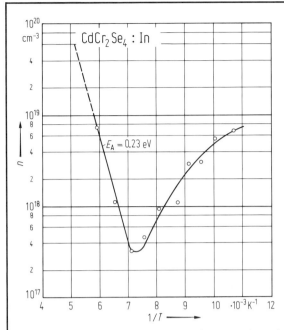

Fig. 22. CdCr$_2$Se$_4$. Electron concentration vs. reciprocal temperature (In doped single crystals) [80K2].

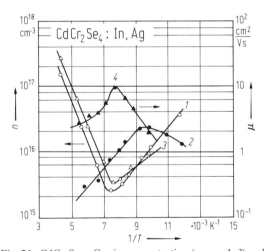

Fig. 24. CdCr$_2$Se$_4$. Carrier concentration (curves *1, 3*) and mobility (*2, 4*) vs. reciprocal temperature in 0.3 at % In (*1, 2*) and 1.2 at % Ag (*3, 4*) doped films [82P].

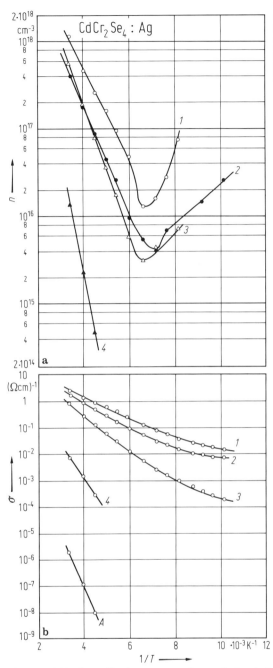

Fig. 23. CdCr$_2$Se$_4$. Carrier concentration and electrical conductivity vs. reciprocal temperature for an undoped (curve *3*) and three Ag doped (curves *1, 2, 4*) single crystals [80K3]. Curve *A* in Fig. b is another undoped sample for comparison.

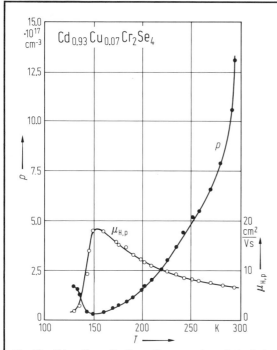

Fig. 25. Cd$_{0.93}$Cu$_{0.07}$Cr$_2$Se$_4$. Concentration of the holes and Hall mobility vs. temperature [79B1].

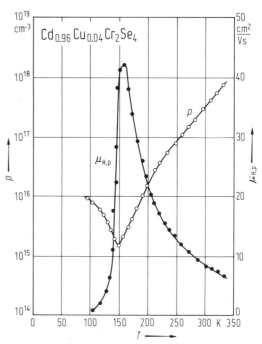

Fig. 26. Cd$_{0.96}$Cu$_{0.04}$Cr$_2$Se$_4$. Hole mobility and hole concentration vs. temperature [78S].

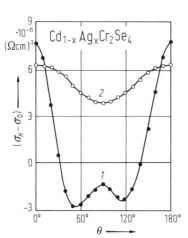

Fig. 27. Cd$_{1-x}$Ag$_x$Cr$_2$Se$_4$. Magnetoconductivity ($\sigma_H - \sigma_0$) vs. angle between the magnetic field direction and the [100] axis. Curve 1: x = 0.005 (n = 9), 2: x = 0.034 (n = 3). H = 3560 Oe, T = 77 K [82B].

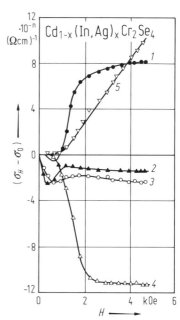

Fig. 28. CdCr$_2$Se$_4$. Magnetoconductivity ($\sigma_H - \sigma_0$) vs. magnetic field (T = 77 K); curves 1, 2, 3: transverse magnetoconductivity with magnetization along [001], [110], [111]; 4: longitudinal magnetoconductivity for $H \| E_{[1\bar{1}0]}$, p-type Cd$_{0.995}$Ag$_{0.005}$Cr$_2$Se$_4$ (n = 9, $\sigma_0 = 319.5 \cdot 10^{-9} \, \Omega^{-1} \, cm^{-1}$); 5: isotropic magnetoconductivity, n-type Cd$_{0.98}$In$_{0.02}$Cr$_2$Se$_4$ (n = 4, $\sigma_0 = 1.2 \cdot 10^{-2} \, \Omega^{-1} \, cm^{-1}$) [82B].

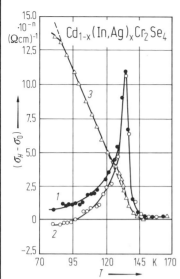

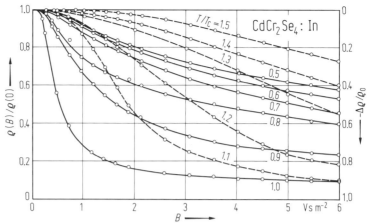

Fig. 30. CdCr$_2$Se$_4$. Field dependence of the normalized resistivity for several temperatures $T \leqq T_C$ (solid lines) and $T \geqq T_C$ (dashed lines); single crystal, In doped sample [73F].

Fig. 29. CdCr$_2$Se$_4$. Magnetoconductivity vs. temperature ($H = 5$ kOe); curves 1, 2: Cd$_{0.995}$Ag$_{0.005}$Cr$_2$Se$_4$ with the magnetization along [001] and [111], respectively, (n = 8); 3: Cd$_{0.98}$In$_{0.02}$Cr$_2$Se$_4$ (n = 4) [82B].

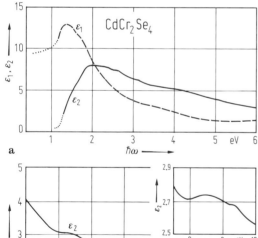

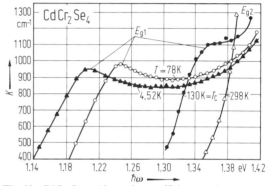

Fig. 32. CdCr$_2$Se$_4$. Absorption coefficient vs. photon energy at various temperatures (single crystal material) [70H]. E_{g_1}: lowest energy absorption edge related to some imperfection; E_{g_2}: higher energy edge.

Fig. 31. CdCr$_2$Se$_4$. Real and imaginary part of the dielectric constant vs. photon energy at 300 K, (a) between 1 and 6 eV, (b) between 5 and 12 eV [79Z1, 79Z2]. Cf. Fig. 38.

Fig.33. CdCr$_2$Se$_4$. Absorption coefficient vs. photon energy for single crystal material in the vicinity of the absorption edge at $T = 300$ K [77B].

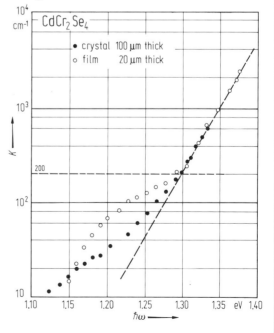

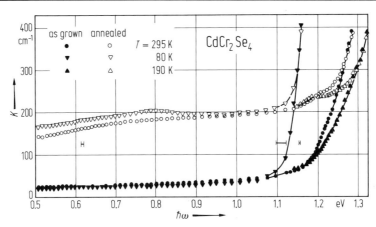

Fig. 34. CdCr$_2$Se$_4$. Optical absorption coefficient vs. photon energy of single crystals as grown and annealed in vacuum for 24 h at $T = 650\,°C$. The weak maxima located at 0.8 eV and at 1.2 eV are not explained [74P].

For Fig. 35, see next page.

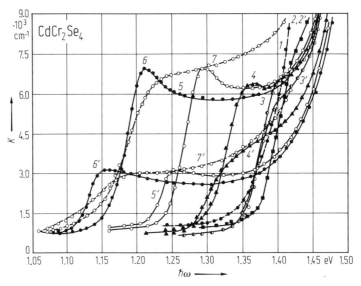

Fig. 36. CdCr$_2$Se$_4$. Absorption coefficient vs. photon energy in the case of unpolarized (curve 1), right-hand ($2 \cdots 7$) and left-hand ($2' \cdots 7'$) circularly polarized light. T [K]: 1: 293; 2, $2'$: 197; 3, $3'$: 162; 4, $4'$: 135; 5, $5'$: 113; 6, $6'$ and 7, $7'$: 16. Curves $1 \cdots 6$ and $2' \cdots 6'$ represent undoped CdCr$_2$Se$_4$; curves 7 and $7'$ represent CdCr$_2$Se$_4$ containing ≈ 4 wt % In [80G].

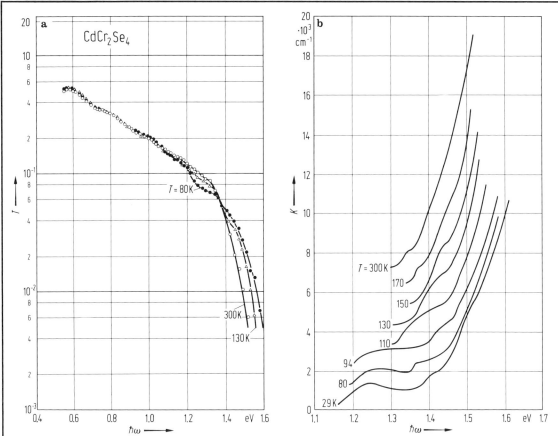

Fig. 35. $CdCr_2Se_4$. (a) Optical transmission vs. photon energy of vacuum-deposited thin polycrystalline film of 2 μm thickness at three temperatures. (b) Optical absorption coefficient vs. photon energy for different temperatures. (The absorption level is displaced by $1000\ cm^{-1}$ for each temperature and the scale of the vertical axis is corresponding to the spectrum at $T = 29$ K) [76S1].

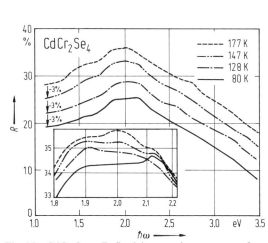

Fig. 37. $CdCr_2Se_4$. Reflectivity vs. photon energy for a single crystal at various temperatures near the Curie temperature in the spectral range 1.15···3.5 eV. The inset shows the enlarged spectra around the main peak band [73I]. Cf. Fig. 42.

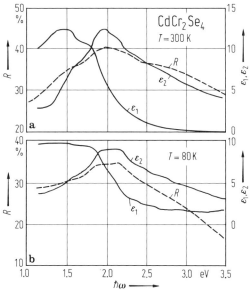

Fig. 38. $CdCr_2Se_4$. Reflectivity and real and imaginary parts of the dielectric constant vs. photon energy for single crystals from the Kramers-Kronig analysis of the data at (a) $T = 300$ K, and (b) $T = 80$ K [73I]. For R at various temperatures, see Fig. 37. See also Figs. 31 and 42.

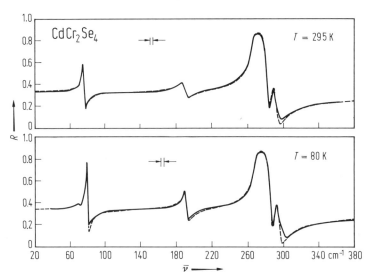

Fig. 39. CdCr$_2$Se$_4$. Reflectivity vs. wavenumber for polycrystalline material in the paramagnetic state (T = 295 K) and in the ferromagnetic state (T = 80 K). Solid: experimental curve; dashed: classical oscillator fit [71W1].

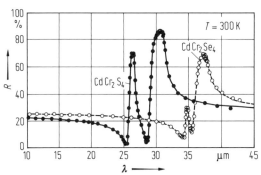

Fig. 40. CdCr$_2$S$_4$, CdCr$_2$Se$_4$. Infrared reflectance vs. wavelength for polycrystalline samples. Typical experimental data and the calculated curves for reflectance at 300 K [71L5].

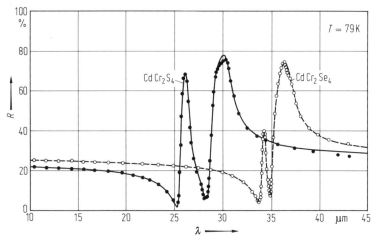

Fig. 41. CdCr$_2$S$_4$, CdCr$_2$Se$_4$. Infrared reflectance vs. wavelength for polycrystalline samples. Typical experimental data and the calculated curves of reflectance at 79 K [71L5].

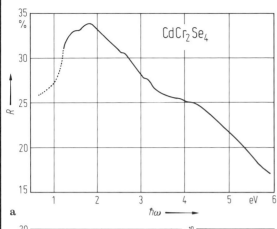

a

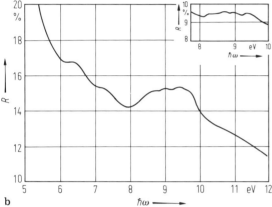

b

Fig. 42. CdCr$_2$Se$_4$. Reflectivity vs. photon energy at 300 K. The insert in b) shows the reflectivity of a chemically etched surface [79Z1].

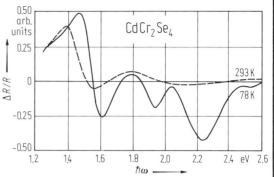

Fig. 43. CdCr$_2$Se$_4$. Wavelength modulated reflectivity vs. photon energy for single crystals at $T = 293$ K (broken line) and 78 K (full line) [80Z].

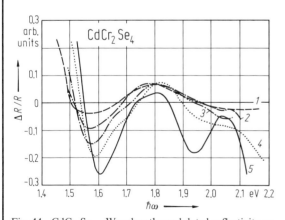

Fig. 44. CdCr$_2$Se$_4$. Wavelength modulated reflectivity vs. photon energy (single crystal sample); curve 1: $T = 293$ K, 2: 233 K, 3: 189 K, 4: 123 K, 5: 78 K [80Z].

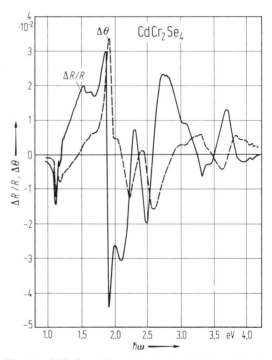

Fig. 45. CdCr$_2$Se$_4$. Magnetic circular dichroism (relative change of reflectance vs. photon energy). The spectrum (solid curve) and the calculated phase shift difference $\Delta\theta$ spectrum (dashed curve) at $T = 4.2$ K $(B = 10$ kG) [77S]. $\frac{\Delta R}{R} = 2 \frac{R_+ - R_-}{R_+ + R_-}$, where R_+ and R_- are the reflectivities for right-handed and left-handed circularly polarized light. $\Delta\theta = \theta^+ - \theta^-$.

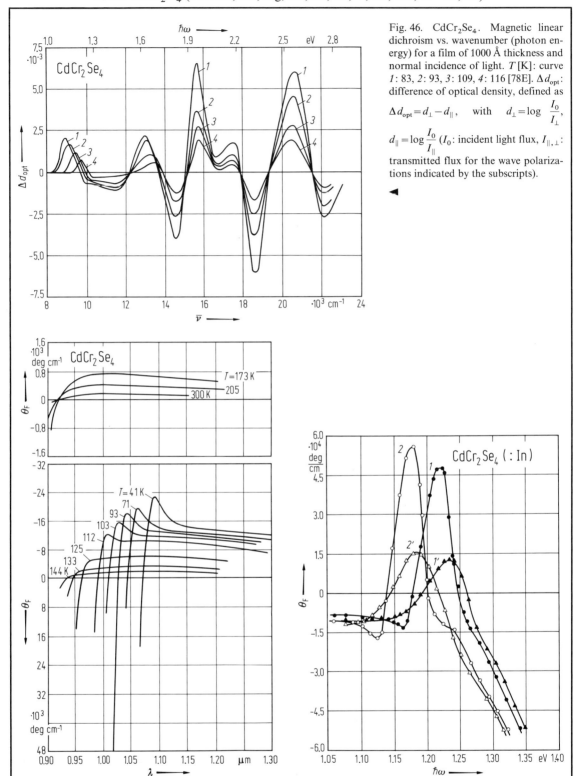

Fig. 46. CdCr₂Se₄. Magnetic linear dichroism vs. wavenumber (photon energy) for a film of 1000 Å thickness and normal incidence of light. T [K]: curve *1*: 83, *2*: 93, *3*: 109, *4*: 116 [78E]. Δd_{opt}: difference of optical density, defined as

$$\Delta d_{opt} = d_\perp - d_\parallel, \quad \text{with} \quad d_\perp = \log \frac{I_0}{I_\perp},$$

$d_\parallel = \log \dfrac{I_0}{I_\parallel}$ (I_0: incident light flux, $I_{\parallel,\perp}$: transmitted flux for the wave polarizations indicated by the subscripts).

◄

Fig. 47. CdCr₂Se₄. Faraday rotation vs. wavelength for various temperatures with an applied magnetic field of 4.6 kOe; single crystal sample [70K2].

Fig. 48. CdCr₂Se₄. Faraday rotation vs. photon energy ($B = 4.5$ kG) T [K]: *1*, *1′*: 80; *2*, *2′*: 16. Curves *1* und *2* represent undoped CdCr₂Se₄; curves *1′* and *2′* represent CdCr₂Se₄ containing ≈4 wt % In [80G]. $B \parallel$ propagation direction of light.

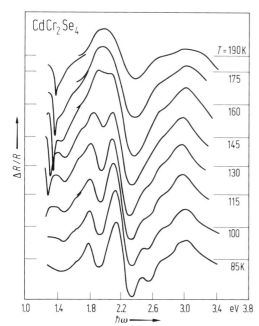

Fig. 49. CdCr$_2$Se$_4$. Thermoreflectance vs. photon energy at various temperatures for an undoped single crystal sample [75T].

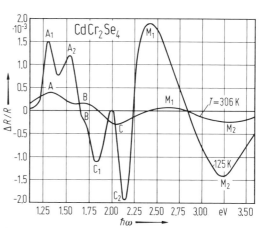

Fig. 50. CdCr$_2$Se$_4$. Thermoreflectance vs. photon energy for polycrystalline material in the paramagnetic (T = 306 K) and ferromagnetic (T = 125 K) regions [75S1]. Structures A and C split near the Curie temperature into a red shifting (A$_1$ and C$_1$) and a blue shifting (A$_2$ and C$_2$) component.

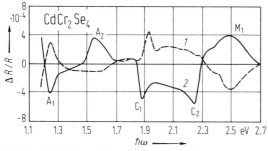

Fig. 51. CdCr$_2$Se$_4$. Piezoreflection vs. photon energy at 77 K: Curve 1: $E \perp X$, 2: $E \parallel X$ [80K4]. X: stress vector.

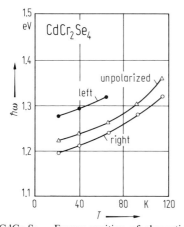

Fig. 53. CdCr$_2$Se$_4$. Energy position of absorption peaks vs. temperature (observed with right and left-handed circularly polarized and unpolarized light) [78K1]. Cf. Fig. 61.

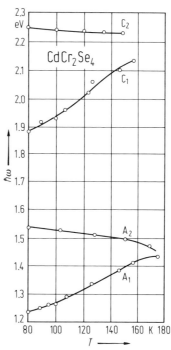

Fig. 52. CdCr$_2$Se$_4$. Piezoreflection peaks vs. temperature (see Fig. 51) [80K4].

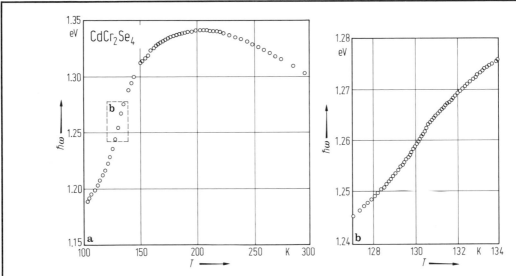

Fig. 54. CdCr$_2$Se$_4$. Temperature dependence of the optical "absorption edge" (a) and the details of this dependence in the vicinity of T_C (b) [77B]. The "absorption edge" is defined by $K = 200$ cm^{-1}.

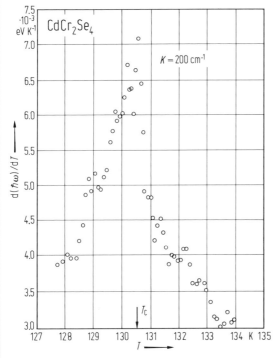

Fig. 55. CdCr$_2$Se$_4$. Temperature derivative of the optical "absorption edge" vs. temperature [77B].

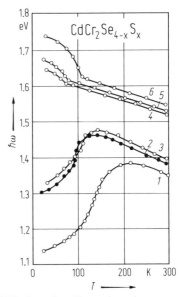

Fig. 56. CdCr$_2$Se$_{4-x}$S$_x$. Absorption edge vs. temperature for various compositions. Curve 1: x = 0, 2: 1.5, 3: 1.7, 4: 3.65, 5: 3.85 6: 4 [76K].

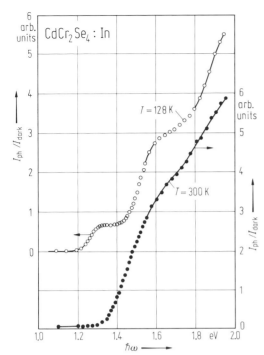

Fig. 57. CdCr₂Se₄. Relative photocurrent vs. photon energy for a 1.4% In doped, p-type single crystal sample at $T = 300$ K and 128 K [71B1].

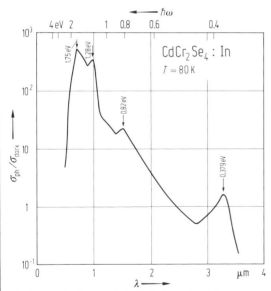

Fig. 59. CdCr₂Se₄. Relative photoconductivity vs. wavelength (photon energy) for an In doped, n-type single crystal at $T = 80$ K [77T2]. $\sigma_{ph} = \sigma - \sigma_{dark}$.

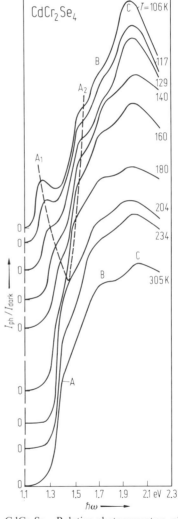

Fig. 58. CdCr₂Se₄. Relative photocurrent vs. photon energy for an undoped, p-type single crystal at various temperatures [76S2]. A and C are assigned to electronic transitions, B to the intra-ion transition $^4A_{2g} \rightarrow {}^4T_{2g}$ of Cr^{3+}. The splitting of A into the "red shifting" component A₁ and the "blue shifting" component A₂ is explained by the splitting of the conduction band into two subbands with opposite spin orientation induced by the magnetic order (cf. Fig. 50).

421

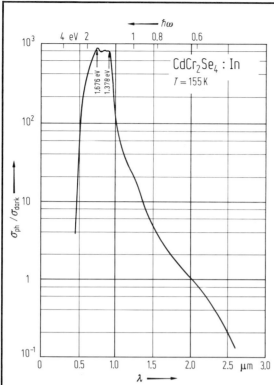

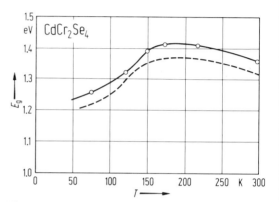

Fig. 60. $CdCr_2Se_4$. Relative photoconductivity vs. wavelength (photon energy) for an In doped, n-type single crystal at $T = 155$ K (above the temperature of the maximum resistivity), (about the same arbitrary units as in Fig. 59). The transport has changed from impurity conduction to band conduction. The electrons are thermally activated from the impurity band to the conduction band. Therefore the peak due to the impurity band has vanished [77T2]. $\sigma_{ph} = \sigma - \sigma_{dark}$.

Fig. 61. $CdCr_2Se_4$. Energy gap derived from photovoltaic measurements on single crystals vs. temperature [68W2]. The dashed line represents results from optical absorption measurements by [66H].

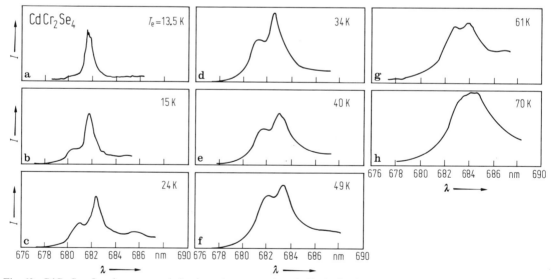

Fig. 62. $CdCr_2Se_4$. Luminescence emission intensity vs. wavelength. Excitation by argon laser of 10 mW power at different temperatures. (a) $T_s = 4.4$ K, $T_e = 13.5$ K; (b) $T_s = 4.9$ K, $T_e = 15$ K; (c) $T_s = T_e = 24$ K; (d) $T_s = T_e = 34$ K; (e) $T_s = T_e = 40$ K; (f) $T_s = T_e = 49$ K; (g) $T_s = T_e = 61$ K; (h) $T_s = T_e = 70$ K [82Y]. T_e: electronic temperature obtained from the high energy tail of the higher energy peak, T_s: sample temperature.

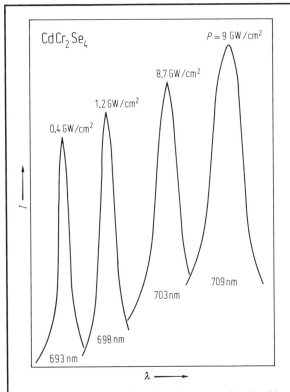

Fig. 63. $CdCr_2Se_4$. Emission intensity vs. wavelength with the excitation power P as parameter; $T = 80$ K [83Y].

For Fig. 66, see next page.

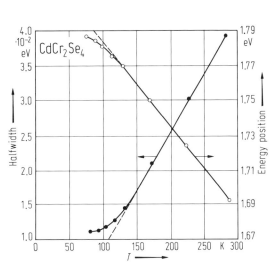

Fig. 64. $CdCr_2Se_4$. Energy position of the luminescence line maximum and halfwidth of luminescence line vs. temperature [75V3]. See also [81Y].

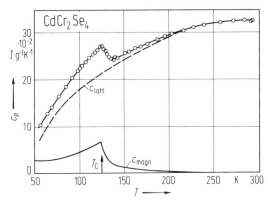

Fig. 65. $CdCr_2Se_4$. Specific heat capacity vs. temperature. The continuous curve represents the magnetic and the dashed curve the lattice contribution [75M1].

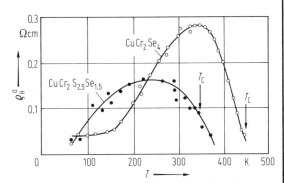

Fig. 67. $CuCr_2S_{2.5}Se_{1.5}$, $CuCr_2Se_4$. Anomalous Hall resistivity $\varrho_H^a = \varrho_H - R_H B$ (ϱ_H = Hall resistivity) vs. temperature [79K2]. $\varrho_H = U_H \dfrac{d}{I}$; d: sample size, U_H: Hall voltage.

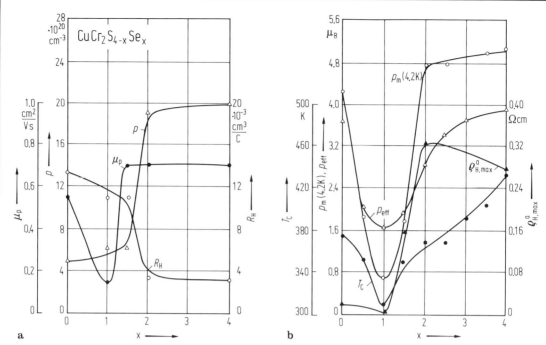

a   b

Fig. 66. CuCr₂S₄₋ₓSeₓ. (a) Hall coefficient, mobility, and hole concentration at 100 K vs. composition (polycrystalline sample). (b) Ferromagnetic Curie point T_C, maximum anomalous Hall resistivity $\varrho^a_{H,max}$, magnetic moment per molecule at 4.2 K p_m(4.2 K) and magnetic moment p_{eff} calculated from C_m vs. composition [79K2].

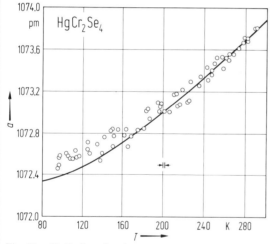

Fig. 68. HgCr₂Se₄. Lattice parameter vs. temperature [76W]. Solid line: theoretical curve calculated from the Debye-Grüneisen theory. Circles: experimental values. Θ_D = 540 K.

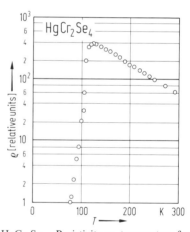

Fig. 69. HgCr₂Se₄. Resistivity vs. temperature for a single crystal [70T1].

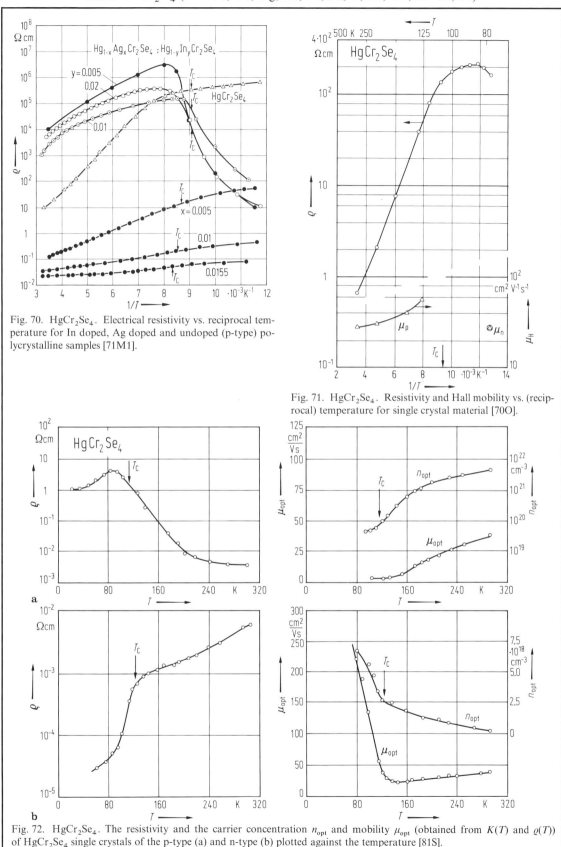

Fig. 70. HgCr$_2$Se$_4$. Electrical resistivity vs. reciprocal temperature for In doped, Ag doped and undoped (p-type) polycrystalline samples [71M1].

Fig. 71. HgCr$_2$Se$_4$. Resistivity and Hall mobility vs. (reciprocal) temperature for single crystal material [70O].

Fig. 72. HgCr$_2$Se$_4$. The resistivity and the carrier concentration n_{opt} and mobility μ_{opt} (obtained from $K(T)$ and $\varrho(T)$) of HgCr$_2$Se$_4$ single crystals of the p-type (a) and n-type (b) plotted against the temperature [81S].

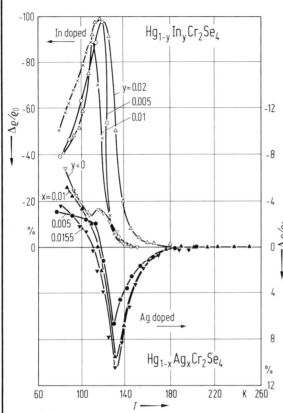

Fig. 73. HgCr₂Se₄. Transverse magnetoresistance at a magnetic flux density $B = 0.7$ V s m^{-2} vs. temperature, for In and Ag doped polycrystalline samples [71M1, 70M1].

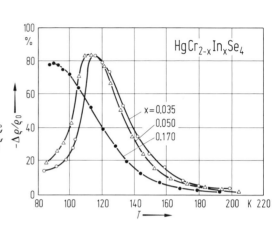

Fig. 74. HgCr₂₋ₓInₓSe₄. Magnetoresistance vs. temperature for single crystals at $B = 0.7$ V s m^{-2} and $B \perp I$ [71T2].

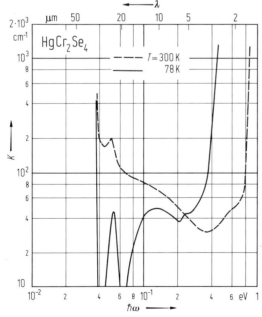

Fig. 76. HgCr₂Se₄. Absorption coefficient vs. photon energy (wavelength) for a single crystal at 78 K and 300 K [71L2].

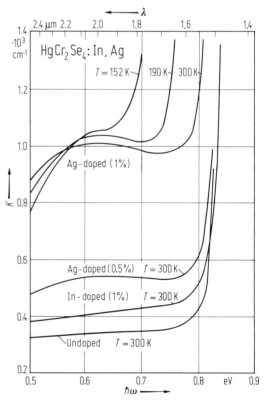

Fig. 75. HgCr₂Se₄. Absorption coefficient vs. photon energy (wavelength) of various single crystal samples without and with Ag or In doping at different temperatures [70M1].

10.2.3 MCr_2X_4 (M = Cd, Cu, Hg, Zn, Fe, Ni, V, Ba, Co; X = Se, Te)

For Fig. 77, see next page.

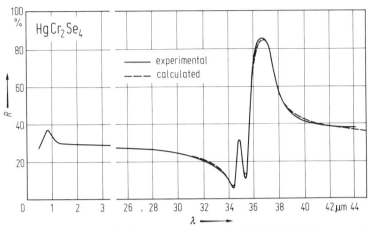

Fig.78. $HgCr_2Se_4$. Reflectance (near normal incidence) vs. wavelength at 300 K [71L2]. Single crystal material.

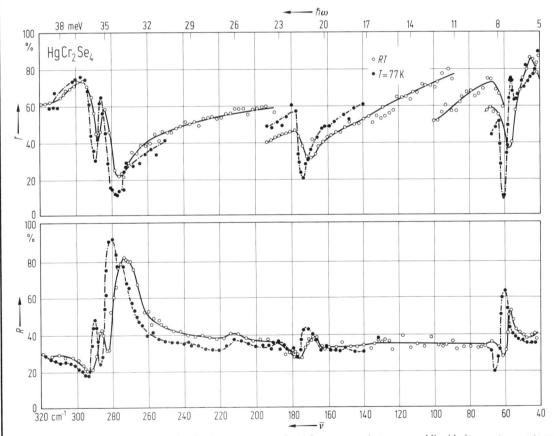

Fig. 79. $HgCr_2Se_4$. Transmission and reflection vs. wavenumber (photon energy) at room and liquid nitrogen temperatures for polycrystalline material [71W2].

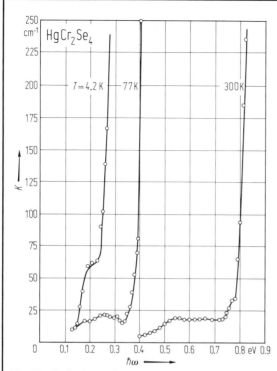

Fig. 77. HgCr₂Se₄. Absorption coefficient at three fixed temperatures, $T = 4.2\,K$, 77 K and 300 K [73A1].

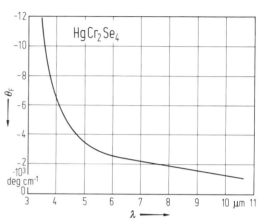

Fig. 80. HgCr₂Se₄. Faraday rotation vs. wavelength for single crystals at $T = 85\,K$ measured with $B = 0.7\,V\,s\,m^{-2}$ [71L2].

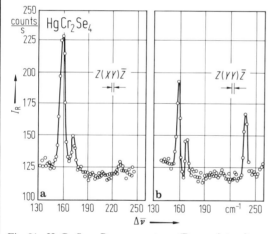

Fig. 81. HgCr₂Se₄. Raman spectrum (Raman intensity vs. Raman shift) at 8 K for (a) crossed and (b) parallel scattering configurations, taken with the 514.5 nm laser line [78I1].

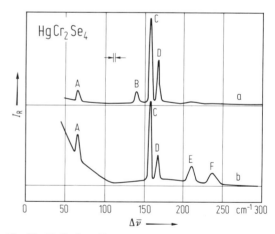

Fig. 82. HgCr₂Se₄. Raman spectrum (intensity vs. Raman shift) with (a) $\lambda = 647.1$ nm and (b) 514.5 nm laser lines at $T = 80\,K$ [78I1].

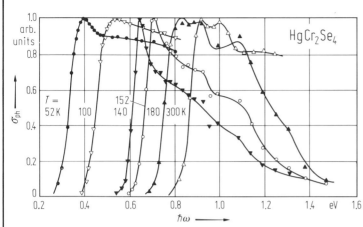

Fig. 83. HgCr$_2$Se$_4$. Photoconductivity vs. photon energy at several temperatures covering $T_C = 106$ K (the peak intensities were normalized) [82W].

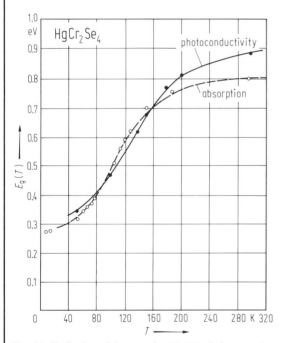

Fig. 84. HgCr$_2$Se$_4$. Edge energies ($E_g(T)$) of photoconductivity and absorption vs. temperature [82W].

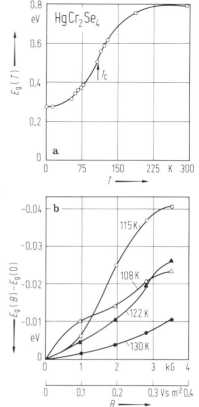

Fig. 85. HgCr$_2$Se$_4$. (a) Temperature dependence of the optical absorption edge, E_g, defined by the value at $K = 240$ cm^{-1}. (b) Differences between the absorption edges obtained with and without magnetic field at a number of fixed temperatures are shown vs. the applied field. Single crystal material [73A1].

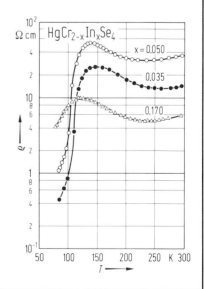

Fig. 86. HgCr$_{2-x}$In$_x$Se$_4$. Electrical resistivity vs. temperature for three single crystals of different composition ($H = 7$ kOe, $I = 1$ mA) [71T2].

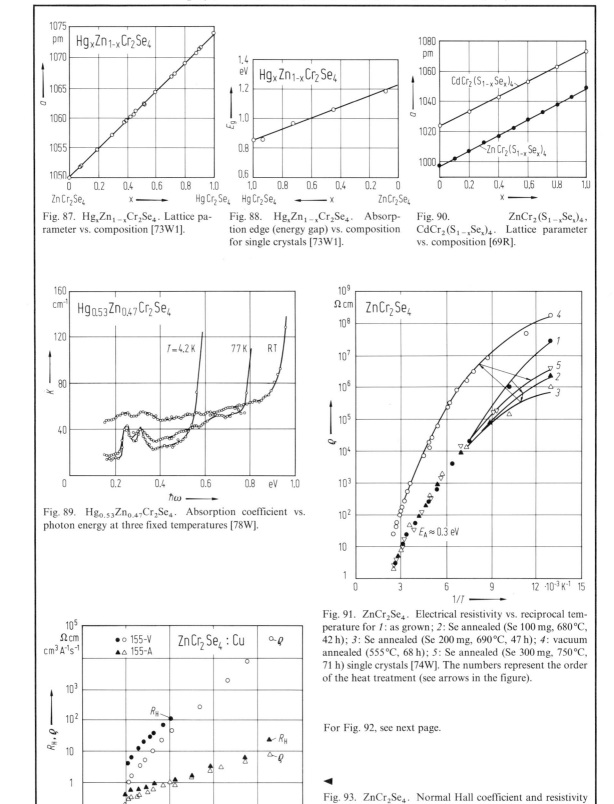

Fig. 87. Hg$_x$Zn$_{1-x}$Cr$_2$Se$_4$. Lattice parameter vs. composition [73W1].

Fig. 88. Hg$_x$Zn$_{1-x}$Cr$_2$Se$_4$. Absorption edge (energy gap) vs. composition for single crystals [73W1].

Fig. 90. ZnCr$_2$(S$_{1-x}$Se$_x$)$_4$, CdCr$_2$(S$_{1-x}$Se$_x$)$_4$. Lattice parameter vs. composition [69R].

Fig. 89. Hg$_{0.53}$Zn$_{0.47}$Cr$_2$Se$_4$. Absorption coefficient vs. photon energy at three fixed temperatures [78W].

Fig. 91. ZnCr$_2$Se$_4$. Electrical resistivity vs. reciprocal temperature for *1*: as grown; *2*: Se annealed (Se 100 mg, 680°C, 42 h); *3*: Se annealed (Se 200 mg, 690°C, 47 h); *4*: vacuum annealed (555°C, 68 h); *5*: Se annealed (Se 300 mg, 750°C, 71 h) single crystals [74W]. The numbers represent the order of the heat treatment (see arrows in the figure).

For Fig. 92, see next page.

◀

Fig. 93. ZnCr$_2$Se$_4$. Normal Hall coefficient and resistivity vs. reciprocal temperature for Cu doped (0.24 wt %) single crystals [74W]. 155-V: vacuum annealed (570°C, 53 h); 155-A: as grown.

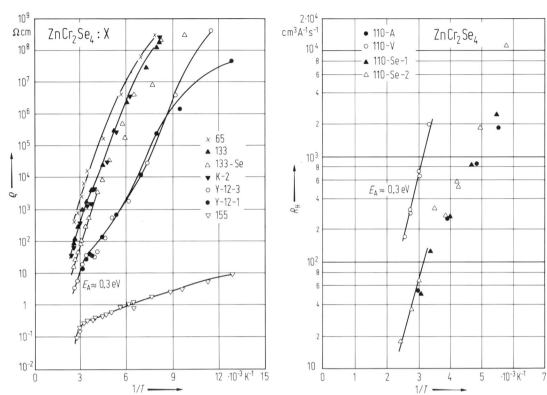

Fig. 92. ZnCr$_2$Se$_4$. Resistivity vs. reciprocal temperature, for differently doped single crystals [74W]. 65: Cd doped (≈ 0.6 wt %); K-2: In doped (≈ 0.4 wt %); 133-Se: Se annealed (Se 200 mg, 540 °C, 53 h); 155: Cu doped (≈ 0.24 wt %); Y-12-1 and Y-12-3: undoped samples obtained from a simple ampoule.

Fig. 94. ZnCr$_2$Se$_4$. Normal Hall coefficient vs. reciprocal temperature for differently treated single crystals [74W]. 110-A: as grown; 110-V: vacuum annealed (555 °C, 68 h); 110-Se-1: Se annealed (Se 100 mg, 680 °C, 42 h); 110-Se-2: Se annealed (Se 200 mg, 690 °C, 47 h).

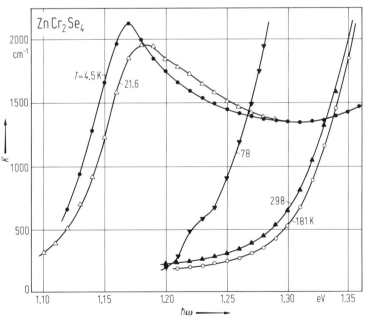

Fig. 95. ZnCr$_2$Se$_4$. Optical absorption coefficient vs. photon energy, at different temperatures with unpolarized light [71L6].

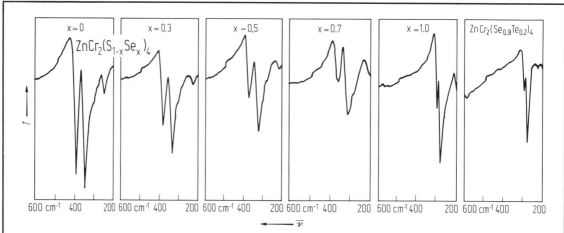

Fig. 96. ZnCr$_2$(S$_{1-x}$Se$_x$)$_4$, ZnCr$_2$(Se$_{0.8}$Te$_{0.2}$)$_4$. Infrared spectra (relative transmission vs. wavenumber) [69R]. For IR spectra, see also [75L].

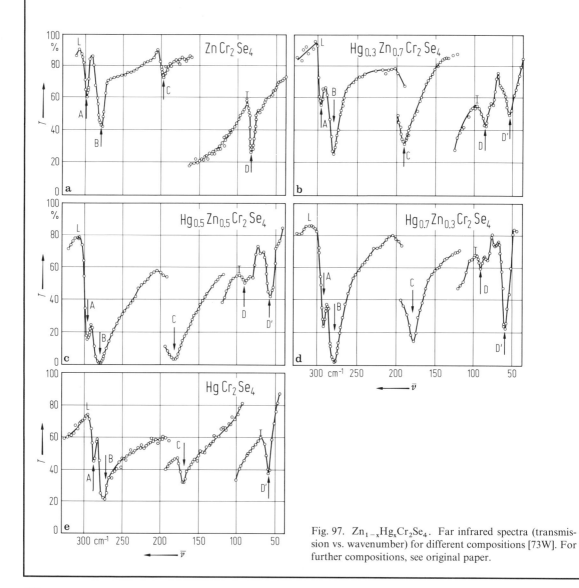

Fig. 97. Zn$_{1-x}$Hg$_x$Cr$_2$Se$_4$. Far infrared spectra (transmission vs. wavenumber) for different compositions [73W]. For further compositions, see original paper.

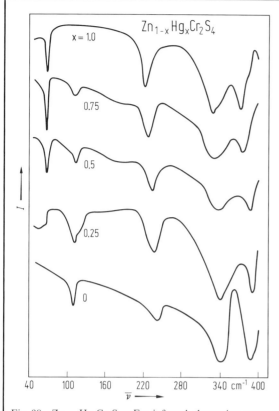

Fig. 98. Zn$_{1-x}$Hg$_x$Cr$_2$S$_4$. Far infrared absorption spectra (relative intensity vs. wavenumber) [75L].

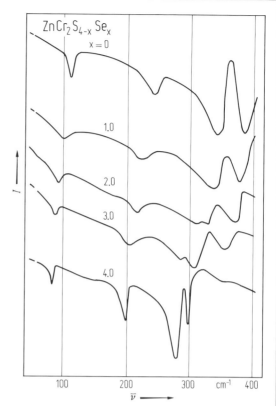

Fig. 99. ZnCr$_2$S$_{4-x}$Se$_x$. Far infrared absorption spectra (relative intensity vs. wavenumber) [75L].

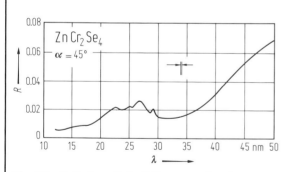

Fig. 100. ZnCr$_2$Se$_4$. Reflectivity vs. wavelength at room temperature measured for s-polarized light at an incidence angle α of 45°. The resolution (0.1 nm) is indicated in the figure [82S].

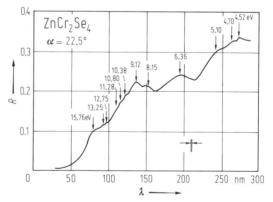

Fig. 101. ZnCr$_2$Se$_4$. Reflectivity vs. wavelength at liquid-nitrogen temperature for p-polarized light at $\alpha = 22.5°$. The resolution (0.4 nm) is indicated in the figure [82S]. α: incidence angle.

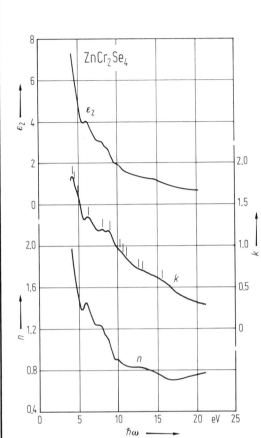

Fig. 102. ZnCr₂Se₄. Dielectric constant (ε_2) and optical constants (n, k) vs. photon energy at liquid-nitrogen temperature. The vertical lines show the energy positions of the reflectance structures for comparison [82S].

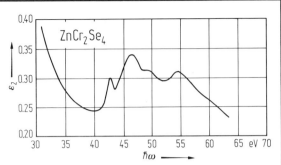

Fig. 103. ZnCr₂Se₄. Dielectric constant vs. photon energy in the $M_{2,3}$ inner-core transition region. The solid curve shows the experimental spectrum [82S].

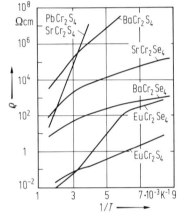

Fig. 105. MCr₂X₄ (M = Pb, Sr, Ba, Eu; X = S, Se). Electrical resistivity vs. reciprocal temperature for needle-type samples [71O].

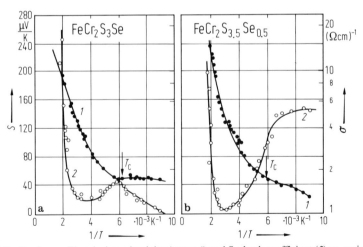

Fig. 104. FeCr₂S₃Se, FeCr₂S₃.₅Se₀.₅. Electrical conductivity (curve 1) and Seebeck coefficient (2) vs. reciprocal temperature for pressed samples [77E].

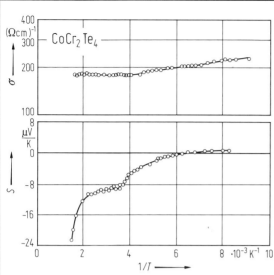

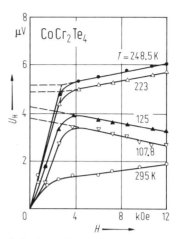

Fig. 106. CoCr$_2$Te$_4$. Electrical conductivity and Seebeck coefficient vs. reciprocal temperature [77V]. Orientation unknown.

Fig. 107. CoCr$_2$Te$_4$. Hall potential vs. magnetic field at various temperatures [77V].

10.2.4 Transition-metal oxides (For tables, see p. 187 ff.)

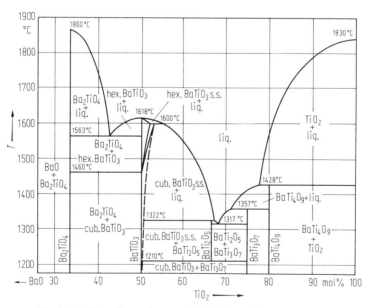

Fig. 1. BaTiO$_3$. Phase diagram of the BaO—TiO$_2$ system [55R1].

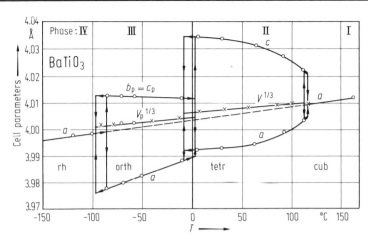

Fig. 2. BaTiO$_3$. Cell parameters vs. temperature [49K]. V: unit cell volume. The index p designates the pseudo-cubic unit cell (see [66C]).

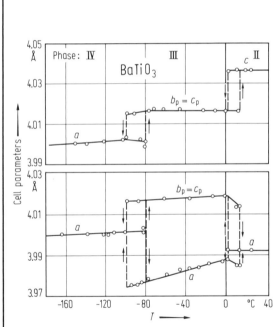

Fig. 3. BaTiO$_3$. Cell parameters vs. temperature [49R]. The symbols are the same as in Fig. 2.

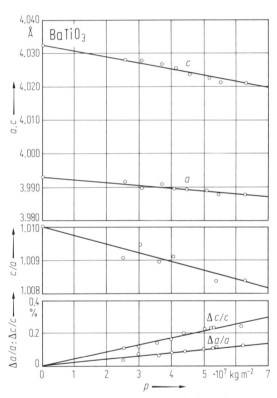

Fig. 4. BaTiO$_3$. Cell parameters vs. hydrostatic pressure [62K]. Data for phase II material.

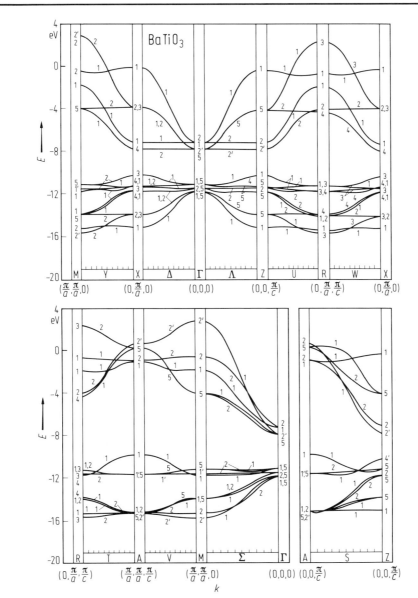

Fig. 5. BaTiO$_3$. Calculated band structure for phase II [73M]. Method: Slater-Koster interpolation scheme with nonorthogonal orbitals.

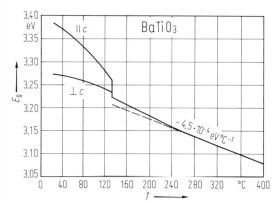

Fig. 6. BaTiO$_3$. Energy gap vs. temperature [70W1]. The band gap energy was estimated by extrapolation of the Urbach edge to the absorption coefficient $K = 3000$ cm^{-1}.

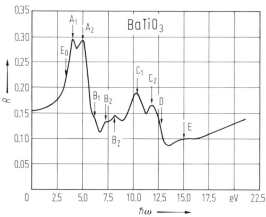

Fig. 7. BaTiO$_3$. Reflectivity vs. photon energy at room temperature (unpolarized light) [65C]. For energies of the fundamental absorption edges, see the tables.

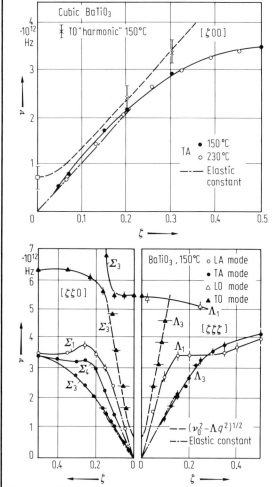

Fig. 8. BaTiO$_3$. Phonon dispersion curves (phonon frequency vs. reduced wave vector coordinate) along three principal directions of the Brillouin zone. For the optical branch in [100] direction the quasi harmonic frequencies deduced from optical (square) or neutron scattering (crosses) experiments are given [71H1]. The dashed line is of the form $v^2 = v_0^2 - \Lambda q^2$.

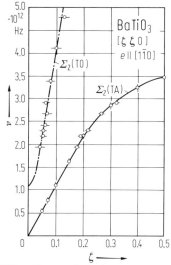

Fig. 9. BaTiO$_3$. Phonon dispersion curves along the [110] direction (frequency vs. reduced wave vector coordinate). Experimental points from neutron scattering at 22°C [70S2]. e: polarization vector.

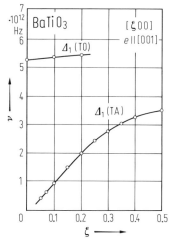

Fig. 10. BaTiO$_3$. Phonon dispersion curves along the [100] direction at RT. Symbols as in Fig. 9 [70S2].

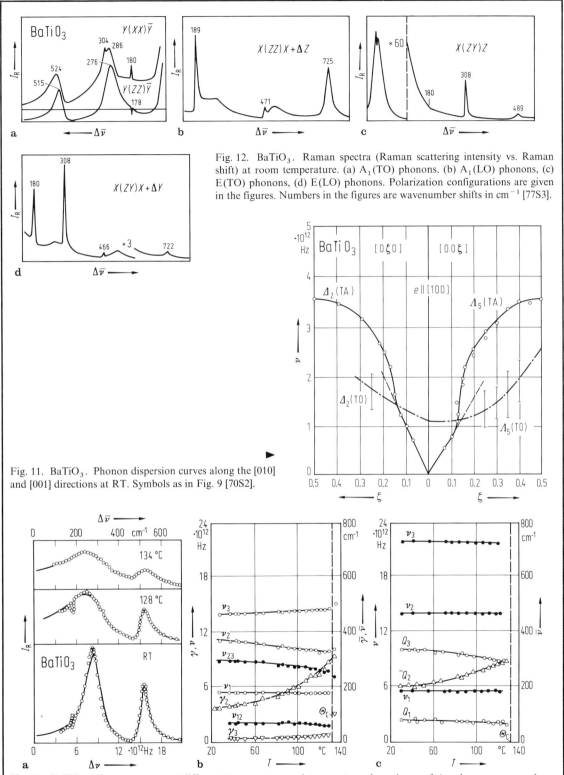

Fig. 12. BaTiO$_3$. Raman spectra (Raman scattering intensity vs. Raman shift) at room temperature. (a) A$_1$(TO) phonons, (b) A$_1$(LO) phonons, (c) E(TO) phonons, (d) E(LO) phonons. Polarization configurations are given in the figures. Numbers in the figures are wavenumber shifts in cm^{-1} [77S3].

Fig. 11. BaTiO$_3$. Phonon dispersion curves along the [010] and [001] directions at RT. Symbols as in Fig. 9 [70S2].

Fig. 13. BaTiO$_3$. Raman spectra at different temperatures and temperature dependence of A$_1$ phonon wavenumbers. (a) Raman intensity vs. Raman shift; dots are experimental points and full lines are the fit of the data to the coupled oscillator model. (b) A$_1$(TO) mode parameters vs. temperature, (c) A$_1$(LO) mode parameters vs. temperature. ν_i: frequency; γ_i: damping parameters of the i-th mode; ν_{ij}: coupling coefficient between the i and j-th mode; Θ_C: ferroelectric Curie temperature; Q_i: effective charge of the i-th mode [77S3].

439

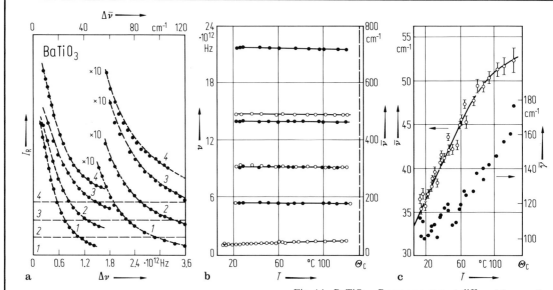

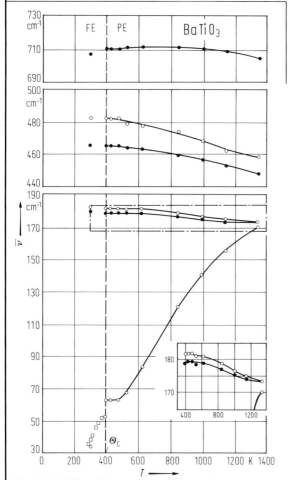

Fig. 14. BaTiO$_3$. Raman spectra at different temperatures and temperature dependence of E phonon wavenumbers. (a) Raman intensity vs. Raman shift at 12.7 °C curve *1*, 40 °C *2*, 78 °C *3*, 121 °C *4*; base lines are shown by broken lines. (b) Mode frequencies (wavenumbers) vs. temperature; open circles: E(TO) modes, full circles: E(LO) mode. (c) Wavenumber ($\bar{\nu}$) and damping parameter ($\bar{\gamma}$) for the overdamped E mode shown in (a) [77S3].

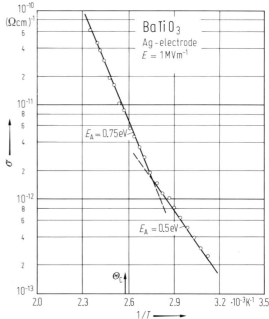

Fig. 15. BaTiO$_3$. Wavenumbers of F$_{1u}$ modes vs. temperature in the ferroelectric (FE) and paraelectric (PE) phases. Open circles: transverse modes, full circles: longitudinal modes, squares: Raman data from [77S3]. [80L].

Fig. 16. BaTiO$_3$. Conductivity vs. reciprocal temperature [58I]. $\sigma \| c$ in the tetragonal phase.

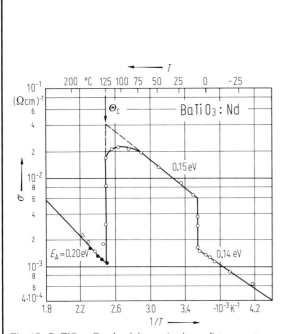

Fig. 17. BaTiO$_3$. Conductivity vs. (reciprocal) temperature for a 0.1% Nd doped sample ($\sigma \| c$) [64B2].

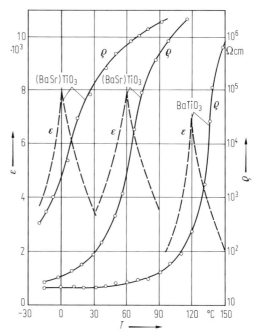

Fig. 18. BaTiO$_3$. Resistivity and dielectric constant vs. temperature for semiconducting ceramics of BaTiO$_3$ and (BaSr)TiO$_3$ [61H].

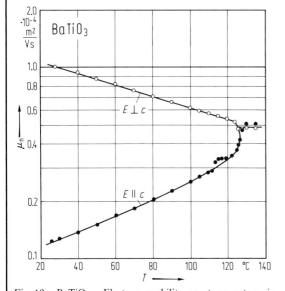

Fig. 19. BaTiO$_3$. Electron mobility vs. temperature in an n-type sample with a donor concentration of $n_d = 8.5 \cdot 10^{18}$ cm^{-3} [67B2]. See also [77F].

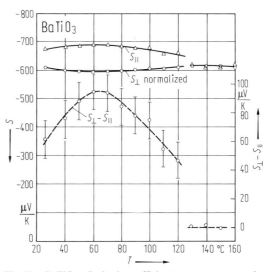

Fig. 20. BaTiO$_3$. Seebeck coefficient vs. temperature for an n-type sample with a donor concentration of $n_d = 6 \cdot 10^{18}$ cm^{-3} [67B2]. See also [71B5].

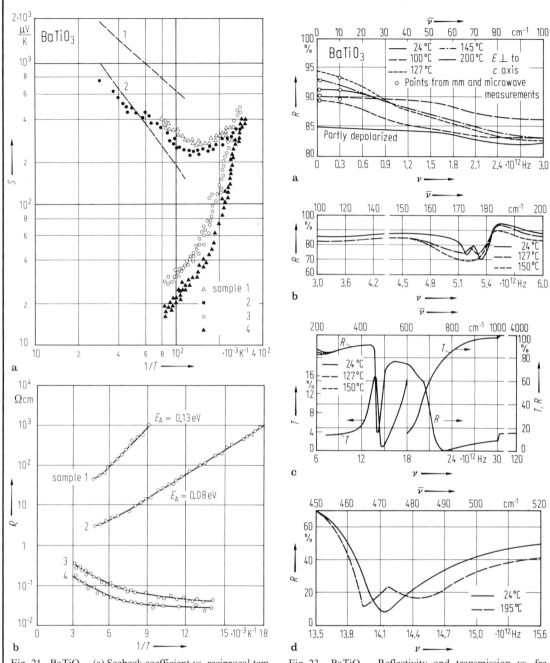

Fig. 21. BaTiO$_3$. (a) Seebeck coefficient vs. reciprocal temperature for four semiconducting ceramic samples reduced in H$_2$ gas at 1100 (sample 1), 1200 (sample 2), 1300 (sample 3) and 1350 °C (sample 4), respectively. The dashed lines show theoretical values for samples 1 and 2. (b) Resistivity vs. reciprocal temperature for the same samples; activation energies are indicated [79T1].

Fig. 23. BaTiO$_3$. Reflectivity and transmission vs. frequency (wavenumber) at various temperatures [64B1]. Note the change in scale at $4.2 \cdot 10^{12}$ Hz in Fig. b.

For Fig. 22, see next page.

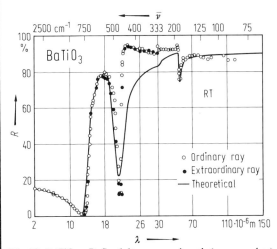

Fig. 22. BaTiO$_3$. Reflectivity vs. wavelength (wavenumber) at room temperature (note change in scale at 30 μm). The solid curve was calculated by using dispersion theory [62S].

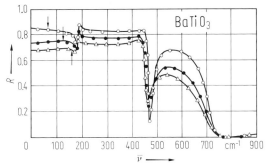

Fig. 24. BaTiO$_3$. Reflectivity vs. wavenumber for cubic material ($\Theta_C = 395$ K) at 400 K (open circles), 850 K (full circles), and 1150 K (triangles) showing F$_{1u}$-type vibrational modes. The curves are the best fit of the dielectric function model to the experimental data. The arrows indicate the soft mode frequency ν_{TO} [80L].

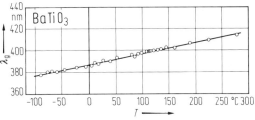

Fig. 26. BaTiO$_3$. Wavelength of absorption edge vs. temperature [54H].

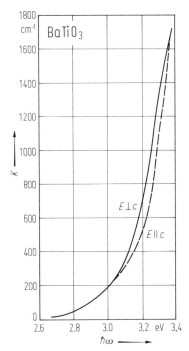

Fig. 25. BaTiO$_3$. Absorption coefficient vs. photon energy at room temperature [59C].

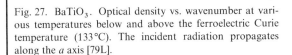

Fig. 27. BaTiO$_3$. Optical density vs. wavenumber at various temperatures below and above the ferroelectric Curie temperature (133 °C). The incident radiation propagates along the a axis [79L].

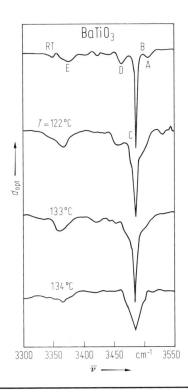

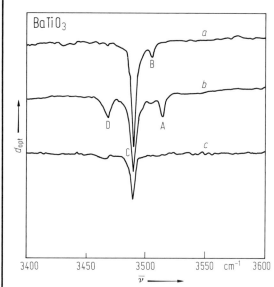

Fig. 28. BaTiO$_3$. Optical density vs. wavenumber for different directions of the radiation k-vector and crystal thickness; curve a: $k \| c$, $d = 5.7$ mm; b: $k \| a$, $d = 5.5$ mm; c: $k \| a$, $d = 2.5$ mm [79L].

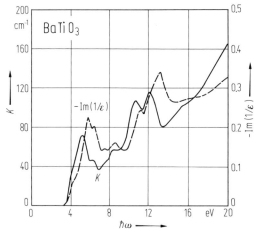

Fig. 30. BaTiO$_3$. Absorption coefficient and Im$(1/\varepsilon)$ vs. photon energy at room temperature (unpolarized light) [65C].

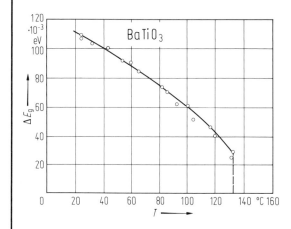

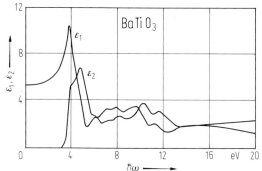

Fig. 29. BaTiO$_3$. Real and imaginary parts of the dielectric constant vs. photon energy at room temperature [65C].

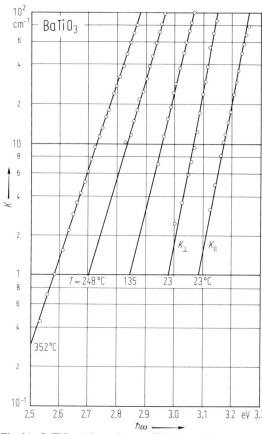

Fig. 31. BaTiO$_3$. Absorption coefficient vs. photon energy at different temperatures. Below the ferroelectric temperature (132 °C) two absorption coefficients correspond to light polarized parallel and perpendicular to the tetragonal c axis, respectively [70W1].

◄

Fig. 32. BaTiO$_3$. Energy separation between the interband absorption edges for light polarized parallel and perpendicular to the tetragonal c axis vs. temperature below the Curie temperature [70W1].

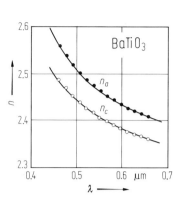

Fig. 33. BaTiO$_3$. Refractive indices in a and c direction vs. wavelength at room temperature [71J].

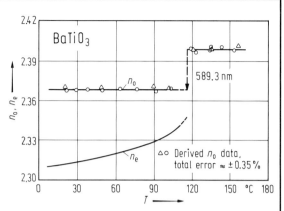

Fig. 34. BaTiO$_3$. Ordinary and extraordinary refractive index vs. temperature at a wavelength of 589.3 nm [64L1].

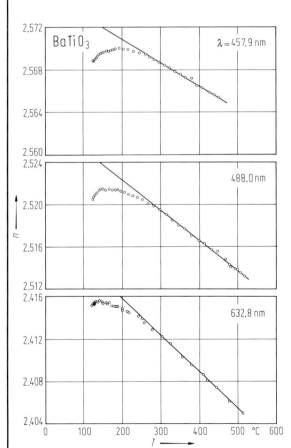

Fig. 35. BaTiO$_3$. Refractive index vs. temperature at three different wavelengths for the cubic phase [82B].

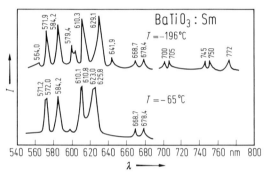

Fig. 36. BaTiO$_3$. Luminescence intensity vs. wavelength for BaTiO$_3$:Sm^{3+} [62M1].

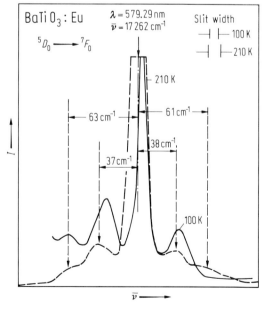

Fig. 37. BaTiO$_3$. Luminescence intensity vs. wavenumber for BaTiO$_3$:Eu^{3+} [67Y].

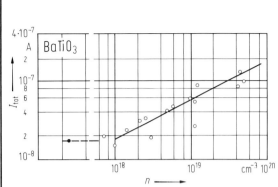

Fig. 38. $BaTiO_3$. Luminescence intensity of various ceramics vs. measured charge carrier density. The luminescence intensity was measured using a fixed primary electron beam of constant spot diameter smaller than the grain size; and the units represent the measured total photomultiplier current. The open point corresponds to an undoped and insulating ceramic [81I]. For luminescence of differently prepared undoped and doped samples, see [81I].

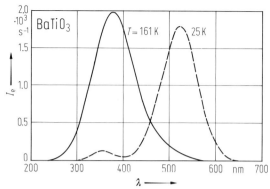

Fig. 39. $BaTiO_3$. Thermoluminescence intensity vs. wavelength at peaks $T = 161$ K and $T = 25$ K [82A1], I_e: emission intensity in 10^3 photons/s.

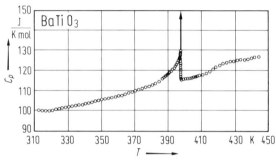

Fig. 40. $BaTiO_3$. Heat capacity vs. temperature [76H1].

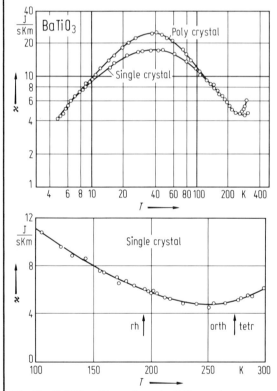

Fig. 41. $BaTiO_3$. Thermal conductivity vs. temperature (for single crystal, $\kappa \parallel \langle 100 \rangle$) [65S5].

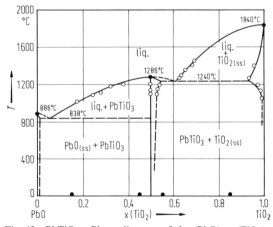

Fig. 42. $PbTiO_3$. Phase diagram of the $(PbO)_{1-x}(TiO_2)_x$ system, refined by mass-loss Knudsen effusion. Full circles: DTA, open circles: mass-loss Knudsen effusion [76H2].

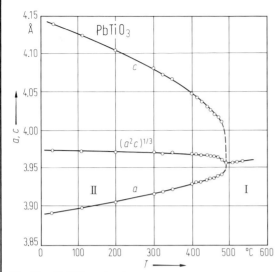

Fig. 43. PbTiO$_3$. Lattice parameters vs. temperature [51S3].

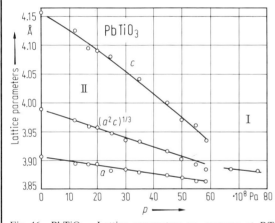

Fig. 46. PbTiO$_3$. Lattice parameters vs. pressure at RT [75I].

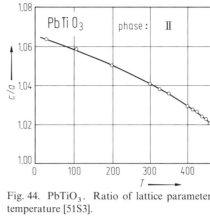

Fig. 44. PbTiO$_3$. Ratio of lattice parameters a and c vs. temperature [51S3].

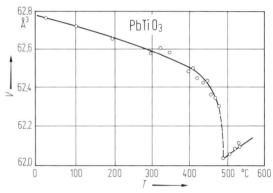

Fig. 45. PbTiO$_3$. Unit cell volume vs. temperature [51S3].

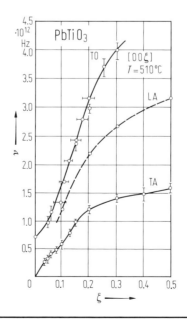

Fig. 47. PbTiO$_3$. Phonon dispersion curves (phonon frequency vs. reduced wave vector coordinate). Results from measurements on a single domain crystal [70S3]. $T = 510\,°C$.

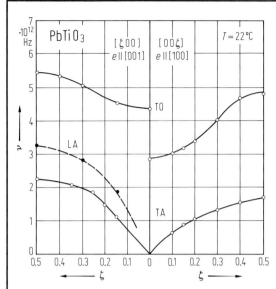

Fig. 48. PbTiO$_3$. Phonon dispersion curves at 22°C (phonon frequency vs. reduced wave vector coordinate) for two directions in the Brillouin zone (cf. Fig. 47) [70S3]. e: polarization vector.

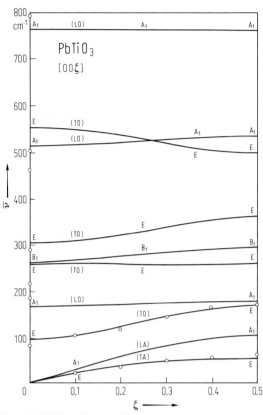

Fig. 49. PbTiO$_3$. Phonon dispersion curves (phonon wavenumber vs. reduced wave vector coordinate) in [100] direction. The dots are values measured by neutron inelastic scattering at room temperature [81F].

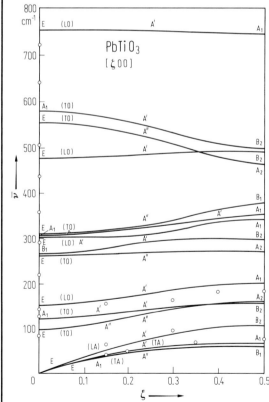

Fig. 50. PbTiO$_3$. Phonon dispersion curves as in Fig. 49 for the [001] direction [81F].

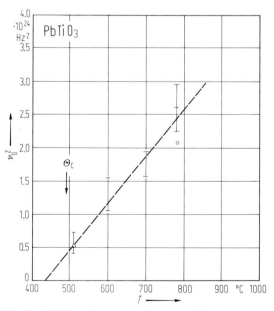

Fig. 51. PbTiO$_3$. Square of the lowest TO phonon frequency at $q=0$ vs. temperature [70S3].

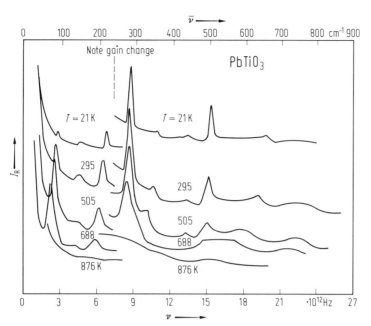

Fig. 52. PbTiO$_3$. Raman intensity vs. Raman frequency (wavenumber) for various temperatures [70T2]. Unpolarized spectra.

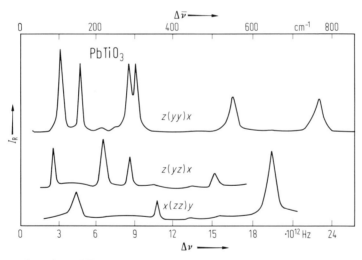

Fig. 53. PbTiO$_3$. Raman intensity at RT vs. Raman shift for various configurations [70T2].

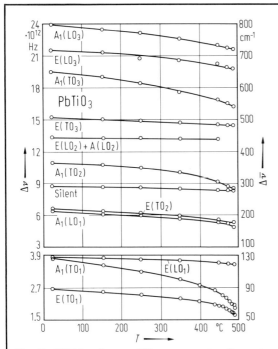

Fig. 54. PbTiO$_3$. Raman shift vs. temperature for various optical modes [73B4].

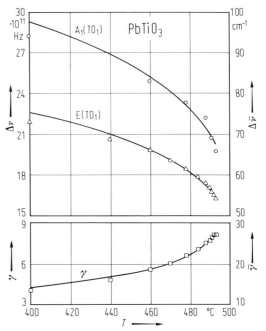

Fig. 55. PbTiO$_3$. Raman shift for the lowest A$_1$ and E modes (upper part) and damping constant of the lowest E mode vs. temperature. Solid lines are calculated using an empirical model [73B4].

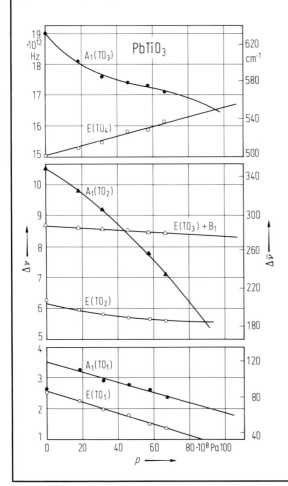

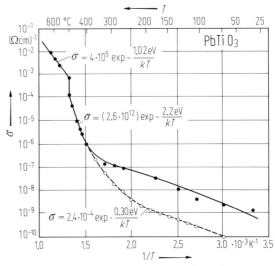

Fig. 57. PbTiO$_3$. Conductivity vs. (reciprocal) temperature. Open circles: dc conductivity, full circles: conductivity measured at 1 kHz which includes the dielectric loss [70R2].

◄

Fig. 56. PbTiO$_3$. Raman shift vs. pressure at RT. Solid lines represent least-square fits to the data [75C].

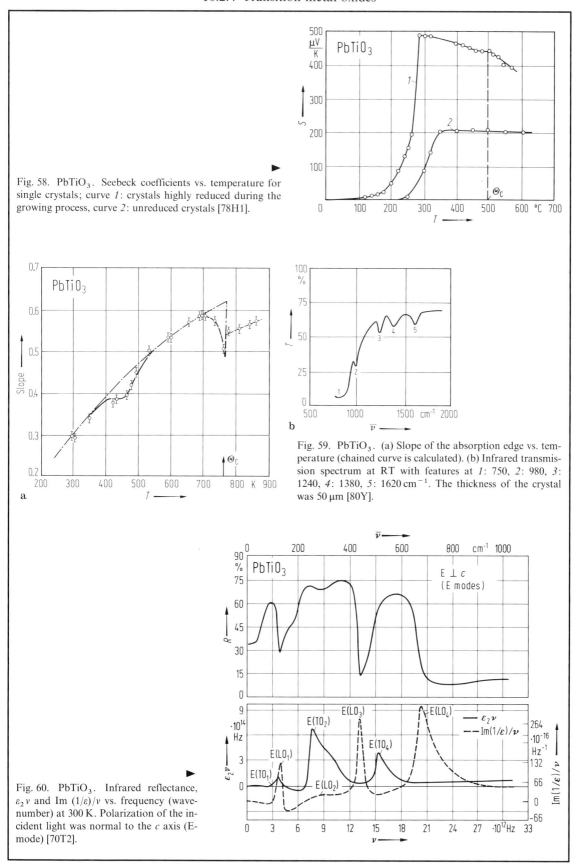

Fig. 58. $PbTiO_3$. Seebeck coefficients vs. temperature for single crystals; curve *1*: crystals highly reduced during the growing process, curve *2*: unreduced crystals [78H1].

Fig. 59. $PbTiO_3$. (a) Slope of the absorption edge vs. temperature (chained curve is calculated). (b) Infrared transmission spectrum at RT with features at *1*: 750, *2*: 980, *3*: 1240, *4*: 1380, *5*: 1620 cm^{-1}. The thickness of the crystal was 50 μm [80Y].

Fig. 60. $PbTiO_3$. Infrared reflectance, $\varepsilon_2 v$ and Im $(1/\varepsilon)/v$ vs. frequency (wavenumber) at 300 K. Polarization of the incident light was normal to the *c* axis (E-mode) [70T2].

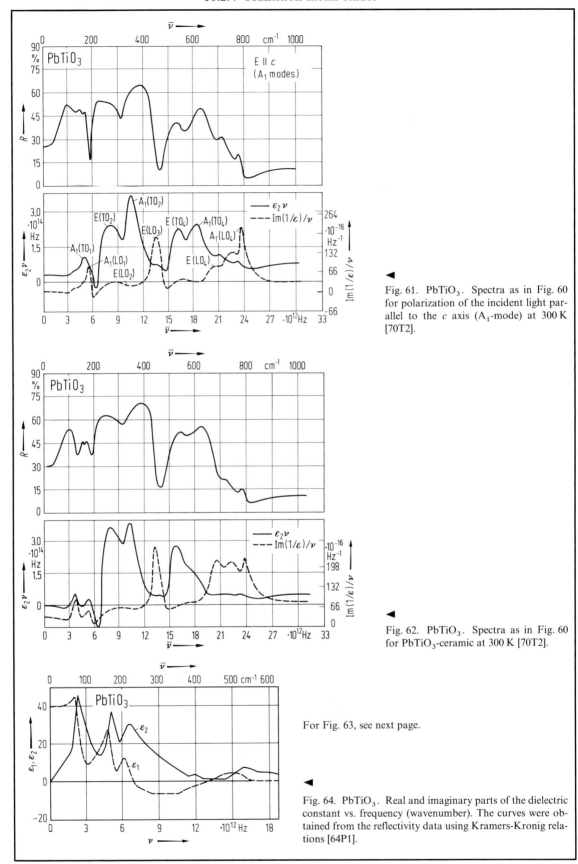

Fig. 61. PbTiO$_3$. Spectra as in Fig. 60 for polarization of the incident light parallel to the c axis (A$_1$-mode) at 300 K [70T2].

Fig. 62. PbTiO$_3$. Spectra as in Fig. 60 for PbTiO$_3$-ceramic at 300 K [70T2].

For Fig. 63, see next page.

Fig. 64. PbTiO$_3$. Real and imaginary parts of the dielectric constant vs. frequency (wavenumber). The curves were obtained from the reflectivity data using Kramers-Kronig relations [64P1].

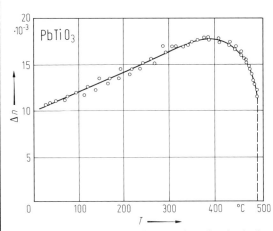

Fig. 63. PbTiO$_3$. Difference between the refractive indices in a and c direction vs. temperature. Na light was used [56S].

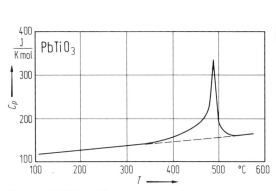

Fig. 65. PbTiO$_3$. Molar heat capacity vs. temperature [51S4].

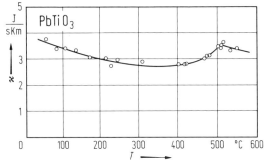

Fig. 66. PbTiO$_3$. Thermal conductivity vs. temperature for a ceramic sample [60Y].

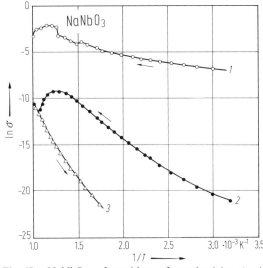

Fig. 67. NaNbO$_3$. Logarithm of conductivity (σ in Ω^{-1} cm^{-1}) vs. reciprocal temperature for single crystals reduced by heating in a vacuum at 800 °C (curve 1) and 600 °C (curve 2) and oxidized by heating in an air atmosphere at normal pressure (curve 3) [79H].

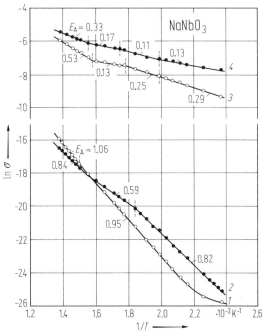

Fig. 68. NaNbO$_3$. Logarithm of conductivity (σ in Ω^{-1} cm^{-1}) vs. reciprocal temperature for samples in four different states of the crystal: nonreduced (curves 1 and 2), reduced (curves 3 and 4), unpolarized (curves 1, 3) and polarized (curves 2, 4). Numerical data are activation energies in eV [79H].

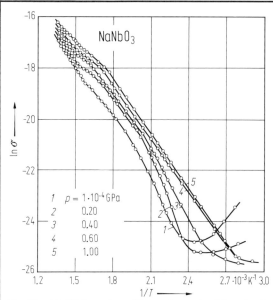

Fig. 69. NaNbO$_3$. Logarithm of conductivity (σ in Ω^{-1} cm^{-1}) vs. reciprocal temperature for selected hydrostatic pressure values [81P].

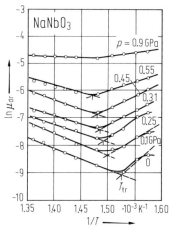

Fig. 71. NaNbO$_3$. Logarithm of drift mobility vs. reciprocal temperature for various pressures [80P]. Orientation not specified. μ_{dr} in cm^2/Vs.

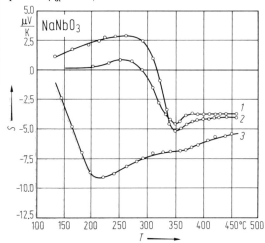

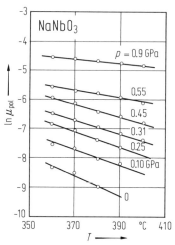

Fig. 70. NaNbO$_3$. Logarithm of small polaron mobility (μ in cm^2/Vs) vs. temperature for various pressures [80P]. Orientation not specified.

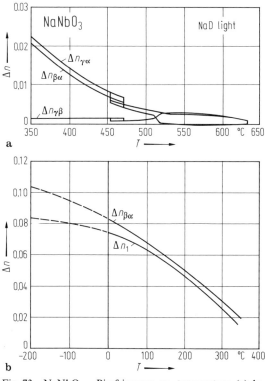

Fig. 73. NaNbO$_3$. Birefringence vs. temperature (a) between 350 °C and 650 °C and (b) below 350 °C. $\Delta n_{\gamma\alpha} = n_\gamma - n_\alpha$ etc., $\Delta n_1 = n_\gamma - n_{(\alpha + \beta)/2}$ = refractive index along the 45° bisector of the axes of the ellipse in the $\alpha\beta$ plane of the indicatrix [55C].

◄

Fig. 72. NaNbO$_3$. Seebeck coefficient vs. temperature; curve 1 for $p=0$, 2: 0.1 GPa, 3: 0.9 GPa [80P]. Orientation not specified.

For Fig. 74, see next page.

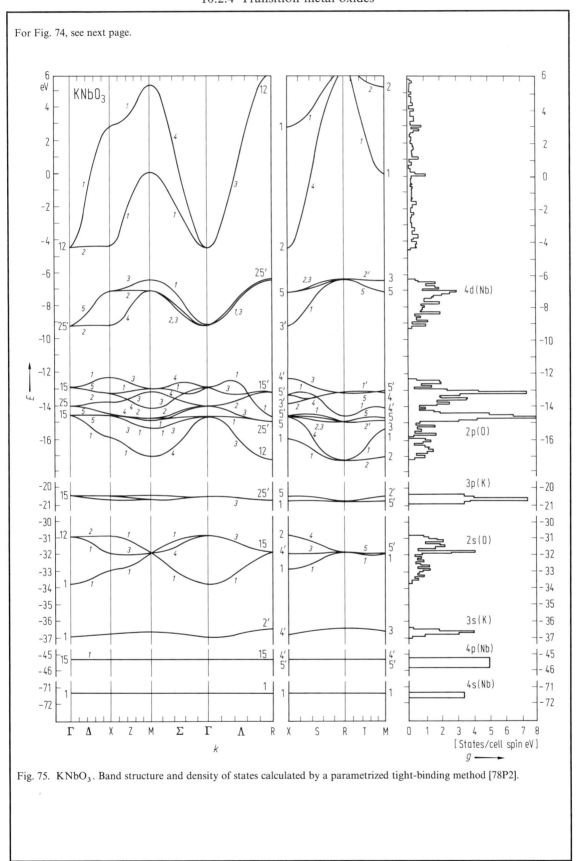

Fig. 75. KNbO$_3$. Band structure and density of states calculated by a parametrized tight-binding method [78P2].

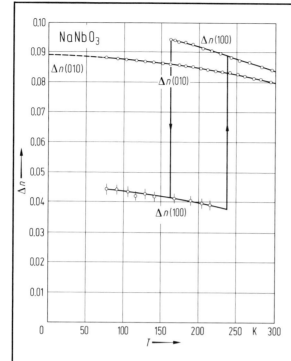

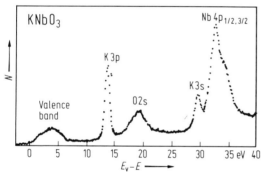

Fig. 76. KNbO$_3$. XPS spectrum (counts per channel vs. binding energy). Incident X-ray: AlKα, resolution 0.55 eV [78P2].

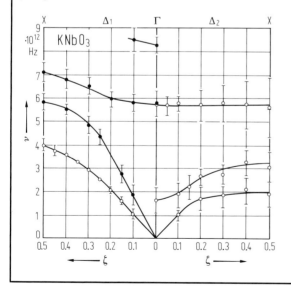

Fig. 74. NaNbO$_3$. Birefringence vs. temperature below RT. Measurements were made on planes correlated to the pseudo-cubic axes ($\lambda = 589$ nm) [78S2].

◀

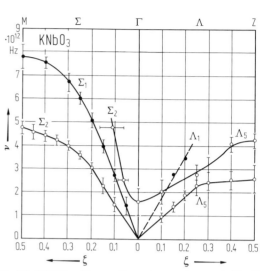

Fig. 78. KNbO$_3$. Phonon dispersion spectrum as in Fig. 77 for the Σ and Λ axes [79F1].

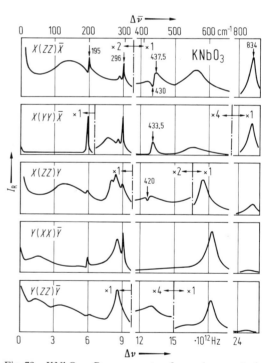

Fig. 79. KNbO$_3$. Raman spectra for various scattering geometries at RT. Peak frequencies are indicated in units of cm^{-1} [76Q].

◀

Fig. 77. KNbO$_3$. Phonon dispersion spectrum (frequency vs. reduced wave vector coordinate) for the Δ_1 and Δ_2 symmetry at 245 °C (tetragonal phase) [79F1].

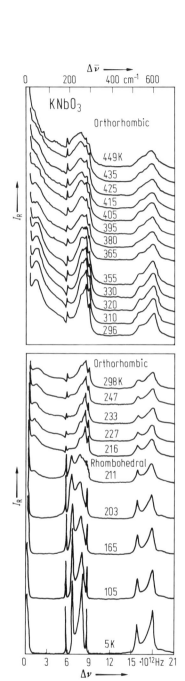

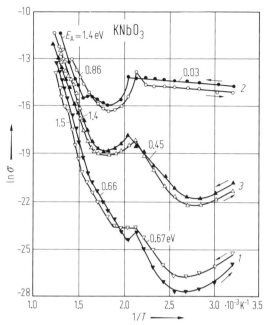

Fig. 81. KNbO$_3$. Logarithm of conductivity (σ in $10^{-8}\,\Omega^{-1}\,cm^{-1}$) vs. reciprocal temperature. Curve 1: full transparent, nonreduced crystal; 2: crystal strongly reduced, nontransparent; 3: crystal partly reduced. Numerical data: activation energies in eV [79H].

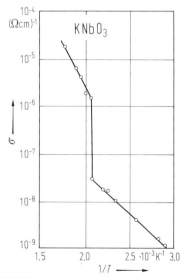

Fig. 80. KNbO$_3$. Temperature dependence of Raman spectra (intensity vs. Raman shift; parameter: temperature). The lowest E(TO$_1$) branch is the temperature dependent ferroelectric mode while E(TO$_2$), E(TO$_3$), and E(TO$_4$) are about 200, 280···290 and 530···550 cm^{-1}, respectively [76P3].

Fig. 82. KNbO$_3$. Conductivity vs. reciprocal temperature. The conductivity was measured by two terminal methods with silver electrodes; applied field strength 1 MV m^{-1}, single crystal of 100 μm thickness [62G].

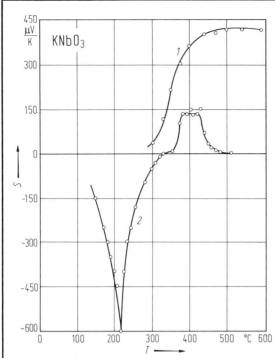

Fig. 83 KNbO$_3$. Seebeck coefficient vs. temperature for a nonreduced (curve *1*) and reduced (*2*) crystal [79H].

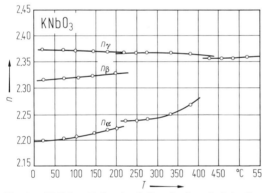

Fig. 85. KNbO$_3$. Refractive indices in the principle directions vs. temperature at $\lambda = 546$ nm [70W2].

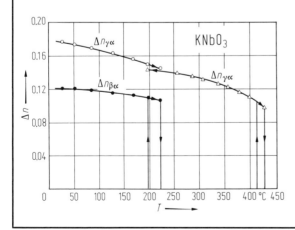

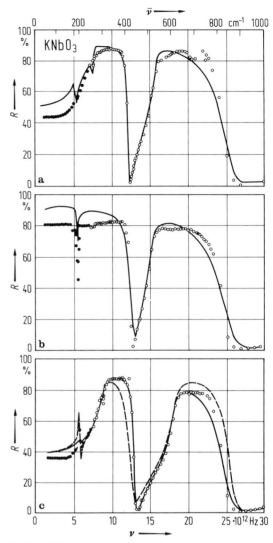

Fig. 84. KNbO$_3$. Reflectivity vs. frequency (wavenumber) at RT. Polarization of the radiation is along x in (a), y in (b) and z in (c). Open circles represent accurate data points. Full circles represent data with uncertain normalization errors. Continuous curves are derived from reflectivity analysis, dashed curve is derived from Raman data [76B2].

◄

Fig. 86. KNbO$_3$. Differences of principal refractive indices vs. temperature at $\lambda = 546$ nm [70W2]. $\Delta n_{\gamma\alpha} = n_\gamma - n_\alpha$, $\Delta n_{\beta\alpha} = n_\beta - n_\alpha$.

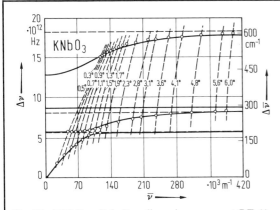

Fig. 87. KNbO$_3$. Polariton dispersion curves at RT (A$_1$ polariton frequency vs. polariton wavenumber). Circles and solid curves are experimental and calculated, respectively. Horizontal broken lines indicate TO modes. Parameter: scattering angle [75F2].

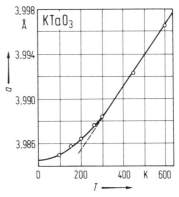

Fig. 88. KTaO$_3$. Lattice parameter vs. temperature [73S3].

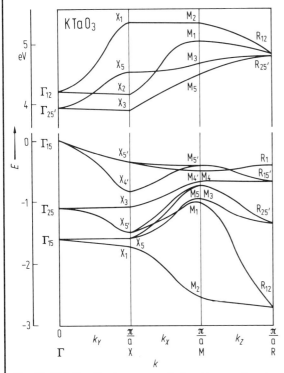

Fig. 89. KTaO$_3$. Band structure calculated by a semiempirical LCAO method [73E].

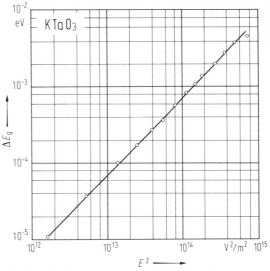

Fig. 90. KTaO$_3$. Increase of the fundamental energy gap vs. square of electric field [67F1].

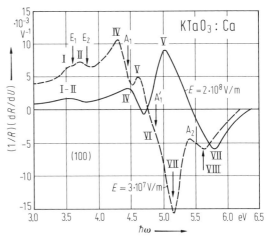

Fig. 91. KTaO$_3$. Electroreflectance vs. photon energy for the (100) surface of a Ca doped sample with donor concentration of $3.8 \cdot 10^{18}$ cm^{-3} (Parameter: field strength E at the (100) interface) [67F1]. U: applied voltage.

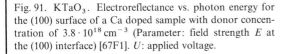

459

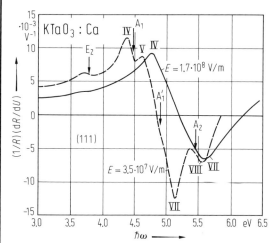

Fig. 92. $KTaO_3$. Electroreflectance vs. photon energy for the (111) surface of a Ca doped sample with donor concentration of $3.5 \cdot 10^{18}$ cm^{-3} [67F1]. Cf. Fig. 91.

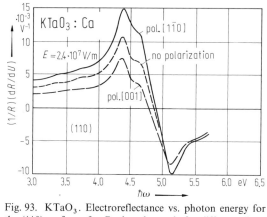

Fig. 93. $KTaO_3$. Electroreflectance vs. photon energy for the (110) surface of a Ca doped sample for different directions of the electric vector of the incident polarized light. E: interface field; cf. Figs. 91, 92 [67F1].

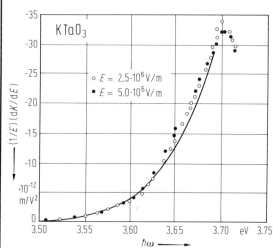

Fig. 94. $KTaO_3$. Electroabsorption vs. photon energy for two values of the electric field. The solid curve gives the slope of the absorption edge $dK/d(\hbar\omega)$ in arbitrary units [67F1].

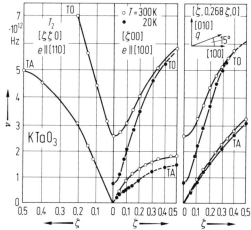

Fig. 95. $KTaO_3$. Phonon dispersion curves (frequency vs. reduced wave vector coordinate) [72C3].

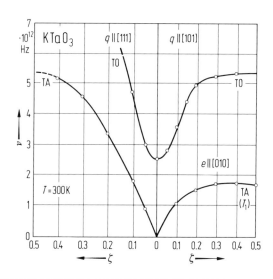

Fig. 96. $KTaO_3$. Phonon dispersion curves as in Fig. 95 for other directions in the Brillouin zone [72C3].

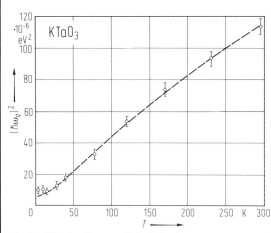

Fig. 97. $KTaO_3$. Square of the TO phonon energy at $q = 0$ (ferroelectric mode) vs. temperature [67S].

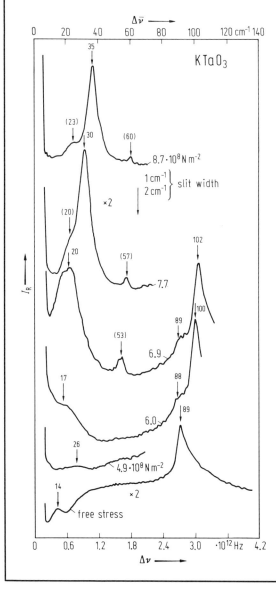

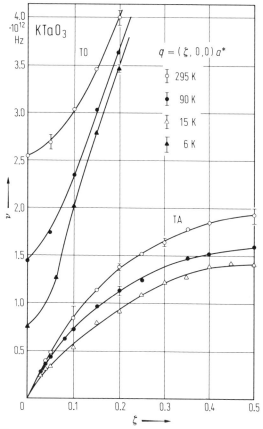

Fig. 98. $KTaO_3$. Frequency of the TA and soft TO phonons vs. reduced wave coordinate along $(\zeta, 0, 0)$ for various temperatures [70A].

For Fig. 99, see next page.

◀

Fig. 100. $KTaO_3$. Raman intensity vs. Raman shift at 2 K for different stress load along the $(010)(Y)$ axis. Peak wavenumbers are given in cm^{-1}. Scattering geometry: $XZ(YY)XZ$ [74Y].

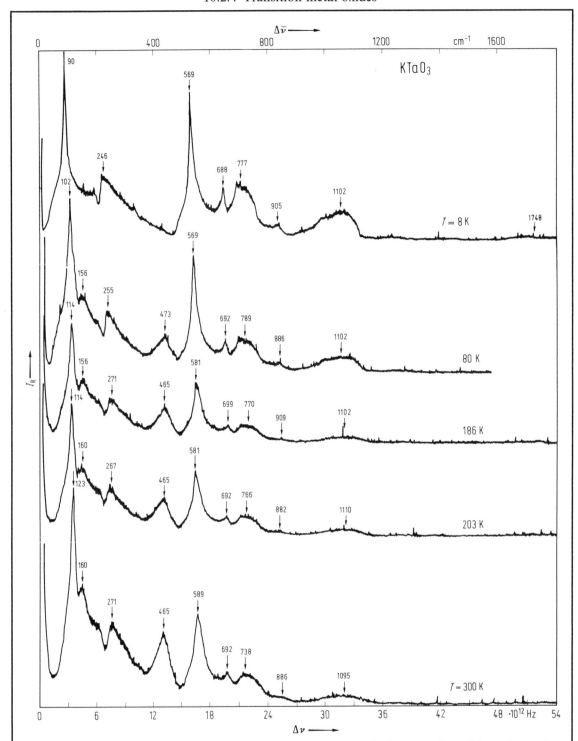

Fig. 99. KTaO₃. Raman intensity vs. Raman shift for various temperatures. Peak wavenumbers of the modes are given in cm⁻¹. Polarization geometry: $X(ZZ)Y$ [67N].

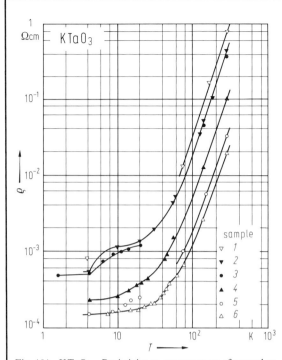

Fig. 101. KTaO₃. Resistivity vs. temperature of several reduced samples with different carrier concentrations from $3.5 \cdot 10^{17}\,cm^{-3}$ (sample *1*) to $1.3 \cdot 10^{19}\,cm^{-3}$ (sample *6*) [65W].

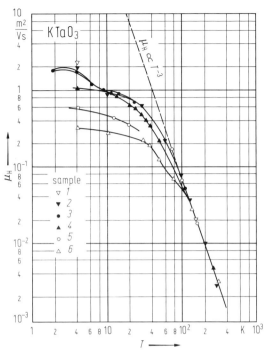

Fig. 102. KTaO₃. Hall mobility vs. temperature for the samples of Fig. 101 [65W].

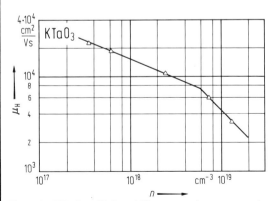

Fig. 103. KTaO₃. Hall mobility vs. carrier concentration at $T = 4.2\,K$ [65W].

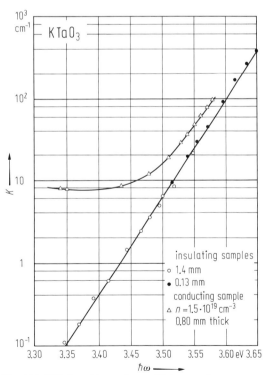

Fig. 104. KTaO₃. Absorption coefficient vs. photon energy near the interband edge for three samples [65W].

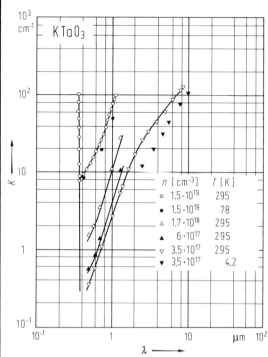

Fig. 105. KTaO$_3$. Absorption coefficient vs. wavelength for various carrier concentrations and temperatures [65W].

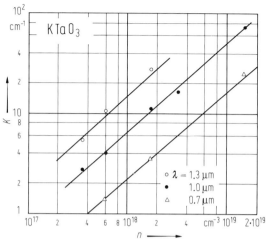

Fig. 106. KTaO$_3$. Absorption coefficient vs. carrier concentration for three wavelengths [65W].

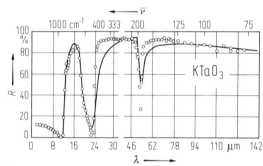

Fig. 107. KTaO$_3$. Reflectivity vs. wavelength (wavenumber) at RT. Note the change of scale at $\lambda = 30\ \mu m$ [63M3].

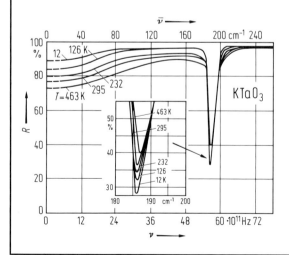

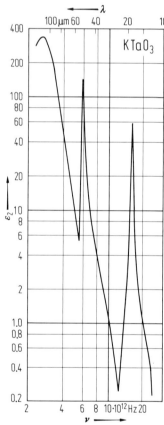

Fig. 109. KTaO$_3$. Imaginary part of the dielectric constant vs. frequency at RT calculated from reflectivity data [63M3].

Fig. 108. KTaO$_3$. Reflectivity vs. frequency (wavenumber) at various temperatures [67P2].

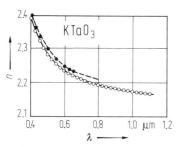

Fig. 110. KTaO$_3$. Refractive index vs. wavelength at 300 K according to [76F3] (solid curve) and [65W] (broken curve).

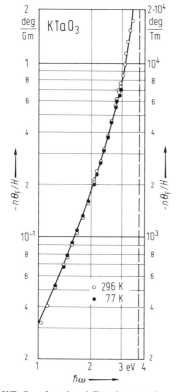

Fig. 111. KTaO$_3$. Interband Faraday rotation vs. photon energy at two temperatures. Solid line: theoretical fit. Dashed line: band gap energy E_g (3.77 eV at 296 K, 3.79 eV at 77 K) [67B1]. n: refractive index, θ_F: Faraday rotation, H: magnetic field strength.

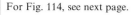

For Fig. 114, see next page.

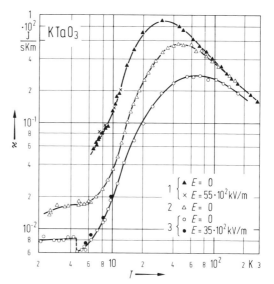

Fig. 112. KTaO$_3$. Thermal conductivity vs. temperature for three samples 1···3 [68S].

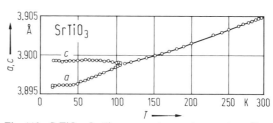

Fig. 113. SrTiO$_3$. Lattice parameters vs. temperature. For comparison between phases I and II the superlattice in phase II below 105 K is ignored [73O].

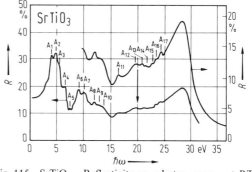

Fig. 115. SrTiO$_3$. Reflectivity vs. photon energy at RT. Data in the energy region below 10 eV from [65C], other data from [78B2]. The vertical arrow marks the expected onset of Sr 4p to conduction band transitions. For peak energies, cf. the tables.

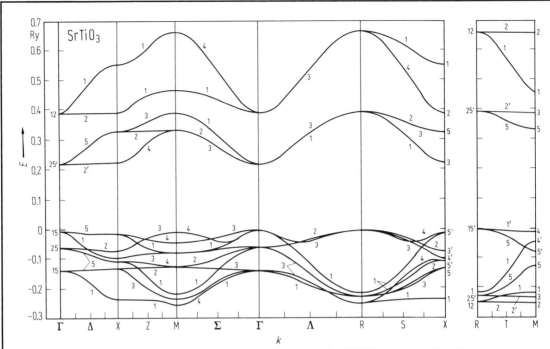

Fig. 114. SrTiO$_3$. Band structure calculated by an adjusted LCAO method [72M6]. 1 Ry $\cong$ 13.606 eV.

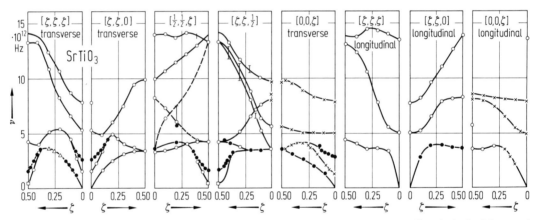

Fig. 116. SrTiO$_3$. Phonon dispersion curves (phonon frequency vs. reduced wave vector coordinate) obtained from various sources [72S4]. Open circles: $T = 90$ K, full circles: $T = 297$ K [72S4]; crosses: $T = 90$ K [64C]; squares: $T = 120$ K [69S2]; triangles: $T = 300$ K [69C].

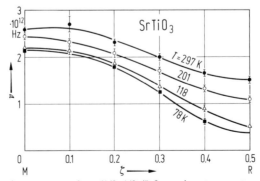

Fig. 117. SrTiO$_3$. Phonon dispersion spectrum along (1/2, 1/2, ζ) for various temperatures [72S4].

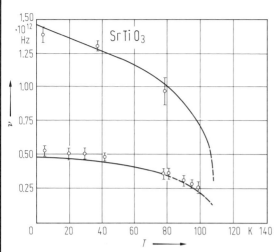

Fig. 118. SrTiO₃. Frequency of the R-phonon ($q=(1.5,$ 1.5, 2.5)) vs. temperature [69S2].

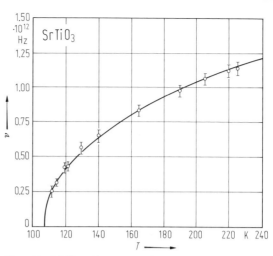

Fig. 119. SrTiO₃. Frequency of the R_{25} (Γ_{25})-phonon vs. temperature [69S2].

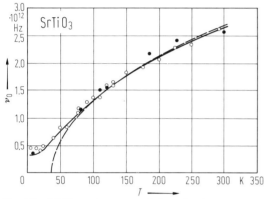

Fig. 120. SrTiO₃. Frequency of the soft phonon at $q=0$ vs. temperature. Full circles: Raman scattering data [67O]; open circles: neutron inelastic scattering [69Y2]. Solid line: $194.4/\varepsilon^{1/2}$. Dashed line: $0.677\,(T-T_0)^{1/2}$; ($T_0=38\,\mathrm{K}$) [69Y2].

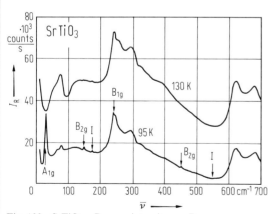

Fig. 122. SrTiO₃. Raman intensity vs. Raman wavenumber in diagonal $X(ZZ)Y$ configuration for two temperatures above and below the cubic-tetragonal transition. The arrows indicate first-order peaks [79T2]. I: impurity induced scattering.

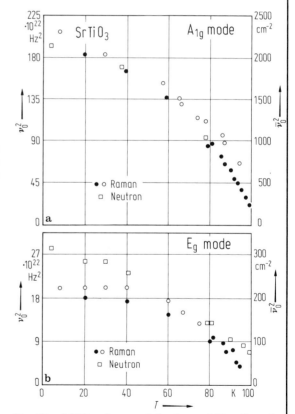

Fig. 121. SrTiO₃. Square of the A_{1g} and E_g soft mode frequencies vs. temperature. Solid circles and open circles are the results from Raman spectra in [74W1] and [68F], respectively. Squares are the result of neutron diffraction experiments in [69S2].

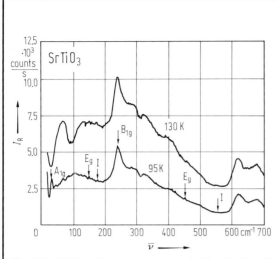

Fig. 123. SrTiO$_3$. Raman spectrum as in Fig. 122 for off-diagonal $X(ZX)Y$ configuration [79T2].

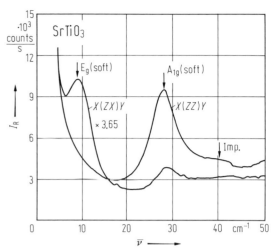

Fig. 125. SrTiO$_3$. Raman spectra as in Fig. 122 for a clear sample at 95 K for two configurations [79T2].

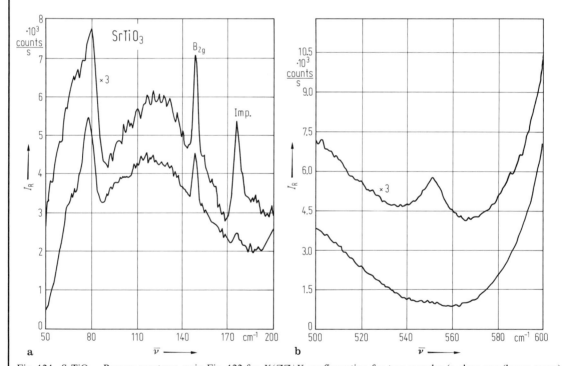

Fig. 124. SrTiO$_3$. Raman spectrum as in Fig. 122 for $X(ZZ)Y$ configuration for two samples (a clear one (lower curve) and a brown one (upper curve)). (a) wavenumber region $50\cdots200$ cm^{-1}, (b) wavenumber region $500\cdots600$ cm^{-1} [79T2].

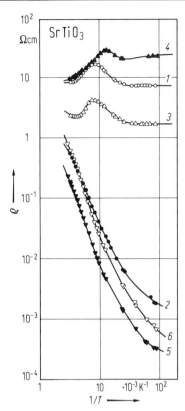

Fig. 126. $SrTiO_3$. Resistivity vs. reciprocal temperature for various differently reduced and doped samples ($n = 1.22 \cdot 10^{17}\,cm^{-3}$ (sample *1*) to $6.0 \cdot 10^{18}\,cm^{-3}$ (sample *5*)) [75L3].

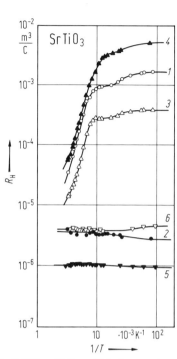

Fig. 127. $SrTiO_3$. Hall coefficient vs. reciprocal temperature for the samples of Fig. 126 [75L3].

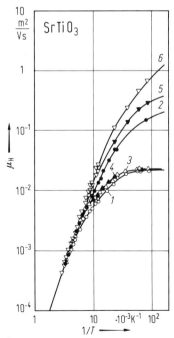

Fig. 128. $SrTiO_3$. Hall mobility vs. reciprocal temperature for the samples of Fig. 126 [75L3].

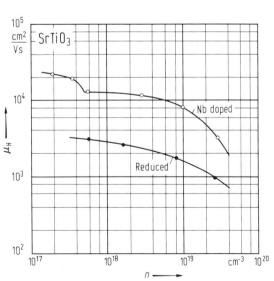

Fig. 129. $SrTiO_3$. Hall mobility vs. electron concentration for a semiconducting single crystal at $T = 2\,K$ [67T].

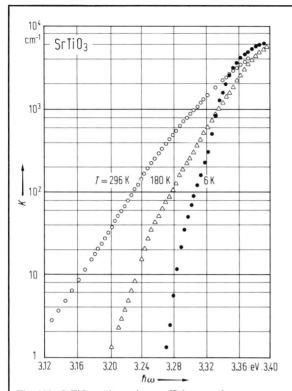

Fig. 130. SrTiO$_3$. Absorption coefficient vs. photon energy at three temperatures (multidomain crystal) [72R].

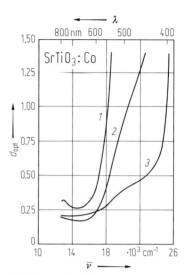

Fig. 131. SrTiO$_3$. Optical density vs. wavenumber (wavelength) for single crystals doped with Co: $6 \cdot 10^{-3}$ wt% (curve 1), $2.5 \cdot 10^{-3}$ wt% (2), $1.2 \cdot 10^{-3}$ wt% (3) [81K2].

For Fig. 133, see next page.

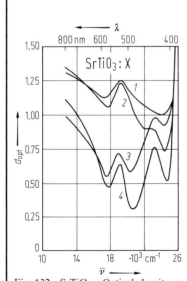

Fig. 132. SrTiO$_3$. Optical density vs. wavenumber for single crystals doped with X = Sm (curve 1), Pr (2), Nd (3) and Er (4) [81K2].

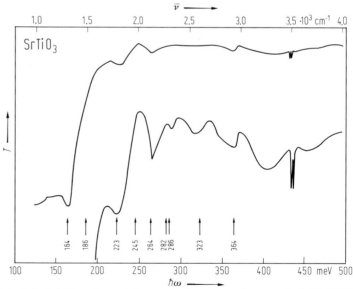

Fig. 134. SrTiO$_3$. Transmission vs. photon energy (wavenumber) ($T = 30$ K). The lower curve is magnified by a factor of 5. The arrows indicate the energies of the defect absorption and of its phonon replicas [81B1].

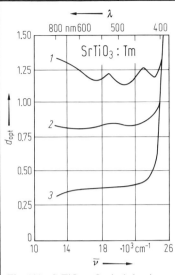

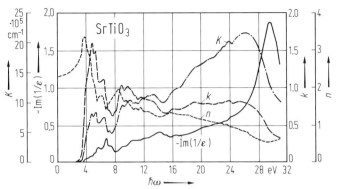

Fig. 137. SrTiO$_3$. Loss function (Im ε^{-1}), refractive index (n), extinction coefficient (k), and absorption coefficient (K) vs. photon energy at RT [78B2].

Fig. 133. SrTiO$_3$. Optical density vs. wavenumber (wavelength) for single crystals with various Tm-concentrations: $5.2 \cdot 10^{-2}$ wt% (curve 1), $3.8 \cdot 10^{-2}$ wt% (2), $2.5 \cdot 10^{-2}$ wt% (3) [81K2].

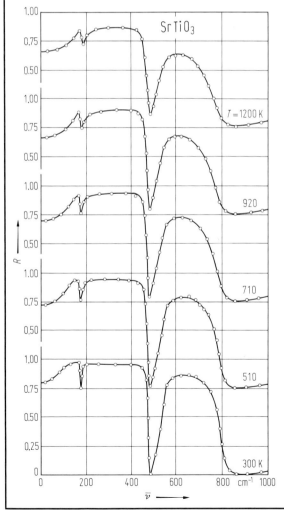

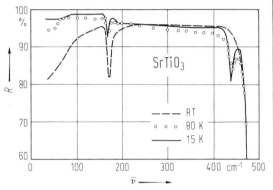

Fig. 136. SrTiO$_3$. Reflectivity vs. wavenumber for three temperatures (unpolarized radiation) [82G].

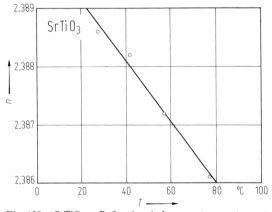

Fig. 138. SrTiO$_3$. Refractive index vs. temperature at 632.8 nm [82S]. Solid line: linear least square fit.

Fig. 135. SrTiO$_3$. Reflectivity vs. wavenumber for various temperatures showing F$_{1u}$ modes [80S].

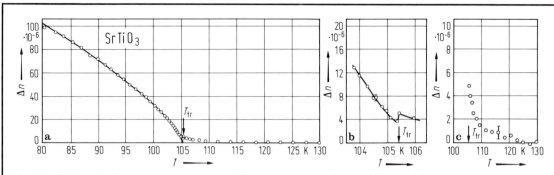

Fig. 139. SrTiO₃. Birefringence vs. temperature. Figs. b and c show details of Fig. a with a vertically expanded scale [72C2].

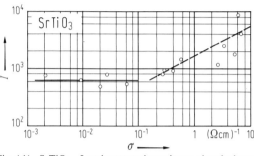

Fig. 141. SrTiO₃. Luminescence intensity vs. electrical conductivity at RT of various ceramics. The intensities were calculated from the areas of the plotted luminescence bands [81I].

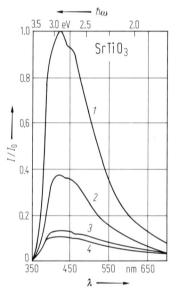

Fig. 140. SrTiO₃. Luminescence intensity vs. wavelength (photon energy) at RT of ceramics sintered in air at 1450°C and afterwards reduced in a dry hydrogen stream at 1100°C for $1/2$ h and doped with 2 mol% La, curve *1*; 1 mol% La, *2*; 0.3 mol% La, *3*; curve *4* is an undoped sample [81I].

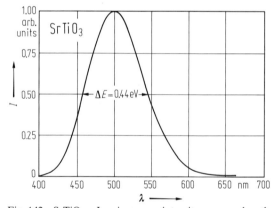

Fig. 142. SrTiO₃. Luminescence intensity vs. wavelength during X-ray irradiation at 15 K [82A2].

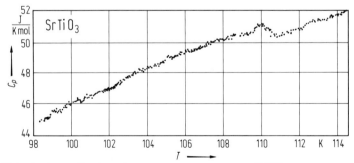

Fig. 143. SrTiO₃. Molar heat capacity at constant pressure vs. temperature [74F1].

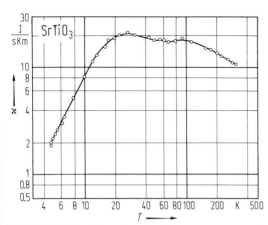

Fig. 144. SrTiO$_3$. Thermal conductivity vs. temperature for a ceramic sample [65S5].

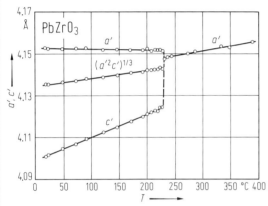

Fig. 146. PbZrO$_3$. Unit cell parameters vs. temperature for ceramic samples. A pseudo-tetragonal unit cell was assumed. $a' \approx a_{mon} = c_{mon}$, $c' \approx b_{mon}$ (mon = monoclinic) [52S].

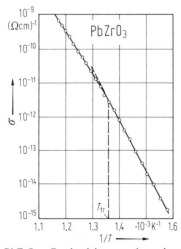

Fig. 148. PbZrO$_3$. Conductivity vs. reciprocal temperature for a ceramic sample [72U].

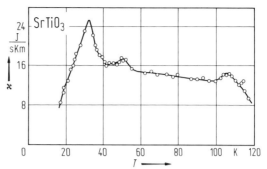

Fig. 145. SrTiO$_3$. Thermal conductivity vs. temperature [66H2].

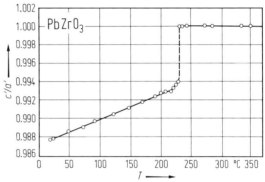

Fig. 147. PbZrO$_3$. c'/a' ratio vs. temperature (see Fig. 146) [53S].

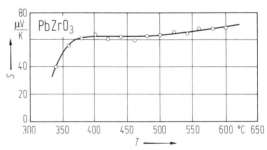

Fig. 149. PbZrO$_3$. Seebeck coefficient vs. temperature for a ceramic sample [72U].

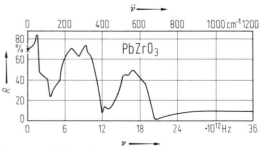

Fig. 150. PbZrO$_3$. Reflectance vs. frequency for a ceramic sample at room temperature [65P].

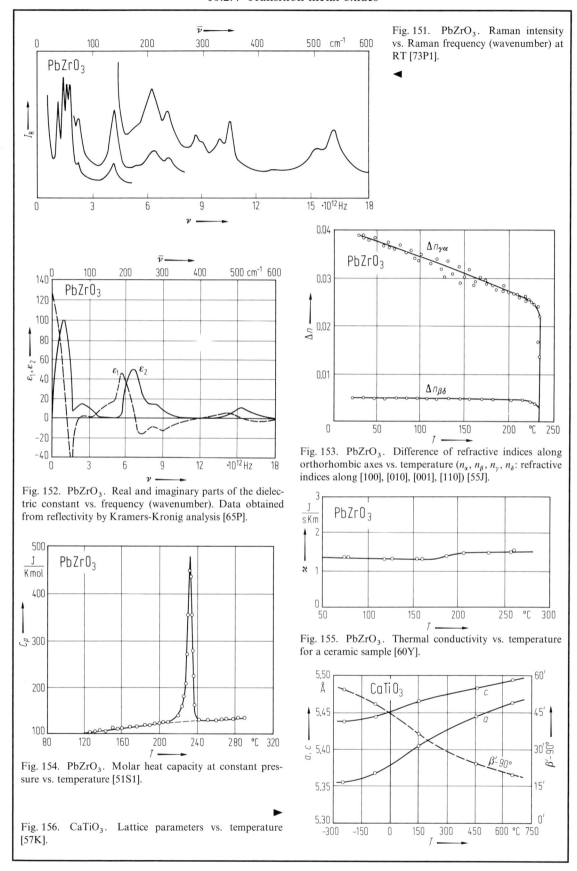

Fig. 151. PbZrO$_3$. Raman intensity vs. Raman frequency (wavenumber) at RT [73P1].

Fig. 152. PbZrO$_3$. Real and imaginary parts of the dielectric constant vs. frequency (wavenumber). Data obtained from reflectivity by Kramers-Kronig analysis [65P].

Fig. 153. PbZrO$_3$. Difference of refractive indices along orthorhombic axes vs. temperature (n_α, n_β, n_γ, n_δ: refractive indices along [100], [010], [001], [110]) [55J].

Fig. 154. PbZrO$_3$. Molar heat capacity at constant pressure vs. temperature [51S1].

Fig. 155. PbZrO$_3$. Thermal conductivity vs. temperature for a ceramic sample [60Y].

Fig. 156. CaTiO$_3$. Lattice parameters vs. temperature [57K].

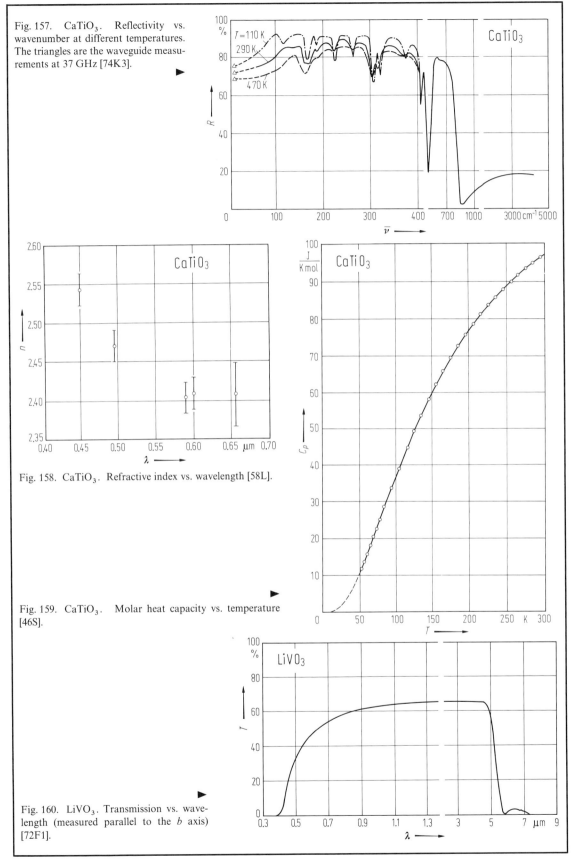

Fig. 157. CaTiO$_3$. Reflectivity vs. wavenumber at different temperatures. The triangles are the waveguide measurements at 37 GHz [74K3].

Fig. 158. CaTiO$_3$. Refractive index vs. wavelength [58L].

Fig. 159. CaTiO$_3$. Molar heat capacity vs. temperature [46S].

Fig. 160. LiVO$_3$. Transmission vs. wavelength (measured parallel to the b axis) [72F1].

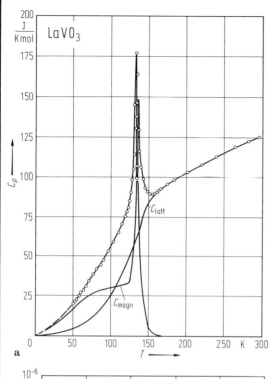

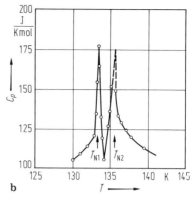

Fig. 161. LaVO$_3$. (a) Heat capacity vs. temperature. The points represent the measured values and the continuous curves are the magnetic C_{magn} and the lattice C_{latt} components. Fig. b shows the T_N range on an expanded scale (T_{N1}: structure transformation temperature, T_{N2}: antiferromagnetic ordering temperature). [74B2].

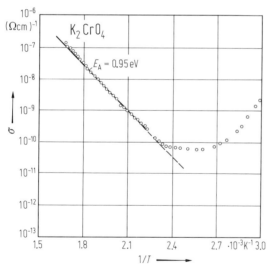

Fig. 162. K$_2$CrO$_4$. Conductivity vs. reciprocal temperature (sample heated in air) [73C].

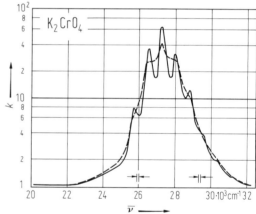

Fig. 163. K$_2$CrO$_4$. Extinction coefficient vs. wavenumber at liquid nitrogen temperature (solid line) and at RT (broken line) [76I].

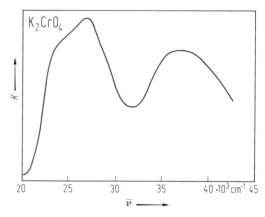

Fig. 164. K$_2$CrO$_4$. Absorption coefficient vs. wavenumbers for a pressed sample [78S1].

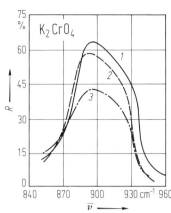

Fig. 165. K_2CrO_4. Reflectance vs. wavenumber at various temperatures T [K]: curve 1: 170, 2: 375, 3: 600 [81T].

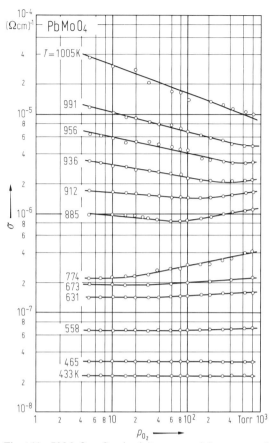

Fig. 166. $PbMoO_4$. Conductivity vs. partial pressure of oxygen at various temperatures [75L1].

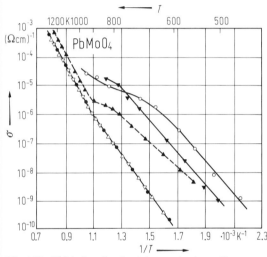

Fig. 167. $PbMoO_4$. Conductivity vs. (reciprocal) temperature. ●●: undoped crystal, equilibrated; ○○: Na doped crystal; ▲▲: nitrogen-annealed crystal, initial run; ▼▼: K doped crystal; ▽▽: Y doped crystal; △△: Bi doped crystal [79G].

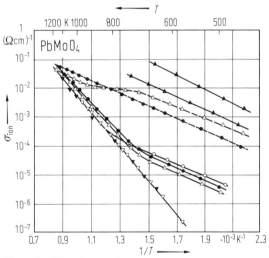

Fig. 168. $PbMoO_4$. Ionic conductivity times temperature vs. (reciprocal) temperature; (dashed line, full circles) undoped, as grown crystal; (solid line, full circles) undoped, equilibrated crystal; (open circles) oxygen annealed crystal; (dashed line, △△) nitrogen annealed crystal, initial run; (solid line, △△) nitrogen annealed crystal, cooling run; (▲▲) Na doped crystal, two concentrations; (▽▽) Bi doped crystal; (▼▼) Y doped crystal [79G].

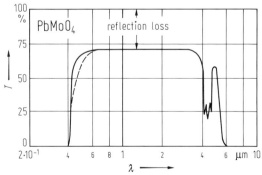

Fig. 169. $PbMoO_4$. Transmission vs. wavelength (dashed curve is for a sample containing impurities) [71C3].

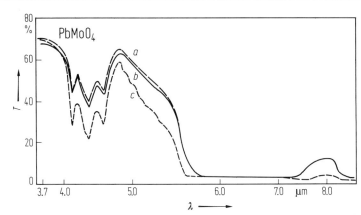

Fig. 170. PbMoO$_4$. Transmission vs. wavelength, curve a: pale yellow sample, oxygen annealed; b: pale yellow sample, unannealed; c: dark yellow, unannealed [74L].

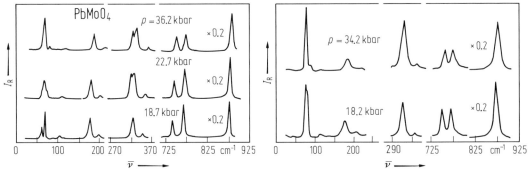

Fig. 171. PbMoO$_4$. Raman intensity vs. wavenumber at various pressures at $T = 80$ K (left hand figure) and at $T = 300$ K (right hand figure) [77G].

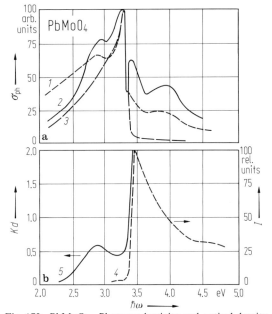

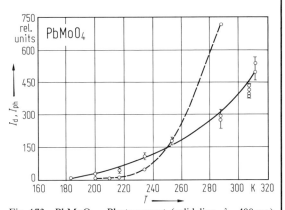

Fig. 172. PbMoO$_4$. Photoconductivity and optical density ($K \cdot d$) vs. photon energy at $T = 80$ K. Curve 1: Maximum values, as-grown crystals; 2: steady-state values, as-grown crystals; 3: annealed crystals; 4: excitation spctrum of the luminescence; 5: absorption spectrum [78G1].

Fig. 173. PbMoO$_4$. Photocurrent (solid line, $\lambda = 400$ nm) and dark current (dashed line) vs. temperature [77B].

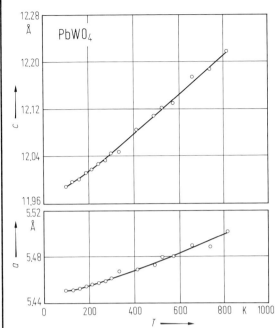

Fig. 174. PbWO$_4$. Lattice parameters vs. temperature [78B1].

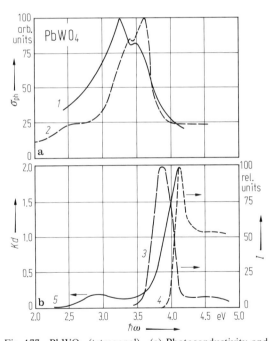

Fig. 177. PbWO$_4$ (tetragonal). (a) Photoconductivity and (b) optical density vs. photon energy at $T = 80$ K. Curve 1: as-grown crystal, 2: annealed crystal, 3: excitation spectrum of the green emission, 4: excitation spectrum of the blue luminescence emission, 5: absorption spectrum [78G1].

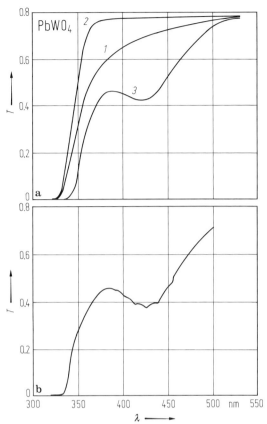

Fig. 175. PbWO$_4$ (tetragonal). Transmission vs. wavelength. (a): Curve 1: $T = 290$ K, crystal grown from Ir crucible in Ar atmosphere; 2: $T = 290$ K, crystal grown from Ir crucible in Ar atmosphere and annealed in air for 48 h; 3: $T = 290$ K, crystal grown from Pt crucible in O$_2$ atmosphere; (b): $T = 4.2$ K, crystal grown from Pt crucible in O$_2$ atmosphere [78O].

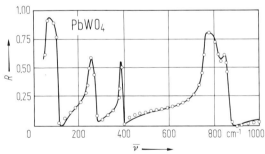

Fig. 176. PbWO$_4$ (tetragonal). Reflectivity vs. wavenumber at RT. $E \parallel c$; circles: best fit obtained with the classical oscillator dispersion equation [76S1].

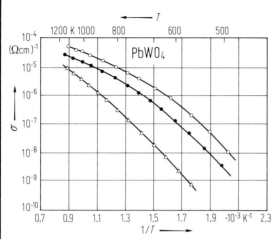

Fig. 178. PbWO$_4$ (tetragonal). Conductivity vs. (reciprocal) temperature (open circles: K doped crystal, full circles: undoped crystal, triangles: Bi doped crystal) [79G].

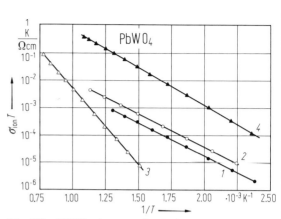

Fig. 179. PbWO$_4$ (tetragonal). Ionic conductivity times temperature vs. reciprocal temperature. Curve 1: undoped, 2: K doped, 3: Bi doped, 4: Y doped crystal [79G].

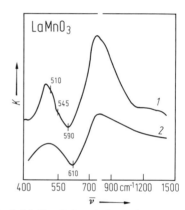

Fig. 181. LaMnO$_3$. Infrared absorption vs. wavenumber. Curve 1: LaMnO$_4$, 2: LaMn$_x^{3+}$Mn$_{1-x}^{4+}$O$_{3+\lambda}$ (x=0.02, 0.05) [74R].

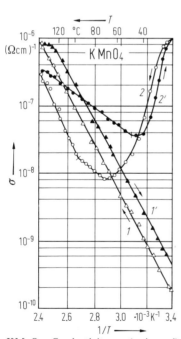

Fig. 180. KMnO$_4$. Conductivity vs. (reciprocal) temperature in dry (curves 1 and 1') and moist (2 and 2') atmosphere. 1, 2: heating; 1', 2': cooling [74K1].

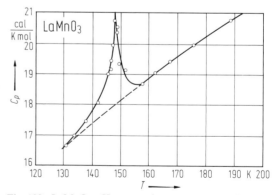

Fig. 182. LaMnO$_3$. Heat capacity vs. temperature [77S2].

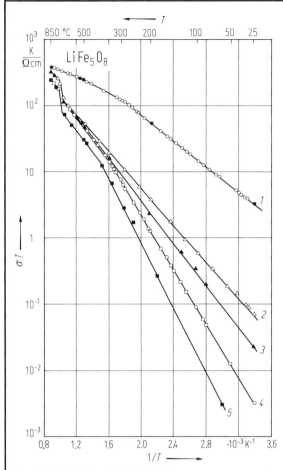

Fig. 183. LiFe$_5$O$_8$. Conductivity times temperature vs. (reciprocal) temperature. Curve 1: dc (●) and ac (○) in Ar, 2: ac in air, 3: dc in air, 4: ac in O$_2$, 5: dc in O$_2$ [77M1].

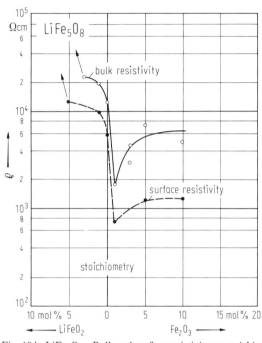

Fig. 184. LiFe$_5$O$_8$. Bulk and surface resistivity vs. stoichiometry of a specimen sintered at 1150°C for 2 h in 1 atm O$_2$ [73B1].

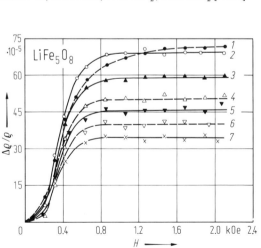

Fig.185. LiFe$_5$O$_8$. Transverse magnetoresistance vs. magnetic field at various temperatures T [K]: Curve 1: 290, 2: 330, 3: 370, 4: 397, 5: 418, 6: 432, 7: 476 [72S1].

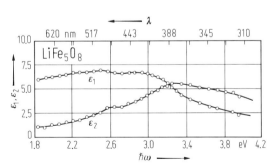

Fig. 186. LiFe$_5$O$_8$. Real and imaginary parts of the dielectric constant vs. photon energy (wavelength) at RT [74M1].

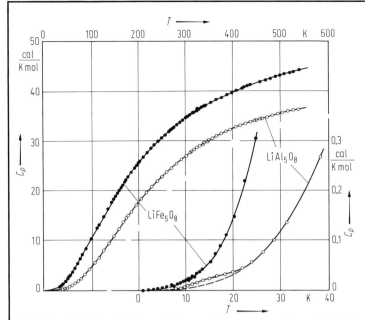

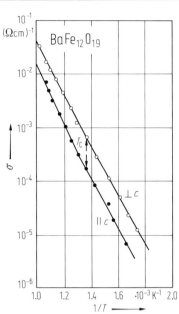

Fig. 187. LiFe$_5$O$_8$, LiAl$_5$O$_8$. Heat capacity vs. temperature. Full circles: LiFe$_5$O$_8$, open circles: LiAl$_5$O$_8$ [75V].

Fig. 188. BaFe$_{12}$O$_{19}$. Conductivity vs. reciprocal temperature measured parallel and perpendicular to the c axis [73B2]. T_C: magnetic Curie temperature.

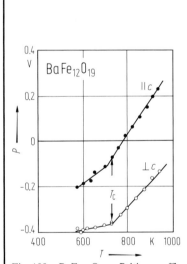

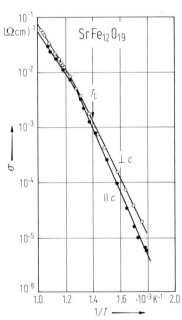

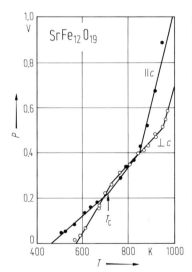

Fig. 189. BaFe$_{12}$O$_{19}$. Peltier coefficient $P = S \cdot T$ vs. temperature measured parallel and perpendicular to the c axis [73B2].

Fig. 190. SrFe$_{12}$O$_{19}$. Conductivity vs. reciprocal temperature measured parallel and perpendicular to the c axis [73B2].

Fig. 191. SrFe$_{12}$O$_{19}$. Peltier coefficient $P = S \cdot T$ vs. temperature measured parallel and perpendicular to the c axis [73B2].

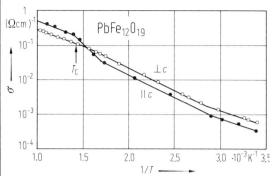

Fig. 192. $PbFe_{12}O_{19}$. Conductivity vs. reciprocal temperature measured parallel and perpendicular to the c axis [73B2].

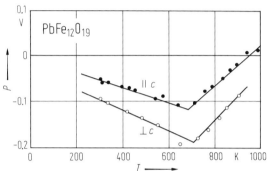

Fig. 193. $PbFe_{12}O_{19}$. Peltier coefficient $P = S \cdot T$ vs. temperature measured parallel and perpendicular to the c axis [73B2].

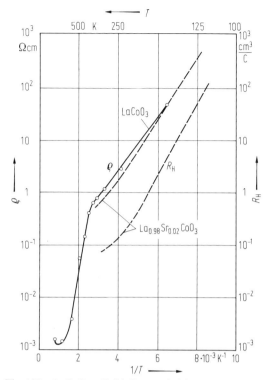

Fig. 195. $LaCoO_3$. Solid line: resistivity vs. (reciprocal) temperature. Broken line: resistivity and Hall coefficient vs. (reciprocal) temperature for a sample with 0.2% Sr [70F2].

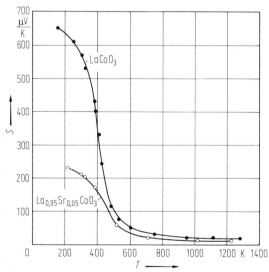

Fig. 194. $LaCoO_3$, $La_{0.95}Sr_{0.05}CoO_3$. Seebeck coefficient vs. temperature [64H].

For Fig. 196, see next page.

▶

Fig. 197. $LaCo_{1-x}Mg_xO_3$. Conductivity vs. (reciprocal) temperature for several compositions [79R]. Activation energies are indicated.

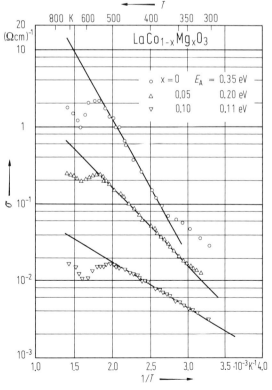

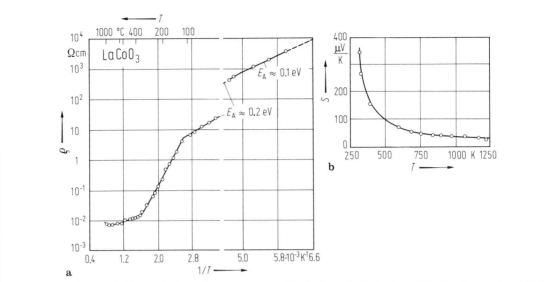

Fig. 196. $LaCoO_3$. Resistivity vs. (reciprocal) temperature (a). Fig. b shows the Seebeck coefficient vs. temperature [72B1].

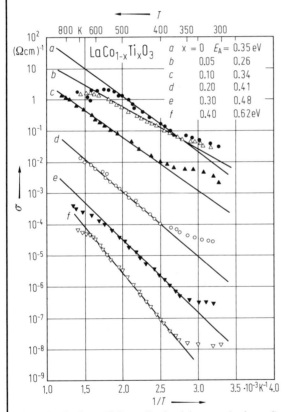

Fig. 198. $LaCo_{1-x}Ti_xO_3$. Conductivity vs. (reciprocal) temperature for several compositions [79R]. Activation energies are indicated.

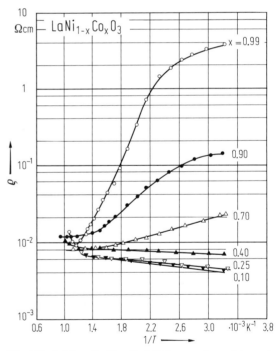

Fig. 199. $LaNi_{1-x}Co_xO_3$. Resistivity vs. reciprocal temperature for samples of different composition [75R].

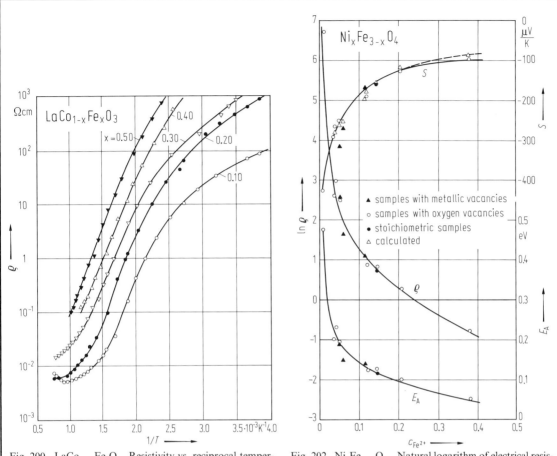

Fig. 200. $LaCo_{1-x}Fe_xO_3$. Resistivity vs. reciprocal temperature for samples of different composition [75R].

Fig. 202. $Ni_xFe_{3-x}O_4$. Natural logarithm of electrical resistivity (ϱ in Ω cm), activation energy of resistivity, and Seebeck coefficient vs. concentration of Fe^{2+} ions per formula unit, $c_{Fe^{2+}}$ ($T = 300$ K) [70N].

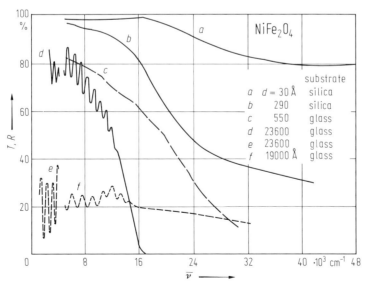

Fig. 201. $NiFe_2O_4$. Transmission (T, curves $a \cdots d$) and reflectance (R, curves e, f) of films of different thickness on glass and fused silica substrates vs. wavenumber. Curves d, e and f show interference fringes on the low frequency side [71W].

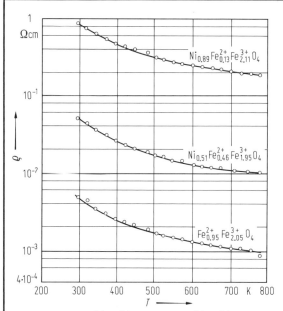

Fig. 203. $Ni_{0.51}Fe_{0.46}^{2+}Fe_{1.95}^{3+}O_4$, $Ni_{0.89}Fe_{0.13}^{2+}Fe_{2.11}^{3+}O_4$. Resistivity vs. temperature [79B]. $Fe_{0.95}^{2+}Fe_{2.05}^{3+}O_4$ for comparison.

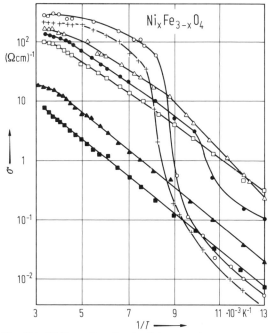

Fig. 204. $Ni_xFe_{3-x}O_4$. Conductivity vs. reciprocal temperature for various compositions. Open circles: x = 0; crosses: x = 0.01; full circles: x = 0.02; open triangles: x = 0.1; open squares: x = 0.2; full triangles: x = 0.4; full squares: x = 0.63. The departure from a straight line for x < 0.2 is the Verwey transition [74C].

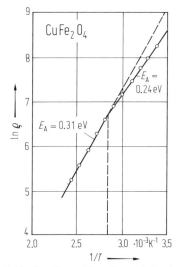

Fig. 205. $CuFe_2O_4$. Natural logarithm of electrical resistivity vs. reciprocal temperature (ϱ in Ω cm) [75B]. Activation energies are given.

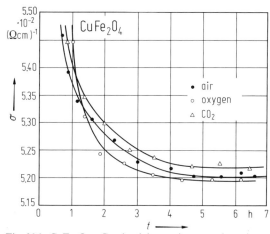

Fig. 206. $CuFe_2O_4$. Conductivity vs. time at various sintering conditions [71C5]. Sintering temperature $T = 1000\,°C$; measuring temperature: $150\,°C$.

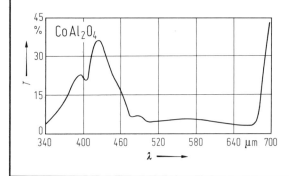

◄

Fig. 207. $CoAl_2O_4$. Transmission vs. wavelength [76Z].

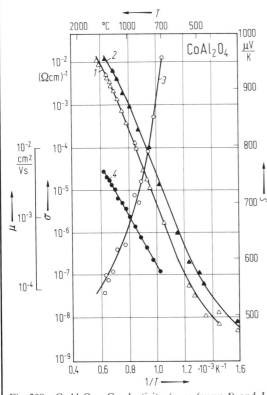

Fig. 208. $CoAl_2O_4$. Conductivity (pure (curve *1*) and Li doped (curve *2*) samples), drift mobility (curve *4*) and Seebeck coefficient (curve *3*) for undoped $CoAl_2O_4$ vs. (reciprocal) temperature [71M].

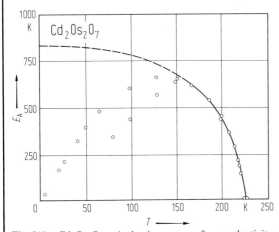

Fig. 210. $Cd_2Os_2O_7$. Activation energy for conductivity vs. temperature [74S2]. 1 K $\cong 8.617 \cdot 10^{-5}$ eV.

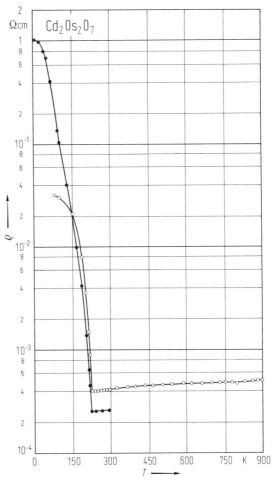

Fig. 209. $Cd_2Os_2O_7$. Resistivity vs. temperature for two differently prepared crystals [74S2].

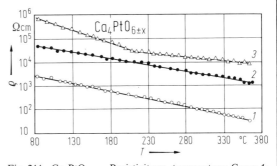

Fig. 211. $Ca_4PtO_{6\pm x}$. Resistivity vs. temperature. Curve *1*: $Ca_4PtO_{5.98}$, *2*: $Ca_4PtO_{6.06}$, *3*: $Ca_4PtO_{5.95}$ [75S2].

10.2.5 Further transition-metal sulfides, selenides and chlorides (For tables, see p. 220 ff.)

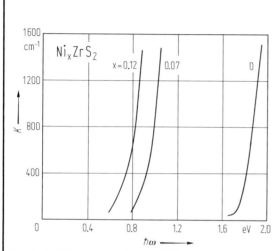

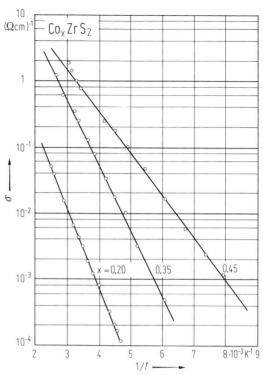

Fig. 1. Ni_xZrS_2. Absorption coefficient vs. photon energy for samples of different composition [79Y].

Fig. 3. Co_xZrS_2. Conductivity vs. reciprocal temperature for powder samples of different composition [75T].

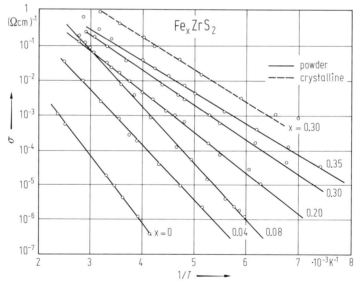

Fig. 2. Fe_xZrS_2. Conductivity vs. reciprocal temperature for samples of different composition [75T].

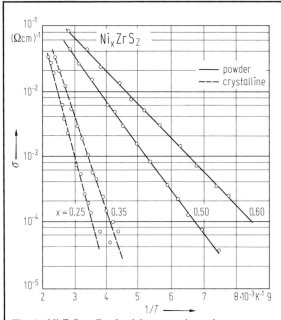

Fig. 4. Ni_xZrS_2. Conductivity vs. reciprocal temperature for samples of different composition [75T].

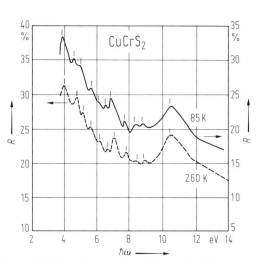

Fig. 5. $CuCrS_2$. Near normal incidence reflectivity vs. photon energy at two different temperatures [80K]. Vertical bars indicate maxima.

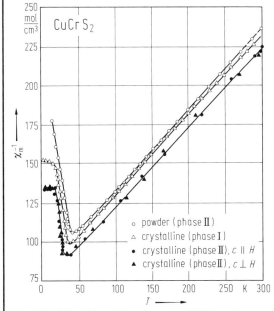

Fig. 6. $CuCrS_2$. Magnetic molar susceptibility vs. temperature (phase I: quenched from 850°C to RT, non-stoichiometric; phase II: slowly cooled; Cu atoms ordered) [79N2]. χ_m in CGS-emu.

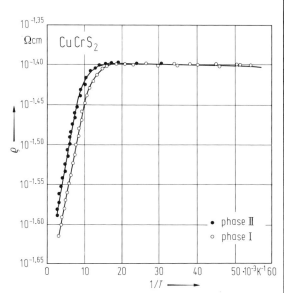

Fig. 7. $CuCrS_2$. Resistivity vs. reciprocal temperature (for phases I, II, see Fig. 6) [79N2]. ($\varrho \perp c$ axis).

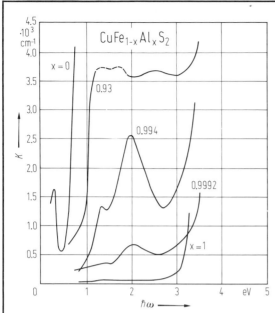

Fig. 8. $CuFe_{1-x}Al_xS_2$. Absorption coefficient vs. photon energy for various compositions [74K].

Fig. 9. $CuFeS_2$. Reflectivity vs. photon energy [80O].

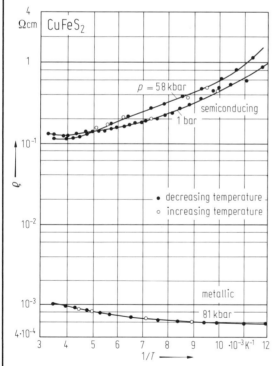

Fig. 10. $CuFeS_2$. Resistivity vs. reciprocal temperature at 1 bar, 58 kbar, and 81 kbar [74P]. ($\varrho \perp c$ axis).

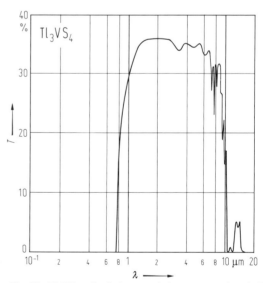

Fig. 11. Tl_3VS_4. Optical transmission vs. wavelength for a sample of 6.75 mm thickness. Curve uncorrected for reflection losses [75I].

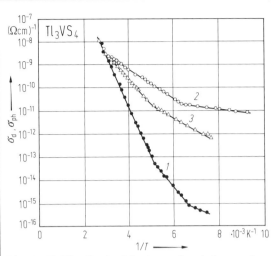

Fig. 12. Tl$_3$VS$_4$. Conductivity (curve *1*) and photoconductivity (*2*, *3*) vs. reciprocal temperature at different illumination levels L=100% (*2*) and L=11% (*3*) [74S]. σ_d: dark conductivity..

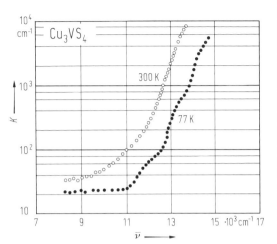

Fig. 13. Cu$_3$VS$_4$. Absorption coefficient vs. wavenumber [81P].

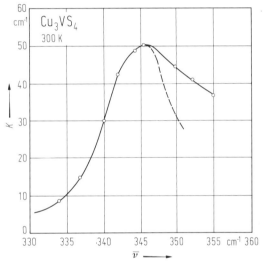

Fig. 14. Cu$_3$VS$_4$. Far-infrared absorption coefficient vs. wavenumber; the dashed line is calculated by assuming anharmonic interaction [81P].

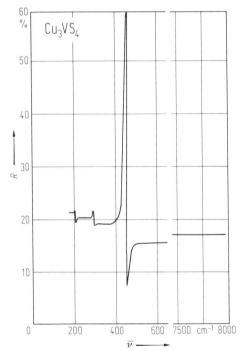

Fig. 15. Cu$_3$VS$_4$. Reflectivity vs. wavenumber. T=300 K [81P].

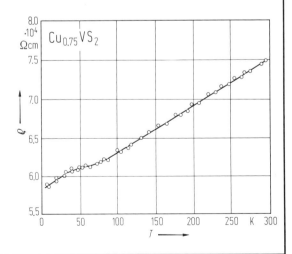

Fig. 16. Cu$_{0.75}$VS$_2$. Resistivity vs. temperature [77L]. $\varrho \perp c$.

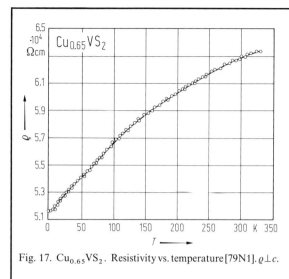

Fig. 17. $Cu_{0.65}VS_2$. Resistivity vs. temperature [79N1]. $\varrho \perp c$.

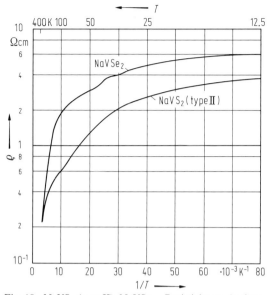

Fig. 18. $NaVS_2$ (type II), $NaVSe_2$. Resistivity vs. (reciprocal) temperature [78B]. Pressed samples.

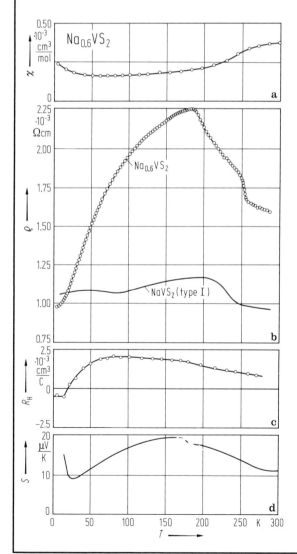

Fig. 20. $NaVS_2$. Molar magnetic susceptibility and reciprocal susceptibility vs. temperature ($H = 8.75$ kOe) [79B]. χ_m in CGS-emu.

◄

Fig. 19. $Na_{0.6}VS_2$. Molar magnetic susceptibility (in CGS-emu) (a), resistivity (b) ($NaVS_2$ (type I) for comparison), Hall coefficient (c), and Seebeck coefficient (d) vs. temperature [78B]. Pressed samples.

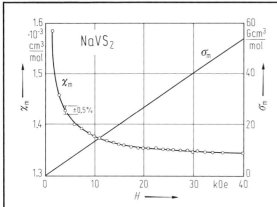

Fig. 21. NaVS$_2$ (type II). Molar magnetization σ_m and susceptibility (in CGS-emu) vs. magnetic field at 4.2 K [78B]. Polycrystalline samples.

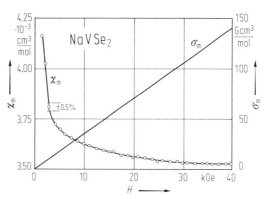

Fig. 22. NaVSe$_2$. Molar magnetization σ_m and susceptibility (in CGS-emu) vs. magnetic field at 4.2 K [78B]. Polycrystalline samples.

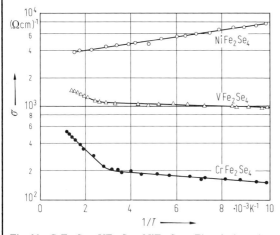

Fig. 23. CrFe$_2$Se$_4$, VFe$_2$Se$_4$, NiFe$_2$Se$_4$. Electrical conductivity vs. reciprocal temperature [74A]. Pressed samples.

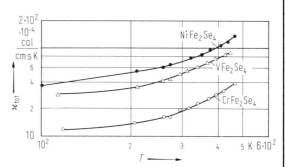

Fig. 24. CrFe$_2$Se$_4$, VFe$_2$Se$_4$, NiFe$_2$Se$_4$. Thermal conductivity vs. temperature [74A]. Pressed samples.

10.3 Ternary rare earth compounds

10.3.1 Oxides LnMO$_3$ with perovskite structure (M = Ti, V, Cr, Mn, Fe, Co, Ni) (For tables, see p. 226 ff.)

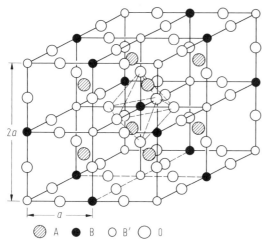

Fig. 1. A$_2$BB′O$_6$. Ordered perovskite crystal structure [79K].

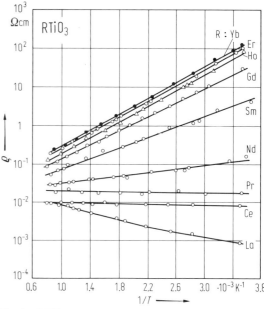

Fig. 2. RTiO$_3$. R = Yb, Er, Ho, Gd, Sm, Nd, Pr, Ce, La. Electrical resistivity vs. reciprocal temperature (polycrystalline samples, orthorhombic modification) [78B].

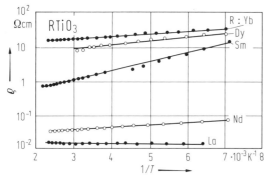

Fig. 3. RTiO$_3$. R = La, Nd, Sm, Dy, Yb. Electrical resistivity vs. reciprocal temperature (polycrystalline samples, cubic modification) [76G].

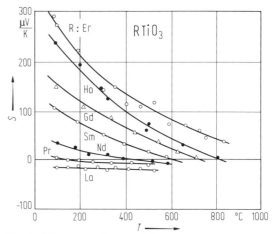

Fig. 4. RTiO$_3$. R = Er, Ho, Gd, Sm, Nd, Pr, La. Seebeck coefficient vs. temperature (polycrystalline samples, orthorhombic modification) [78B].

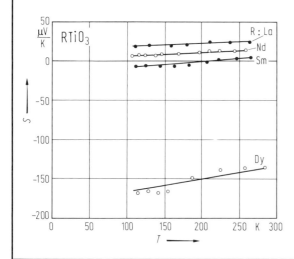

◄

Fig. 5. RTiO$_3$. R = La, Nd, Sm, Dy. Seebeck coefficient vs. temperature (polycristalline samples, cubic modification) [76G].

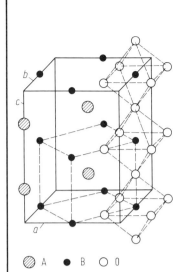

Fig. 6. ABO$_3$. Orthorhombic distortion of crystal structure [79K].

⊘ A   ● B   ○ O

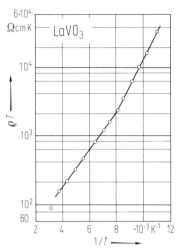

Fig. 7. LaVO$_3$. Electrical resistivity times temperature vs. reciprocal temperature (pressed polycrystalline pellet) [76S1].

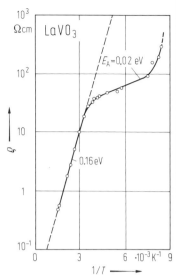

Fig. 8. LaVO$_3$. Temperature dependence of electrical resistivity for single crystals [66R]. Activation energies are indicated. Orientation not specified.

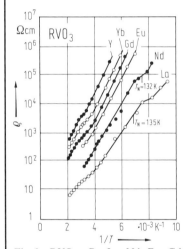

Fig. 9. RVO$_3$. R = La, Nd, Eu, Gd, Yb, Y. Temperature dependence of electrical resistivity (polycrystalline samples) [76G].

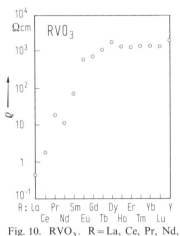

Fig. 10. RVO$_3$. R = La, Ce, Pr, Nd, Sm, Eu, Gd, Tb, Dy, Ho, Er, Tm, Yb, Lu, Y. Electrical resistivity at room temperature (pressed polycrystalline pellets) [76S1].

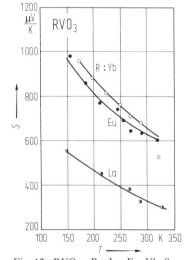

Fig. 12. RVO$_3$. R = La, Eu, Yb. Seebeck coefficient vs. temperature (polycrystalline samples) [76G].

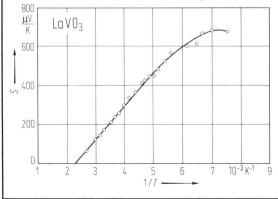

◄

Fig. 11. LaVO$_3$. Temperature dependence of Seebeck coefficient (polycrystalline sample) [75S].

495

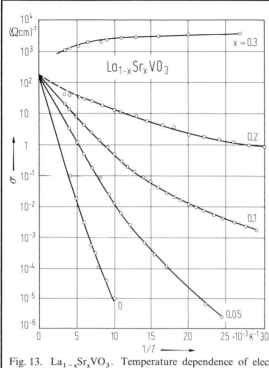

Fig. 13. La$_{1-x}$Sr$_x$VO$_3$. Temperature dependence of electrical conductivity for different concentrations of Sr (polycrystalline samples) [75S].

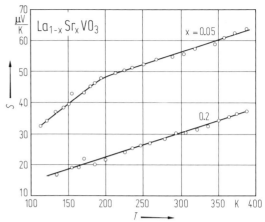

Fig. 14. La$_{1-x}$Sr$_x$VO$_3$. Seebeck coefficient vs. temperature for x = 0.05 and x = 0.2 (polycrystalline samples) [75S].

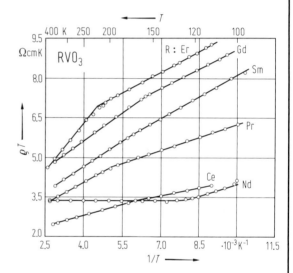

Fig. 15. RVO$_3$. R = Ce, Pr, Nd, Sm, Gd, Er. Electrical resistivity times temperature vs. (reciprocal) temperature (pressed polycrystalline pellets) [76S1].

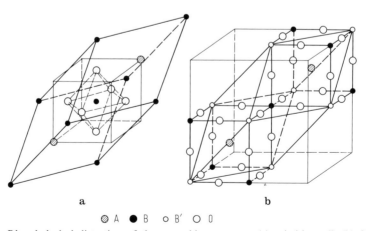

Fig. 16. A$_2$BB′O$_3$. Rhombohedral distortion of the perovskite structure, (a) primitive cell, (b) face-centered cell and primitive cell [79K].

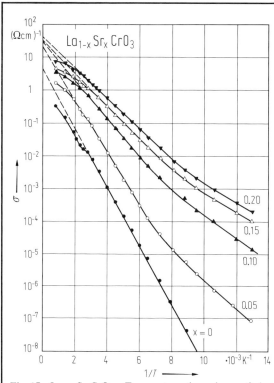

Fig. 17. La$_{1-x}$Sr$_x$CrO$_3$. Temperature dependence of electrical conductivity (pressed polycrystalline pellets) [77W].

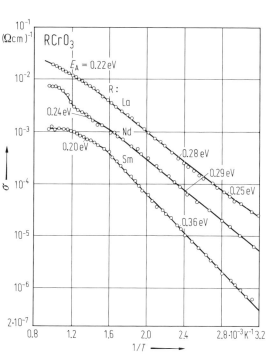

Fig. 18. RCrO$_3$. R = La, Nd, Sm. Temperature dependence of ac conductivity at 10³ Hz (pressed polycrystalline pellets) [80T1]. Activation energies are indicated.

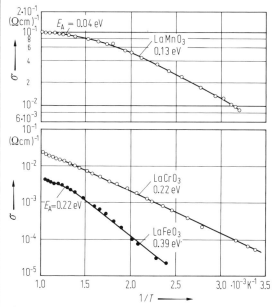

Fig. 19. LaMnO$_3$, LaCrO$_3$, LaFeO$_3$. Temperature dependence of ac conductivity ($f = 1$ kHz) with corresponding activation energies (pressed polycrystalline pellets) [71S].

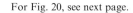

For Fig. 20, see next page.

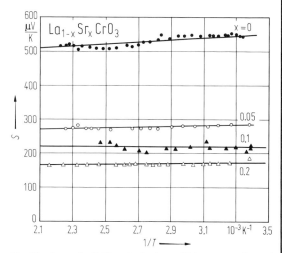

Fig. 21. La$_{1-x}$Sr$_x$CrO$_3$. Seebeck coefficient vs. reciprocal temperature for x = 0, 0.05, 0.1, 0.2 (pressed polycrystalline pellets) [77W].

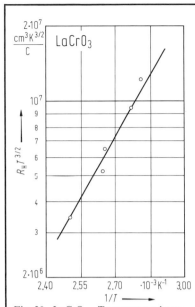

Fig. 20. LaCrO$_3$. Temperature dependence of Hall coefficient (polycrystalline sample) [67R1].

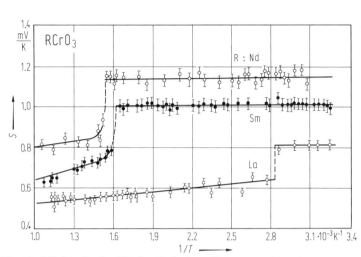

Fig. 22. RCrO$_3$. R = La, Nd, Sm. Seebeck coefficient vs. reciprocal temperature (pressed polycrystalline pellets) [80T1].

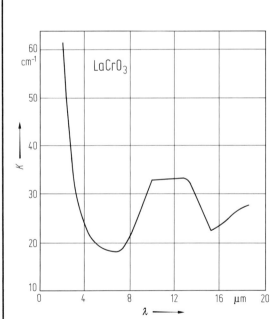

Fig. 23. LaCrO$_3$. Absorption coefficient vs. wavelength (polycrystalline sample) [67R1].

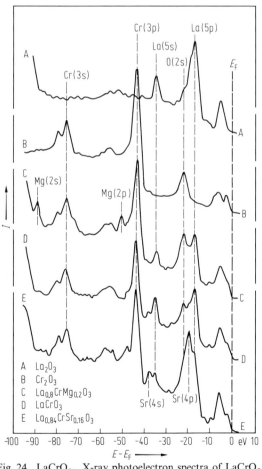

A   La$_2$O$_3$
B   Cr$_2$O$_3$
C   La$_{0,8}$CrMg$_{0,2}$O$_3$
D   LaCrO$_3$
E   La$_{0,84}$CrSr$_{0,16}$O$_3$

Fig. 24. LaCrO$_3$. X-ray photoelectron spectra of LaCrO$_3$ and related compounds in the lower binding energy range (polycrystalline samples) [80H].

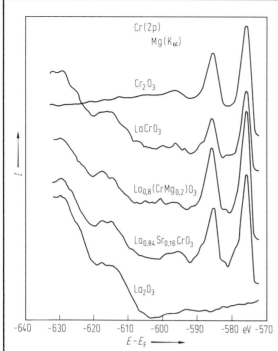

Fig. 25. LaCrO₃. X-ray photoelectron spectra of LaCrO₃ and related compounds with Cr (2p) doublet and satellite structure (polycrystalline sample) [80H]. Mg(Kα): X-ray source used.

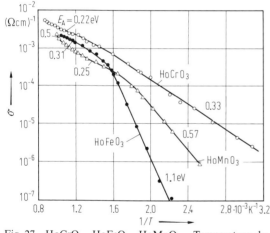

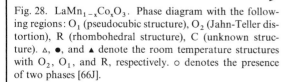

Fig. 27. HoCrO₃, HoFeO₃, HoMnO₃. Temperature dependence of ac conductivity (pressed polycrystalline pellets) [71S]. Activation energies are indicated. f = 1 kHz.

Fig. 28. LaMn₁₋ₓCoₓO₃. Phase diagram with the following regions: O₁ (pseudocubic structure), O₂ (Jahn-Teller distortion), R (rhombohedral structure), C (unknown structure). △, ●, and ▲ denote the room temperature structures with O₂, O₁, and R, respectively. ○ denotes the presence of two phases [66J].

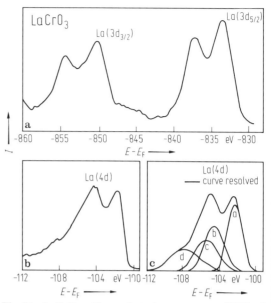

Fig. 26. LaCrO₃. X-ray photoelectron spectra [80H]. a) La(3d), b) La(4d), c) resolved La (4d) curve.

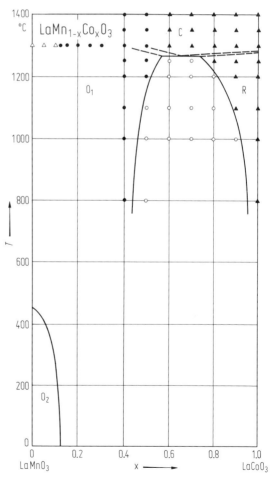

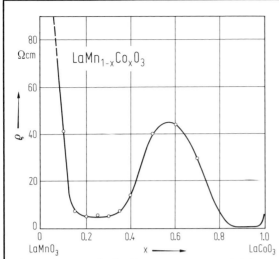

Fig. 29. LaMn$_{1-x}$Co$_x$O$_3$. Room temperature resistivity vs. composition (polycrystalline samples) [66J].

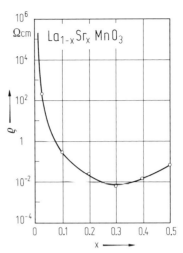

Fig. 30. La$_{1-x}$Sr$_x$MnO$_3$. Room temperature resistivity vs. composition (polycrystalline samples) [54J].

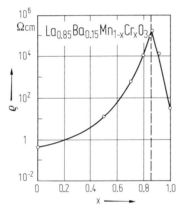

Fig. 31. La$_{0.85}$Ba$_{0.15}$Mn$_{1-x}$Cr$_x$O$_3$. Room temperature resistivities vs. composition (polycrystalline samples) [54J].

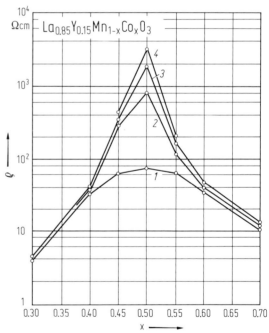

Fig. 33. La$_{0.85}$Y$_{0.15}$Mn$_{1-x}$Co$_x$O$_3$. Room temperature resistivities vs. composition for polycrystals after different heat treatments: *1*: samples prepared at 1350 °C in air; *2*: after 3 days annealing at 900 °C; *3*: after 6 days annealing at 800 °C in addition to *2*; *4*: after 2 days annealing at 1000 °C, 3 days at 900 °C, and 5 days at 800 °C in addition to *3* [66J].

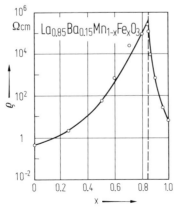

Fig. 32. La$_{0.85}$Ba$_{0.15}$Mn$_{1-x}$Fe$_x$O$_3$. Room temperature resistivities vs. composition (polycrystalline samples) [54J].

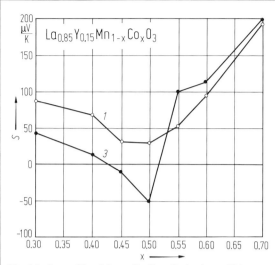

Fig. 34. La$_{0.85}$Y$_{0.15}$Mn$_{1-x}$Co$_x$O$_3$. Seebeck coefficient vs. composition (polycrystalline samples) [66J]. *1*: unannealed, *3*: annealed (compare *3* in Fig. 33).

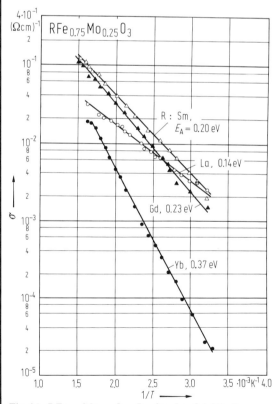

Fig. 35. RFe$_{0.75}$Mo$_{0.25}$O$_3$. R = Sm, La, Gd, Yb. Temperature dependence of electrical conductivity (pressed polycrystalline pellets) [80S]. Activation energies are indicated.

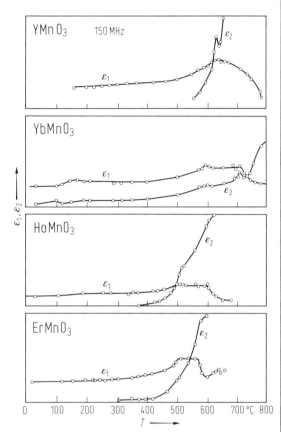

Fig. 34α. RMnO$_3$. R = Y, Yb, Ho, Er. Dielectric constants ε_1, ε_2 vs. temperature [66C].

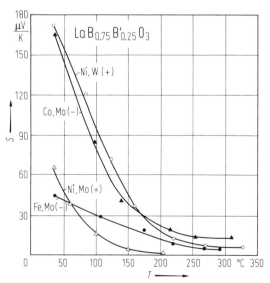

Fig. 36. LaB$_{0.75}$B$'_{0.25}$O$_3$. Seebeck coefficient vs. temperature (pressed polycrystalline pellets). Elements B,B$'$ and the sign of S are indicated [80S].

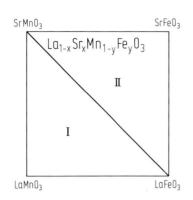

Fig. 37. $La_{1-x}Sr_xMn_{1-y}Fe_yO_3$. Phase diagram. I = region of Mn conductivity, II = region of Fe conductivity [54J]. The diagonal from $SrMnO_3$ to $LaFeO_3$ gives the nonconducting compositions.

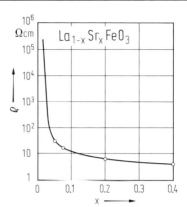

Fig. 38. $La_{1-x}Sr_xFeO_3$. Resistivity vs. composition (polycrystalline samples) [54J].

For Fig. 40, see next page.

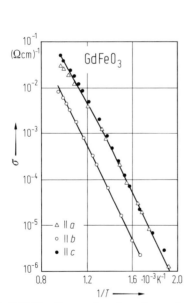

Fig. 39. $GdFeO_3$. Temperature dependence of ac conductivity [71S].

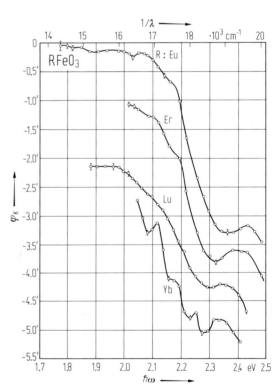

Fig. 41. $RFeO_3$. R = Eu, Er, Lu, Yb. Low-energy tails of polar Kerr rotation φ_K [69K].

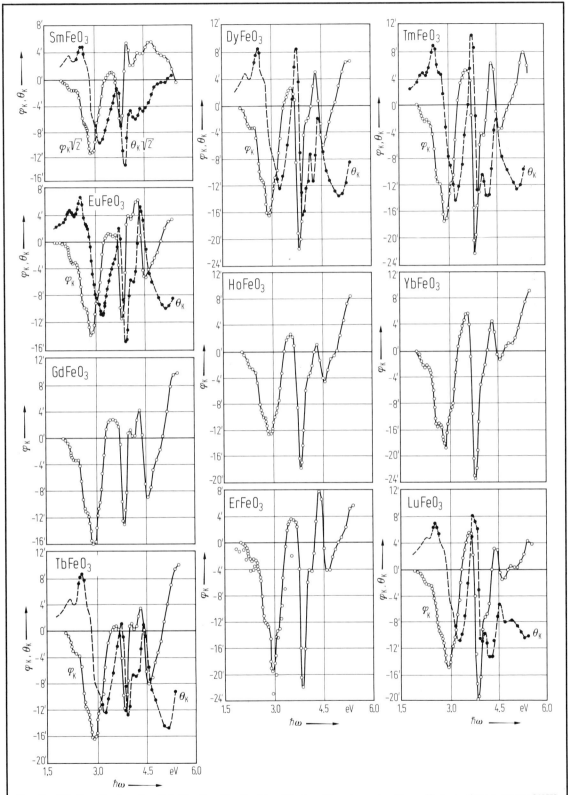

Fig. 40. RFeO₃. R = Sm, Eu, Gd, Tb, Dy, Ho, Er, Tm, Yb, Lu. Complex polar Kerr effect vs. photon energy [69K]. Complex polar Kerr effect is defined by $\Phi_K = \varphi_K + i\,\theta_K$ with φ_K: rotation angle of plane and θ_K: ellipticity.

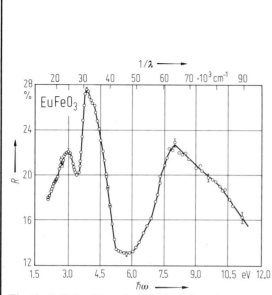

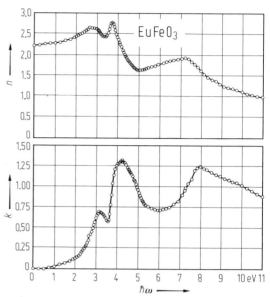

Fig. 42. EuFeO$_3$. Normal reflectivity at (001) facet vs. photon energy [69K].

Fig. 43. EuFeO$_3$. Complex index of refraction vs. photon energy [69K]. $N = n + ik$.

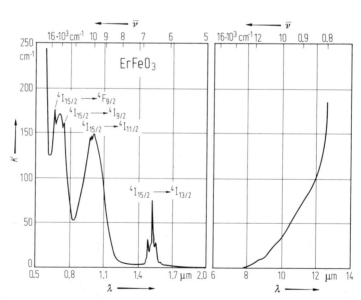

Fig. 44. ErFeO$_3$. Absorption coefficient vs. wavelength (wavenumber) [69W].

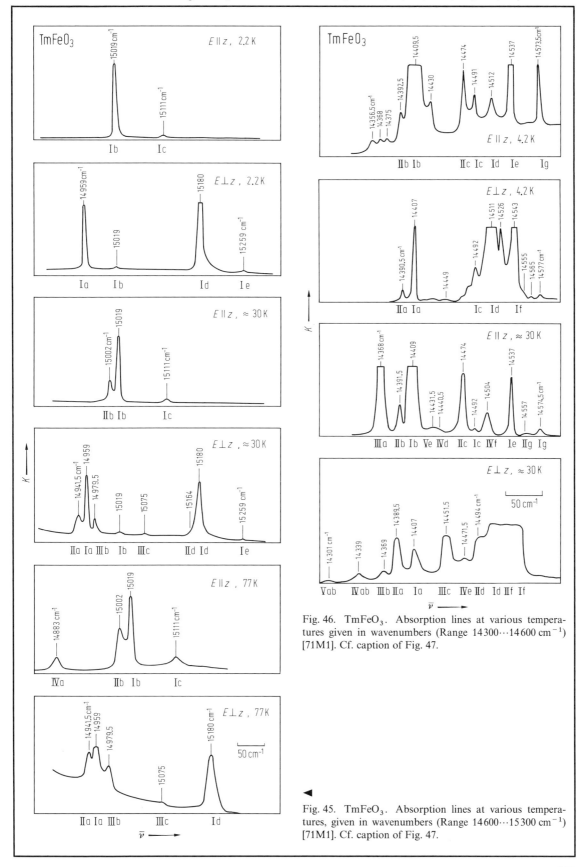

Fig. 46. TmFeO₃. Absorption lines at various temperatures given in wavenumbers (Range 14300···14600 cm⁻¹) [71M1]. Cf. caption of Fig. 47.

◄

Fig. 45. TmFeO₃. Absorption lines at various temperatures, given in wavenumbers (Range 14600···15300 cm⁻¹) [71M1]. Cf. caption of Fig. 47.

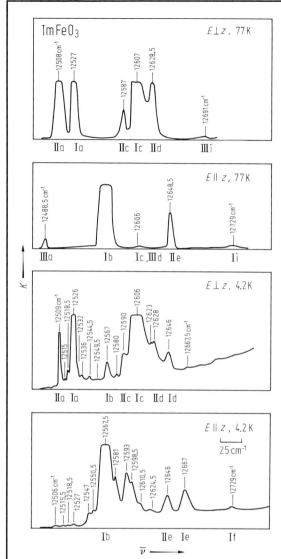

Fig. 47. TmFeO₃. Absorption lines at 4.2 K and 77 K given in wavenumbers (Range 12400···12800 cm⁻¹) [71M1]. Most of the lines can be identified as single-ion transitions from the ground state multiplet, labeled I, II, III etc., to the excited state multiplet, labelled a, b, c etc. in order of increasing energy.

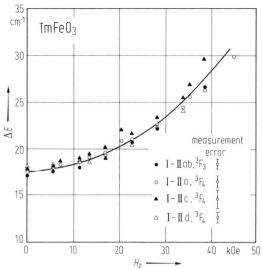

Fig. 48. TmFeO₃. Zeeman splitting of ground state doublet at 10 K with magnetic field in z-direction [71M1]. See caption of Fig. 47.

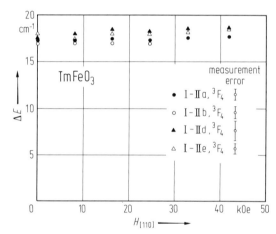

Fig. 49. TmFeO₃. Zeeman splitting of ground state doublet at 10 K with magnetic field in [110]-direction [71M1]. See caption of Fig. 47.

For Fig. 50, see next page.

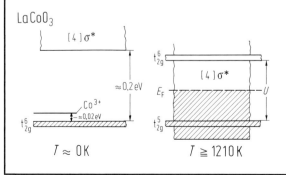

Fig. 51. LaCoO₃. Band scheme in the low- and high-temperature region [72B1].

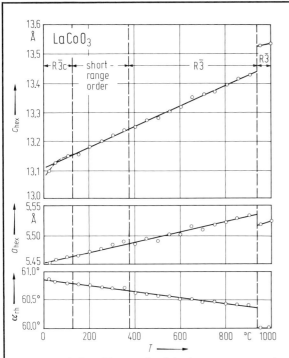

Fig. 50. LaCoO$_3$. Hexagonal cell lattice parameters and rhombohedral angle vs. temperature [67R2].

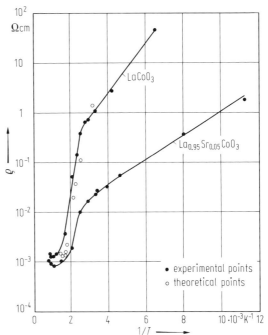

Fig. 52. LaCoO$_3$, La$_{0.95}$Sr$_{0.05}$CoO$_3$. Electrical resistivity vs. reciprocal temperature [64H].

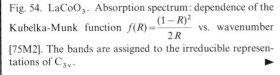

Fig. 54. LaCoO$_3$. Absorption spectrum: dependence of the Kubelka-Munk function $f(R) = \dfrac{(1-R)^2}{2R}$ vs. wavenumber [75M2]. The bands are assigned to the irreducible representations of C$_{3v}$. ▶

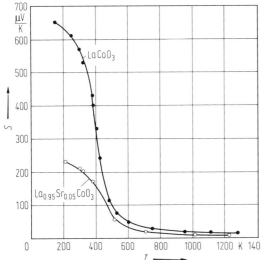

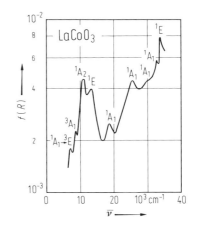

Fig. 53. LaCoO$_3$, La$_{0.95}$Sr$_{0.05}$CoO$_3$. Seebeck coefficient vs. temperature [64H].

▶

Fig. 55. LaCoO$_3$. ^{5}A$_1$-splitting of Co^{3+} due to spin-orbit coupling and due to magnetic field [80M]. β: Bohr magneton. $\boldsymbol{H} = (0, 0, H)$.

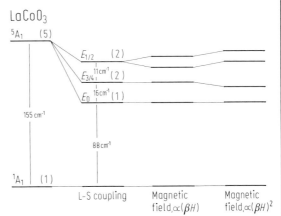

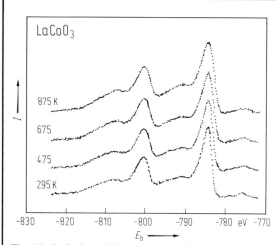

Fig. 56. LaCoO₃. XPS spectra at different temperatures [81M]. Binding energy scale has not been corrected for sample charging.

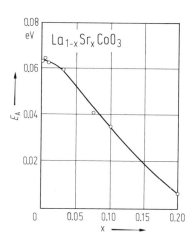

Fig. 57. La₁₋ₓSrₓCoO₃. Activation energy vs. Sr content [62G].

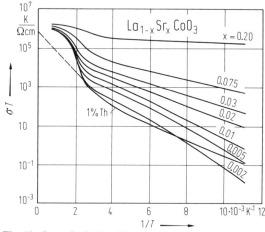

Fig. 58. La₁₋ₓSrₓCoO₃. Temperature dependence of electrical conductivity at different Sr concentrations (structure not specified) [62G]. Curve with 1% Th doped LaCoO₃ for comparison.

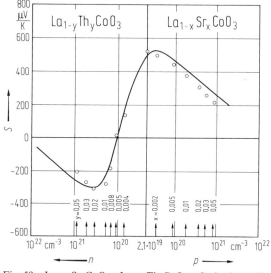

Fig. 59. La₁₋ₓSrₓCoO₃, La₁₋yThyCoO₃. Seebeck coefficient at different doping levels of Sr and Th vs. carrier density [62G].

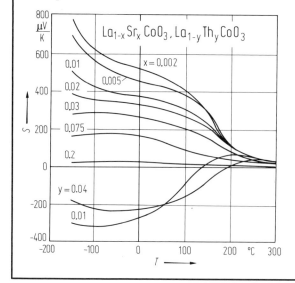

Fig. 60. La₁₋ₓSrₓCoO₃, La₁₋yThyCoO₃. Seebeck coefficient vs. temperature [62G].

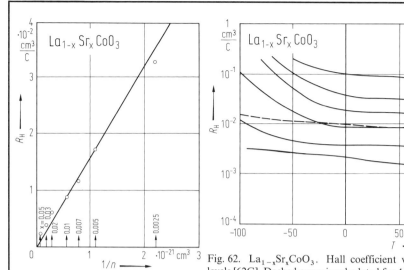

Fig. 61. La$_{1-x}$Sr$_x$CoO$_3$. Hall coefficient vs. reciprocal carrier density at room temperature [62G].

Fig. 62. La$_{1-x}$Sr$_x$CoO$_3$. Hall coefficient vs. temperature at different doping levels [62G]. Dashed curve is calculated for 1% Sr with an energy gap of 0.35 eV.

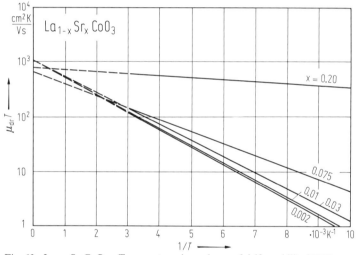

Fig. 63. La$_{1-x}$Sr$_x$CoO$_3$. Temperature dependence of drift mobility [62G].

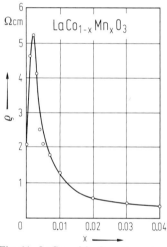

Fig. 64. LaCo$_{1-x}$Mn$_x$O$_3$. Room-temperature resistivity vs. Mn content (polycrystalline samples) [66J]. Cf. Fig. 29.

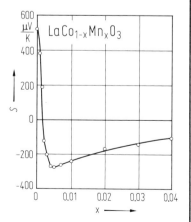

Fig. 65. LaCo$_{1-x}$Mn$_x$O$_3$. Seebeck coefficient vs. Mn content (polycrystalline samples) [66J].

10.3.2 Other oxides with transition-metal elements (For tables, see p. 248 ff.)

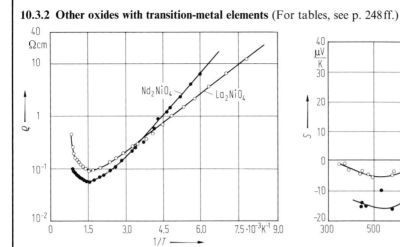

Fig. 1. La_2NiO_4, Nd_2NiO_4. Temperature dependence of electrical resistivity (polycrystalline samples) [73G1].

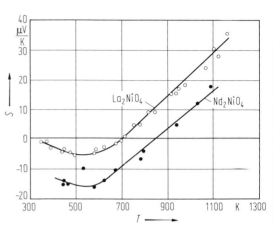

Fig. 2. La_2NiO_4, Nd_2NiO_4. Seebeck coefficient vs. temperature (polycrystalline samples) [73G1].

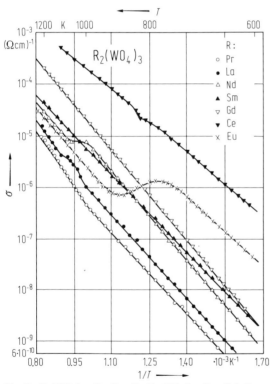

Fig. 3. $R_2(WO_4)_3$. R = La, Ce, Pr, Nd, Sm, Eu, Gd. Temperature dependence of electrical conductivity (pressed polycrystalline pellets) [79D].

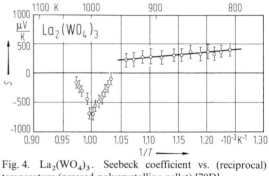

Fig. 4. $La_2(WO_4)_3$. Seebeck coefficient vs. (reciprocal) temperature (pressed polycrystalline pellet) [79D].

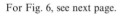

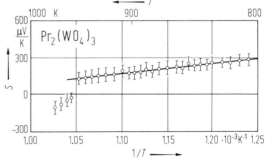

Fig. 5. $Ce_2(WO_4)_3$. Seebeck coefficient vs. (reciprocal) temperature (pressed polycrystalline pellet) [79D].

For Fig. 6, see next page.

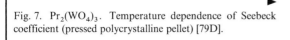

►

Fig. 7. $Pr_2(WO_4)_3$. Temperature dependence of Seebeck coefficient (pressed polycrystalline pellet) [79D].

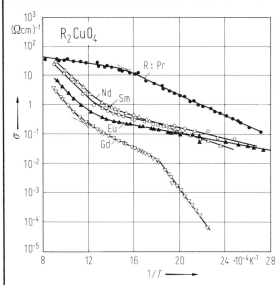

Fig. 6. R_2CuO_4. R = Pr, Nd, Sm, Eu, Gd. Temperature dependence of electrical conductivity (pressed polycrystalline pellets) [74G1].

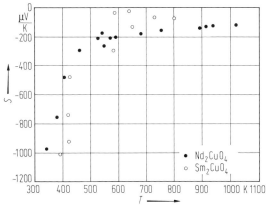

Fig. 9. Nd_2CuO_4, Sm_2CuO_4. Seebeck coefficient vs. temperature (polycrystalline samples) [73G1].

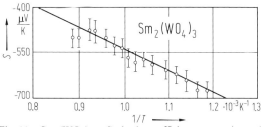

Fig. 11. $Sm_2(WO_4)_3$. Seebeck coefficient vs. reciprocal temperature (pressed polycrystalline pellet) [79D].

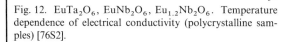

Fig. 12. $EuTa_2O_6$, $EuNb_2O_6$, $Eu_{1.2}Nb_2O_6$. Temperature dependence of electrical conductivity (polycrystalline samples) [76S2].

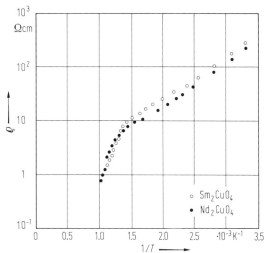

Fig. 8. Nd_2CuO_4, Sm_2CuO_4. Temperature dependence of electrical resistivity (polycrystalline sample) [73G1].

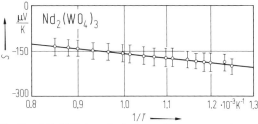

Fig. 10. $Nd_2(WO_4)_3$. Seebeck coefficient vs. reciprocal temperature (pressed polycrystalline pellet) [79D].

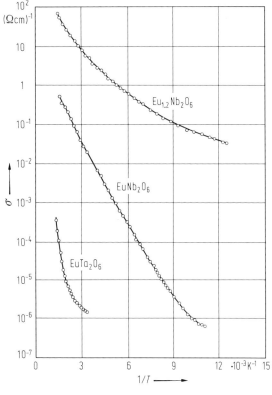

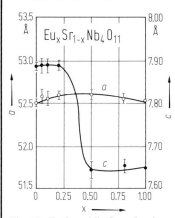

Fig. 13. $Eu_xSr_{1-x}Nb_4O_{11}$. Lattice parameters vs. composition [77S1].

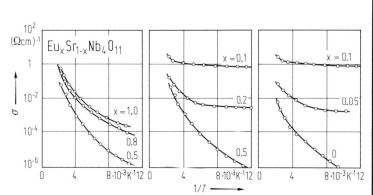

Fig. 14. $Eu_xSr_{1-x}Nb_4O_{11}$. Temperature dependence of electrical conductivity (pressed polycrystalline pellet) [77S1].

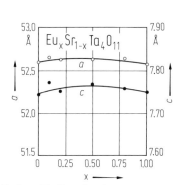

Fig. 15. $Eu_xSr_{1-x}Ta_4O_{11}$. Lattice parameters vs. composition [77S1].

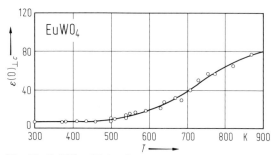

Fig. 17. $EuWO_4$. Dielectric constant vs. temperature measured at 450 kHz [74L].

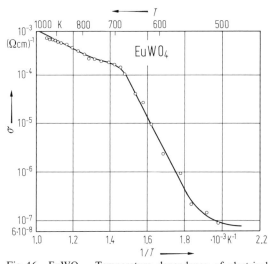

Fig. 16. $EuWO_4$. Temperature dependence of electrical conductivity perpendicular to the c axis [74L].

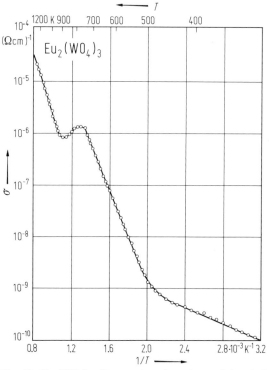

Fig. 18. $Eu_2(WO_4)_3$. Temperature dependence of electrical conductivity parallel to the c axis [76L].

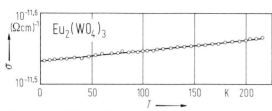

Fig. 19. $Eu_2(WO_4)_3$. Temperature dependence of electrical conductivity (single crystal) [76L]. $\sigma \parallel c$.

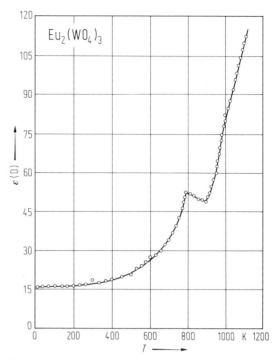

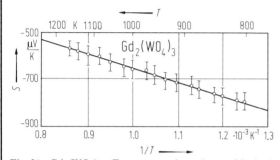

Fig. 21. $Gd_2(WO_4)_3$. Temperature dependence of Seebeck coefficient (polycrystalline sample) [79D].

Fig. 20. $Eu_2(WO_4)_3$. Dielectric constant ($\varepsilon_{\parallel c}$) vs. temperature measured at 1.542 kHz [76L].

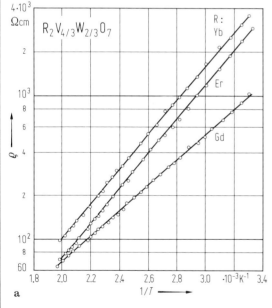

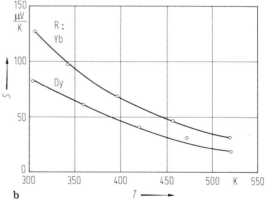

Fig. 22. $R_2V_{4/3}W_{2/3}O_7$. Electrical resistivity (R = Yb, Er, Gd) and Seebeck coefficient (R = Yb, Dy) vs. temperature (polycrystalline samples) [79S].

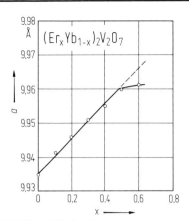

Fig. 23. $(Er_xYb_{1-x})_2V_2O_7$. Lattice constant vs. composition [77S2].

Fig. 24. $Lu_2V_2O_7$, $Yb_2V_2O_7$, $Tm_2V_2O_7$. Electrical conductivity vs. reciprocal temperature (pressed polycrystalline pellets) [77S2].

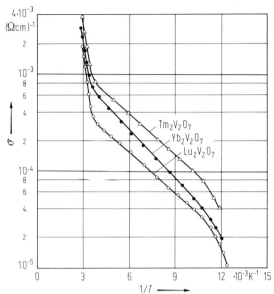

10.3.3 RE oxynitrides (For tables, see p. 255)

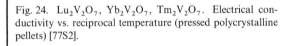

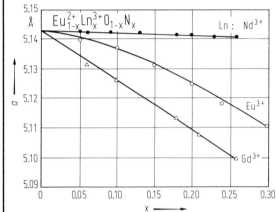

Fig. 1. $Eu_{1-x}Ln_xO_{1-x}N_x$ (Ln = Nd, Eu, Gd). Lattice parameters vs. composition [80E].

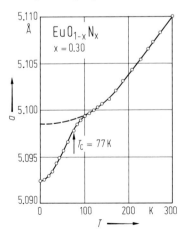

Fig. 2. $EuO_{1-x}N_x$ (x = 0.3). Lattice parameter vs. temperature [80E].

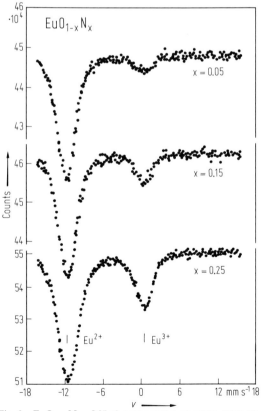

Fig. 3. $EuO_{1-x}N_x$. Mössbauer spectra at room temperature [80E].

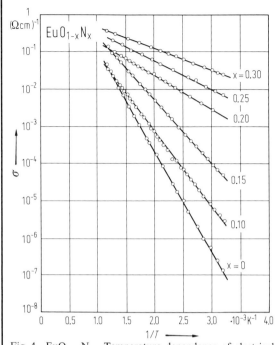

Fig. 4. $EuO_{1-x}N_x$. Temperature dependence of electrical conductivity (polycrystalline samples) [80E].

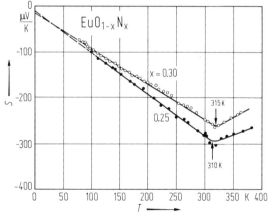

Fig. 5. $EuO_{1-x}N_x$. Seebeck coefficient vs. temperature (polycrystalline samples) [80E].

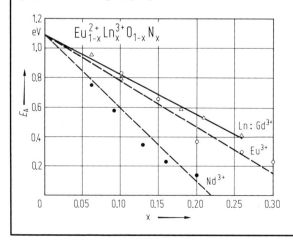

Fig. 7. $Eu_{1-x}Ln_xO_{1-x}N_x$. Ln = Nd, Eu, Gd. Activation energy vs. composition [80E].

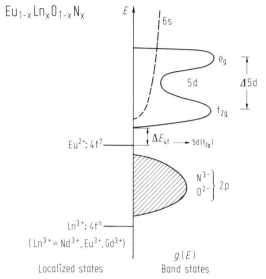

Fig. 6. $Eu_{1-x}Ln_xO_{1-x}N_x$. Ln = Nd, Eu, Gd. Qualitative energy level diagram of oxynitrides [80E].

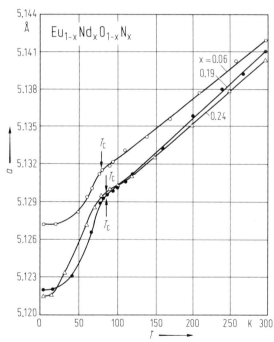

Fig. 9. $Eu_{1-x}Nd_xO_{1-x}N_x$. Lattice parameter vs. temperature for different compositions [80E]. Curie temperature is indicated.

For Fig. 8, see next page.

◄

515

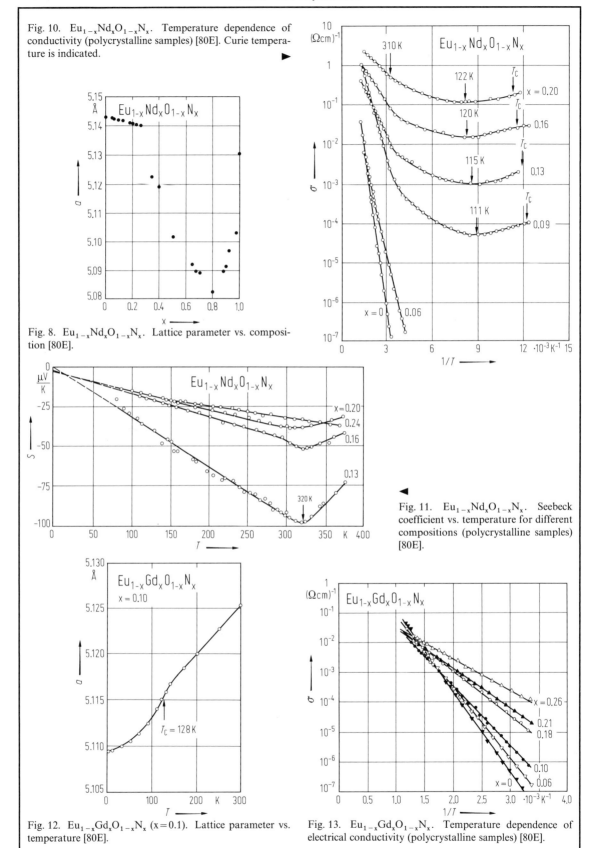

Fig. 10. $Eu_{1-x}Nd_xO_{1-x}N_x$. Temperature dependence of conductivity (polycrystalline samples) [80E]. Curie temperature is indicated. ▶

Fig. 8. $Eu_{1-x}Nd_xO_{1-x}N_x$. Lattice parameter vs. composition [80E].

◀ Fig. 11. $Eu_{1-x}Nd_xO_{1-x}N_x$. Seebeck coefficient vs. temperature for different compositions (polycrystalline samples) [80E].

Fig. 12. $Eu_{1-x}Gd_xO_{1-x}N_x$ (x = 0.1). Lattice parameter vs. temperature [80E].

Fig. 13. $Eu_{1-x}Gd_xO_{1-x}N_x$. Temperature dependence of electrical conductivity (polycrystalline samples) [80E].

10.3.4 RE sulfides, selenides, tellurides (For tables, see p. 256 ff.)

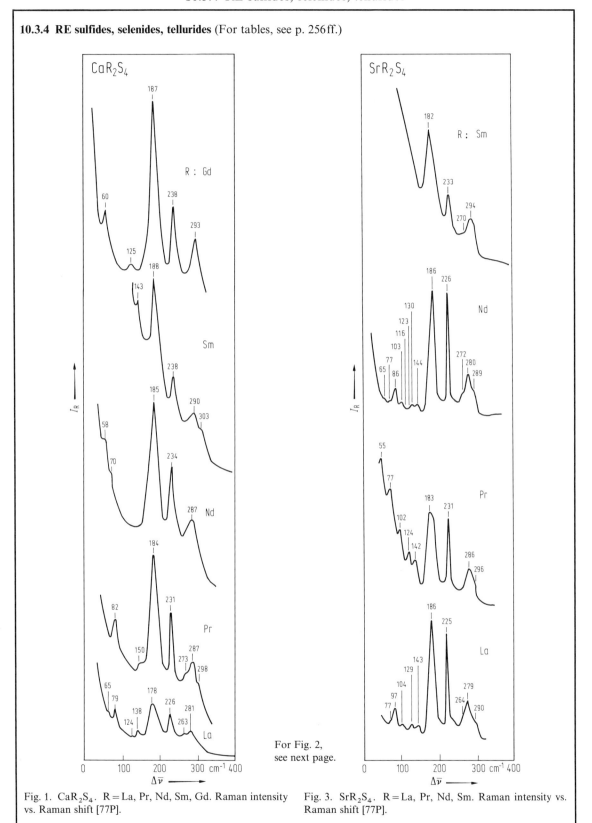

For Fig. 2,
see next page.

Fig. 1. CaR$_2$S$_4$. R = La, Pr, Nd, Sm, Gd. Raman intensity vs. Raman shift [77P].

Fig. 3. SrR$_2$S$_4$. R = La, Pr, Nd, Sm. Raman intensity vs. Raman shift [77P].

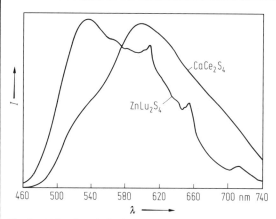

Fig. 2. ZnLu$_2$S$_4$, CaCe$_2$S$_4$. Room temperature cathodoluminescence at 12 kV excitation vs. wavelength [73Y].

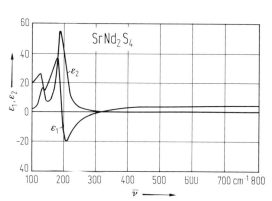

Fig. 4. SrNd$_2$S$_4$. Real and imaginary parts of the dielectric constant vs. wavenumber [77P].

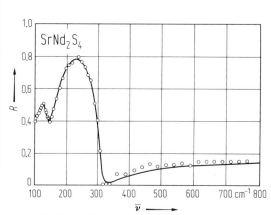

Fig. 5. SrNd$_2$S$_4$. Experimental reflectance data points and theoretically fitted curve vs. wavenumber [77P].

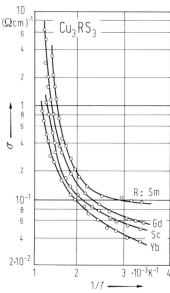

Fig. 6. Cu$_3$RS$_3$. R = Sm, Gd, Sc, Yb. Temperature dependence of electrical conductivity [81R].

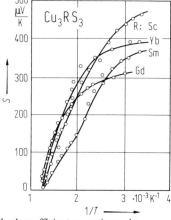

Fig. 7. Cu$_3$RS$_3$. R = Sm, Gd, Sc, Yb. Seebeck coefficient vs. reciprocal temperature [81R].

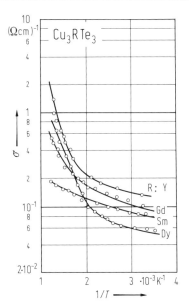

Fig. 9. Cu_3RTe_3. $R = Y$, Gd, Sm, Dy. Temperature dependence of electrical conductivity [81R].

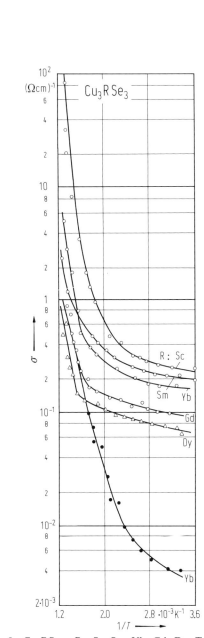

Fig. 8. Cu_3RSe_3. $R = Sc$, Sm, Yb, Gd, Dy. Temperature dependence of electrical conductivity [81R].

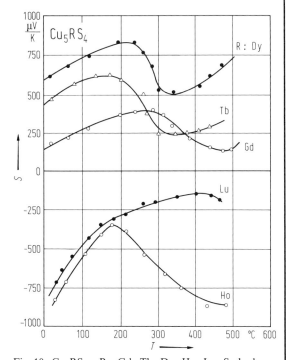

Fig. 10 Cu_5RS_4. $R = Gd$, Tb, Dy, Ho, Lu. Seebeck coefficient vs. temperature [76R]. Orientation not specified.

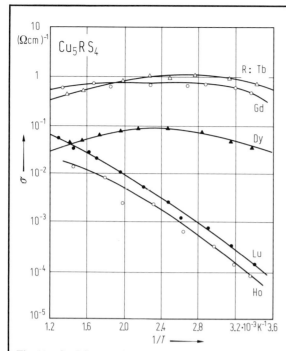

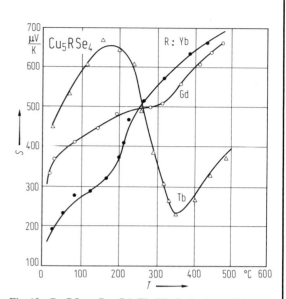

Fig. 11. Cu_5RS_4 R = Tb, Gd, Dy, Lu, Ho. Temperature dependence of electrical conductivity [76R]. (Note deviation from values in the corresponding tables.) Orientation not specified.

Fig. 12. Cu_5RSe_4. R = Gd, Tb, Yb. Seebeck coefficient vs. temperature [76R]. Orientation not specified.

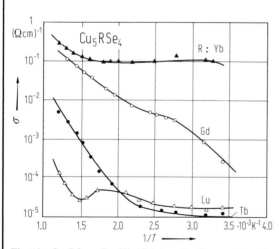

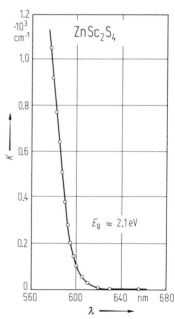

Fig. 13. Cu_5RSe_4. R = Yb, Gd, Lu, Tb. Temperature dependence of electrical conductivity. (Note deviation from the values in the corresponding tables) [76R]. Orientation not specified.

Fig. 14. $ZnSc_2S_4$. Absorption coefficient vs. wavelength at room temperature [73Y].

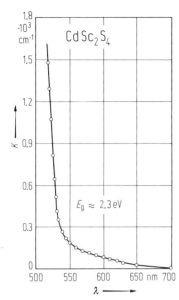

Fig. 15. $CdSc_2S_4$. Absorption coefficient vs. wavelength at room temperature [73Y].

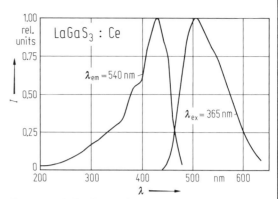

Fig. 16. $LaGaS_3$:Ce. Emission and excitation spectra vs. wavelength [80T2].

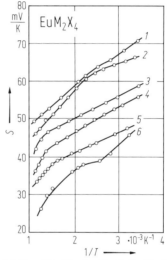

Fig. 18. EuM_2X_4. M = Ga, In; X = S, Se, Te. Seebeck coefficient vs. reciprocal temperature for 1: $EuGa_2S_4$, 2: $EuGa_2Se_4$, 3: $EuGa_2Te_4$, 4: $EuIn_2S_4$, 5: $EuIn_2Se_4$, 6: $EuIn_2Te_4$ [76A]. (Note deviation from values in the corresponding tables.)

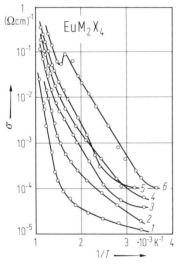

Fig. 17. EuM_2X_4. M = Ga, In; X = S, Se, Te. Temperature dependence of electrical conductivity for 1: $EuGa_2S_4$, 2: $EuGa_2Se_4$, 3: $EuGa_2Te_4$, 4: $EuIn_2S_4$, 5: $EuIn_2Se_4$, 6: $EuIn_2Te_4$ [76A]. (Note deviation from values in the corresponding tables.)

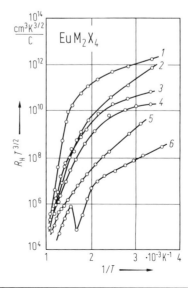

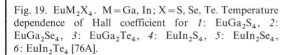

Fig. 19. EuM_2X_4. M = Ga, In; X = S, Se, Te. Temperature dependence of Hall coefficient for 1: $EuGa_2S_4$, 2: $EuGa_2Se_4$, 3: $EuGa_2Te_4$, 4: $EuIn_2S_4$, 5: $EuIn_2Se_4$, 6: $EuIn_2Te_4$ [76A].

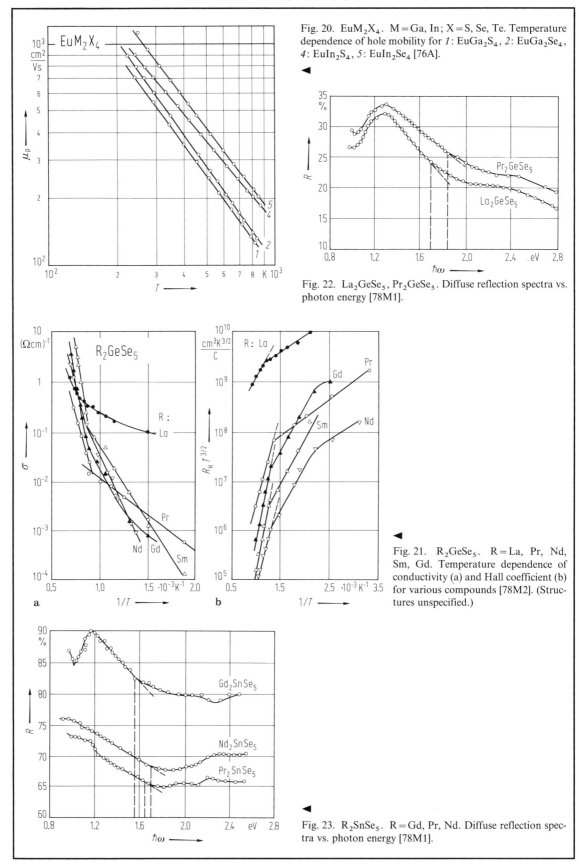

Fig. 20. EuM_2X_4. $M = Ga$, In; $X = S$, Se, Te. Temperature dependence of hole mobility for 1: $EuGa_2S_4$, 2: $EuGa_2Se_4$, 4: $EuIn_2S_4$, 5: $EuIn_2Se_4$ [76A].

Fig. 22. La_2GeSe_5, Pr_2GeSe_5. Diffuse reflection spectra vs. photon energy [78M1].

Fig. 21. R_2GeSe_5. $R = La$, Pr, Nd, Sm, Gd. Temperature dependence of conductivity (a) and Hall coefficient (b) for various compounds [78M2]. (Structures unspecified.)

Fig. 23. R_2SnSe_5. $R = Gd$, Pr, Nd. Diffuse reflection spectra vs. photon energy [78M1].

10.4 Further ternary compounds

10.4.1 I_x–IV_y–VI_z compounds (For tables, see p. 292 ff.)

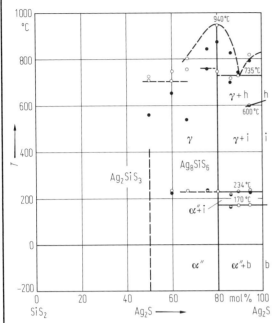

Fig. 1 Ag_8SiS_6. Phase diagram of the system SiS_2—Ag_2S [68G]. Open circles: decreasing T, full circles: increasing T. b, i and h are the phases of Ag_2X, called α, β and γ in sect. 9.3 (Vol. III/17e).

Fig. 2. Ag_2GeS_6. Phase diagram of the system GeS_2—Ag_2S [68G]. See caption of Fig. 1.

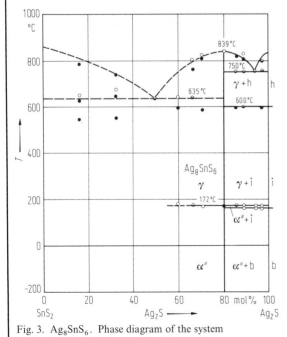

Fig. 3. Ag_8SnS_6. Phase diagram of the system SnS_2—Ag_2S [68G]. See caption of Fig. 1.

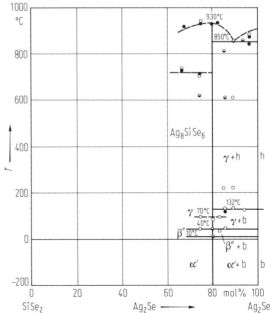

Fig. 4. Ag_8SiSe_6. Phase diagram of the system $SiSe_2$—Ag_2Se [68G]. See caption of Fig. 1.

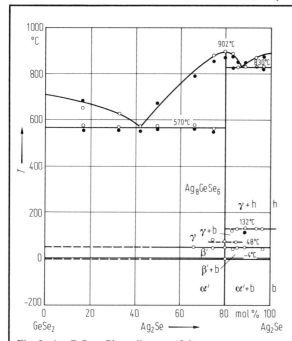

Fig. 5. Ag$_8$GeSe$_6$. Phase diagram of the system
GeSe$_2$–Ag$_2$Se [68G]. See caption of Fig. 1.

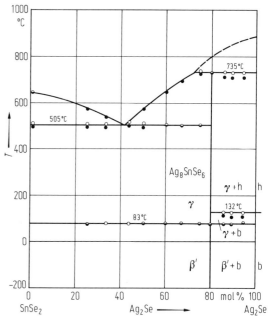

Fig. 6. Ag$_8$SnSe$_6$. Phase diagram of the system
SnSe$_2$–Ag$_2$Se [68G]. See caption of Fig. 1.

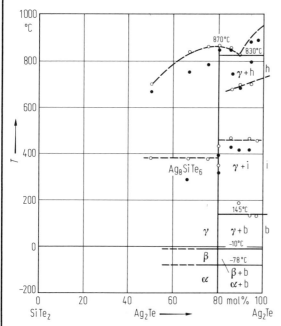

Fig. 7. Ag$_8$SiTe$_6$. Phase diagram of the system
SiTe$_2$–Ag$_2$Te [68G]. See caption of Fig. 1.

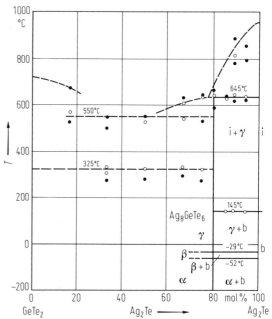

Fig. 8. Ag$_8$GeTe$_6$. Phase diagram of the system
GeTe$_2$–Ag$_2$Te [68G]. See caption of Fig. 1.

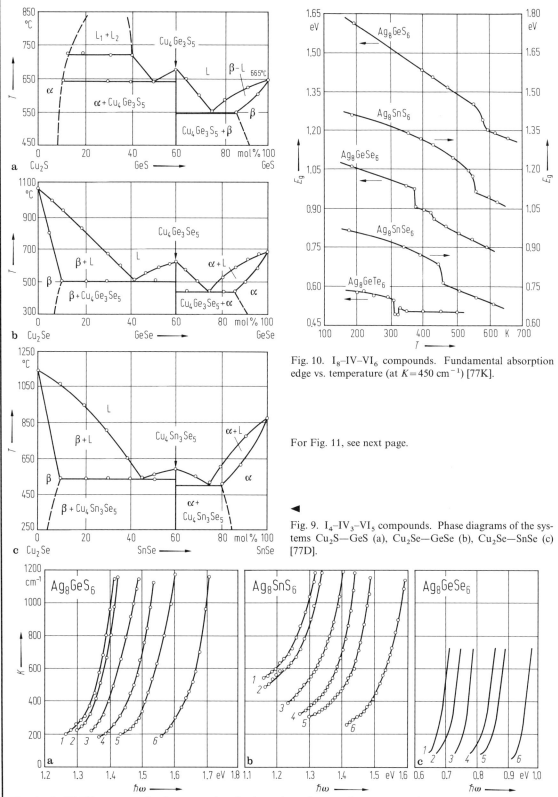

Fig. 10. I_8–IV–VI_6 compounds. Fundamental absorption edge vs. temperature (at $K = 450$ cm^{-1}) [77K].

For Fig. 11, see next page.

◄

Fig. 9. I_4–IV_3–VI_5 compounds. Phase diagrams of the systems Cu$_2$S—GeS (a), Cu$_2$Se—GeSe (b), Cu$_2$Se—SnSe (c) [77D].

Fig. 12. I_8–IV–VI_6 compounds. Long wavelength absorption edge vs. photon energy for (a) Ag$_8$GeS$_6$, (b) Ag$_8$SnS$_6$ and (c) Ag$_8$GeSe$_6$ at different temperatures. The curves *1* to *6* belong to measurements (a) at $T = 510$, 495, 485, 375, 295, 110 K [75O], (b) $T = 505$, 455, 438, 375, 295, 110 K [75O], (c) $T = 510$, 430, 357, 344, 300, 77 K [76K], respectively.

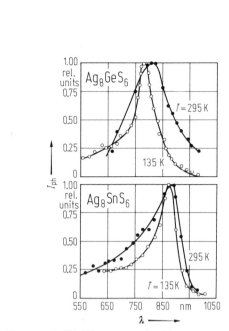

Fig. 11. I_8–IV–VI_6 compounds. Photocurrent vs. wavelength for Ag_8GeS_6 and Ag_8SnS_6 [72O].

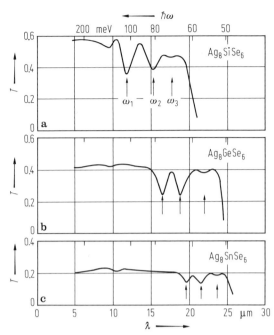

Fig. 13. I_8–IV–VI_6 compounds. Transmission vs. wavelength (photon energy) in the infrared region at 330 K, (a) Ag_8SiSe_6 (thickness of the sample 274 μm), (b) Ag_8GeSe_6 (330 μm), (c) Ag_8SnSe_6 (430 μm) [75B].

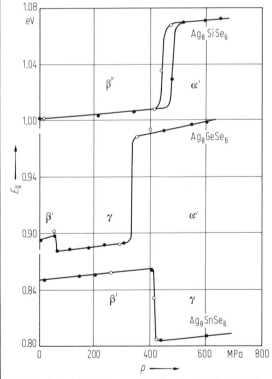

Fig. 14. I_8–IV–VI_8 compounds. Fundamental absorption edge vs. pressure for Ag_8SiSe_6 ($T=293$ K), Ag_8GeSe_6 (315 K), Ag_8SnSe_6 (350 K) [76B2]. Open circles: decreasing p, full circles: increasing p.

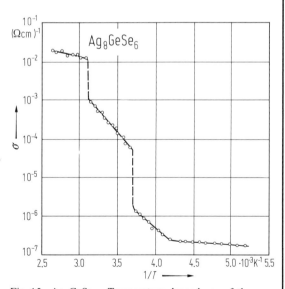

Fig. 15. Ag_8GeSe_6. Temperature dependence of the conductivity showing steps in the conductivity [77O].

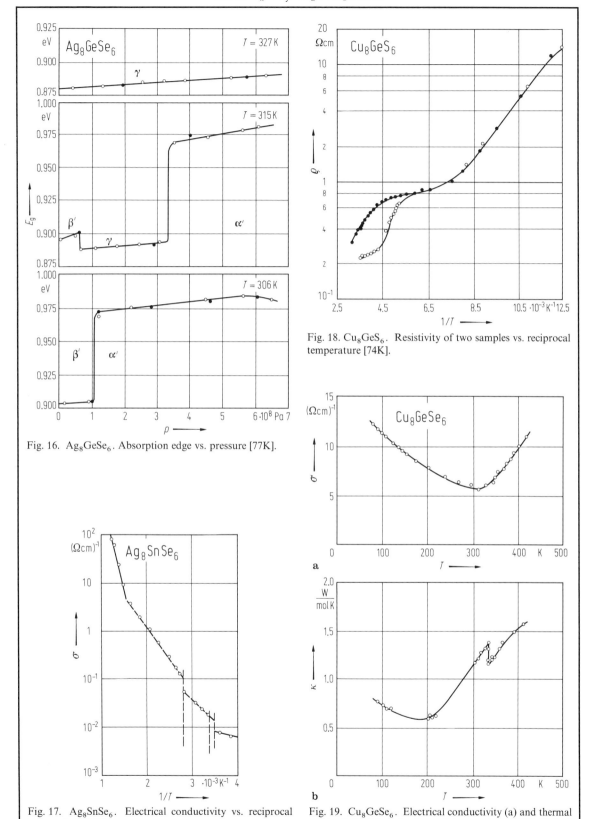

Fig. 16. Ag_8GeSe_6. Absorption edge vs. pressure [77K].

Fig. 17. Ag_8SnSe_6. Electrical conductivity vs. reciprocal temperature [76P].

Fig. 18. Cu_8GeS_6. Resistivity of two samples vs. reciprocal temperature [74K].

Fig. 19. Cu_8GeSe_6. Electrical conductivity (a) and thermal conductivity (b) vs. (reciprocal) temperature [82A]. Unit of thermal conductivity κ as given in the original paper.

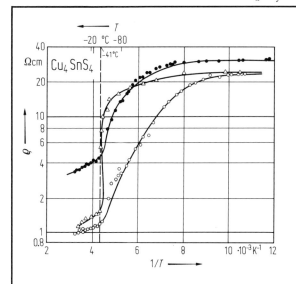

Fig. 20. Cu$_4$SnS$_4$. Resistivity of three samples vs. (reciprocal) temperature [74K].

◀

10.4.2 I$_x$–V$_y$–VI$_z$ compounds (For tables, see p. 296 ff.)

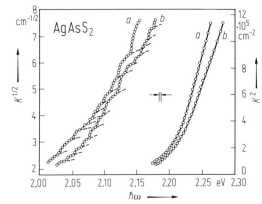

Fig. 1. AgAsS$_2$. Fine structure of the indirect absorption edge (left) and direct absorption edge (right) (square root and square of the absorption coefficient vs. photon energy, respectively). Curve a: polarization perpendicular to the crystal axis, b: polarization parallel to the crystal axis; T = 295 K [74G].

Fig.4. AgAsS$_2$. Electroabsorption vs. photon energy at 295 K for a single crystal [74G].

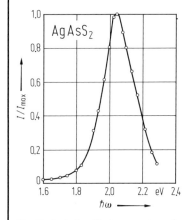

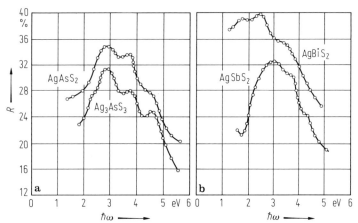

Fig. 2. AgAsS$_2$. Photocurrent vs. photon energy [75G1].

Fig. 3. AgAsS$_2$, Ag$_3$AsS$_3$, AgSbS$_2$, AgBiS$_2$. Reflectivity vs. photon energy at 295 K for various I–V–VI compounds [75G1].

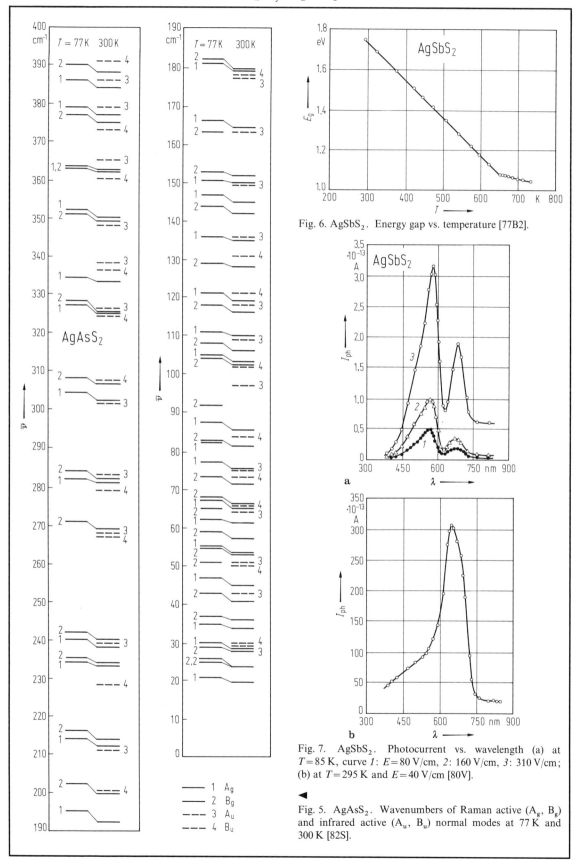

Fig. 6. AgSbS$_2$. Energy gap vs. temperature [77B2].

Fig. 7. AgSbS$_2$. Photocurrent vs. wavelength (a) at $T = 85$ K, curve 1: $E = 80$ V/cm, 2: 160 V/cm, 3: 310 V/cm; (b) at $T = 295$ K and $E = 40$ V/cm [80V].

◄

Fig. 5. AgAsS$_2$. Wavenumbers of Raman active (A$_g$, B$_g$) and infrared active (A$_u$, B$_u$) normal modes at 77 K and 300 K [82S].

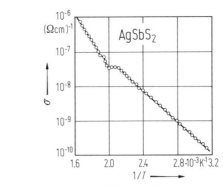

Fig. 8. AgSbS$_2$. Conductivity vs. reciprocal temperature [79V].

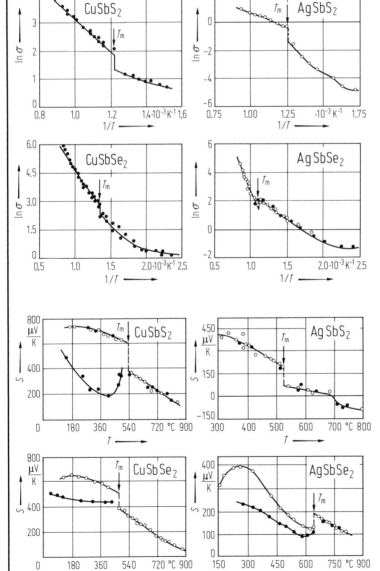

Fig. 9. CuSbS$_2$, CuSbSe$_2$, AgSbS$_2$, AgSbSe$_2$. Conductivity vs. temperature in the solid and liquid phases of various I–V–VI compounds. Open circles: annealed specimens, full circles: cast specimens [68A]. σ in Ω^{-1} cm^{-1}.

Fig. 10. CuSbS$_2$, CuSbSe$_2$, AgSbS$_2$, AgSbSe$_2$. Seebeck coefficient vs. temperature in the solid and liquid phases of various I–V–VI compounds. Open circles: annealed specimens, full circles: cast specimens [68A].

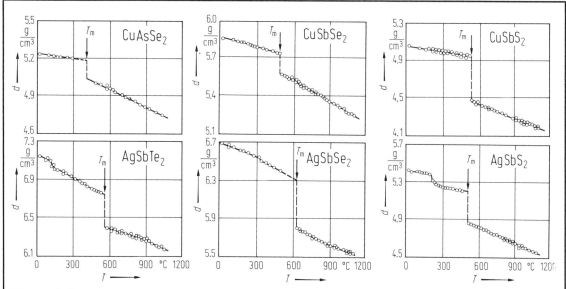

Fig. 11. $CuSbS_2$, $CuSbSe_2$, $CuAsSe_2$, $AgSbS_2$, $AgSbSe_2$, $AgSbTe_2$. Density vs. temperature in the solid and liquid phases [68K].

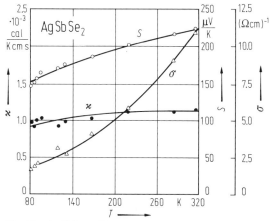

Fig. 12. $AgSbSe_2$. Temperature dependence of thermal conductivity, electrical conductivity, and Seebeck coefficient of a sample slowly cooled from the melting point to room temperature [62P].

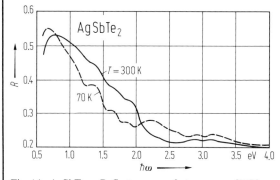

Fig. 14. $AgSbTe_2$. Reflectance vs. photon energy [80B].

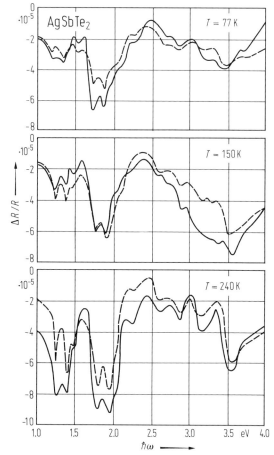

Fig. 13. $AgSbTe_2$. Thermoreflectance vs. photon energy. Full lines: experimental, dashed lines: calculated [80B].

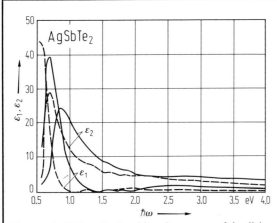

Fig. 15. $AgSbTe_2$. Real and imaginary parts of the dielectric constant vs. photon energy at $T = 70$ K (dashed line) and 300 K (full line) [80B].

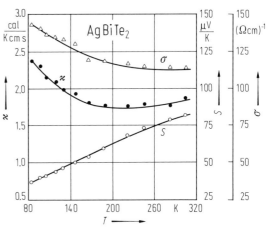

Fig. 16. $AgBiTe_2$. Temperature dependence of electrical conductivity, thermal conductivity and Seebeck coefficient of a quenched sample [62P].

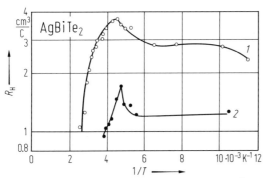

Fig. 17. $AgBiTe_2$. Hall coefficient vs. temperature for an annealed (curve 1) and a quenched (2) sample [62P].

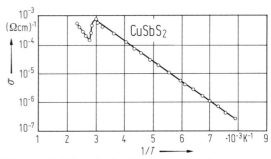

Fig. 18. $CuSbS_2$. Conductivity vs. reciprocal temperature for a p-type sample [57W].

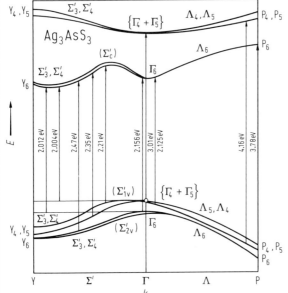

Fig. 19. Ag_3AsS_3. Band structure along two principal axes in the Brillouin zone estimated by group theoretical considerations and the experimental results of Figs. 21 and 22 [73D].

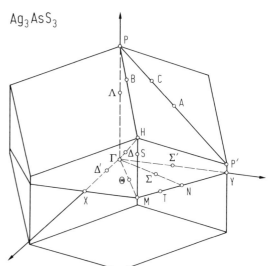

Fig. 20. Ag_3AsS_3. Brillouin zone of the proustite structure [73D].

532

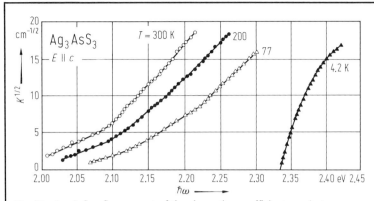

Fig. 21. Ag_3AsS_3. Square root of the absorption coefficient vs. photon energy. Electrical field vector parallel to the c axis [71D].

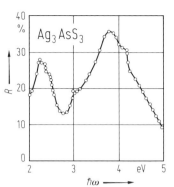

Fig. 22. Ag_3AsS_3. Reflection coefficient vs. photon energy at $T = 300$ K [71D]; see also Fig. 3.

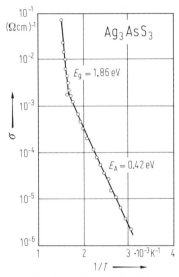

Fig. 23. Ag_3AsS_3. Electrical conductivity vs. reciprocal temperature for a single crystal showing activation energies of $E_A = 0.42$ eV and $E_g = 1.86$ eV [75G2].

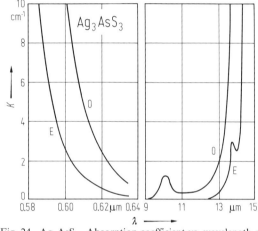

Fig. 24. Ag_3AsS_3. Absorption coefficient vs. wavelength at 293 K (O: ordinary ray, E: extraordinary ray). In the interval between 0.65 μm and 9 μm there are no absorption bands stronger than 0.1 cm^{-1} [67H].

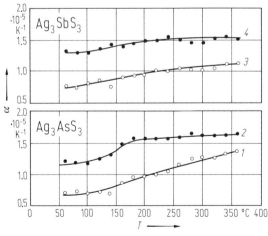

Fig. 25. Ag_3AsS_3, Ag_3SbS_3. Coefficient of linear thermal expansion vs. temperature for α parallel to the c axis (curves 1, 3) and perpendicular to the c axis (curves 2, 4) [74B1].

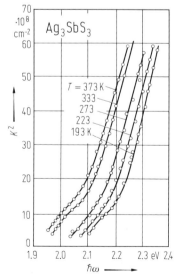

Fig. 26. Ag_3SbS_3. Square of the absorption coefficient vs. photon energy for films of the thickness of 0.72 μm prepared on a glass substrate [76G2].

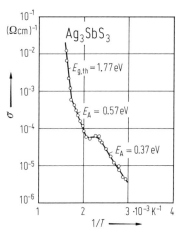

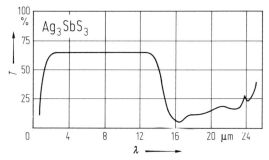

Fig. 27. Ag$_3$SbS$_3$. Electrical conductivity vs. reciprocal temperature for a single crystal showing activation energies of $E_A = 0.37$ eV and 0.57 eV and $E_{g,th} = 1.77$ eV [75G2].

Fig. 28. Ag$_3$SbS$_3$. Transmission vs. wavelength for a single crystal of 1.2 mm thickness (without correction for refraction) [73G].

10.4.3 II$_x$–III$_y$–VI$_z$ compounds (For tables, see p. 304 ff.)

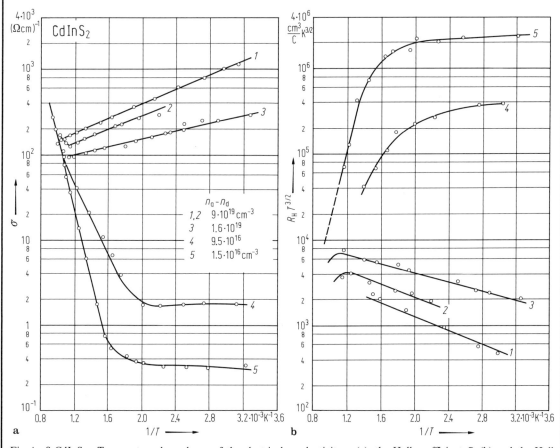

Fig. 1. β-CdInS$_2$. Temperature dependence of the electrical conductivity σ (a), the Hall coefficient R_H (b) and the Hall mobility $\mu_{H,p}$ (c) of five samples with impurity concentrations of 10^{16} to 10^{20} cm^{-3} [69G1].

For Fig. 1c, see next page.

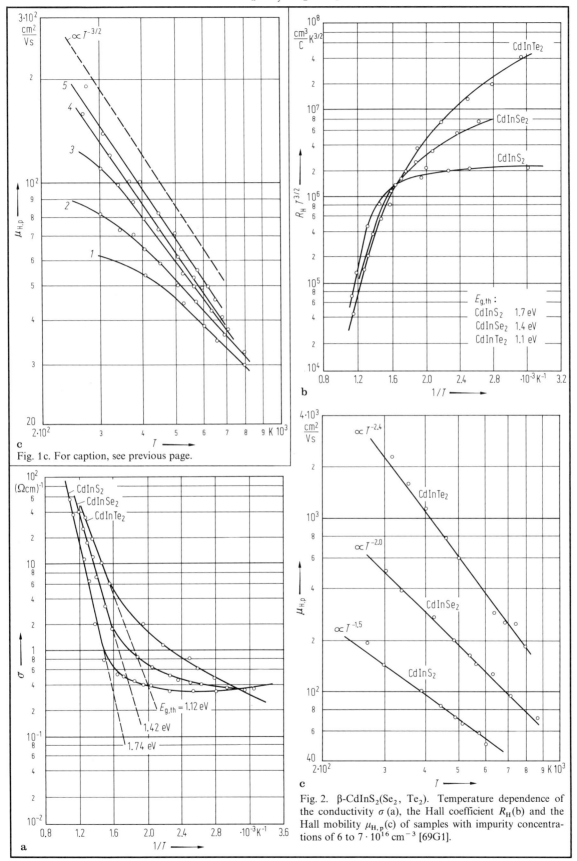

Fig. 1c. For caption, see previous page.

Fig. 2. β-CdInS$_2$(Se$_2$, Te$_2$). Temperature dependence of the conductivity σ (a), the Hall coefficient R_H(b) and the Hall mobility $\mu_{H,p}$(c) of samples with impurity concentrations of 6 to $7 \cdot 10^{16}$ cm^{-3} [69G1].

535

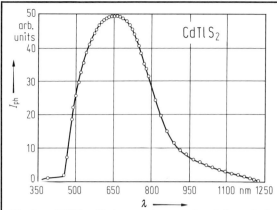

Fig. 3. α-CdTlS$_2$. Photocurrent vs. wavelength for a single crystal at 300 K [67G].

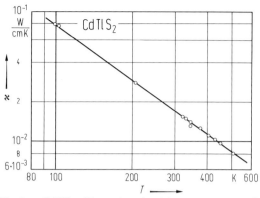

Fig. 5. α-CdTlS$_2$. Thermal conductivity vs. temperature for a p-type sample [67G].

For Fig. 6, see next page.

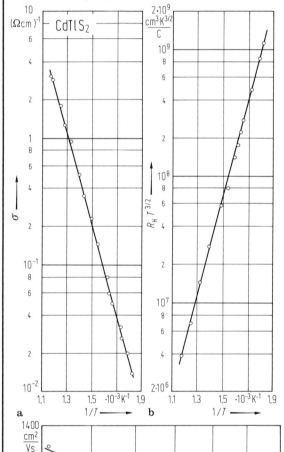

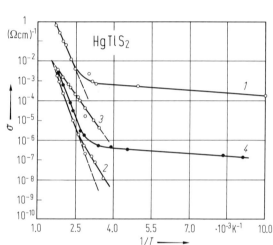

Fig. 7. HgTlS$_2$. Electrical conductivity vs. reciprocal temperature for four different samples (1, 2 single crystals, 3 polycrystalline, 4 polycrystalline with partially oriented crystals) [68G].

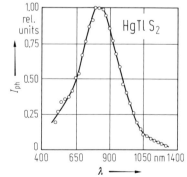

Fig. 4. α-CdTlS$_2$. Temperature dependence of the electrical conductivity σ (a), the Hall coefficient R_H(b) and the Hall mobility $\mu_{H,p}$(c) [67G].

Fig. 8. HgTlS$_2$. Photoconductivity vs. wavelength of a single crystal at $T = 300$ K [68G].

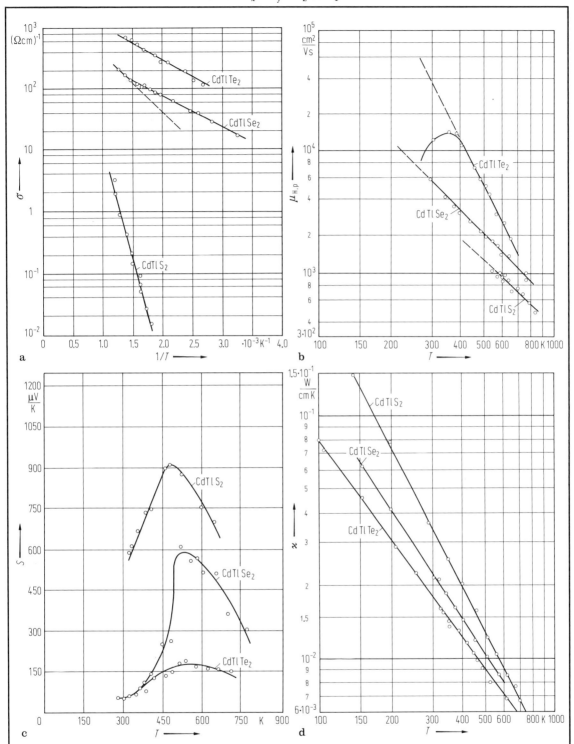

Fig. 6. α-CdTlS$_2$(Se$_2$, Te$_2$). Temperature dependence of the electrical conductivity σ (a), the Hall mobility $\mu_{H,p}$(b), the thermoelectric power S(c) and the thermal conductivity κ(d) of p-type samples [69G2].

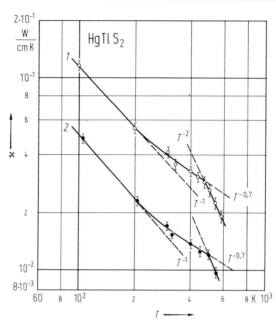

Fig. 9. $HgTlS_2$. Thermal conductivity of two samples vs. temperature [68G].

10.4.4 $III_x–V_y–VI_z$ compounds (For tables, see p. 306ff.)

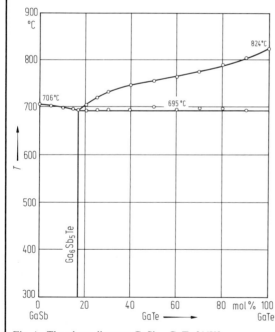

Fig. 1. The phase diagram GaSb—GaTe [66K].

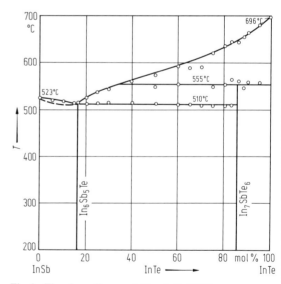

Fig. 2. The phase diagram InSb—InTe [66K].

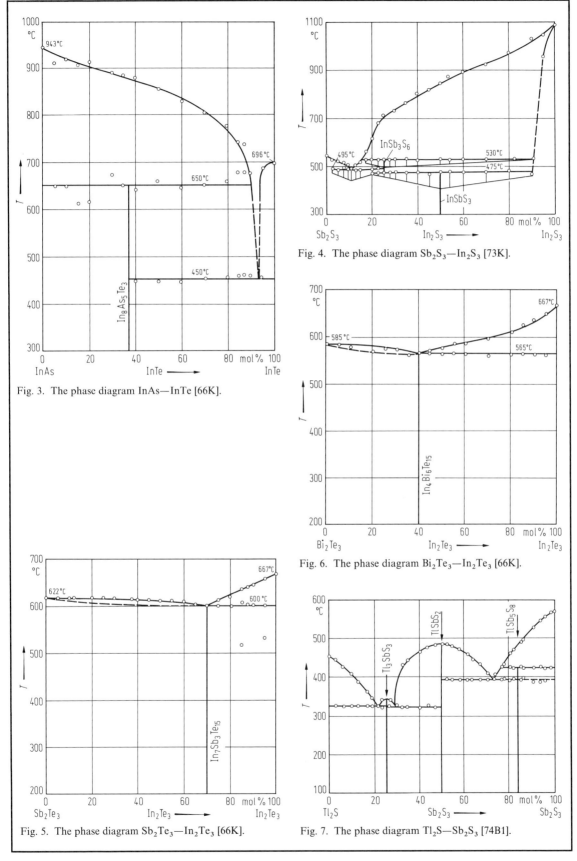

Fig. 3. The phase diagram InAs—InTe [66K].

Fig. 4. The phase diagram Sb$_2$S$_3$—In$_2$S$_3$ [73K].

Fig. 5. The phase diagram Sb$_2$Te$_3$—In$_2$Te$_3$ [66K].

Fig. 6. The phase diagram Bi$_2$Te$_3$—In$_2$Te$_3$ [66K].

Fig. 7. The phase diagram Tl$_2$S—Sb$_2$S$_3$ [74B1].

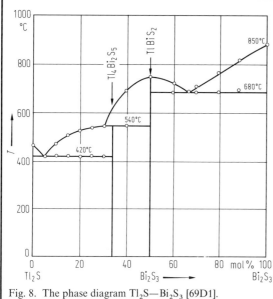

Fig. 8. The phase diagram Tl$_2$S—Bi$_2$S$_3$ [69D1].

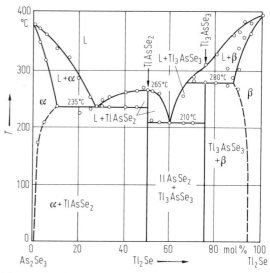

Fig. 9. The phase diagram As$_2$Se$_3$—Tl$_2$Se [69D1]; see also [74R].

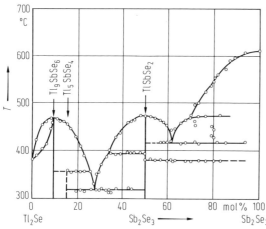

Fig. 10. The phase diagram Tl$_2$Se—Sb$_2$Se$_3$ [73G], see also [75B].

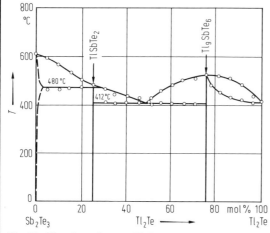

Fig. 11. The phase diagram Sb$_2$Te$_3$—Tl$_2$Te [77B1].

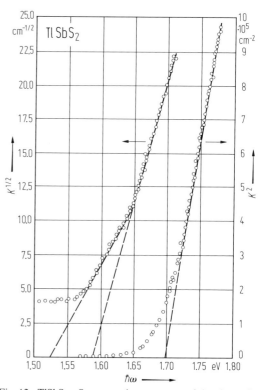

Fig. 12. TlSbS$_2$. Square and square root of the absorption coefficient vs. photon energy in the region of the absorption edge showing the threshold energies for the direct transitions with emission or absorption of a photon [70S].

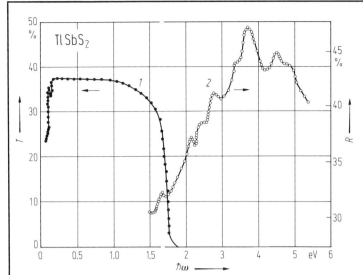

Fig. 13. TlSbS$_2$. Transmission *1* and reflection coefficient *2* vs. photon energy at room temperature [70S].

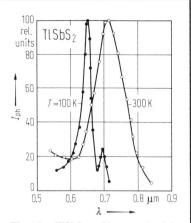

Fig. 15. TlSbS$_2$. Photoconductivity spectra of a single crystal at 300 K and 100 K [76C].

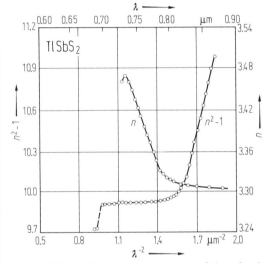
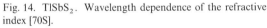

Fig. 14. TlSbS$_2$. Wavelength dependence of the refractive index [70S].

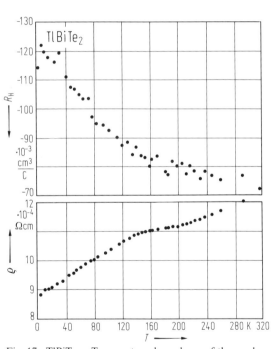

Fig. 17. TlBiTe$_2$. Temperature dependence of the conductivity and the Hall coefficient [72J].

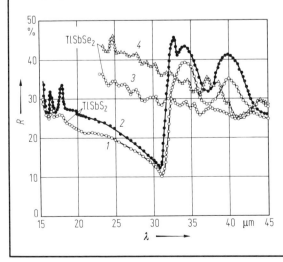

Fig. 16. TlSbS$_2$, TlSbSe$_2$. Reflection spectra at 300 K, curves *1, 3*, and 80 K, *2, 4* [75S].

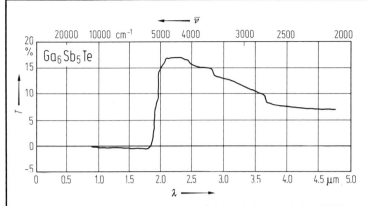

Fig. 18. Ga_6Sb_5Te. Spectral dependence of the optical transmission [66K].

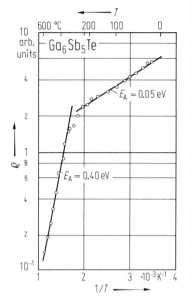

Fig. 19. Ga_6Sb_5Te. Resistivity vs. (reciprocal) temperature [66K]. Activation energies are indicated.

10.4.5 IV_x–V_y–VI_z compounds (For tables, see p. 310ff.)

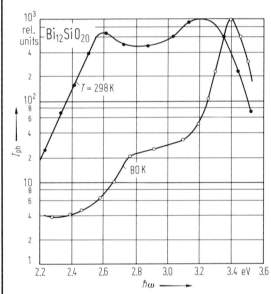

Fig. 1. $Bi_{12}SiO_{20}$. Photocurrent vs. photon energy of an undoped crystal at two temperatures [73H].

For Fig. 2, see next page.

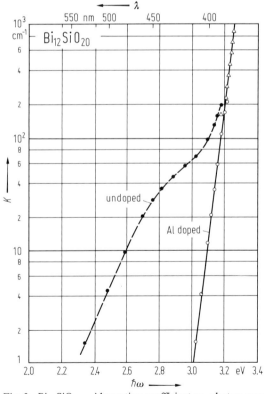

Fig. 3. $Bi_{12}SiO_{20}$. Absorption coefficient vs. photon energy (wavelength) of an undoped and a heavily Al-doped single crystal. The absorption below 3 eV belongs to an extrinsic absorption band [73H].

Fig. 4. $Bi_{12}SiO_{20}$. Index of refraction vs. wavelength. The index of refraction of $Bi_{12}GeO_{20}$ is identical with that of $Bi_{12}SiO_{20}$ within the experimental error (± 0.005) [71A]. ▶

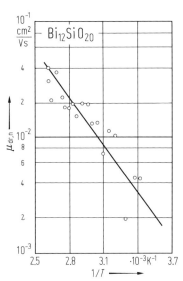

Fig. 2. $Bi_{12}SiO_{20}$. Drift mobility of carriers vs. reciprocal temperature in a field of $2.3 \cdot 10^4$ V/cm [80K].

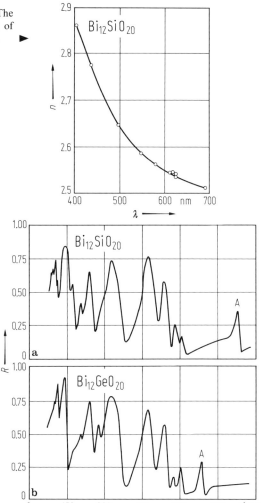

Fig. 6. $Bi_{12}SiO_{20}$, $Bi_{12}GeO_{20}$. Reflectivity vs. wavenumber at 300 K for (a) $Bi_{12}SiO_{20}$ and (b) $Bi_{12}GeO_{20}$ [79W].

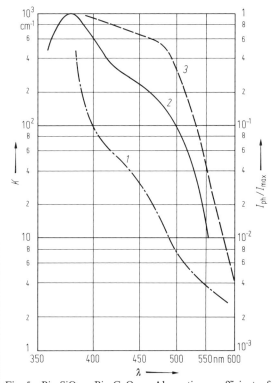

Fig. 5. $Bi_{12}SiO_{20}$, $Bi_{12}GeO_{20}$. Absorption coefficient of $Bi_{12}SiO_{20}$ 1 and spectral response of photoconductivity in $Bi_{12}GeO_{20}$ 2 and $Bi_{12}SiO_{20}$ 3 vs. wavelength. Results are normalized to constant light intensity [71A].

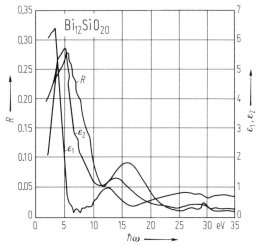

Fig. 7. $Bi_{12}SiO_{20}$. Reflectivity, real and imaginary parts of the dielectric constant vs. photon energy at 300 K [80E].

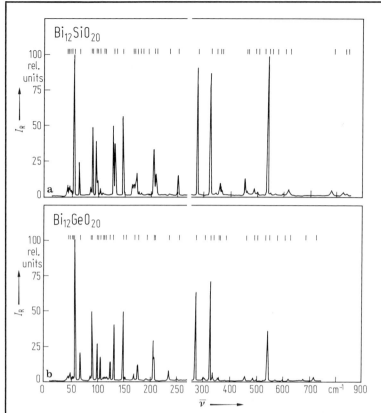

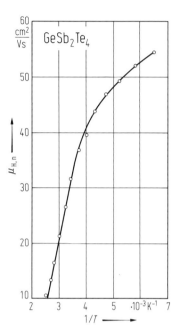

Fig. 8. Bi$_{12}$SiO$_{20}$, Bi$_{12}$GeO$_{20}$. Raman intensity vs. wavenumber at 10 K for (a) Bi$_{12}$SiO$_{20}$ and (b) Bi$_{12}$GeO$_{20}$ [82B2]. Maxima are indicated.

Fig. 10. GeSb$_2$Te$_4$. Hall mobility of an n-type sample vs. reciprocal temperature [72F].

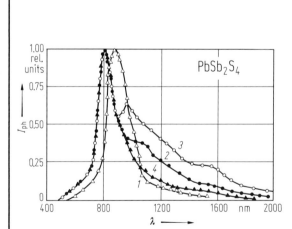

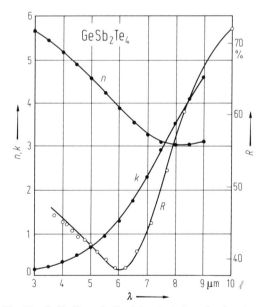

Fig. 9. PbSb$_2$S$_4$. Photocurrent vs. wavelength at *1*: 300 K, *2*, *3*: 100 K obtained after illumination for 15 and 30 min, respectively, *4*: 100 K [74D2].

Fig. 11. GeSb$_2$Te$_4$. Reflectivity R, index of refraction n and extinction coefficient k vs. wavelength. Open circles mark the experimental points, solid circles and lines represent theoretical values obtained from the experimental data [72F].

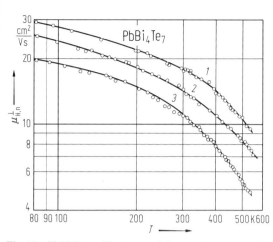

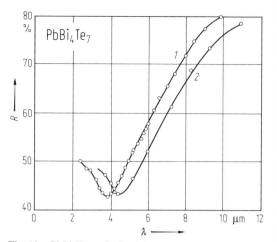

Fig. 12. PbBi$_4$Te$_7$. Electron mobility vs. temperature in three samples with *1*: $\sigma_\perp = 1.1 \cdot 10^3\,\Omega^{-1}\,cm^{-1}$, *2*: $\sigma_\perp = 1.3 \cdot 10^3\,\Omega^{-1}\,cm^{-1}$, *3*: $\sigma_\perp = 2.0 \cdot 10^3\,\Omega^{-1}\,cm^{-1}$, respectively [71F].

Fig. 13. PbBi$_4$Te$_7$. Reflectivity vs. wavelength near the plasma edge for two samples [71F].

10.4.6 V–VI–VII compounds (For tables, see p. 313 ff.)

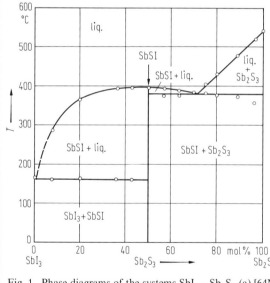

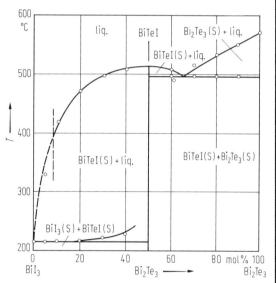

Fig. 1. Phase diagrams of the systems SbI$_3$—Sb$_2$S$_3$ (a) [64M] and BiI$_3$—Bi$_2$Te$_3$ (b) [68H2]. S: solid.

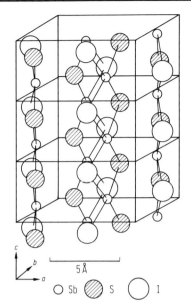

Fig. 2. SbSI. Crystal structure in the paraelectric phase as an example for the D_{2h}^{16}—Pnam lattice [67K].

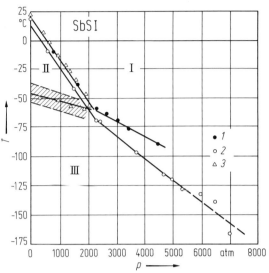

Fig. 3. SbSI. Phase diagram: I paraelectric phase, II ferroelectric phase, III C_2^2 phase, measurements on three samples. The low temperature transition region is shaded [72S4].

For Figs. 4 and 5, see next page.

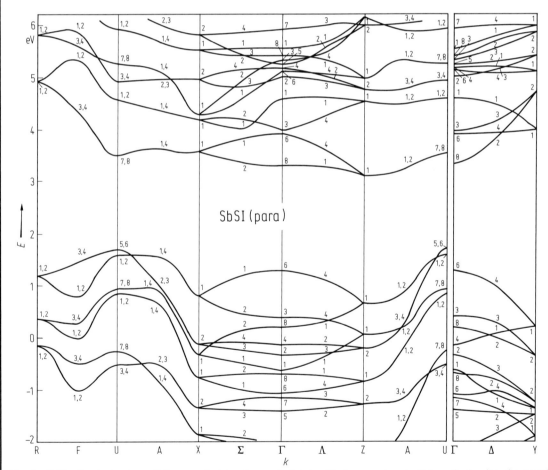

Fig. 6. SbSI. Band structure of the paraelectric phase calculated with the pseudopotential method using the atomic positions at 35 °C as parameters [73N].

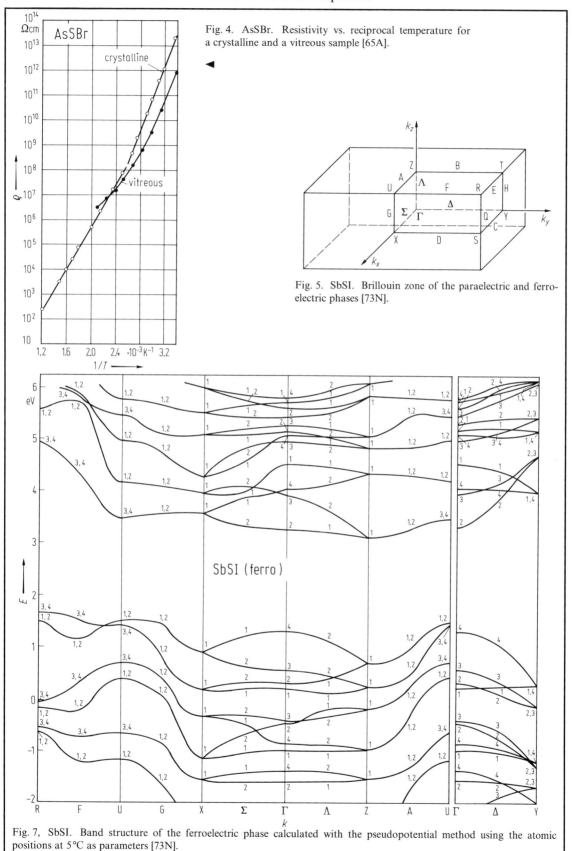

Fig. 4. AsSBr. Resistivity vs. reciprocal temperature for a crystalline and a vitreous sample [65A].

Fig. 5. SbSI. Brillouin zone of the paraelectric and ferro-electric phases [73N].

SbSI (ferro)

Fig. 7, SbSI. Band structure of the ferroelectric phase calculated with the pseudopotential method using the atomic positions at 5°C as parameters [73N].

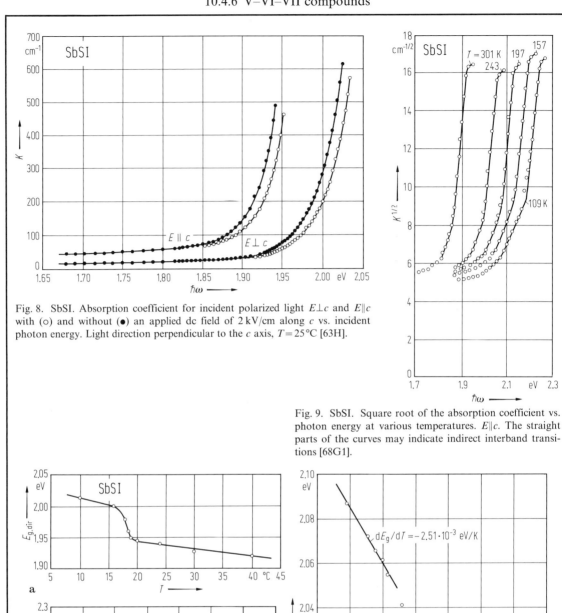

Fig. 8. SbSI. Absorption coefficient for incident polarized light $E \perp c$ and $E \parallel c$ with (o) and without (•) an applied dc field of 2 kV/cm along c vs. incident photon energy. Light direction perpendicular to the c axis, $T = 25\,°C$ [63H].

Fig. 9. SbSI. Square root of the absorption coefficient vs. photon energy at various temperatures. $E \parallel c$. The straight parts of the curves may indicate indirect interband transitions [68G1].

Fig. 10. SbSI. Temperature dependence of the energy gap determined (a) from the absorption edge for $T > 10\,°C$ [66F], (b) from transmission ($E \parallel c$) for $-180\,°C < T < 20\,°C$ [78I2], (c) from thermoabsorption near the phase transition temperature [82G].

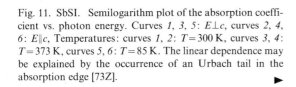

Fig. 11. SbSI. Semilogarithm plot of the absorption coefficient vs. photon energy. Curves *1, 3, 5*: $E \perp c$, curves *2, 4, 6*: $E \| c$, Temperatures: curves *1, 2*: $T = 300$ K, curves *3, 4*: $T = 373$ K, curves *5, 6*: $T = 85$ K. The linear dependence may be explained by the occurrence of an Urbach tail in the absorption edge [73Z]. ▶

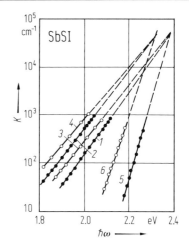

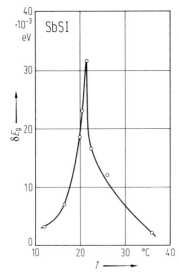

Fig. 12. SbSI. Shift of absorption edge under a dc field of 2 kV/cm applied along the *c* axis vs. temperature around Θ_C [63H].

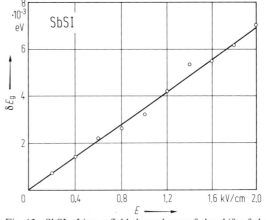

Fig. 13. SbSI. Linear field dependence of the shift of the absorption edge in the ferroelectric phase. $T = 16.6\,°C$. Electric dc field applied along the *c* axis [63H].

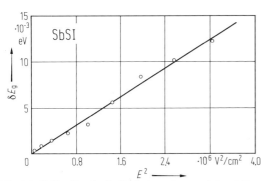

Fig. 14. SbSI. Quadratic field dependence of the shift of the absorption edge in the paraelectric phase. $T = 25.5\,°C$ [63H]. dc field along the *c* axis.

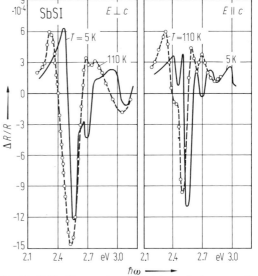

Fig. 15. SbSI. Temperature modulated reflectance vs. photon energy (dashed lines [76E]) and wavelength modulated reflectance vs. photon energy (solid lines [74F]).

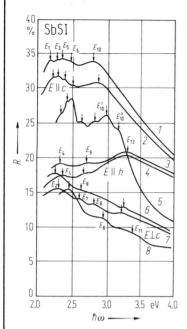

Fig. 16. SbSI. Reflectivity vs. photon energy in the range 2 to 4 eV. Temperatures: curves *1, 3, 6*: $T = 300$ K, curves *2, 4, 7*: $T = 273$ K, curves *5, 8*: $T = 90$ K. Configurations: curves *1, 2, 5*: $E\|c$, curves *3, 4*: $E\|b$, curves *6, 7, 8*: $E\perp c$. Critical points $E_1 \cdots E_{12}$ are indicated [71B3].

For Fig. 19, see next page.

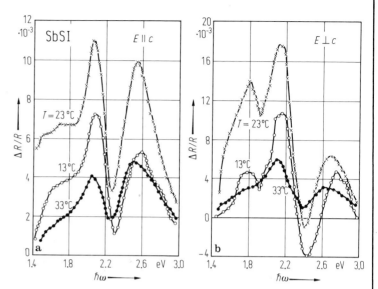

Fig. 17. SbSI. Electroreflectance vs. incident photon energy (a) with $E\|c$, (b) with $E\perp c$. [71G].

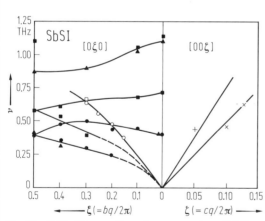

Fig. 18. SbSI. Phonon dispersion curves along [0ζ0] and [00ζ]. Open symbols indicate longitudinal modes, closed symbols transverse modes polarized along [001] for $q\|[0\zeta0]$ and along [010] for $q\|[00\zeta]$. Ferroelectric phase: □ and ■ at 10.3 °C. Paraelectric phase: △ and ▲ at 18.6 °C, ○ and ● at 23.3 °C. The symbols + and × coincide for both phases, measured at 10.3 and 23.3 °C [77P].

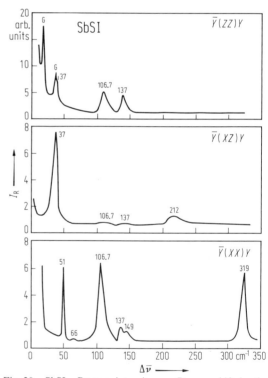

Fig. 20. SbSI. Raman intensity vs. Raman shift in the paraelectric phase ($T = 299$ K) for three scattering geometries. The peaks labeled G are grating ghosts [70P].

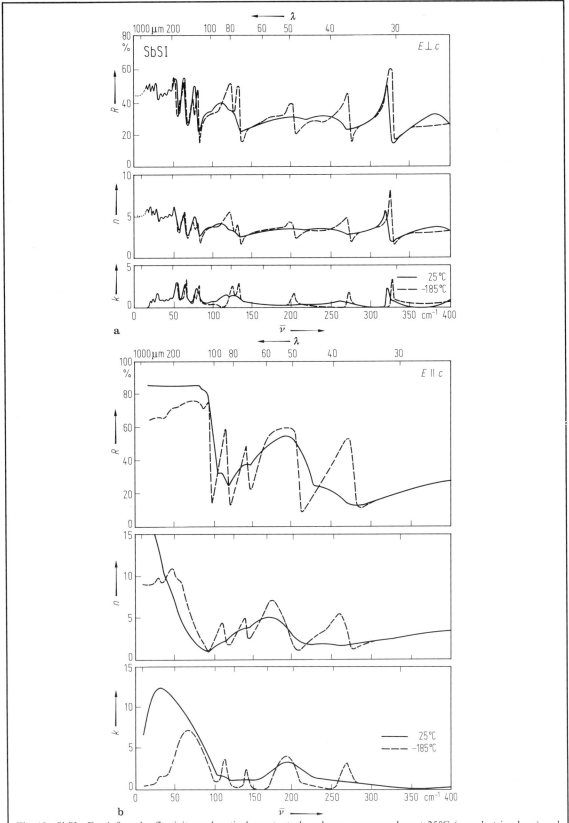

Fig. 19. SbSI. Far infrared reflectivity and optical constants k and n vs. wavenumber at 25°C (paraelectric phase) and −185°C (ferroelectric phase) (a) for $E \perp c$, (b) for $E \| c$ [69P1].

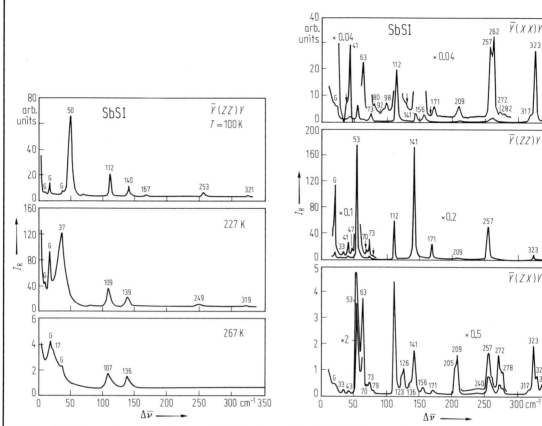

Fig. 21. SbSI, Raman intensity vs. Raman shift in the ferroelectric phase in the $\bar{Y}(ZZ)Y$ configuration for $T = 100$ K, 227 K, 267 K [70P]. The peaks labeled G are grating ghosts.

Fig. 22. SbSI. Raman intensity vs. Raman shift at 15 K in the three scattering geometries [71A].

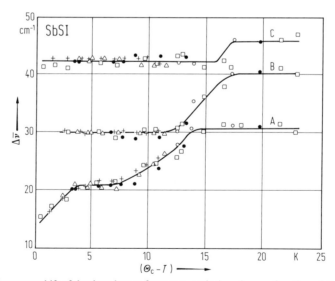

Fig. 23. SbSI. Raman frequency shift of the three lowest frequency optical modes as a function of $\Theta_C - T$ in the ferroelectric phase. A "soft mode" is seen to decrease by approaching Θ_C interacting with two "hard modes" at 30 cm^{-1} and 41 cm^{-1}. The data have been obtained at constant temperature by changing Θ_C by pressure. Different symbols represent measurements at 0 °C (●), 1.5 °C (△) and 3 °C (+, ○). Squares indicate the data obtained at $p = 200 \cdot 10^5$ Pa and changing the temperature [72T].

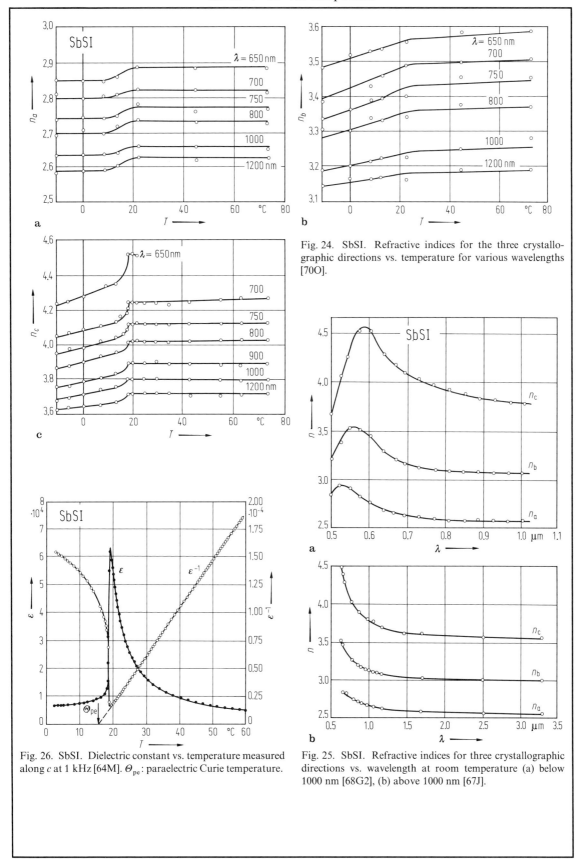

a

b

c

Fig. 24. SbSI. Refractive indices for the three crystallographic directions vs. temperature for various wavelengths [70O].

a

b

Fig. 26. SbSI. Dielectric constant vs. temperature measured along c at 1 kHz [64M]. Θ_{pe}: paraelectric Curie temperature.

Fig. 25. SbSI. Refractive indices for three crystallographic directions vs. wavelength at room temperature (a) below 1000 nm [68G2], (b) above 1000 nm [67J].

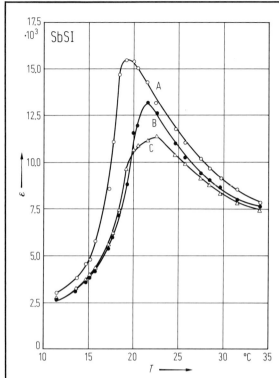

Fig. 27. SbSI. Dielectric constant vs. temperature measured along the c axis (f=1 kHz): A – in the dark, B – in the dark with an electric field of 1.0 kV/cm present, C – with additional illumination [67U].

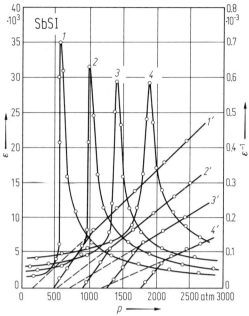

Fig. 28. SbSI. Dielectric constant vs. pressure measured along the c axis (f=1 kHz) for T=271 K (curve 1), 254 K (2), 239 K (3), 220 K (4) [69G]. Primed numbers refer to reciprocal dielectric constant.

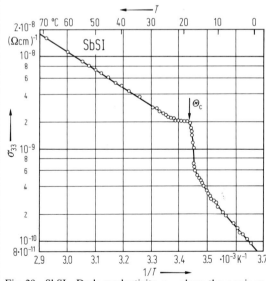

Fig. 29. SbSI. Dark conductivity σ_{33} along the c axis vs. (reciprocal) temperature [77I].

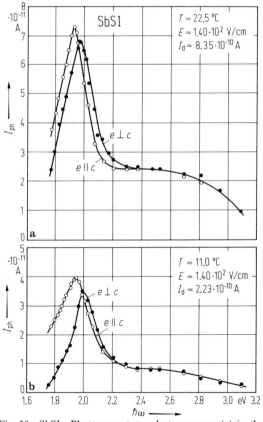

Fig. 30. SbSI. Photocurrent vs. photon energy (a) in the paraelectric phase (T=22.5 °C), (b) in the ferroelectric phase (T=11.0 °C) [73I]. I_d: dark current, e: polarization vector. Electric field E‖c.

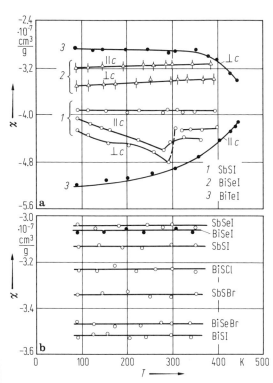

Fig. 31. SbSI and other V–VI–VII compounds. Magnetic susceptibility vs. temperature (a) according to [76B], (b) according to [76M]; powdered sample. χ in CGS-emu.

Fig. 32. SbSI. Relative change in the length of a specimen along the c axis vs. temperature (curve A). The length also varies by application of a dc field (1.5 kV/cm in curve B). A further change in length is caused by illumination (not shown in the figure) [66T].

For Fig. 33, see next page.

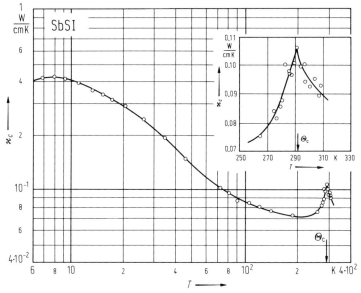

Fig. 34. SbSI. Thermal conductivity along the c axis vs. temperature [68S2].

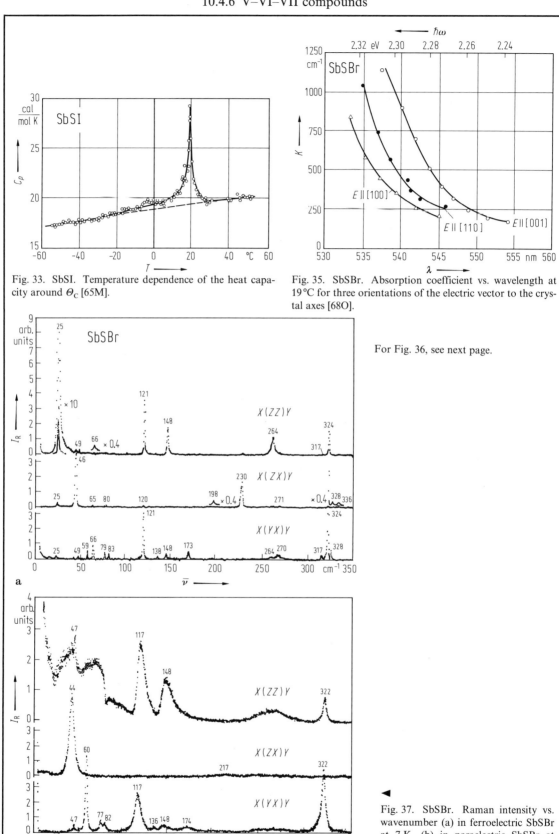

Fig. 33. SbSI. Temperature dependence of the heat capacity around Θ_C [65M].

Fig. 35. SbSBr. Absorption coefficient vs. wavelength at 19°C for three orientations of the electric vector to the crystal axes [68O].

For Fig. 36, see next page.

Fig. 37. SbSBr. Raman intensity vs. wavenumber (a) in ferroelectric SbSBr at 7 K, (b) in paraelectric SbSBr at 300 K [82I].

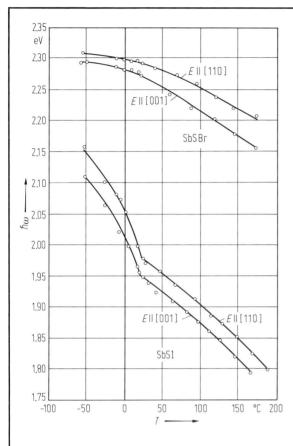

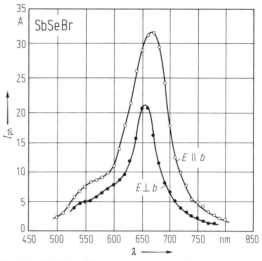

Fig. 38. SbSeBr. Photocurrent vs. wavelength at 300 K [68H3].

Fig. 36. SbSBr. Photon energy at an absorption of $K = 500 \, \text{cm}^{-1}$ vs. temperature for different orientations compared with the same measurements on SbSI [68O].

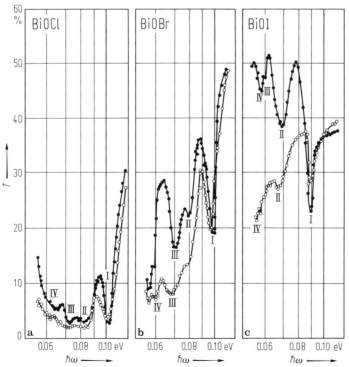

Fig. 39. BiOCl(Br, I). Infrared transmission vs. photon energy ($E \perp c$) of single crystals of (a) BiOCl, (b) BiOBr, (c) BiOI at 80 K (o) and 290 K (●). The structure denoted by I, II, III and IV results from two-phonon processes [73P].

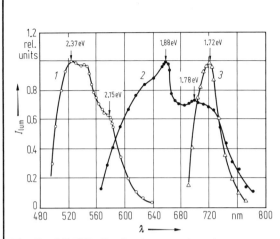

Fig. 40. BiOCl(Br, I). Luminescence intensity vs. wavelength at 130 K for *1* BiOCl, *2* BiOBr, *3* BiOI. Spectral resolution about 0.02 eV [72S2].

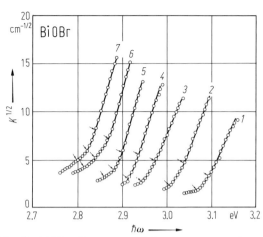

Fig. 42. BiOBr. Square root of the absorption coefficient vs. photon energy in the region of indirect transitions. The arrows indicate kinks corresponding to phonon emission and absorption threshold energies. The curves *1* to *7* correspond to $T = 90, 150, 203, 293, 370, 417, 473$ K, respectively [72S3].

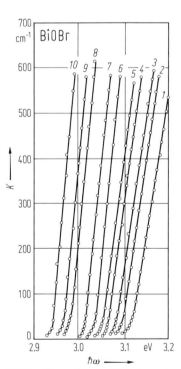

Fig. 41. BiOBr. Absorption coefficient vs. photon energy. The curves *1* to *10* correspond to $T = 100, 173, 200, 230, 260, 295, 325, 378, 400, 436$ K, respectively [72S3].

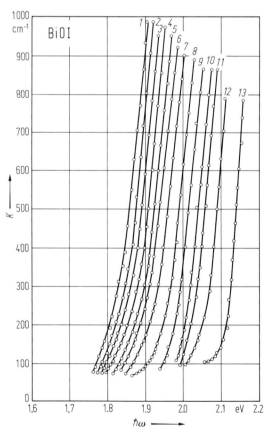

Fig. 43. BiOI. Absorption coefficient vs. photon energy. The curves *1* to *13* correspond to $T = 470, 450, 420, 398, 370, 350, 320, 257, 263, 230, 203, 168, 90$ K, respectively [72D3].

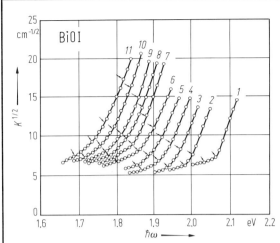

Fig. 44. BiOI. Square root of the absorption coefficient vs. photon energy in the region of indirect transitions. The arrows indicate kinks corresponding to phonon emission and absorption threshold energies. The curves *1* to *11* correspond to $T = 90$, 173, 203, 233, 263, 293, 330, 355, 380, 415, 445 K, respectively [72S3].

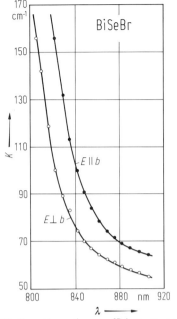

Fig. 45. BiSeBr. Absorption coefficient vs. wavelength [68H3].

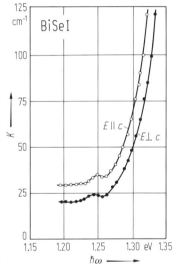

Fig. 46. BiSeI. Absorption coefficient vs. photon energy [68C].

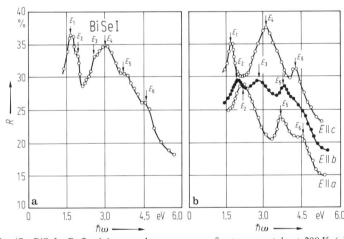

Fig. 47. BiSeI. Reflectivity vs. photon energy of n-type crystals at 298 K (a) in unpolarized light, (b) in polarized light [74B].

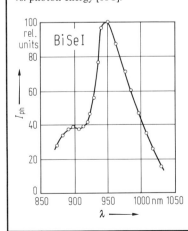

Fig. 48. BiSeI. Photocurrent ($I\|c$) vs. wavelength [68C].

559

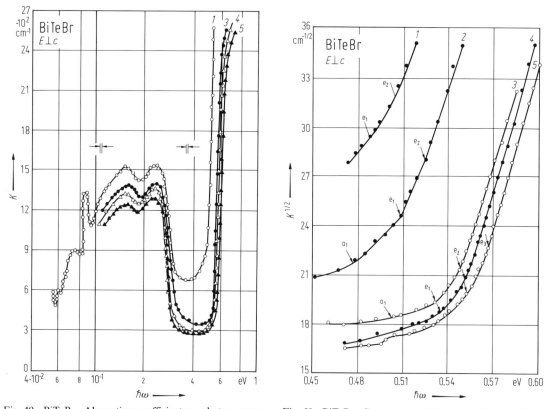

Fig. 49. BiTeBr. Absorption coefficient vs. photon energy between 0.06 eV and 1 eV for $E \perp c$ at $T = 295$, 190, 80, 22, 5 K (curves *1* to *5*) [73B2].

Fig. 50. BiTeBr. Square root of the absorption coefficient shown in Fig. 49 vs. photon energy ($E \perp c$). Arrows indicate the onset of transitions with emission (e) or absorption (a) of phonons [73B2].

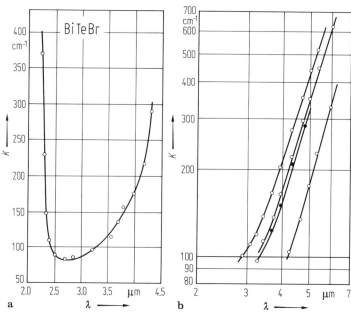

Fig. 51. BiTeBr. (a) Absorption coefficient vs. wavelength at room temperature and (b) long wavelength part of the absorption edge for 4 n-type single crystals [80M].

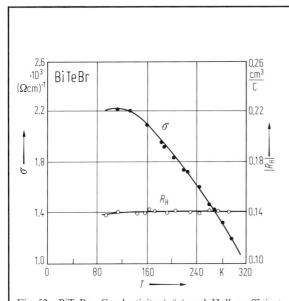

Fig. 52. BiTeBr. Conductivity ($\sigma \| c$) and Hall coefficient ($B \perp c$) vs. temperature for an n-type single crystal [80M].

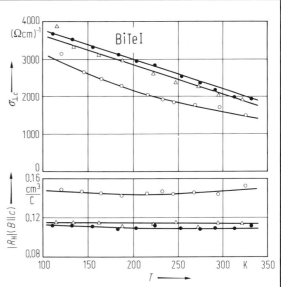

Fig. 53. BiTeI. Conductivity ($\sigma \perp c$) and Hall coefficient ($B \| c$) vs. temperature for three samples [78D].

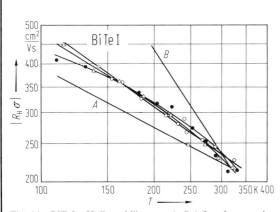

Fig. 54. BiTeI. Hall mobility $\mu_H = |\sigma R_H|$ for the samples of Fig. 53. Theoretical curves A for optical phonon scattering, B for acoustic phonon scattering [78D].

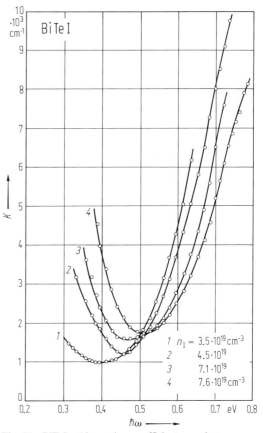

Fig. 55. BiTeI. Absorption coefficient vs. photon energy at RT for four samples with different impurity concentrations. The shift of the absorption edge is due to the Burstein effect [70H1].

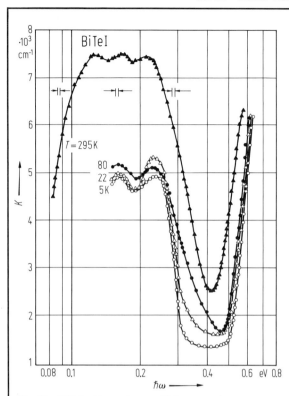

Fig. 56. BiTeI. Absorption coefficient vs. photon energy at different temperatures [74P].

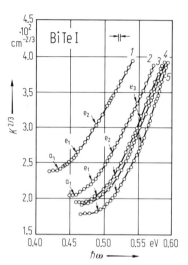

Fig. 57. BiTeI. 2/3-power of the absorption coefficient vs. photon energy at $T = 295$, 173, 80, 22, 5 K (curves $1\cdots5$, respectively). The arrows indicate threshold energies for indirect transitions with phonon emission (e) or absorption (a) [74P].

10.4.7 Other ternary compounds (For tables, see p. 326 ff.)

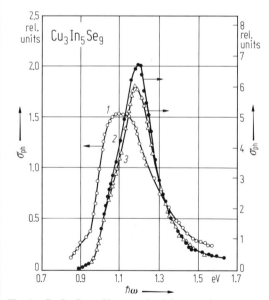

Fig. 1. $Cu_3In_5Se_9$. Photoconductivity vs. photon energy at 300 K (curve 1) and 77 K (2, 3). Curve 3 was obtained after preliminary illumination [80T2].

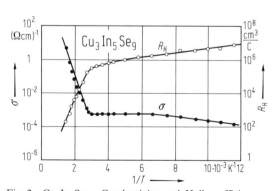

Fig. 2. $Cu_3In_5Se_9$. Conductivity and Hall coefficient vs. reciprocal temperature [80T1].

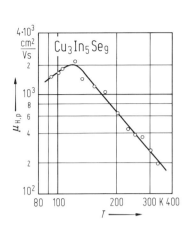

Fig. 3. Cu$_3$In$_5$Se$_9$. Hall mobility of holes vs. temperature [80T2].

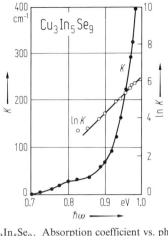

Fig. 4. Cu$_3$In$_5$Se$_9$. Absorption coefficient vs. photon energy plotted on ordinary and logarithmic scales [80T2].

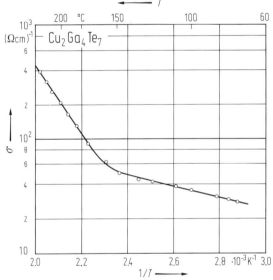

Fig. 5. Cu$_2$Ga$_4$Te$_7$. Conductivity vs. (reciprocal) temperature [73C].

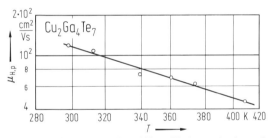

Fig. 6. Cu$_2$Ga$_4$Te$_7$. Hall mobility of holes vs. (reciprocal) temperature [73C].

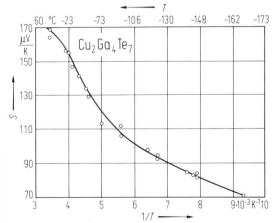

Fig. 7. Cu$_2$Ga$_4$Te$_7$. Seebeck coefficient vs. (reciprocal) temperature [73C].

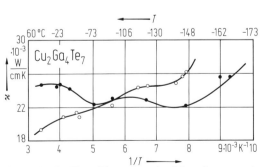

Fig. 8. Cu$_2$Ga$_4$Te$_7$. Thermal conductivity of two samples vs. (reciprocal) temperature [73C].

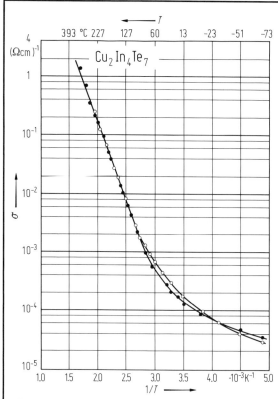

Fig. 9. $Cu_2In_4Te_7$. Conductivity of two samples vs. (reciprocal) temperature [72C].

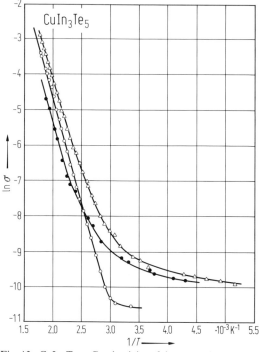

Fig. 13. $CuIn_3Te_5$. Conductivity of three samples vs. reciprocal temperature [71C]. σ in $\Omega^{-1}\,cm^{-1}$.

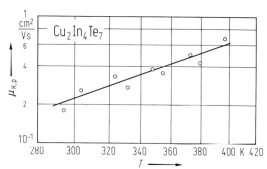

Fig. 10. $Cu_2In_4Te_7$. Hall mobility of holes vs. temperature [72C].

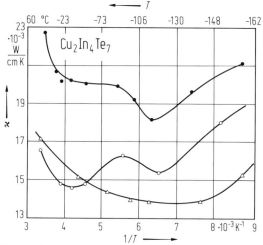

Fig. 11. $Cu_2In_4Te_7$. Thermal conductivity of three samples vs. (reciprocal) temperature [72C].

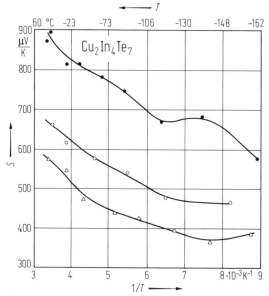

Fig. 12. $Cu_2In_4Te_7$. Seebeck coefficient of three samples vs. (reciprocal) temperature [72C].

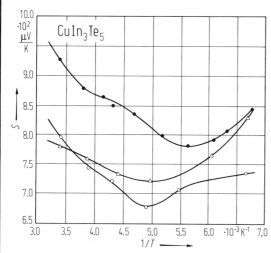

Fig. 14. $CuIn_3Te_5$. Seebeck coefficient of the samples of Fig. 13 vs. reciprocal temperature [71C].

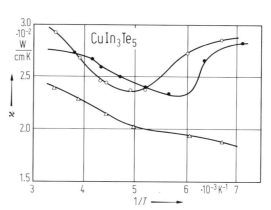

Fig. 15. $CuIn_3Te_5$. Thermal conductivity of the samples of Fig. 13 vs. reciprocal temperature [71C].

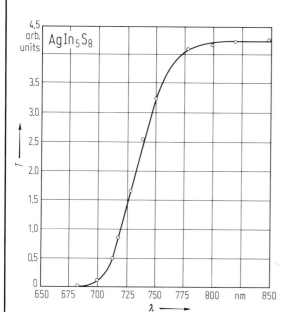

Fig. 16. $AgIn_5S_8$. Optical transmission vs. wavelength of a single crystal at 300 K [77P].

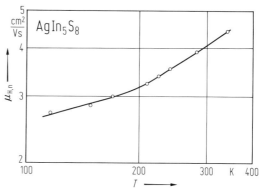

Fig. 17. $AgIn_5S_8$. Electron Hall mobility of a single crystal vs. temperature [77P].

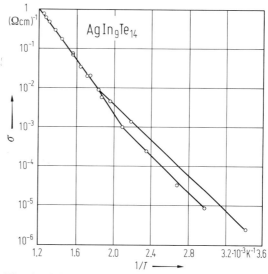

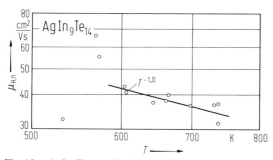

Fig. 19. $AgIn_9Te_{14}$. Carrier mobility vs. temperature [67C1].

Fig. 18. $AgIn_9Te_{14}$. Conductivity vs. reciprocal temperature [67C1]. The two curves at lower temperatures correspond to the first and second heating cycle.